Engineering Design with
SolidWorks 2012

and MultiMedia DVD

A Step-by-Step Project Based Approach Utilizing 3D Solid Modeling

David C. Planchard & Marie P. Planchard CSWP

ISBN: 978-1-58503-697-4

Schroff Development Corporation

www.SDCpublications.com

Schroff Development Corporation
P.O. Box 1334
Mission KS 66222
(913) 262-2664
www.SDCpublications.com

Publisher: Stephen Schroff

Trademarks and Disclaimer

software for design, analysis and product data management applications. Microsoft Windows®, Microsoft Office® and its family of products are registered trademarks of the Microsoft Corporation. Other software applications and parts described in this book are trademarks or registered trademarks of their respective owners.

The publisher and the authors make no representations or warranties with respect to the accuracy or completeness of the contents of this work and specifically disclaim all warranties, including without limitation warranties of fitness for a particular purpose. No warranty may be created or extended by sales or promotional materials. Dimensions of parts are modified for illustration purposes. Every effort is made to provide an accurate text. The authors and the manufacturers shall not be held liable for any parts, components, assemblies or drawings developed or designed with this book or any responsibility for inaccuracies that appear in the book. Web and company information was valid at the time of this printing.

Examination Copies

Teacher evaluation copies for 2012, 2011, 2010, 2009, 2008 and 2007 SolidWorks books are available with classroom support materials (PowerPoint presentations, files, Term Projects, labs, quizzes and more) with initial and final SolidWorks models. Books received as examination copies are for review purposes only. Examination copies *are not intended for student* use. Resale of examination copies or student use is prohibited.

Electronic Files

Any electronic files associated with this book are licensed to the original user only. These files may not be transferred to any other party.

Introduction

Engineering Design with SolidWorks 2012 and Video Instruction is written to assist students, designers, engineers and professionals. The book provides a solid foundation in SolidWorks by utilizing projects with step-by-step instructions for the beginning to intermediate SolidWorks user. Explore the user interface, CommandManager, menus, toolbars and modeling techniques to create parts, assemblies and drawings in an engineering environment.

Follow the step-by-step instructions and develop multiple parts and assemblies that combine machined, plastic and sheet metal components. Formulate the skills to create, modify and edit sketches and solid features. Learn the techniques to reuse features, parts and assemblies through symmetry, patterns, copied components, design tables, Bills of Materials, Custom Properties and Configurations. Address various SolidWorks analysis tools: *SimulationXpress, Sustainability / SustainabilityXpress* and *DFMXpress* and Intelligent Modeling techniques. Learn by doing, not just by reading!

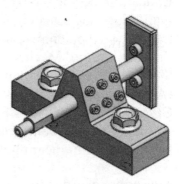

GUIDE-ROD Assembly
Project 1 and Project 2

Desired outcomes and usage competencies are listed for each project. Know your objective up front. Follow the steps in Projects 1 - 8 to achieve the design goals. Work between multiple documents, features, commands and custom properties that represent how engineers and designers utilize SolidWorks in industry.

Review individual features, commands and tools with the enclosed Video Instruction DVD. The projects contain exercises. The exercises analyze and examine usage competencies. Collaborate with leading industry suppliers such as SMC Corporation of America, Boston Gear and 80/20 Inc.

FLASHLIGHT Assembly
Project 4 and Project 5

Collaborative information translates into numerous formats such as paper drawings, electronic files, rendered images and animations. On-line intelligent catalogs guide designers to the product that meets both their geometric requirements and performance functionality.

The authors developed the industry scenarios by combining their own industry experience with the knowledge of engineers, department managers, vendors and manufacturers. These professionals are directly involved with SolidWorks every day. Their responsibilities go far beyond the creation of just a 3D model.

The book is designed to compliment the SolidWorks Tutorials contained in SolidWorks 2012. There are over 2.5 hours of Video Instruction on the enclosed DVD.

About the Cover

Displayed on the front cover is the GUIDE-CYLINDER assembly manufactured by SMC Corporation of America, Indianapolis, IN, USA. All parts are used with permission.

About the Authors

David Planchard is the founder of D&M Education LLC. Before starting D&M Education, he spent over 27 years in industry and academia holding various engineering, marketing, and teaching positions and degrees. He holds five U.S. patents and one international patent. He has published and authored numerous papers on Machine Design, Product Design, Mechanics of Materials, and Solid Modeling. He is an active member of the SolidWorks Users Group and the American Society of Engineering Education (ASEE). David holds a BSME, MSM with the following Professional Certifications: CCAI, CCNA, CCNP, CSWA, and CSWP. David is a SolidWorks Solution Partner, an Adjunct Faculty member and the SAE advisor at Worcester Polytechnic Institute in the Mechanical Engineering department.

Marie Planchard is the Director of World Education Markets at DS SolidWorks Corp. Before she joined SolidWorks, Marie spent over 10 years as an engineering professor at Mass Bay College in Wellesley Hills, MA. She has 14 plus years of industry software experience and held a variety of management and engineering positions. Marie holds a BSME, MSME and a Certified SolidWorks Professional (CSWP) Certification. She is an active member of the American Society of Mechanical Engineers (ASME) and the American Society for Engineering Education (ASEE).

David and Marie Planchard are co-authors of the following books:

- **A Commands Guide for SolidWorks® 2012**, 2011, 2010, 2009 and 2008

- **A Commands Guide Reference Tutorial for SolidWorks® 2007**

- **Assembly Modeling with SolidWorks® 2010,** 2008, 2006, 2005-2004, 2003 and 2001Plus

- **Drawing and Detailing with SolidWorks® 2011**, 2010, 2009, 2008, 2007, 2006, 2005, 2004, 2003, 2002 and 2001/2001Plus

- **Engineering Design with SolidWorks® and Video Instruction 2012**, 2011, 2010, 2009, 2008, 2007, 2006, 2005, 2004, 2003, 2001Plus, 2001 and 1999

- **Engineering Graphics with SolidWorks and Video Instruction 2012**, 2011, 2010

- **SolidWorks® The Basics with Multimedia CD** 2009, 2008, 2007, 2006, 2005, 2004 and 2003

- **SolidWorks® Tutorial with and Video Instruction 2012**, 2011, 2010, 2009, 2008, 2007, 2006, 2005, 2004, 2003 and 2001/2001Plus

- **The Fundamentals of SolidWorks®: Featuring the VEXplorer robot, 2008** and 2007

- **Official Certified SolidWorks® Associate Examination Guide, Version 3; 2011, 2010, 2009,** Version 2; 2010, 2009, 2008, Version 1; 2007

- **Official Certified SolidWorks® Professional (CSWP) Certification Guide with Multimedia DVD, 2011, 2010**

- **Applications in Sheet Metal Using Pro/SHEETMETAL & Pro/ENGINEER**

Acknowledgments

Writing this book was a substantial effort that would not have been possible without the help and support of my loving family and of my professional colleagues. I would like to thank Professor John Sullivan and Robert Norton and the community of scholars at Worcester Polytechnic Institute who have enhanced my life, my knowledge, and helped to shape the approach and content of this book.

The author is greatly indebted to my colleagues from Dassault Systèmes SolidWorks Corporation for their help and continuous support: Jeremy Luchini, Avelino Rochino, and Mike Puckett.

Thanks also to Professor Richard L. Roberts of Wentworth Institute of Technology, Professor Dennis Hance of Wright State University, and Professor Jason Durfess of Eastern Washington University who provided insight and invaluable suggestions.

Finally to my wife, who is infinitely patient for her support and encouragement and to our loving daughter Stephanie who supported me during this intense and lengthy project.

Contact the Authors

This is the 13th edition of Engineering Design with SolidWorks. We realize that keeping software application books current is imperative to our customers. We value the hundreds of professors, students, designers, and engineers that have provided us input to enhance our book. We value your suggestions and comments. Please contact us or visit our website www.dmeducation.net with any comments, questions or suggestions on this book or any of our other SolidWorks books. David Planchard, D & M Education, LLC, dplanchard@msn.com or planchard@wpi.edu.

Table of Contents

Introduction   **I-1**
About the Cover   I-2
About the Authors   I-2
Acknowledgements   I-3
Contact the Authors   I-3
Note to Instructors   I-3
Trademarks, Disclaimer and Copyrighted Material   I-3
References   I-4
Table of Contents   I-6
Overview of Projects   I-14
What is SolidWorks?   I-19
About the Book   I-21
Windows Terminology in SolidWorks   I-22

Project 1 - Fundamentals of Part Modeling   **1-1**
Project Objective   1-3
Project Situation   1-4
Project Overview   1-6
File Management   1-7
Start a SolidWorks 2012 Session   1-9
Understand the SolidWorks User Interface and CommandManager   1-10
  Menu bar toolbar   1-10
  Menu bar menu   1-10
  Drop-down menu   1-11
  Right-click   1-11
  Consolidated toolbar   1-12
  System feedback icons   1-12
  Confirmation Corner   1-12
  Heads-up View toolbar   1-13
  SolidWorks CommandManager   1-15
  FeatureManager Design Tree   1-19
  Fly-out FeatureManager   1-21
  Task Pane   1-22
    Design Library   1-22
    File Explore   1-23
    Search   1-23
    View Palette   1-23
    Appearances, Scenes, and Decals   1-24
    Custom Properties   1-24
    Document Recovery   1-24
  Motion Study tab   1-25
MotionManager   1-25
  Animation   1-25
  Basic Motion   1-25
System Options   1-31
Part Document Template and Document Properties   1-32
PLATE Part Overview   1-35
PLATE Part-New SolidWorks Document   1-37

PLATE Base Feature 1-38
Machined Part 1-39
Reference Planes and Orthographic Projection 1-40
PLATE Part-Extruded Boss/Base Feature 1-44
PLATE Part-Modify Dimensions and Rename 1-52
Display Modes, View Modes, View tools, and Appearances ... 1-54
Fasteners 1-56
PLATE Part-Extruded Cut Feature 1-58
PLATE Part-Fillet Feature 1-63
PLATE Part-Hole Wizard 1-65
ROD Part Overview 1-68
ROD Part-Extruded Boss/Base Feature 1-70
ROD Part-Hole Wizard Feature 1-72
ROD Part-Chamfer Feature 1-73
ROD Part-Extruded Cut Feature & Convert Entities Sketch Tool ... 1-74
ROD Part-View Orientation, Named Views & Viewport option ... 1-79
ROD Part-Copy/Paste Function 1-80
ROD Part-Design Changes with Rollback Bar 1-81
ROD Part-Recover from Rebuild Errors 1-83
ROD Part-Edit Part Color 1-87
GUIDE Part Overview 1-89
GUIDE Part-Extruded Boss/Base Feature and Dynamic Mirror Feature ... 1-91
GUIDE Part-Extruded Cut Slot Profile 1-94
GUIDE Part-Mirror Feature 1-98
GUIDE Part-Holes 1-99
GUIDE PART-Linear Pattern Feature 1-102
GUIDE Part-Materials Editor and Mass Properties 1-104
Manufacturing Considerations 1-106
Sketch Entities and Sketch Tools 1-109
Project Summary 1-110
Project Terminology 1-110
Questions / Exercises 1-114

Project 2 - Fundamentals of Assembly Modeling   **2-1**
Project Objective 2-3
Project Situation 2-4
Project Overview 2-5
Bottom-up Assembly Modeling Approach 2-5
Linear Motion and Rotational Motion 2-6
GUIDE-ROD assembly 2-7
GUIDE-ROD assembly - Insert Components 2-11
FeatureManager Syntax 2-13
Mate Types 2-16
  Standard Mates 2-16
  Advanced Mates 2-17
  Mechanical Mates 2-17
GUIDE-ROD Assembly - Mate the ROD Component 2-18
GUIDE-ROD Assembly - Mate the PLATE Component 2-23
GUIDE-ROD Assembly - Mate Errors 2-27
Collision Detection 2-30
Modify Component Dimension 2-31

SolidWorks Design Library   2-32
GUIDE-ROD Assembly - Inert Mates for Flange bolts   2-35
Socket Head Cap Screw Part   2-39
SmartMates   2-44
Coincident/Concentric SmartMate   2-44
Tolerance and Fit   2-47
Exploded View   2-51
Section View   2-56
Analyze an Interference Problem   2-58
Save As Copy Option   2-59
GUIDE-ROD Assembly-Feature Driven Component Pattern   2-62
Redefining Mates and Linear Components Pattern   2-64
Folders and Suppressed Components   2-68
Make-Buy Decision   2-69
CUSTOMER Assembly   2-71
Copy the CUSTOMER Assembly - Pack and Go   2-77
Project Summary   2-79
Project Terminology   2-80
Questions / Exercises   2-87

Project 3 - Fundamentals of Drawing   **3-1**
Project Objective   3-3
Project Situation   3-4
Project Overview   3-4
Drawing Template and Sheet Format   3-5
Sheet Format and Title Block   3-12
Company Logo   3-17
Save Sheet Format and Save As Drawing Template   3-19
GUIDE Part-Modify   3-22
GUIDE Part - Drawing   3-23
Move Views and Properties of the Sheet   3-26
Auxiliary View, Section View and Detail View   3-29
Auxiliary View   3-30
Section View   3-31
Detail View   3-32
Partial Auxiliary View - Crop View   3-33
Display Modes and Performance   3-35
Detail Drawing   3-37
Move Dimensions in the Same View   3-40
Partial Auxiliary View-Crop View   3-40
Move Dimensions to a Different View   3-44
Dimension Holes and the Hole Callout   3-45
Center Marks and Centerlines   3-48
Modify the Dimension Scheme   3-50
GUIDE Part-Insert an Additional Feature   3-54
General Notes and Parametric Notes   3-56
Revision Table   3-59
Part Number and Document Properties   3-61
Exploded View   3-67
Balloons   3-69
Bill of Materials   3-71

Project Summary   3-76
Project Terminology   3-77
Questions / Exercises   3-80

Project 4 - Extrude and Revolve Features   **4-1**
Project Objective   4-3
Project Overview   4-4
Design Intent   4-6
Project Situation   4-9
Part Template   4-11
BATTERY Part   4-15
BATTERY Part - Extruded Boss/Base Feature   4-17
BATTERY Part - Fillet Feature Edge   4-22
BATTERY Part - Extruded Cut Feature   4-23
BATTERY Part - Fillet Feature Face   4-25
BATTERY Part - Extruded Boss/Boss Feature   4-27
Injection Molded Process   4-32
BATTERYPLATE Part   4-33
Save As, Delete, Modify and Edit Feature   4-34
BATTERYPLATE Part - Extruded Boss/Base Feature   4-36
BATTERYPLATE Part - Fillet Features-Full Round, options   4-37
Multi-body Parts and the Extruded Boss/Base Feature   4-40
LENS Part   4-42
LENS Part-Revolved Base Feature   4-43
LENS Part-Shell Feature   4-46
Extruded Boss Feature and Convert Entities Sketch tool   4-47
LENS Part-Hole Wizard   4-48
LENS Part - Revolved Boss Thin Feature   4-51
LENS Part - Extruded Boss/Boss Feature and Offset Entities   4-53
LENS Part - Extruded Boss/Boss Feature and Transparent Optical Property   4-55
BULB Part   4-57
BULB Part - Revolved Base Feature   4-58
BULB Part - Revolved Boss Feature and Spline Sketch tool   4-60
BULB Part - Revolved Cut Thin Feature   4-62
BULB Part - Dome Feature   4-64
BULB Part - Circular Pattern Feature   4-65
Customizing Toolbars and Short Cut Keys   4-69
Design Checklist and Goals before Plastic Manufacturing   4-71
Mold Base   4-73
Applying SolidWorks Features for Mold Tooling Design   4-74
Manufacturing Design Issues   4-83
Project Summary   4-84
Project Terminology   4-85
Questions / Exercises   4-89

Project 5 - Swept, Lofted and Additional Features   **5-1**
Project Objective   5-3
Project Overview   5-4
Project Situation   5-5
O-RING Part - Swept Base Feature   5-7

O-RING Part - Design Table   5-9
SWITCH Part - Lofted Base Feature   5-13
SWITCH Part - Dome Feature   5-18
Four Major Categories of Solid Features   5-20
LENSCAP Part   5-20
LENSCAP Part - Extruded Boss/Base, Extruded Cut and Shell Features   5-21
LENSCAP Part - Revolved Cut Thin Feature   5-24
LENSCAP Part - Thread, Swept Feature and Helix/Spiral Curve   5-25
HOUSING Part   5-31
HOUSING Part - Lofted Boss Feature   5-34
HOUSING Part - Second Extruded Boss/Base Feature   5-38
HOUSING Part - Shell Feature   5-39
HOUSING Part - Third Extruded Boss/Base Feature   5-40
HOUSING Part - Draft Feature   5-41
HOUSING Part - Thread with Swept Feature   5-43
HOUSING Part - Handle with Swept Feature   5-48
HOUSING Part - Extruded Cut Feature with Up To Surface   5-53
HOUSING Part - First Rib and Linear Pattern Feature   5-55
HOUSING Part - Second Rib Feature   5-58
HOUSING Part - Mirror Feature   5-61
FLASHLIGHT Assembly   5-64
Assembly Template   5-65
LENSANDBULB Sub-assembly   5-65
BATTERYANDPLATE Sub-assembly   5-70
CAPANDLENS Sub-assembly   5-72
FLASHLIGHT Assembly   5-76
Addressing Interference Issues   5-82
Export Files and eDrawings   5-83
Project Summary   5-86
Project Terminology   5-86
Questions / Exercises   5-89

Project 6 - Top-Down Assembly Modeling and Sheet Metal   **6-1**
Project Objective   6-3
Project Situation   6-4
Top-Down Assembly Modeling   6-5
BOX Assembly Overview   6-8
InPlace Mates and In-Context features   6-10
Part Template and Assembly Template   6-12
Box Assembly and Layout Sketch   6-13
Global Values and Equations   6-17
MOTHERBOARD - Insert Component   6-20
POWERSUPPLY - Insert Component   6-26
Sheet Metal Overview   6-34
Bends   6-34
Relief   6-37
CABINET - Insert Component   6-37
CABINET - Rip Feature and Sheet Metal Bends   6-40
CABINET - Edge Flange   6-42
CABINET - Hole Wizard and Linear Pattern   6-45
CABINET - Sheetmetal Library Feature   6-49

CABINET - Louver Forming tool   6-53
Manufacturing Considerations   6-54
Additional Pattern Options   6-60
CABINET - Formed and Flat States   6-62
CABINET - Sheet Metal Drawing with Configurations   6-64
PEM Fasteners and IGES Components   6-70
Derived Component Pattern   6-74
MOTHERBOARD - Assembly Hole Feature   6-76
Assembly FeatureManager and External References   6-77
Replace Components   6-79
Equations   6-82
Design Tables   6-86
BRACKET Part - Sheet Metal Features   6-89
BRACKET Part - In-Content Features   6-91
BRACKET Part - Edge, Tab, Break Corner and Miter Features   6-93
BRACKET Part - Mirror Component   6-98
MirrorBRACKET Part - Bends, Fold, Unfold and Jog Features   6-102
Project Summary   6-106
Project Terminology   6-107
Questions / Exercises   6-109

Project 7 - SimulationXpress, Sustainability and DFMXpress   **7-1**
Project Objective   7-3
SolidWorks SimulationXpress   7-3
SolidWorks SimulationXpress Wizard   7-7
    Welcome   7-7
    Fixtures   7-7
    Loads   7-8
    Materials   7-8
    Run   7-8
    Results   7-8
    Optimize   7-8
Analyze the MGPMRod Part   7-9
Review of SolidWorks SimulationXpress   7-12
SolidWorks Sustainability   7-13
Life Cycle Assessment   7-14
    Key Elements   7-15
    Carbon Footprint   7-16
    Energy Consumption   7-16
    Air Acidification   7-16
    Water Eutrophication   7-16
SustainabilityXpress Wizard   7-15
    Material Class   7-15
    Material Name   7-15
    Manufacturing Process   7-16
    Generate a Report   7-17
References - Sustainability   7-18
Methodology   7-18
SolidWorks DFMXpress   7-26

DFMXpress Wizard   7-26
  Run   7-27
  Settings   7-27
  Close   7-27
  Help   7-28
Project Summary   7-28
Questions / Exercises   7-28

Project 8 - Intelligent Modeling Techniques   **8-1**
Project Objective   8-3
Design Intent   8-4
Sketch   8-4
  Geometric relations   8-4
  Full Defined Sketch tool   8-5
  SketchXpert   8-8
Equations   8-11
  Dimension driven equations   8-11
  Equation Driven Curve   8-14
  Explicit Driven Equation Curve   8-14
  Parametric Driven Equation Curve   8-16
Curves   8-18
  Curve Through XYZ Points   8-19
  Projected Composite Curves   8-21
Feature - End Conditions   8-23
Along A Vector   8-26
FeatureXpert (Constant Radius)   8-27
Symmetry   8-28
  Bodies to mirror   8-28
Planes   8-30
  Conic Sections   8-31
Assembly   8-32
  Assembly Visualization   8-32
  SolidWorks Sustainability   8-33
  MateXpert   8-34
Drawing   8-34
  DimXpert   8-35
Summary   8-35

Appendix   **A-1**
ECO Form   A-1
Types of Decimal Dimensions (ASME Y14.5M)   A-1
SolidWorks Keyboard Shortcuts   A-3
Windows Shortcuts   A-3
CSWA Certification Introduction   A-5
Intended Audience   A-9
Helpful On-Line information   A-10

Index   **I-1**

The Instructors DVD contains the following PowerPoint presentations, Adobe files along with avi files, Term Projects, quizzes with initial and final models.

Term Project folder
avi folder
Gears using SolidWorks
SolidWorks Drawings Documents in General
General Tolerancing and Fits
Non-Standard Drawing View Types
Assemblies and Mates in General
Weldments
Surface Finish
Design Intent
Part and Drawing Dimensioning
Drafting and Dimensioning Standards
Fundamental ASME Y14.5 Dimensioning Rules
Fastners in General
Alphabet of lines and Precedent of Line Types
Multiview_Visualization
Annotations in Drawings
Tolerance, Weld and Texture Symbols
Open a Drawing Document 2012
Open an Assembly Document 2012
Boolean Operation
Open an Assembly Document 2011
Split Line tool for a Static load analysis using SimulationXpress
Planes, Measurement tool, Equations, Design Tables and Configurations
3D Modeling Features and Strategy
Basic Sketching
SolidWorks Basic Concepts

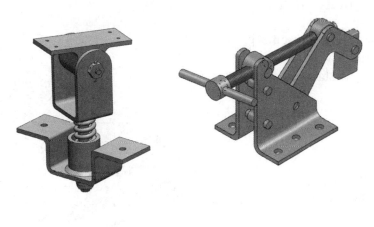

Overview of Projects

Project 1: Fundamentals of Part Modeling

Project 1 introduces the basic concepts behind SolidWorks and the SolidWorks 2012 user interface.

Create file folders and sub-folders to manage projects. Apply System Options and Document Properties. Develop a Custom Part Template. Create three parts: PLATE, ROD and GUIDE.

Utilize the following features: Extruded Boss/Base, Instant3D, Extruded Cut, Fillet, Mirror, Chamfer, Hole Wizard and Linear Pattern. Apply materials and appearances. Learn the SolidWorks interface, how to select the correct Sketch plane, edit sketches and features and copy and paste features.

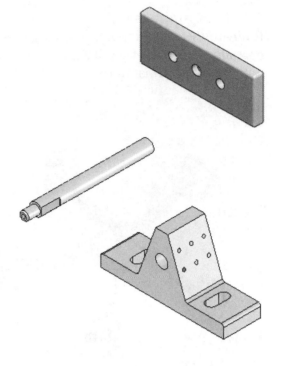

Project 2: Fundamentals of Assembly Modeling

Project 2 introduces the fundamentals of Assembly Modeling (Bottom-up) method along with creating Standard (Coincident, Concentric, Distance, Parallel, Tangent) and SmartMates.

Create an Assembly template. Review the Assembly FeatureManager syntax.

Create two assemblies: GUIDE-ROD and CUSTOMER. Edit component dimensions and address tolerance and fit.

Incorporate design changes into an assembly. Obtain additional SolidWorks parts using 3D ContentCentral and the SolidWorks Toolbox.

All templates, logos, and needed models are included on the enclosed DVD. Copy the models from the DVD to your local hard drive. Work directly from your local hard drive. View the Multi-media files for additional help.

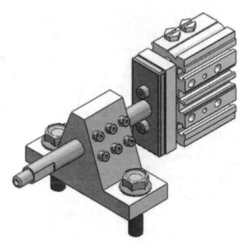

Documents library
ENGDESIGN-W-SOLIDWORKS

Name

- Chapter 1 - Homework
- Chapter 2 - Homework
- Chapter 3 - Homework
- Chapter 4 - Homework
- Chapter 5 - Homework
- Chapter 6 - Homework
- Chapter 7 - Homework
- Chapter 8 - Intelligent Modeling
- COMPONENTS
- DFMXpress
- Extra Model Files - Templates
- LOGO
- SimulationXpress
- SustainabilityXpress
- VENDOR-COMPONENTS

Project 3: Fundamentals of Drawing

Project 3 covers the development of a customized drawing template.

Learn the two Sheet Format modes: 1.) *Edit Sheet Format* 2.) *Edit Sheet*

Develop and insert a Company logo from a bitmap or picture file.

Create the GUIDE drawing with Custom Properties, various drawing views and a Revision table.

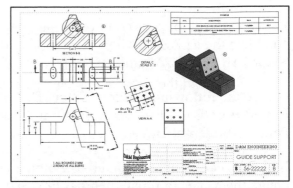

Create the GUIDE-ROD drawing with Custom Properties, an Exploded Isometric view with a Bill of Materials and balloons.

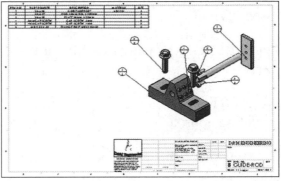

Project 4: Extrude and Revolve Feature

Project 4 focuses on the customer's design requirements. Create four key FLASHLIGHT components: BATTERY, BATTERYPLATE, LENS, and BULB. Develop an ANSI - IPS Part Template.

Create the BATTERY and BATTERYPLATE part with the Extruded Boss/Base feature and the Instant3D tool.

Create the LENS and BULB with the Revolved Boss/Base feature. Utilize the following features: Extruded Boss/Base, Extruded Cut, Revolved Base, Revolved Cut, Dome, Shell, Fillet and Circular Pattern.

Utilize the Mold tools to create the cavity plate for the BATTERYPLATE.

Tangent edges are displayed for educational purposes in the book.

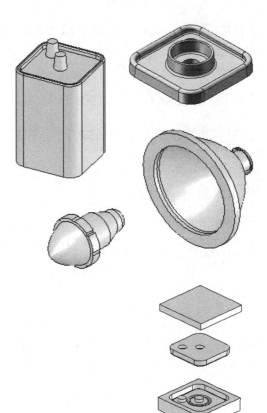

Project 5: Swept, Lofted and Additional Features

Project 5 develops four additional components to complete the FLASHLIGHT assembly: O-RING, SWITCH, LENSCAP and HOUSING.

Utilize the following features: Swept Boss/Base, Lofted Boss/Base, Rib, Linear Pattern, Circular Pattern, Draft and Dome.

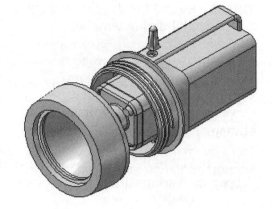

Insert the components and sub-assemblies and create the Mates to finish the final FLASHLIGHT assembly.

Project 6: Top-Down assembly and Sheet Metal parts

Project 6 focuses on the Top-Down assembly modeling approach. Develop a Layout Sketch.

Create components and modify them In-Context of the assembly.

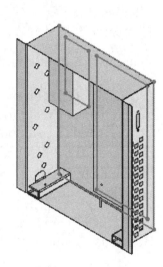

Create Sheet metal features. Utilize the following features: Rip, Insert Sheet metal Bends, Base Flange, Edge Flange, Miter Flange, Break Corners, Hem and more.

Utilize the Die Cut Feature and Louver Form tool.

Add IGES format part files from the Internet.

Replace fasteners in the assembly and redefine mates.

Utilize equations, global values and a Design Table to create multiple configurations of the BOX assembly.

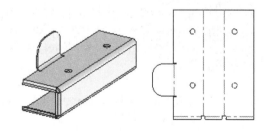

☀ The book is designed to expose the new user to many tools, techniques and procedures. It may not always use the most direct tool or process.

Project 7: SimulationXpress, Sustainability/SustainabilityXpress and DFMXpress

Project 7 introduces a few general SolidWorks analysis tools: SimulationXpress, Sustainability/ SustainabilityXpress and DFMXpress.

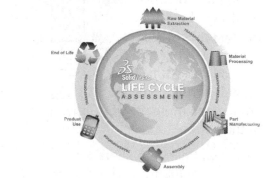

Execute a SolidWorks SimulationXpress analysis on a part. Determine if the part can support an applied load under a static load condition.

Perform a SustainabilityXpress analysis on a part. View the environmental impact calculated in four key areas: *Carbon Footprint, Energy Consumption, Air Acidification* and *Water Eutrophication*.

Material and Manufacturing process region and Usage region are used as input variables. Compare similar materials and environmental impacts to the base line design.

Implement DFMXpress on a part. DFMXpress is an analysis tool that validates the manufacturability of SolidWorks parts. Use DFMXpress to identify design areas that may cause problems in fabrication or increase the costs of production.

Perform short stand-alone step-by-step tutorials to practice and reinforce the subject matter and objectives.

Project 8: Intelligent Modeling Techniques

Project 8 introduces some of the available tools in SolidWorks to perform intelligent modeling. Intelligent modeling is incorporating design intent into the definition of the sketch, feature, part, and assembly or drawing document. Intelligent modeling is most commonly addressed through design intent.

Perform short stand-alone step-by-step tutorials to practice and reinforce the subject matter and objectives. Learn by doing, not just by reading! All needed models for this chapter are located on the DVD in the book!

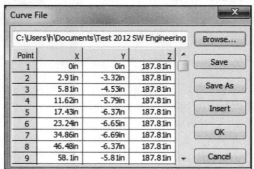

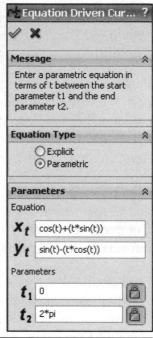

Cover and perform short tutorials on the following topics: *Fully Defined Sketch tool, SketchXpert, Equations, Explicit Equation Driven Curve tool, Parametric Equation Driven Curve tool, Curve Through XYZ Points tool, Feature- End Condition options, Curve Through XYZ Points tool, FeatureXpert, Symmetry, Assemblies, Drawings and more.*

The Instructors DVD contains the following PowerPoint presentations, Adobe files along with avi files, Term Projects, quizzes with initial and final models.

What is SolidWorks?

SolidWorks® is a mechanical design automation software package used to build parts, assemblies and drawings that takes advantage of the familiar Microsoft® Windows graphical user interface.

SolidWorks is an easy to learn design and analysis tool, (SolidWorks SimulationXpress, SolidWorks Motion, SolidWorks Flow Simulation, etc.) which makes it possible for designers to quickly sketch 2D and 3D concepts, create 3D parts and assemblies and detail 2D drawings.

In SolidWorks, you create 2D and 3D sketches, 3D parts, 3D assemblies and 2D drawings. The part, assembly and drawing documents are related. Additional information on SolidWorks and its family of products can be obtained at their URL, www.SolidWorks.com.

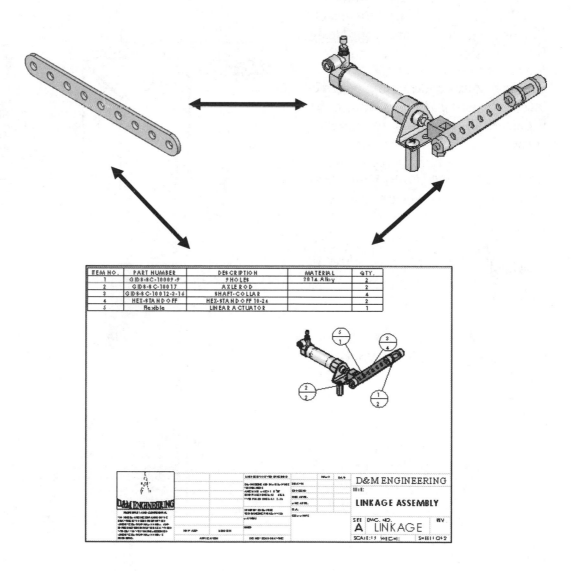

Features are the building blocks of parts. Use feature tools such as: Extruded Boss/Base, Extruded Cut, Fillet, etc. from the Features tab in the CommandManager to create 3D parts.

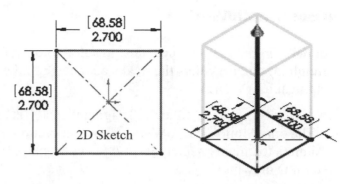

Extruded features begin with a 2D sketch created on a Sketch plane.

The 2D sketch is a profile or cross section. Use sketch tools such as: Line, Center Rectangle, Slot, Circle, etc. from the Sketch tab in the CommandManager to create a 2D sketch. Sketch the general shape of the profile. Add geometric relationships and dimensions to control the exact size of the geometry and your Design Intent. Design for change!

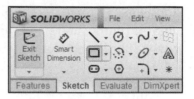

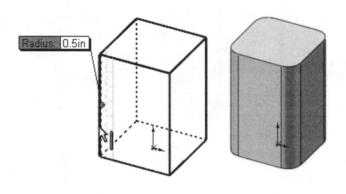

Create features by selecting edges or faces of existing features, such as a Fillet. The Fillet feature rounds sharp corners.

Dimensions drive features. Change a dimension, and you change the size of the part.

Use Geometric relationships: Vertical, Horizontal, Parallel, etc. and various End Conditions to maintain the Design Intent.

Create a hole that penetrates through a part (Through All). SolidWorks maintains relationships through the change.

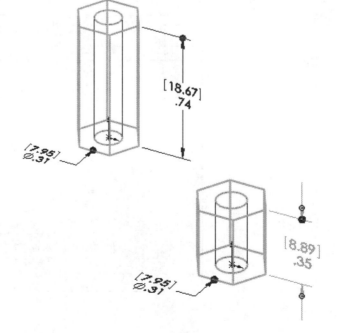

The step-by-step approach used in this text allows you to create, edit and modify parts, assemblies and drawings. Change is an integral part of design!

About the Book

You will find a wealth of information in this book. The following conventions are used throughout the text:

- The term document is used to refer a SolidWorks part, drawing or assembly file.

- The list of items across the top of the SolidWorks interface is the Main menu. Each item in the Main menu has a pull-down menu. When you need to select a series of commands from these menus, the following format is used; Click **Insert**, **Reference Geometry**, **Plane** from the Main bar. The Plane PropertyManager is displayed.

- Screen shots in the book were made using SolidWorks 2012 SP0 running Windows® 7.

- The ANSI overall drafting standard and Third Angle projection is used as the default setting in this text. IPS (inch, pound, second) and MMGS (millimeter, gram, second) unit systems are used.

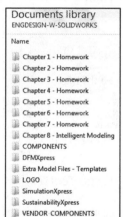

- The book is organized into eight Projects. Each Project is focus on a specific subject or feature.

- Use the enclosed Multi-Media DVD in the book to review features, procedures and to obtain models used in the text.

The book is design to expose the new user to many tools, techniques and procedures. It may not always use the most direct tool or process. Learn by doing, not just by reading!

The following command syntax is used throughout the text. Commands that require you to perform an action are displayed in **Bold** text.

Format:	Convention:	Example:
Bold	<ul><li>All commands actions.</li><li>Selected icon button.</li><li>Selected geometry: line, circle.</li><li>Value entries.</li></ul>	<ul><li>Click **Options** from the Menu bar toolbar.</li><li>Click the **Extruded Boss/Base** feature. 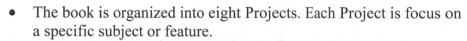</li><li>Click **Corner Rectangle** from the Consolidated Sketch toolbar. 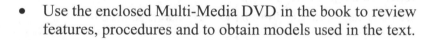</li><li>Select the **centerpoint**.</li><li>Enter **3.0** for Radius.</li></ul>
Capitalized	<ul><li>Filenames.</li><li>First letter in a feature name.</li></ul>	<ul><li>Save the **FLASHLIGHT** assembly.</li><li>Click the **Fillet** feature.</li></ul>

Windows Terminology in SolidWorks

The mouse buttons provide an integral role in executing SolidWorks commands. The mouse buttons execute commands, select geometry, display Shortcut menus and provides information feedback.

A summary of mouse button terminology is displayed below:

Item:	Description:
Click	Press and release the left mouse button.
Double-click	Double press and release the left mouse button.
Click inside	Press the left mouse button. Wait a second, and then press the left mouse button inside the text box. Use this technique to modify Feature names in the FeatureManager design tree.
Drag	Point to an object, press and hold the left mouse button down. Move the mouse pointer to a new location. Release the left mouse button.
Right-click	Press and release the right mouse button. A Shortcut menu is displayed. Use the left mouse button to select a menu command.
ToolTip	Position the mouse pointer over an Icon (button). The tool name is displayed below the mouse pointer.
Large ToolTip	Position the mouse pointer over an Icon (button). The tool name and a description of its functionality are displayed below the mouse pointer.
Mouse pointer feedback	Position the mouse pointer over various areas of the sketch, part, assembly or drawing. The cursor provides feedback depending on the geometry.

A mouse with a center wheel provides additional functionality in SolidWorks. Roll the center wheel downward to enlarge the model in the Graphics window. Hold the center wheel down. Drag the mouse in the Graphics window to rotate the model. Review various Windows terminology that describe; menus, toolbars and commands that constitute the graphical user interface in SolidWorks.

Visit the SolidWorks website: http://www.solidworks.com /sw/support/hardware.html to view their supported operating systems and hardware requirements.

SolidWorks					
Operating Systems	SolidWorks 2009	SolidWorks 2010	SolidWorks 2011	SolidWorks 2012	(SolidWorks 2013)
Windows 7	✗	✔ *	✔	✔	✔
Windows Vista	✔	✔	✔	✔	✔
Windows XP	✔	✔	✔	✔	✗
Minimum Hardware	Configuring a SolidWorks Workstation				
RAM	2 GB or more				
Disk Space	5 GB or more				
Video Card	Certified cards and drivers				
Processor	Intel or AMD with SSE2 support. 64-bit operating system recommended				
Install Media	DVD Drive or Broadband Internet Connection				

All models were created using SolidWorks 2012 SP0.

The book does not cover starting a SolidWorks session in detail for the first time. A default SolidWorks installation presents you with several options.

All templates, logos and needed models for this book are included on the enclosed DVD. Copy the information from the DVD to your local hard drive. Work from your local hard drive. DO NOT WORK directly from the DVD.

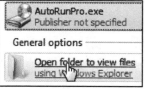

View the Multi-Media files on the DVD for additional help.

The book is designed to expose the new user to numerous tools and procedures. It may not always use the simplest and most direct process.

The book does not cover starting a SolidWorks session in detail for the first time. A default SolidWorks installation presents you with several options. For additional information for an Education Edition, visit the following sites: http://www.solidworks.com/goedu and http://www.solidworks.com/sw/education/ 6443_ENU_HTML.htm.

Documents library
ENGDESIGN-W-SOLIDWORKS

Name

- Chapter 1 - Homework
- Chapter 2 - Homework
- Chapter 3 - Homework
- Chapter 4 - Homework
- Chapter 5 - Homework
- Chapter 6 - Homework
- Chapter 7 - Homework
- Chapter 8 - Intelligent Modeling
- COMPONENTS
- DFMXpress
- Extra Model Files - Templates
- LOGO
- SimulationXpress
- SustainabilityXpress
- VENDOR-COMPONENTS

goEDU

Installation Instructions

Education User License Agreement (EULA)

System Requirements

Product Description

Instructions to Access Instructors' Curriculum

Workgroup PDM Installation Instructions

Workgroup PDM Video Tutorials

Notes:

Project 1
Fundamentals of Part Modeling

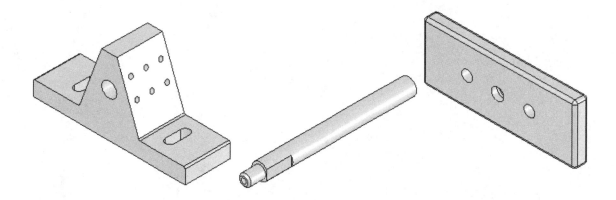

Below are the desired outcomes and usage competencies based on the completion of Project 1.

Project Desired Outcomes:	Usage Competencies:
• Understand and navigate the SolidWorks User Interface (UI) and CommandManager.	• Start a SolidWorks session. • Use the default SolidWorks User Interface, CommandManager, Menus and Toolbars.
• Address File Management.	• Create file folders for various Project Models and Document Templates.
• PART-ANSI-MM Part Template.	• Apply and modify System Options and Document Properties.

• Three Parts:
  o PLATE
  o ROD
  o GUIDE
	• Create 2D Sketch profiles on the correct Sketch plane.
	• Knowledge to create and modify the following SolidWorks 3D features: Extruded Boss/Base, Extruded Cut, Instant3D, Fillet, Mirror, Chamfer, Hole Wizard and Linear Pattern.

Notes:

Project 1 - Fundamentals of Part Modeling

Project Objective

Provide a comprehensive understanding of the SolidWorks 2012 default User Interface and CommandManager: *Menu bar toolbar, Menu bar menu, Drop-down menu, Right-click Pop-up menus, Context toolbars / menus, Fly-out FeatureManager, System feedback icons, Confirmation Corner, Heads-up View toolbar and an understanding of System Options, Document Properties, Part templates, File management and more.*

Create three folders: PROJECTS, MY-TEMPLATES and VENDOR COMPONENTS.

Start a SolidWorks session. Create the PART-MM-ANSI Part template utilized for the parts in this project.

Create three parts:

- PLATE

- ROD

- GUIDE

On the completion of this project, you will be able to:

- Start a SolidWorks 2012 session.

- Understand and navigate through the SolidWorks (UI) and CommandManager.

- Create file folders to manage and organize projects.

- Apply and understand System Options and Document Properties.

- Create a Part template.

- Open, Save and Close Part documents and templates.

- Create 2D sketch profiles on the proper Sketch plane.

- Utilize the following Sketch tools: Line, Corner Rectangle, Circle, Arc, Trim Entities, Convert Entities, Straight Slot and Mirror Entities.

- Apply and edit sketch dimensions.

- Insert the following Geometric relations: Horizontal, Vertical, Symmetric, Equal and Coradial.

- Create and modify 3D parts.

- Create and modify the following SolidWorks features: Extruded Boss\Base, Extruded Cut, Fillet, Mirror, Hole Wizard, Chamfer and Linear Pattern.

☼ Screen shots and illustrations in the book display the SolidWorks user default setup.

Project Situation

You receive an email or a concept sketch from a customer. The customer is retooling an existing assembly line.

You are required to design and manufacture a ROD. The ROD part is 10mm in diameter x 100mm in length.

One end of the ROD connects to an existing customer GUIDE CYLINDER assembly.

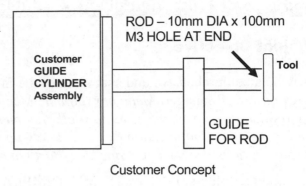

Customer Concept

The other end of the ROD connects to the customer's tool. The ROD contains a 3mm hole and a key-way to attach the tool.

The ROD requires a support GUIDE. The ROD travels through the support GUIDE. The GUIDE-ROD assembly is the finished customer product.

The GUIDE-ROD assembly is a component used in a low volume manufacturing environment. Investigate a few key design issues:

- How will the customer use the GUIDE-ROD assembly?

- How are the parts: PLATE, ROD and GUIDE used in the GUIDE-ROD assembly?

- Does the GUIDE-ROD assembly affect other components?

- Identify design requirements for load, structural integrity or other engineering properties.

- Identify cost effective materials.

- How are parts manufactured?

- What are their critical design features?

- How will each part behave when modified?

You may not have access to all of the required design information. Placed in a concurrent engineering situation, you are dependent on others and are ultimately responsible for the final design.

Dimensions for this project are in millimeters. Design information is provided from various sources. Ask questions. Part of the learning experience is to know which questions to ask and who to ask.

The ROD part requires support. During the manufacturing operations, the ROD exhibits unwanted deflection.

The engineering group calculates working loads on test samples. Material test samples include 6061-0(SS), AISI 304 steel and a machinable plastic Nylon 101.

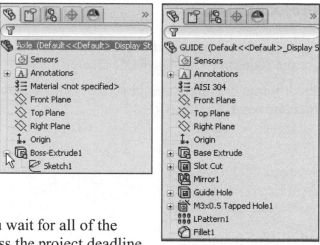

The engineering group recommends AISI 304 steel for the ROD and GUIDE parts.

In the real world, there are numerous time constraints. The customer requires a quote, design sketches and a delivery schedule, YESTERDAY! If you wait for all of the required design information, you will miss the project deadline.

You create a rough concept sketch with notes in a design review meeting.

Your colleagues review and comment on the concept sketch.

The ROD cannot mount directly to the customer's GUIDE CYLINDER assembly without a mounting PLATE part.

The PLATE part mounts to the customer's PISTON PLATE part in the GUIDE CYLINDER assembly.

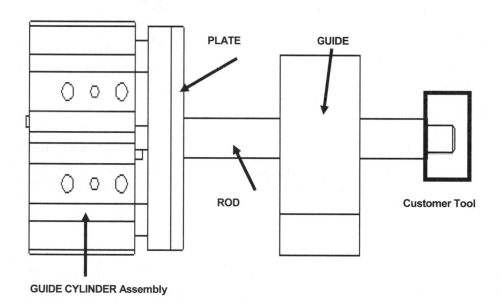

Concept Sketch and Design
Review Notes

Project Overview

A key goal in this project is to create three individual parts: PLATE, ROD and GUIDE for the requested GUIDE-ROD assembly.

Incorporate the three individual parts into the GUIDE-ROD assembly in Project 2 using Standard mates.

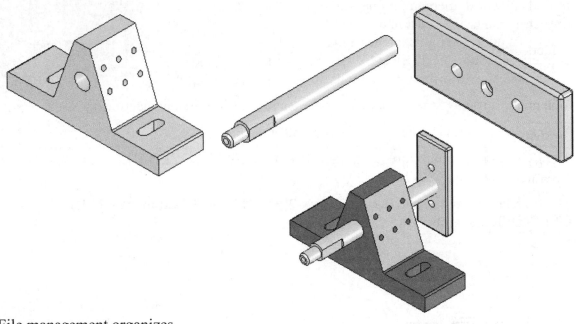

File management organizes documents and folders. The GUIDE-ROD assembly consists of numerous documents.

Create file folders to organize parts, assemblies, drawings, templates and vendor components.

Drafting standards such as ANSI or ISO and unit systems such as IPS (inch, pound, second) or MMGS (millimeter, gram, second) are defined in Document Properties and are stored in the Part, Assembly or Drawing template.

Plan file organization and templates before you create parts, assemblies or drawings. Create a PART-MM-ANSI Part template for the metric (MMGS) parts required for the GUIDE-ROD assembly.

File Management

File management organizes parts, assemblies and drawings. Why do you need file management? A large assembly contains hundreds or even thousands of parts.

Parts and assemblies are distributed between team members to save time. Design changes occur frequently in the development process. How do you manage and control changes? Answer: Through file management. File management is a very important tool in the development process.

The GUIDE-ROD assembly consists of many files. Utilize file folders to organize projects, vendor parts and assemblies, templates and various libraries.

Folders exist on the local hard drive, example C:\. Folders also exist on a network drive, example Z:\. The letters C:\ and Z:\ are used as examples for a local drive and a network drive respectfully. The following example utilizes the folder, "Documents" to contain the folders for your projects.

Activity: File Management

Create a New file folder in Windows. Note: The procedure will be different depending on your operating system.

1)   Click **Start** from the Windows taskbar.

2)   Click **Documents** in the Windows dialog box.

3)   Click **New folder** from the Windows Main menu.

Enter the new folder name.

4)   Enter **ENGDESIGN-W-SOLIDWORKS**. Note: In this book, all models, assemblies and templates are saved to the ENGDESIGN-W-SOLIDWORKS folder.

During the initial SolidWorks installation, you are requested to select either the ISO or ANSI drafting standard. ISO is typically a European overall drafting standard and uses First Angle Projection. The book was written using the ANSI (US) overall drafting standard and Third Angle Projection for drawings.

The below procedure can be different if you are on a network system or using a different operating system. The goal of this section is to organize files and folders. This is very important when you start creating assemblies and drawings with internal and external references and using different computers.

Create the first sub-folder. Note: The procedure will be different depending on your operating system.

5)   Double-click on the **ENGDESIGN-W-SOLIDWORKS** file folder.

6) Click **New folder**. A new folder icon is displayed.

7) Enter **MY-TEMPLATES** for the folder name.

Create the second sub-folder.
8) Click **New folder**.

9) Enter **PROJECTS** for the second sub-folder name.

Create the third sub-folder.
10) Click **New folder**.

11) Enter **VENDOR-COMPONENTS** for the third sub-folder name.

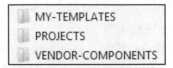

Return to the ENGDESIGN-W-SOLIDWORKS folder.
12) Click the **Back arrow** button to return to the ENGDESIGN-W-SOLIDWORKS folder.

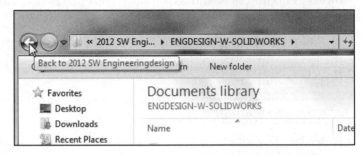

Utilize the MY-TEMPLATES folder and the PROJECTS folder throughout the text. Store the Part template, Assembly template and Drawing template in the MY-TEMPLATES folder.

Store the parts, assemblies and drawings that you create in the PROJECTS folder.

Store parts and assemblies, which you download from the internet or 3D ContentCentral in the VENDOR-COMPONENTS folder.

Start a SolidWorks Session

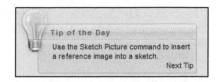

The SolidWorks application is located in the Programs folder. SolidWorks displays the Tip of the Day at the bottom of the Task Pane. Read the Tip of the Day to obtain additional information on SolidWorks.

Create a New part. Click File, New from the Menu bar menu or click New from the Menu bar toolbar. There are two options for new documents: *Novice* and *Advanced*. Select the Advanced option. The Advanced option provides the ability to create additional tabs for custom templates. Select the Part document.

Activity: Start a SolidWorks 2012 Session

Start a SolidWorks 2012 session.

13) Click **Start** on the Windows taskbar.

14) Click **All Programs**.

15) Click the **SolidWorks 2012** folder.

16) Click **SolidWorks 2012** application. The SolidWorks program window opens. Note: Do not open a document at this time!

If available, double-click the SolidWorks icon on the Windows Desktop to start a SolidWorks session.

Read the Tip of the Day dialog box.

17) If you do not see this screen, click the **SolidWorks Resources** tab on the right side of the Graphics window located in the Task Pane.

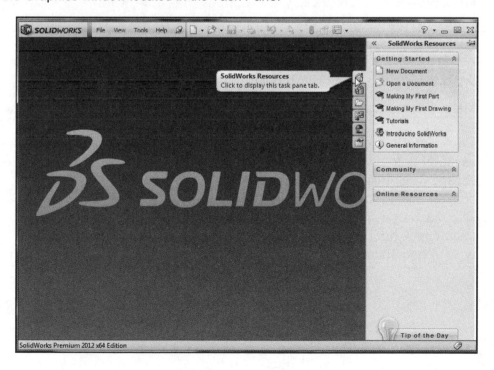

Activity: Understand the SolidWorks User Interface and CommandManager

Menu bar toolbar

The SolidWorks 2012 (UI) is designed to make maximum use of the Graphics window. The Menu Bar toolbar contains a set of the most frequently used tool buttons from the Standard toolbar. The available default tools are:

- **New** ▢ - Creates a new document.

- **Open** ▱ - Opens an existing document.

- **Save** ▣ - Saves an active document.

- **Print** ▱ - Prints an active document.

- **Undo** ↺ - Reverses the last action.

- **Select** ▱ ▾ - Selects Sketch entities, components and more.

- **Rebuild** ▯ - Rebuilds the active part, assembly or drawing.

- **File Properties** ▱ - Shows the summary information on the active document.

- **Options** ▤ - Changes system options and Add-Ins for SolidWorks.

🔆 The Search *Knowledge Base* and *Community Forum* options from the Menu Bar toolbar requires Internet access.

Menu bar menu

Click SolidWorks in the Menu bar toolbar to display the Menu bar menu. SolidWorks provides a context-sensitive menu structure. The menu titles remain the same for all three types of documents, but the menu items change depending on which type of document is active.

Example: The Insert menu includes features in part documents, mates in assembly documents, and drawing views in drawing documents. The display of the menu is also dependent on the workflow customization that you have selected. The default menu items for an active document are: *File*, *Edit*, *View*, *Insert*, *Tools*, *Window*, *Help* and *Pin*.

The Pin option displays the Menu bar toolbar and the Menu bar menu as illustrated. Throughout the book, the Menu bar menu and the Menu bar toolbar are referred to as the Menu bar.

Drop-down menu

SolidWorks takes advantage of the familiar Microsoft® Windows user interface. Communicate with SolidWorks through drop-down menus, Context sensitive toolbars, Consolidated toolbars or the CommandManager tabs.

A command is an instruction that informs SolidWorks to perform a task.

To close a SolidWorks drop-down menu, press the Esc key. You can also click any other part of the SolidWorks Graphics window or click another drop-down menu.

Right-click

Right-click in the Graphics window on a model, or in the FeatureManager on a feature or sketch to display the Context-sensitive toolbar. If you are in the middle of a command, this toolbar displays a list of options specifically related to that command.

The most commonly used tools are located in the Pop-up Context toolbar and CommandManager.

Press the **s** key to view/access previous command tools in the Graphics window.

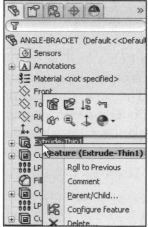

Consolidated toolbar

Similar commands are grouped together in the CommandManager. Example: Variations of the Rectangle sketch tool are grouped in a single fly-out button as illustrated.

If you select the Consolidated toolbar button without expanding:

- For some commands such as Sketch, the most commonly used command is performed. This command is the first listed and the command shown on the button.

- For commands such as rectangle, where you may want to repeatedly create the same variant of the rectangle, the last used command is performed. This is the highlighted command when the Consolidated toolbar is expanded.

System feedback

SolidWorks provides system feedback by attaching a symbol to the mouse pointer cursor.

The system feedback symbol indicates what you are selecting or what the system is expecting you to select.

As you move the mouse pointer across your model, system feedback is displayed in the form of a symbol, riding next to the cursor as illustrated. This is a valuable feature in SolidWorks.

Confirmation Corner

When numerous SolidWorks commands are active, a symbol or a set of symbols are displayed in the upper right hand corner of the Graphics window. This area is called the Confirmation Corner.

When a sketch is active, the confirmation corner box displays two symbols. The first symbol is the sketch tool icon. The second symbol is a large red X. These two symbols supply a visual reminder that you are in an active sketch. Click the sketch symbol icon to exit the sketch and to save any changes that you made.

When other commands are active, the confirmation corner box provides a green check mark and a large red X. Use the green check mark to execute the current command. Use the large red X to cancel the command.

Heads-up View toolbar

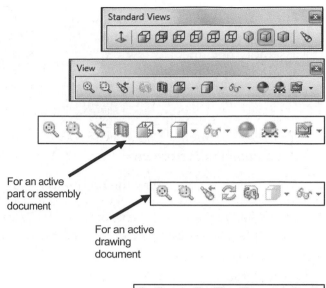

SolidWorks provides the user with numerous view options from the Standard Views, View and Heads-up View toolbar.

The Heads-up View toolbar is a transparent toolbar that is displayed in the Graphics window when a document is active.

For an active part or assembly document

For an active drawing document

You can hide, move or modify the Heads-up View toolbar. To modify the Heads-up View toolbar: right-click on a tool and select or deselect the tools that you want to display.

The following views are available: Note: *The available views are document dependent.*

- *Zoom to Fit* : Zooms the model to fit the Graphics window.

- *Zoom to Area* : Zooms to the areas you select with a bounding box.

- *Previous View* : Displays the previous view.

- *Section View* : Displays a cutaway of a part or assembly, using one or more cross section planes.

- *View Orientation* : Provides the ability to select a view orientation or the number of viewports. The available options are: *Top, Isometric, Trimetric, Dimetric, Left, Front, Right, Back, Bottom, Single view, Two view - Horizontal, Two view - Vertical Four view.*

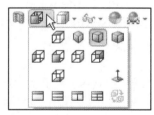

- *Display Style* : Provides the ability to display the style for the active view: The available options are: *Wireframe, Hidden Lines Visible, Hidden Lines Removed, Shaded, Shaded With Edges.*

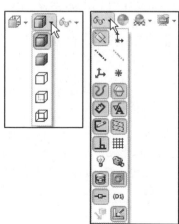

- *Hide/Show Items* : Provides the ability to select items to hide or show in the Graphics window. Note: *The available items are document dependent.*

- *Edit Appearance* : Provides the ability to edit the appearance of entities of the model.

- *Apply Scene* 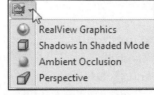: Provides the ability to apply a scene to an active part or assembly document. View the available options.

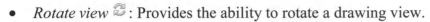

- *View Setting* : Provides the ability to select the following settings: *RealView Graphics*, *Shadows in Shaded Mode*, *Ambient Occlusion* and *Perspective*.

- *Rotate view* : Provides the ability to rotate a drawing view.

- *3D Drawing View* : Provides the ability to dynamically manipulate the drawing view to make a selection.

☼ The default part and document setting displays the grid. To deactivate the grid, click **Options**, **Document Properties** tab. Click **Grid/Snaps**, uncheck the **Display grid** box.

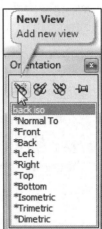

☼ To add a custom view to the Heads-up View toolbar, press the **space** key. The Orientation dialog box is displayed. Click the **New View** tool. The Name View dialog box is displayed. Enter a new **named** view. Click **OK**.

☼ Press the **g** key to activate the Magnifying glass tool. Use the Magnifying glass tool to inspect a model and make selections without changing the overall view.

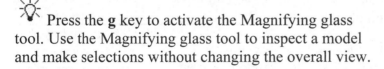

☼ The book does not cover starting a SolidWorks session in detail for the first time. A default SolidWorks installation presents you with several options. For additional information for an Education Edition, visit the following sites: http://www.solidworks.com/goedu and http://www.solidworks.com/sw/education/6443_ENU_HTML.htm.

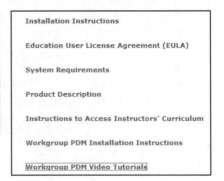

goEDU

Installation Instructions

Education User License Agreement (EULA)

System Requirements

Product Description

Instructions to Access Instructors' Curriculum

Workgroup PDM Installation Instructions

Workgroup PDM Video Tutorials

SolidWorks CommandManager

The SolidWorks CommandManager is a *Context-sensitive toolbar* that automatically updates based on the toolbar you want to access. By default, it has toolbars embedded in it based on your active document type. When you click a tab below the CommandManager, it updates to display that toolbar. Example, if you click the Sketch tab, the Sketch toolbar is displayed. The default Part tabs are: *Features, Sketch, Evaluate, DimXpert* and *Office Products*.

If you have SolidWorks, SolidWorks Professional, or SolidWorks Premium; the Office Products tab appears on the CommandManager.

Below is an illustrated CommandManager for a default Part document.

Select the Add-In directly from the Office Products tab.

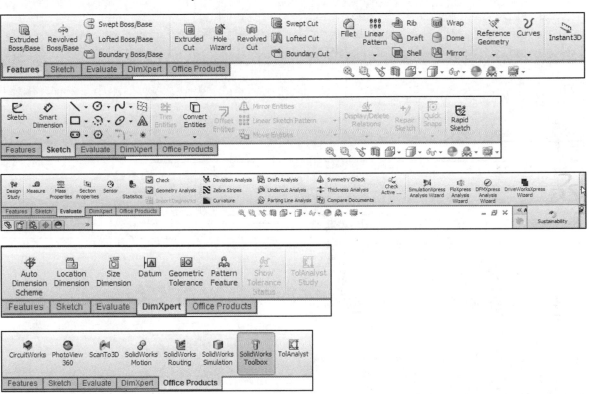

To customize the CommandManager, right-click on a tab and select Customize CommandManager.

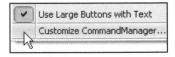

Below is an illustrated CommandManager for a default Drawing document. The default Drawing tabs are: *View Layout*, *Annotation*, *Sketch*, *Evaluate* and *Office Products*.

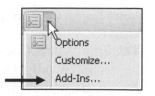

If you have SolidWorks, SolidWorks Professional, or SolidWorks Premium, the Office Products tab appears on the CommandManager.

 Double-clicking the CommandManager when docked will make it float. Double-clicking the CommandManager when it is floating will return it to its last position in the Graphics window.

Select the Add-In directly from the Office Products tab.

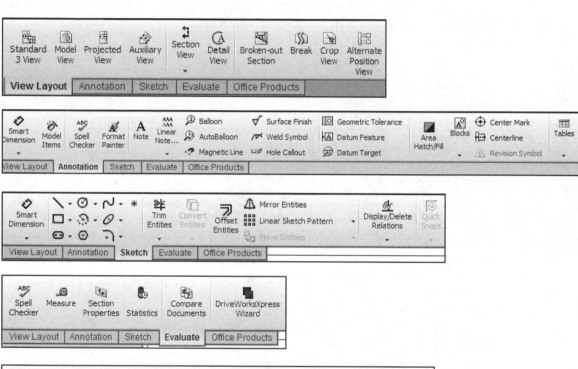

To add a custom tab to your CommandManager, right-click on a tab and click Customize CommandManager from the drop-down menu. The Customize dialog box is displayed. You can also select to add a blank tab as illustrated and populate it with custom tools from the Customize dialog box.

Below is an illustrated CommandManager for a default Assembly document. The default Assembly tabs are: *Assembly*, *Layout*, *Sketch*, *Evaluate* and *Office Products*.

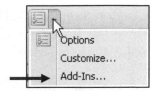

If you have SolidWorks, SolidWorks Professional, or SolidWorks Premium, the Office Products tab appears on the CommandManager.

 Select the Add-In directly from the Office Products tab.

 The Instant3D and Rapid Sketch tool is active by default.

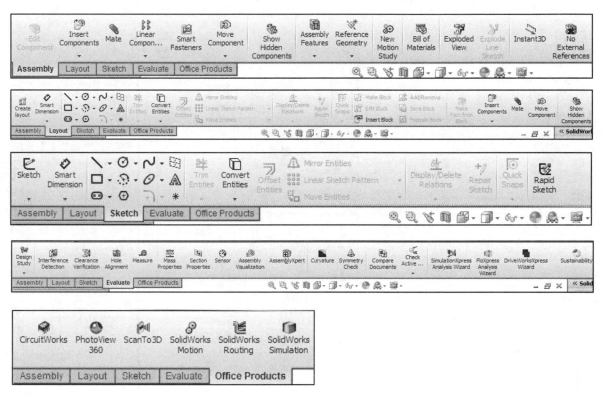

By default, the illustrated options are selected in the Customize box for the CommandManager. Right-click on an existing tab, and click Customize CommandManager to view your options.

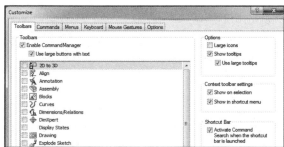

Drag or double-click the CommandManager and it becomes a separate floating window. Once it is floating, you can drag the CommandManager anywhere on or outside the SolidWorks window.

To dock the CommandManager when it is floating, do one of the following:

- While dragging the CommandManager in the SolidWorks window, move the pointer over a docking icon - ◸ Dock above, ◂ Dock left, ▸ Dock right and click the needed command.
- Double-click the floating CommandManager to revert the CommandManager to the last docking position.

☼ Screen shots in the book were made using SolidWorks 2012 SP0 running Windows® 7 Ultimate.

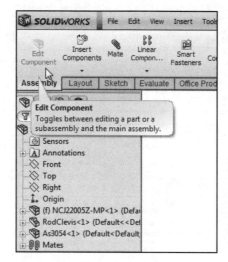

Save space in the CommandManager, right-click in the CommandManager and uncheck the Use Large Buttons with Text box. This eliminates the text associated with the tool.

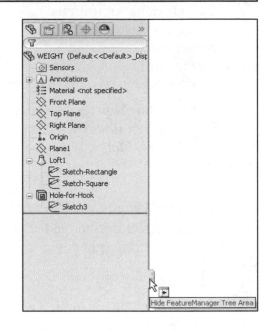

FeatureManager Design Tree

The FeatureManager design tree is located on the left side of the SolidWorks Graphics window. The FeatureManager provides a summarized view of the active part, assembly, or drawing document. The tree displays the details on how the part, assembly or drawing document was created.

Understand the FeatureManager design tree to troubleshoot your model. The FeatureManager is used extensively throughout this book.

The FeatureManager consists of five default tabs:

- *FeatureManager design tree*

- *PropertyManager*

- *ConfigurationManager*

- *DimXpertManager*

- *DisplayManager*

Select the Hide FeatureManager Tree Area arrows 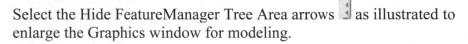 as illustrated to enlarge the Graphics window for modeling.

The Sensors tool ⌖ Sensors located in the FeatureManager monitors selected properties in a part or assembly and alerts you when values deviate from the specified limits. There are four sensor types: *Mass properties*, *Measurement*, *Interference Detection* and *Simulation data*.

DimXpert provides the ability to graphically check if the model is fully dimensioned and toleranced. DimXpert automatically recognizes manufacturing features. Manufacturing features are *not SolidWorks features*. Manufacturing features are defined in 1.1.12 of the ASME Y14.5M-1994 Dimensioning and Tolerancing standard. See SolidWorks Help for additional information.

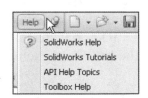

Various commands provide the ability to control what is displayed in the FeatureManager design tree. They are:

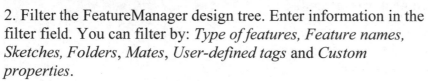
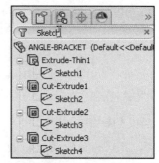

1. Show or Hide FeatureManager items.

⌖ Click **Options** 🗐 from the Menu bar. Click **FeatureManager** from the System Options tab. **Customize** your FeatureManager from the Hide/Show Tree Items dialog box.

2. Filter the FeatureManager design tree. Enter information in the filter field. You can filter by: *Type of features, Feature names, Sketches, Folders*, *Mates, User-defined tags* and *Custom properties*.

⌖ Tags are keywords you can add to a SolidWorks document to make them easier to filter and to search. The Tags ⌀ icon is located in the bottom right corner of the Graphics window.

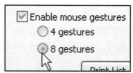

⌖ To collapse all items in the FeatureManager, **right-click** and select **Collapse items**, or press the **Shift +C** keys.

The FeatureManager design tree and the Graphics window are dynamically linked. Select sketches, features, drawing views, and construction geometry in either pane.

Split the FeatureManager design tree and either display two FeatureManager instances, or combine the FeatureManager design tree with the ConfigurationManager or PropertyManager.

Move between the FeatureManager design tree, PropertyManager, ConfigurationManager, and DimXpertManager by selecting the tabs at the top of the menu.

⌖ Right-click and drag in the Graphics area to display the Mouse Gesture wheel. You can customize the default commands for a sketch, part, assembly or drawing.

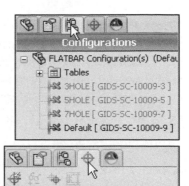

The ConfigurationManager is located to the right of the FeatureManager. Use the ConfigurationManager to create, select and view multiple configurations of parts and assemblies.

The icons in the ConfigurationManager denote whether the configuration was created manually or with a design table.

The DimXpertManager tab provides the ability to insert dimensions and tolerances manually or automatically. The DimXpertManager provides the following selections: *Auto Dimension Scheme* , *Show Tolerance Status* , *Copy Scheme* and *TolAnalyst Study* .

TolAnalyst is available in SolidWorks Premium.

Fly-out FeatureManager

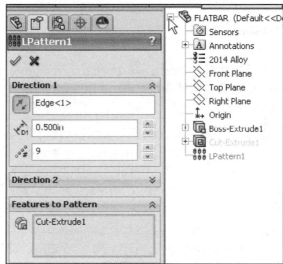

The fly-out FeatureManager design tree provides the ability to view and select items in the PropertyManager and the FeatureManager design tree at the same time.

Throughout the book, you will select commands and command options from the drop-down menu, fly-out FeatureManager, Context toolbar or from a SolidWorks toolbar.

Another method for accessing a command is to use the accelerator key. Accelerator keys are special key strokes, which activate the drop-down menu options. Some commands in the menu bar and items in the drop-down menus have an underlined character.

Press the Alt key followed by the corresponding key to the underlined character activates that command or option.

Press the **s** key to view the Shortcut toolbar. Shortcut menus provide convenient access to previous applied tools and commands.

Illustrations may vary depending on your SolidWorks version and operating system.

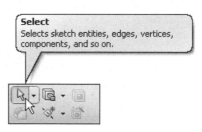

Task Pane

The Task Pane is displayed when a SolidWorks session starts. The Task Pane displays the following states: *visible or hidden, expanded or collapsed, pinned or unpinned, docked or floating.* The Task Pane contains the following default tabs: *SolidWorks Resources* 🏠 , *Design Library* 🗒 , *File Explorer* 🗁 , *View Palette* 🗒 , *Appearances, Scenes, and Decals* 🌐 and *Custom Properties* 🗒 .

SolidWorks Resources

The basic SolidWorks Resources 🏠 menu displays the following default selections: *Getting Started, Community, Online Resources* and *Tip of the Day.*

Other user interfaces are available during the initial software installation selection: *Machine Design, Mold Design* or *Consumer Products Design.*

Design Library

The Design Library 🗒 contains reusable parts, assemblies, and other elements, including library features.

The Design Library tab contains four default selections. Each default selection contains additional sub categories. The default selections are: *Design Library, Toolbox, 3D ContentCentral* and *SolidWorks Content.*

☀ Activate the SolidWorks Toolbox. Click Tools, Add-Ins.., from the Main menu; check the SolidWorks Toolbox box and SolidWorks Toolbox Browser box from the Add-ins dialog box.

To access the Design Library folders in a non-network

environment, click Add File Location 🗒 and browse to the needed path. Paths will vary depending on your SolidWorks version and Windows setup. In a network environment, contact your IT department for system details.

File Explorer

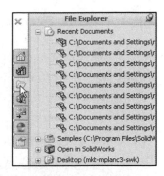

File Explorer ⌂ duplicates Windows Explorer from your local computer and displays *Recent Documents*, *directories*, and the *Open in SolidWorks* and *Desktop* folders.

Search

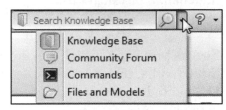

The SolidWorks Search box is displayed in the upper right corner of the SolidWorks Graphics window. Enter the text or key words to search.

In addition to searching for files and models, you can search the SolidWorks *Knowledge Base*, *Community Forum* and *Commands*. Internet access is required for the Community Forum and Knowledge Base.

View Palette

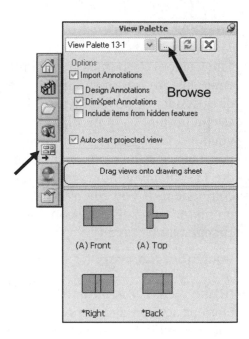

The View Palette tab located in the Task Pane provides the ability to insert drawing views of an active document, or click the Browse button to locate the desired document.

Drag and drop the view from the View Palette into an active drawing sheet to create a drawing view.

The selected model in the illustration is View Palette 13-1. The **(A) Front** and **(A) Top** drawing views are displayed with DimXpert Annotations which was applied at the part level.

Appearances, Scenes, and Decals

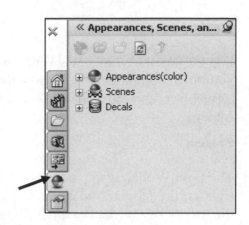

Appearances, Scenes, and Decals ⬤ provide a simplified way to display models in a photo-realistic setting using a library of Appearances, Scenes, and Decals.

An appearance defines the visual properties of a model, including color and texture. Appearances do not change physical properties, which are defined by materials.

Scenes provide a visual backdrop behind a model. In SolidWorks, they provide reflections on the model. Drag and drop a selected appearance, scene or decal on a part or assembly.

Custom Properties

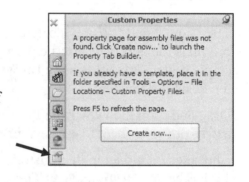

The Custom Properties 🖺 tool provides the ability to enter custom and configuration specific properties directly into SolidWorks files. In assemblies, you can assign properties to multiple parts at the same time. If you select a lightweight component in an assembly, you can view the component's custom properties in the Task Pane without resolving the component. If you edit a value, you are prompted to resolve the component so the change can be saved.

Document Recovery

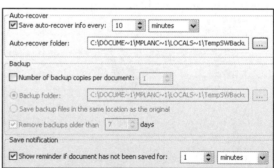

If auto recovery is initiated in the System Options section and the system terminates unexpectedly with an active document, the saved information files are available on the Task Pane Document Recovery tab the next time you start a SolidWorks session.

🔆 Run DFMXpress from the Evaluate tab or from Tools, DFMXpress in the Menu bar menu. The DFMXpress icon is displayed in the Task Pane.

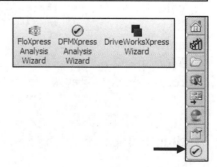

Motion Study tab

Motion Studies are graphical simulations of motion for an assembly. Access MotionManager from the Motion Study tab. The Motion Study tab is located in the bottom left corner of the Graphics window.

Incorporate visual properties such as lighting and camera perspective. Click the Motion Study tab to view the MotionManager. Click the Model tab to return to the FeatureManager design tree.

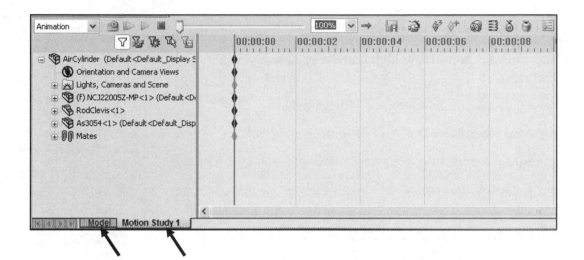

The MotionManager display a timeline-based interface, and provide the following selections from the drop-down menu as illustrated:

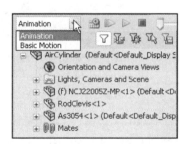

- *Animation:* Apply Animation to animate the motion of an assembly. Add a motor and insert positions of assembly components at various times using set key points. Use the Animation option to create animations for motion that do **not** require accounting for mass or gravity.

- *Basic Motion:* Apply Basic Motion for approximating the effects of motors, springs, collisions and gravity on assemblies. Basic Motion takes mass into account in calculating motion. Basic Motion computation is relatively fast, so you can use this for creating presentation animations using physics-based simulations. Use the Basic Motion option to create simulations of motion that account for mass, collisions or gravity.

If the Motion Study tab is not displayed in the Graphics window, click **View, MotionManager** from the Menu bar.

For older assemblies created before 2008, the Animation1 tab maybe displayed. View the Assembly Chapter for additional information.

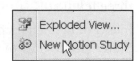

To create a new Motion Study, click **Insert, New Motion Study** from the Menu bar.

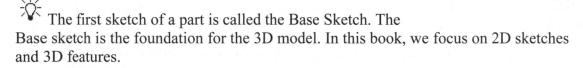

View SolidWorks Help for additional information on Motion Study.

Activity: Create a new Part

A part is a 3D model, which consist of features. What are features?

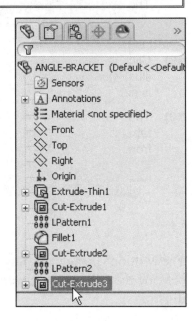

- Features are geometry building blocks.

- Most features either add or remove material.

- Some features do not affect material (Cosmetic Thread).

- Features are created either from 2D or 3D sketched profiles or from edges and faces of existing geometry.

- Features are an individual shape that combined with other features, makes up a part or assembly. Some features, such as bosses and cuts, originate as sketches. Other features, such as shells and fillets, modify a feature's geometry.

- Features are displayed in the FeatureManager as illustrated (Extrude-Thin1, Cut-Extrude1, LPattern1, Fillet1, Cut-Extrude2, Lpatern2, and Cut-Extrude3).

You can suppress a feature. A suppressed feature is displayed in light gray.

The first sketch of a part is called the Base Sketch. The Base sketch is the foundation for the 3D model. In this book, we focus on 2D sketches and 3D features.

During the initial SolidWorks installation, you are requested to select either the ISO or ANSI drafting standard. ISO is typically a European-drafting standard and uses First Angle Projection. The book is written using the ANSI (US) overall drafting standard and Third Angle Projection for drawings.

There are two modes in the New SolidWorks Document dialog box: *Novice* and *Advanced*. The *Novice* option is the default option with three templates. The *Advanced* mode contains access to additional templates and tabs that you create in system options. Use the *Advanced* mode in this book.

Create a New part.

18) Click **New** ⬜ from the Menu bar. The New SolidWorks Document dialog box is displayed.

Select the Advanced mode.

19) Click the **Advanced** button as illustrated. The Advanced mode is set.

20) The Templates tab is the default tab. Part is the default template from the New SolidWorks Document dialog box. Click **OK** from the New SolidWorks Document dialog box.

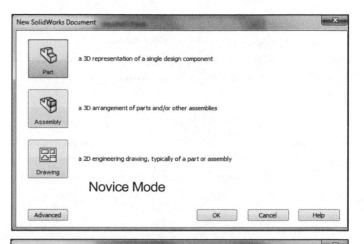

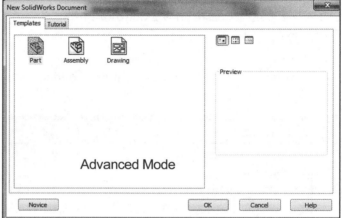

🔆 Illustrations may vary depending on your SolidWorks version and operating system.

The *Advanced* mode remains selected for all new documents in the current SolidWorks session. When you exit SolidWorks, the *Advanced* mode setting is saved.

The default SolidWorks installation contains two tabs in the New SolidWorks Document dialog box: *Templates* and *Tutorial*. The *Templates* tab corresponds to the default SolidWorks templates. The *Tutorial* tab corresponds to the templates utilized in the SolidWorks Tutorials.

Note: The MY-TEMPLATES tab is not displayed in the New SolidWorks Document dialog box when the folder is empty. Save your templates to the MY-TEMPLATES folder. In some network installations, you are required to reference the MY-TEMPLATES folder at the beginning of each SolidWorks session. The System Option, File Locations, Document Templates option sets the pathname to additional template folders. File Locations are explored in this Project.

🔆 Click **View**, **Origins** from the Menu bar menu to display the Origin in the Graphics window.

Part1 is displayed in the FeatureManager and is the name of the document. Part1 is the default part window name. The Menu bar, CommandManager, FeatureManager, Heads-up View toolbar, SolidWorks Resources, SolidWorks Search, Task Pane and the Origin are displayed in the Graphics window.

The Part Origin ⚓ is displayed in blue in the center of the Graphics window. The Origin represents the intersection of the three default reference planes: *Front Plane*, *Top Plane* and *Right Plane*. The positive X-axis is horizontal and points to the right of the Origin in the Front view. The positive Y-axis is vertical and point upward in the Front view. The FeatureManager contains a list of features, reference geometry, and settings utilized in the part.

🔆 You can now Edit the Document units directly from the Graphics window.

🔆 Grid/Snaps are deactivated in the Graphics window for improved modeling clarity.

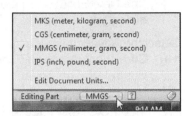

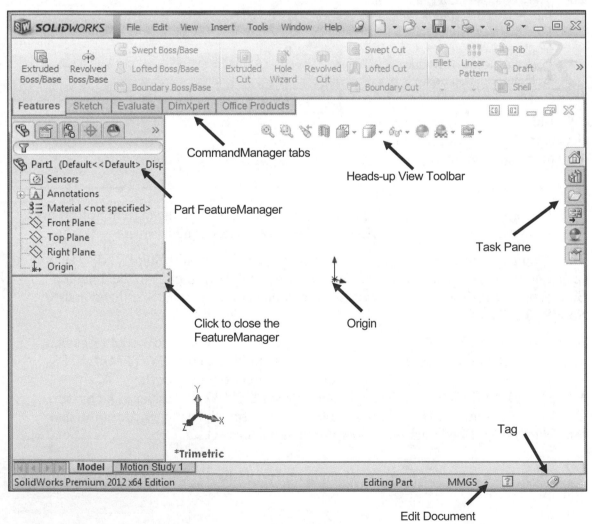

Activity: Menu Bar toolbar, Menu Bar menu, Heads-up View Toolbar

Maximize the Graphics window.

21) Click the **Maximize** button in the top right corner of the SolidWorks window.

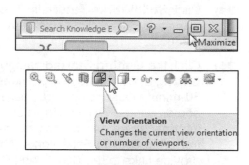

Display tools and tool tips.

22) Position the **mouse pointer** over the Heads-up View toolbar and view the tool tips.

23) **Read** the large tool tip. Select the **drop-down arrow** ▾ to view the available view orientation tools.

Display the View toolbar and the Menu bar.

24) Right-click in the **gray area** of the Menu bar.

25) Click **View**. The View toolbar is displayed.

26) If required, **click and drag** the View toolbar off the Graphics window.

27) Click **SolidWorks** to expand the Menu bar menu.

28) **Pin** the Menu bar as illustrated. Use both the Menu bar menu and the Menu bar toolbar in this book.

☀ The SolidWorks Help Topics contains step-by-step instructions for various commands. The Help ? icon is displayed in the dialog box or in the PropertyManager for each feature. Display the SolidWorks Help Home Page. Use SolidWorks Help to locate information on What's New, Sketches, Features, Assemblies and more.

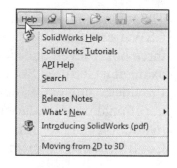

29) Click **Help** from the Menu bar.

30) Click **SolidWorks Help**. The SolidWorks Help Home Page is displayed by default.

☀ SolidWorks Web Help is active by default under Help in the Main menu.

View your options and features.

31) Click the **Home Page** ⌂ icon to return to the Home Page.

32) **Close** ⊠ the SolidWorks Home Page dialog box.

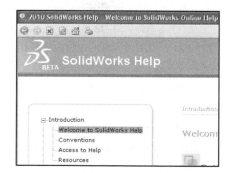

Display and explore the SolidWorks tutorials.

33) Click **Help** from the Menu bar.

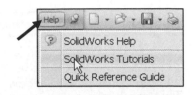

34) Click **SolidWorks Tutorials**. The SolidWorks Tutorials are displayed. The SolidWorks Tutorials are presented by category.

35) Click the **Getting Started** category. The Getting Started category provides three 30-minute lessons on parts, assemblies, and drawings. This section also provides information for new users who are switching from AutoCAD to SolidWorks. The tutorials provide links to the CSWP and CSWA Certification programs and a new What's New Tutorials for 2012.

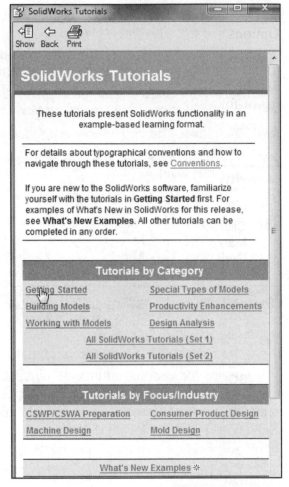

SolidWorks Corporation offers various levels of certification representing increasing levels of expertise in 3D CAD design as it applies to engineering.

The *Certified SolidWorks Associate* CSWA certification indicates a foundation in and apprentice knowledge of 3D CAD design and engineering practices and principles.

The main requirement for obtaining the CSWA certification is to take and pass the three hour, seven question on-line proctored exam at a Certified SolidWorks CSWA Provider; "university, college, technical, vocational or secondary educational institution" and to sign the SolidWorks Confidentiality Agreement.

Passing this exam provides students the chance to prove their working knowledge and expertise and to be part of a worldwide industry certification standard.

36) **Close** the ⊠ Online Tutorial dialog box. Return to the SolidWorks Graphics window.

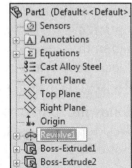

Rename a feature or sketch. Slowly click the feature or sketch name twice and enter the new name when the old one is highlighted.

System Options

System Options are stored in the registry of the computer. System Options are not part of the document. Changes to the System Options affect current and future documents. Note: This can be different in a network environment. Ask your IT administrator.

Review and modify the System Options. If you work on a local drive C:\, the System Options are stored on the computer.

If you work on a network drive Z:\, set System Options for each SolidWorks session.

Set the System Options before you start a project. The File Locations Option contains a list of folders referenced during a SolidWorks session.

Add the ENGDESIGN-W-SOLIDWORKS\MY-TEMPLATES folder path name to the Document Templates File Locations list.

Activity: Set System Options

Set the System Options.

37) Click **Options** ⊞ from the Menu bar. The System Options General dialog box is displayed.

38) Click **File Locations** to set the folder path for custom Document Templates.

39) Click the **Add** button.

40) Select the **ENGDESIGN-W-SOLIDWORKS\MY-TEMPLATES** folder in the Browse For Folder dialog box.

41) Click **OK**. Click **Yes**.

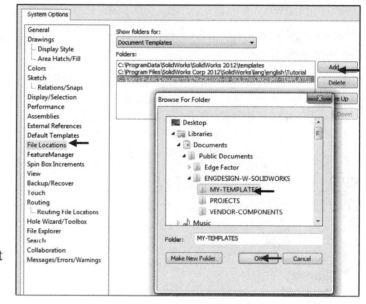

Each folder listed in the System Options, File Locations, Document Templates, Show Folders For option produces a corresponding tab in the New SolidWorks Document dialog box.

The templates used for this project are located in the following folder:
Desktop\Libraries\Documents\ENGDESIGN-W-SolidWorks\MY-TEMPLATES.

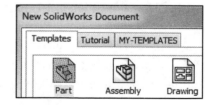

Verify the General options.
42) Click the **General** option. Review the default values.

Part Document Template and Document Properties

The Part template is the foundation for a SolidWorks part. Part1 displayed in the FeatureManager utilizes the *Part* .prtdot default template located in the New dialog box.

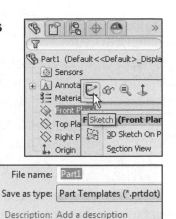

Document Properties contain the default settings for the Part template. The Document Properties include the drafting / dimensioning standard, units, dimension decimal display, grids, note font and line styles.

There are hundreds of document properties. Modify the following Document Properties: *Drafting standard*, *Units* and *Decimal Places*.

The Drafting standard determines the display of dimension text, arrows, symbols and spacing. Units are the measurement of physical quantities. Millimeter (MMGS) dimensioning and decimal inch (IPS) dimensioning are the two most common unit types specified for engineering parts and drawings.

Document Properties are stored with the document. Apply the Document Properties to the Part template.

Create a Part template named PART-MM-ANSI from the default Part template. Save the Custom Part template in the ENGDESIGN-W-SOLIDWORKS\MY-TEMPLATES folder. Utilize the PART-MM-ANSI Part template for all metric parts.

Conserve modeling time. Set the Document Properties and create the required templates before starting a project.

The Overall Drafting standard options are: *ANSI, ISO, DIN, JIS, BSI, GOST* and *GB*.

Overall Drafting standard options:	Abbreviation:	Description:
ANSI	ANSI	American National Standards Institute.
ISO	ISO	International Standards Organization
DIN	DIN	Deutsche Institute für Normumg (German)
JIS	JIS	Japanese Industry Standard
BSI	BSI	British Standards Institution
GOST	GOST	Gosndarstuennye State Standard (Russian)
GB	GB	Guo Biao (Chinese)

Display the parts, assemblies and drawings created in the GUIDE-ROD assembly in the ANSI drafting standard.

The Units Document Property assists the designer by defining the Unit System, Length unit, Angular unit, Density unit and Force unit of measurement for the Part, Assembly or Drawing document.

The Decimals option displays the number of decimal places for the Length and Angle unit of measurement.

Activity: Create a Part Document Template and apply Document Properties

---

Set Document Properties for the model.

43) Click the **Document Properties** tab.

44) Select **ANSI** from the Overall drafting standard drop-down menu. Various Detailing options are available depending on the selected standard.

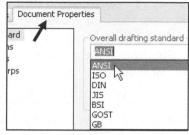

Set document units and decimal places.

45) Click **Units**.

46) Click **MMGS (millimeter, grams, second)** for Unit system.

47) Select **.12** (two decimal places) for Length basic units.

48) Select **None** for Angle decimal places.

49) Click **OK** from the Document Properties - Units dialog box.

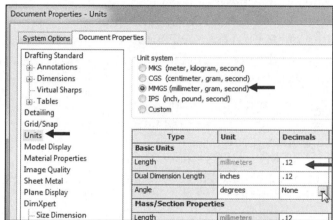

Save the Part template.

50) Click **Save As** from the Menu bar.

51) Select **Part Templates *.prtdot** from the Save as type box. The default Part Templates folder is displayed.

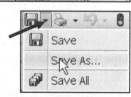

52) Click the **drop-down arrow** in the Save in box.

53) Select the **ENGDESIGN-W-SOLIDWORKS\MY-TEMPLATES** folder.

54) Enter **PART-MM-ANSI** in the File name box.

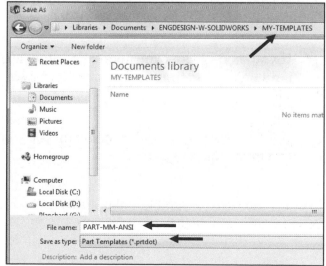

55) Click **Save**. PART-MM-ANSI is displayed in the FeatureManager.

Close the PART-MM-ANSI Part template.

56) Click **File**, **Close** from the Menu bar. All SolidWorks documents are closed. If required, click **Window**, **Close All** from the Menu bar to close all open documents.

Additional details on System Options and Document Properties are available using the Help option. Index keyword: System Options, Document Templates, units and properties.

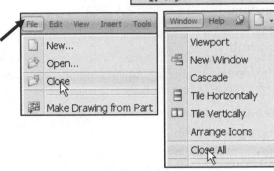

Review of the File Folders and Custom Part Template

You created the following folders: *PROJECTS*, *MY-TEMPLATES* and *VENDOR-COMPONENTS* to manage the documents and templates for this project.

System Options, File Locations directed SolidWorks to open Document Templates from the MY-TEMPLATES folder. System Options are stored in the registry of the computer.

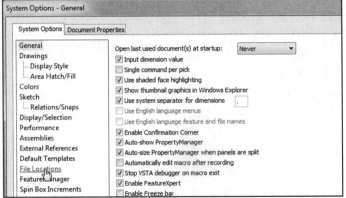

You created a custom Part template, PART-MM-ANSI from the default Part document. Document Properties control the drafting standard, units, tolerance and other properties. Document Properties are stored in the current document. The Part template file extension is .prtdot.

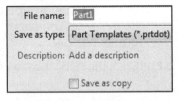

If you modify a document property from an Overall drafting standard, a modify message is displayed as illustrated.

Illustrations may vary depending on your SolidWorks version and operating system.

PLATE Part Overview

Determine the functional and geometric requirements of the PLATE.

Functional:

The PLATE part fastens to the customer's PISTON PLATE part in the GUIDE-CYLINDER assembly.

Fasten the PLATE part to the ROD part with a countersunk screw.

Geometric:

The dimensions of the PISTON PLATE part are 56mm x 22mm.

The dimensions of the PLATE part are 56mm x 22mm.

Locate the 4mm mounting holes with respect to the PISTON PLATE mounting holes. The mounting holes are 23mm apart on center.

The PLATE part requires a Countersink Hole to fasten the ROD part to the PLATE part.

Review the mating part dimensions before creating the PLATE part.

The GUIDE-CYLINDER assembly dimensions referenced in this project are derived from the SMC Corporation of America (www.smcusa.com).

☀ Click **View**, **Origins** from the Menu bar menu to display the Origin in the Graphics window.

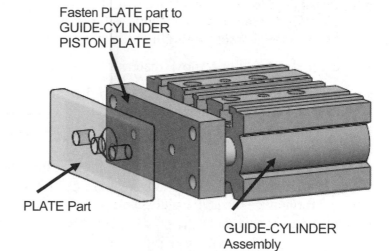

Fasten PLATE part to GUIDE-CYLINDER PISTON PLATE

PLATE Part

GUIDE-CYLINDER Assembly

Courtesy of SMC Corporation of America

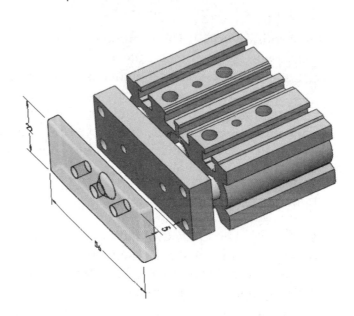

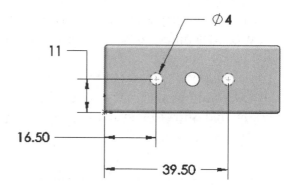

∅**4**

11

16.50

39.50

Start the translation of the initial design functional and geometric requirements into SolidWorks features.

What are features?

- Features are geometry building blocks.

- Features can add or remove material.

- Features are created from sketched profiles or from edges and faces of existing geometry.

Utilize the following features to create the PLATE part:

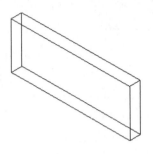

- ***Extruded Boss/Base*** : The Extruded Boss/Base feature adds material to the part. The Boss-Extrude1 (Base) feature is the first feature of the PLATE. An extrusion extends a profile along a path normal to the profile plane for a specified distance. The movement along that path becomes the solid 3D model. Sketch the 2D rectangle (close profile) on the Front plane. Fully define the Base Sketch with Geometric relations and dimensions.

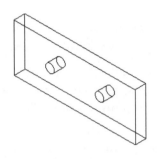

- ***Extruded Cut*** : The Extruded Cut feature removes material. The Extruded Cut starts with a 2D circle sketched on the front face. The front face is your Sketch plane. Copy the sketch circle to create the second circle. Utilize the Through All Depth End Condition (Design Intent). The holes extend through the entire Boss-Extrude1 feature. Note: Use the Hole Wizard feature when you need to incorporate complex or blind holes in a part. The Hole Wizard (Call out) information can be imported directly into a drawing and provides the correct hole annotations.

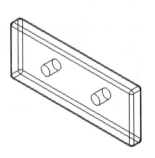

- ***Fillet*** : The Fillet feature removes sharp edges of the PLATE. Add Fillets to a solid, not the sketch. Group small corner edge fillets together. In this exercise, group Tangent edge fillets together.

- ***Hole Wizard*** : The Hole Wizard feature creates the Countersink hole at the center of the PLATE. The Hole Wizard requires a Sketch plane. Select the back face of the PLATE for your Sketch plane. Note: In this book, we focus on 2D Sketches vs. 3D Sketches.

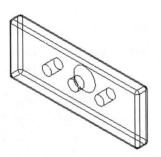

A goal of this book is to expose the new user to various SolidWorks design tools and features.

Activity: Create the PLATE Part using a Part Template

Create a New part.

57) Click **New** 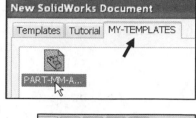 from the Menu bar. The New SolidWorks Document dialog box is displayed.

58) Click the **MY-TEMPLATES** tab.

59) Double click the **PART-MM-ANSI** icon. The Part FeatureManager is displayed.

In the New SolidWorks Document, Advanced option, Large icons are displayed by default. Utilize the List option or List Detail option to view the complete template name.

🔆 Set the MY-TEMPLATES tab with **Options** ⊟, **System Options**, **File Locations**, **Document Templates** option. The Templates tab is displayed in the New SolidWorks Document Advanced option. If the *Novice* option is displayed, select the *Advanced* button.

🔆 The first system default Part filename is: Part1. The system attaches the .sldprt suffix to the created part. The second created part in the same session, increments to the filename: Part2.

🔆 There are numerous ways to manage a part. Use Project Data Management (PDM) systems to control, manage and document file names and drawing revisions. Use appropriate filenames that describe the part.

Save the Empty part.

60) Click **Save** 💾.

61) Select the **ENGDESIGN-W-SOLIDWORKS\PROJECTS** folder.

62) Enter **PLATE** for File name.

63) Enter **PLATE 56MM x 22MM** for Description.

🔆 Create a smart part to create a smart assembly and drawing document.

64) Click **Save**.

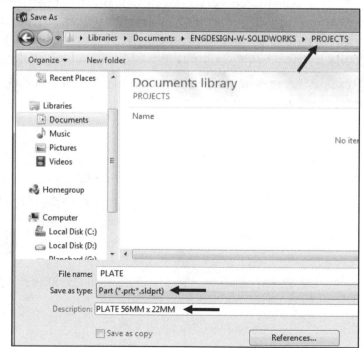

PLATE is displayed in the FeatureManager design tree. The Part 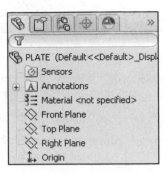 icon is displayed at the top of the FeatureManager.

☼ Click **View**, **Origins** from the Menu bar menu to display the Origin in the Graphics window.

Base Feature

What is a Base feature? The Base feature (Boss-Extrude1) is the first feature that is created. The Base feature is the foundation of the part. Keep the Base feature <u>simple!</u>

The Base feature geometry for the PLATE is an extrusion. Rename the base feature Extruded-Base. Note: The default name is Boss-Extrude1.

How do you create a 3D Extruded Boss/Base feature?

- Select a Sketch plane.

- Sketch a 2D profile on the Sketch plane.

- Apply all needed Geometric relations and dimensions to fully define the sketch.

- Apply the Extruded Boss/Base feature tool. Extend the profile perpendicular ($\perp$) to the Sketch plane.

☼ It is considered very poor design practice not to fully define a sketch. You will be tempted in order to save design time, but keep in mind that the extra couple of minutes you take to do something right the first time will save you additional time in the end. Fully defined sketches are required to manufacture the part.

Review the part manufacturing, materials, reference planes and orthographic projection before creating a Base feature.

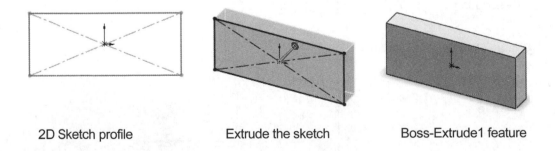

2D Sketch profile   Extrude the sketch   Boss-Extrude1 feature

Machined Part

In earlier conversations with manufacturing, a decision was made that the part would be machined.

Your material supplier stocks raw material in rod, sheet, plate, angle and block forms.

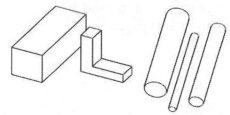

Block, Angle and Rod Stock

You decide to start with a standard plate form. A standard plate form will save time and money.

Select the best profile for the extrusion. The best profile is a simple 2D rectangle.

The boss and square internal cuts are very costly to manufacture from machined bar stock.

As a designer, review your manufacturing options.

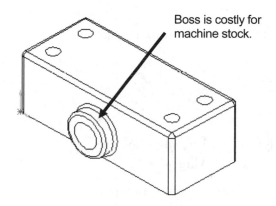

Boss is costly for machine stock.

Utilize standard material thickness and hole sizes in the PLATE design. Utilize standard stock in the ROD design.

Utilize slot cuts in the GUIDE design. Machined parts require dimensions to determine the overall size and shape. Datum planes determine the location of referenced dimensions. Add geometric relations and dimensions to fully define the geometry Extruded Boss/Base feature.

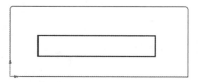

Internal Square Cut (expensive)

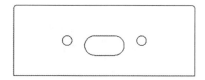

Holes and Slots (less expensive)

Understanding reference planes and views is important as you design machined parts. Before you create the PLATE, review the next topic on Reference Planes and Orthographic Projection.

Reference Planes and Orthographic Projection

The three default ⊥ Reference planes represent infinite 2D planes in 3D space:

- Front

- Top

- Right

Planes have no thickness or mass. Orthographic projection is the process of projecting views onto parallel planes with ⊥ projectors.

Default ⊥ datum Planes are:

- Primary

- Secondary

- Tertiary

Use the following planes in manufacturing:

- Primary datum plane: Contacts the part at a minimum of three points.

- Secondary datum plane: Contacts the part at a minimum of two points.

- Tertiary datum plane: Contacts the part at a minimum of one point.

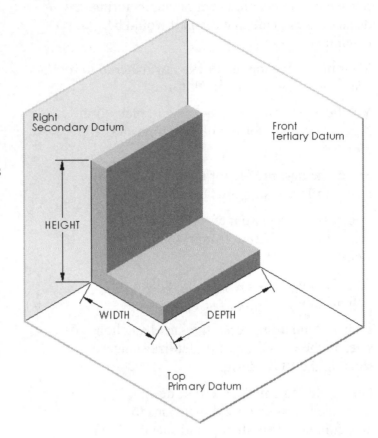

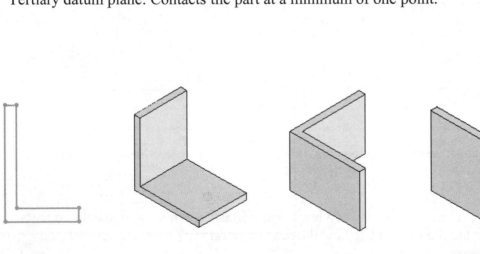

2D Profile   Front Plane   Top Plane   Right Plane

The part view orientation is dependent on the first feature Sketch plane. Compare the available default Sketch planes in the FeatureManager: *Front Plane, Top Plane and Right Plane.*

Each Extruded feature above was created with an L-shaped 2D Sketch profile. The six principle views of Orthographic projection listed in the ASME Y14.3M standard are:

- Top
- Front
- Right side
- Bottom
- Rear
- Left side

SolidWorks Standard view names correspond to these Orthographic projection view names.

ASME Y14.3M Principle View Name:	SolidWorks Standard View:
Front	Front
Top	Top
Right side	Right
Bottom	Bottom
Rear	Back
Left side	Left

The most common standard drawing views in Third angle Orthographic projection are:

- Front
- Top
- Right
- Isometric

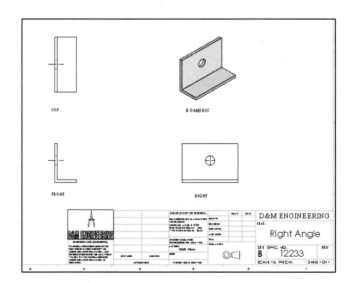

There are two Orthographic projection drawing systems. The first Orthographic projection system is called: Third Angle Projection. The second Orthographic projection system is called: First Angle Projection. The two drawing systems derive from positioning a 3D object in the Third or First quadrant as illustrated.

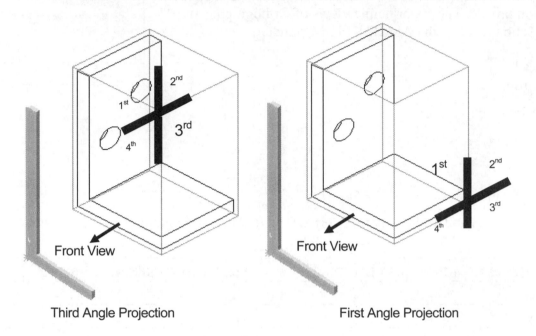

Third Angle Projection   First Angle Projection

Third Angle Projection

Third angle projection is a method of creating a 2D drawing of a 3D object. The 3D object is position in the third quadrant. The 2D projection planes are located between the viewer and the part. The projected views are placed on a drawing.

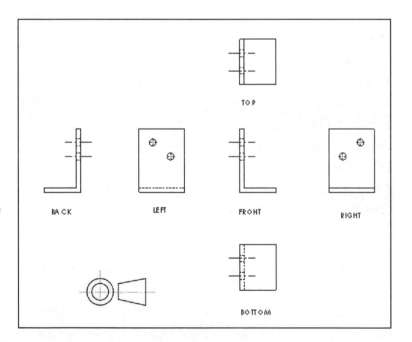

First Angle Projection

First-angle projection is a method of creating a 2D drawing of a 3D object. It is mainly used in Europe and Asia. The 3D object is position in the first quadrant. Views are projected onto the planes located behind the part. The projected views are placed on a drawing.

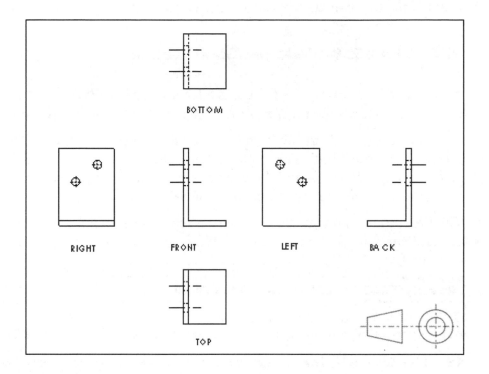

Third Angle Projection is primarily used in the U.S. & Canada and is based on the ASME Y14.3M multi and sectional view drawings standard. Designers should have knowledge and understanding of both systems.

There are numerous multi-national companies. Example: A part is designed in the U.S., manufactured in Japan and destined for a European market.

Third Angle Projection is used in this text. A truncated cone symbol should appear on the drawing to indicate the Projection system.

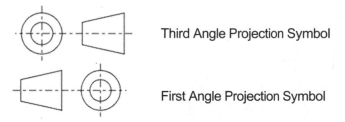

Third Angle Projection Symbol

First Angle Projection Symbol

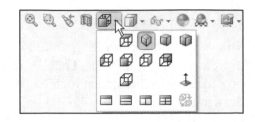

Utilize the *Heads-up View* toolbar in the Graphics window to orient or set view style for your model.

Before incorporating your design intent into the Sketch plane, ask a question: How will the part be oriented in the assembly? Answer: Orient or align the part to assist in the final assembly. Utilize the Front Plane to create the Extruded Base feature for the PLATE part.

PLATE Part-Extruded Boss/Base Feature

An Extruded Boss/Base feature is a feature in SolidWorks that utilizes a sketched profile and extends the profile perpendicular ($\perp$) to the Sketch plane.

In SolidWorks, a 2D profile is called a sketch. A sketch requires a Sketch plane and a 2D profile. The sketch in this example uses the Front Plane. The 2D profile is a rectangle. Geometric relationships and dimensions define the exact size of the rectangle. The rectangle is extruded perpendicular to the Sketch plane.

The Front Plane's imaginary boundary is represented on the screen with four visible edges. Planes are flat and infinite. Planes are used as the primary sketch surface for creating Extruded Boss/Base and Extruded Cut features.

Activity: PLATE Part-Extruded-Base Feature

Select the Sketch plane. Create a 2D sketch.

65) Right-click **Front Plane** from the FeatureManager. The Context toolbar is displayed.

66) Click **Sketch** from the Context toolbar. The Sketch toolbar is displayed. Front Plane is your Sketch Plane. Note: The grid is de-activated to improve picture clarity.

The plane or face you select for the Base sketch determines the orientation of the part.

Click **View**, **Origins** from the Menu bar menu to display the Origin in the Graphics window.

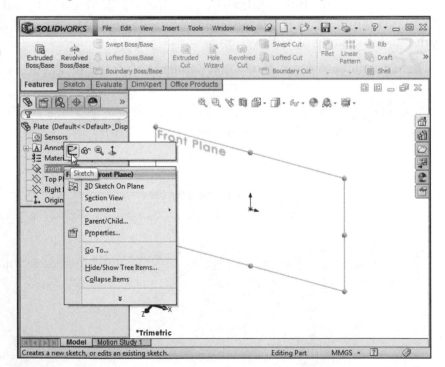

You can also click the Front Plane from the FeatureManager and click the Sketch tab from the CommandManager. Front Plane is the Sketch plane and the Sketch toolbar is displayed.

The Origin ⊥ represents the intersection of the Front, Top and Right Planes. The left corner point of the rectangle is Coincident with the origin. The Origin is displayed in red.

Sketch a Corner rectangle.

67) Click the **Corner Rectangle** ▢ tool from the Sketch toolbar. The sketch opens in the Front

view. The mouse pointer displays the Corner Rectangle feedback symbol ▢. The Front Plane feedback indicates the current Sketch plane. The red dot feedback indicates the Origin point location.

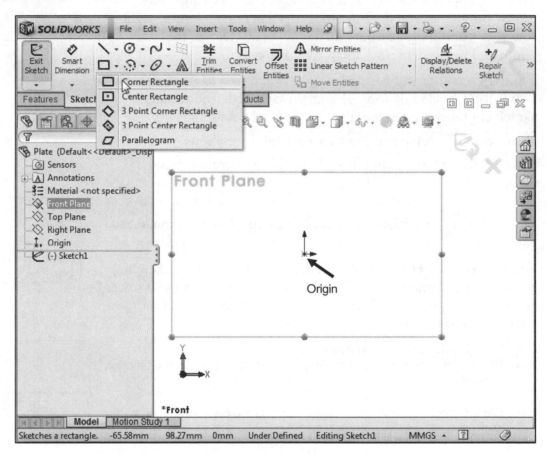

The Rectangle-based tool uses a Consolidated Rectangle PropertyManager. The SolidWorks application defaults to the last used rectangle type.

68) Click the **Origin** ⊥. This is the first point of the rectangle. It is very important that you always reference the sketch to the origin. This helps to fully define the sketch.

69) Drag the **mouse pointer** up and to the right.

70) Release the **mouse button** to create the second point and the rectangle. The first point is Coincident to the Origin.

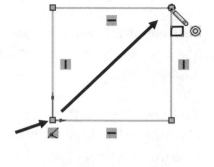

View the illustrated sketch relations in the Graphics window.

71) Click **View**, **Sketch Relations** from the Menu bar. The sketch relations are displayed in the Graphics window.

To deactivate the illustrated sketch relations, click **View**; uncheck **Sketch Relations** from the Menu bar.

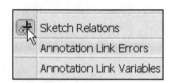

The X-Y coordinates of the rectangle are displayed above the mouse pointer as you drag the mouse pointer up and to the right. The X-Y coordinates display different values. Define the exact width and height with the Smart dimension tool.

The Corner Rectangle tool ▭ remains selected. The rectangle sketch contains predefined geometric sketch relations: two Horizontal ▬ relations and two Vertical | relations. The first point of the rectangle contains a Coincident ⊀ relation at the Origin.

If you make a mistake, select the **Undo** ⟲ icon in the Menu bar.

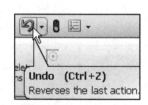

SolidWorks uses color and cursor feedback to aid in the sketching process. The Corner Rectangle tool remains active until you select another tool or right-click the Select. The right mouse button contains additional tools.

Deactivate the Corner Rectangle tool.

72) Right-click a **position** in the Graphics window. The Context toolbar is displayed.

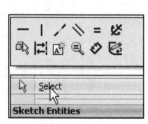

73) Click **Select**. The mouse pointer displays the Select ⍜ icon and the Corner Rectangle Sketch tool is deactivated.

The phrase, "Right-click Select" means **right-click** the mouse in the Graphics window. Click **Select**. This will de-select the current tool that is active.

Size the geometry of the rectangle.

74) Click and drag the **top horizontal line** of the rectangle upward.

75) Release the **mouse button**.

The Line Properties PropertyManager is displayed to the left of the Graphics window. The selected light blue line displays a Horizontal relation in the Existing Relations box.

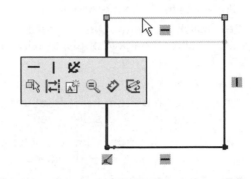

The rectangular sketch is displayed in three colors: Light blue, blue and black. The geometry consists of four lines and four points. The selected top horizontal line is displayed in light blue. The right vertical line is displayed in blue. The left vertical line and the bottom horizontal line are displayed in black.

💡 Context menus and toolbars save time. Commands and tools vary depending on the mouse position in the SolidWorks window and the active Sketch tool and history.

💡 Rename a feature or sketch. Slowly click the feature or sketch name twice and enter the new name when the old one is highlighted.

💡 Design Intent is how your part reacts as parameters are modified. Example: If you have a hole in a part that must always be .125≤ from an edge, you would dimension to the edge rather than to another point on the sketch. As the part size is modified, the hole location remains .125≤ from the edge.

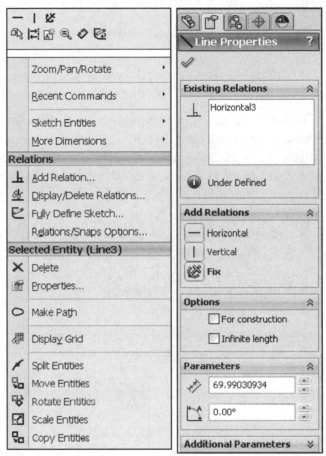

Dimension the bottom horizontal line.

76) Click the **Smart Dimension** tool from the Sketch toolbar. The pointer displays the dimension feedback symbol .

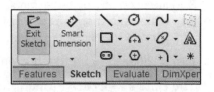

77) Click the **bottom horizontal line** of the rectangle. Note: A dimension value is displayed.

78) Click a **position** below the bottom horizontal line. The Modify dialog box is displayed.

79) Enter **56**mm in the Modify dialog box.

80) Click the **Green Check mark** in the Modify dialog box.

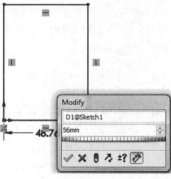

You can either press the enter key or click the Green Check mark to accept the dimension value.

The Dimension PropertyManager is displayed when a dimension is selected. Modify Dimension properties for the active model such as *Tolerance/ Precision*, *Primary Value*, *Dimension Text* and *Dual Dimension* display directly from the Dimension PropertyManager.

The Smart Dimension tool uses the Dimension PropertyManager. The Smart Dimension PropertyManager provides the ability to select the **Value**, **Leaders** or **Other** tab. Each tab provides a separate menu.

The Value tab is selected by default.

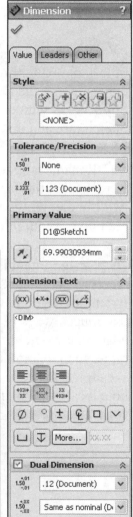

Dimension the vertical line.

81) Click the **left vertical line**.

82) Click a **position** to the left of the vertical line.

83) Enter **22**mm.

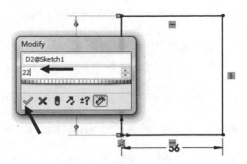

84) Click the **Green Check mark** in the Modify dialog box. The sketch is fully defined. All lines and vertices are displayed in black.

To flip the arrow direction, click the control point as illustrated.

Deactivate the sketch relations and select the dimension value.

85) Click **View**; uncheck **Sketch Relations** from the Menu bar.

86) Right-click **Select** in the Graphics window.

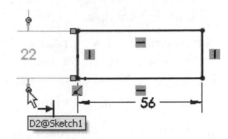

87) Position the **mouse pointer** over the dimension text. The pointer changes to a linear dimension symbol, with a displayed text box. D2@Sketch1 represents the second linear dimension created in Sketch1.

Modify the dimension text.

88) Double-click the **22** dimension text in the Graphics window. The Modify dialog box is displayed.

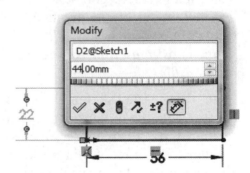

89) Enter **42**mm in the Modify dialog box.

90) Click the **Green Check mark** from the Modify dialog box.

Click the **Spin Box Arrows** to increase or decrease dimensional values.

Return to the original vertical dimension.

91) Double-click the **42**mm dimension text in the Graphics window.

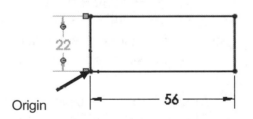

92) Enter **22**mm.

93) Click the **Green Check mark** from the Modify dialog box.

Click the **Undo** tool to return to the original dimension.

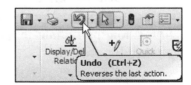

Options in the Modify dialog box:

- Green Check mark ✓ - Saves the current value and exits the Modify dialog box.

- Restore ✗ - Restores the original value and exits.

- Rebuild 🔒 - Rebuilds the model with the current value.

- Reset ±? - Modifies the spin box increment.

- Mark for Drawing 🖉 - Set by default. Inserts part dimensions into the drawing.

The System displays Under Defined in the Status bar located in the lower right corner of the Graphics window. In an under defined Sketch, the entities that require position, dimensions or Sketch relations are displayed in blue.

In a fully defined sketch, all entities are displayed in black. The Status bar displays Fully Defined. In machining practices, parts require fully defined sketches. *A fully defined sketch has defined positions, dimensions and/or relationships.*

In an over defined sketch, there is geometry conflict with the dimensions and or relationships. In an over defined sketch, entities are displayed in red. SolidWorks provides the SketchXpert tool to correct an over defined sketch. The SketchXpert PropertyManager is displayed by default for an over defined sketch. Click the Diagnose button and accept an option to correct the over defined sketch.

🔆 Although dimensions are not required to create features in SolidWorks, dimensions provide location and size information. Models require dimensions for manufacturing. Dimension the rectangle with horizontal and vertical dimensions.

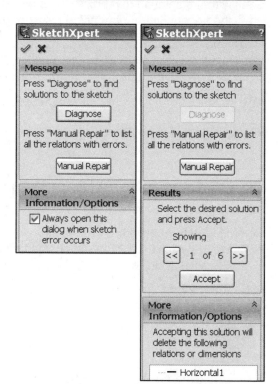

Insert an Extruded Boss/Base feature. This is the first (Base) feature of the model. Utilize the default Blind End Condition.

94) Click the **Features** tab from the CommandManager.

95) Click the **Extruded Boss/Base** tool from the Features toolbar. The Boss-Extrude PropertyManager is displayed. The extruded sketch is previewed in a Trimetric view. The preview displays the direction of the extrude feature.

Reverse the direction of the extruded depth.

96) Click the **Reverse Direction** box. Note the location of the Origin in the Graphics window. Blind is the default End Condition for Direction1.

97) Enter **10**mm for Depth in Direction1.

98) Click **OK** from the Boss-Extrude PropertyManager. The Boss-Extrude1 feature is displayed in the Graphics window. The name Boss-Extrude1 is displayed in the PLATE FeatureManager

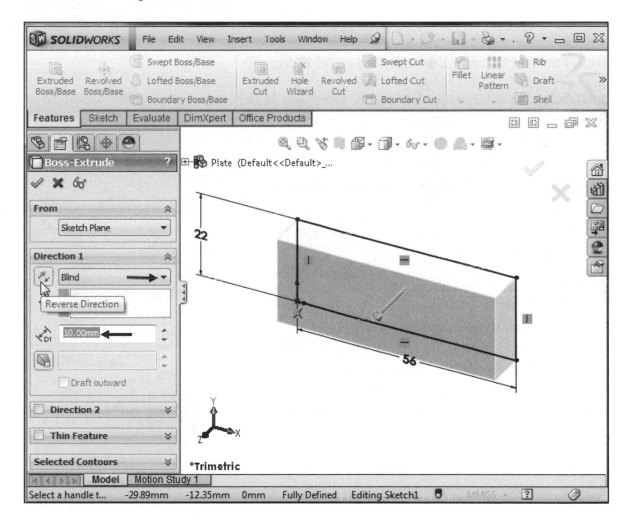

🔆 The OK ✔ button accepts and completes the Feature process. The Tool tip displays OK in the PropertyManager for Feature tools.

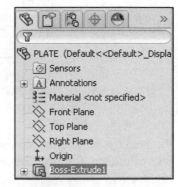

🔆 The OK ✔ button also displays different Tool tips such as; *Apply, Close Dialog, Close Accept* and *Exit* for different Sketch tools.

🔆 If you exit the sketch before selecting the Extruded Boss/Base tool from the Features toolbar, Sketch1 is displayed in the FeatureManager. To create the Extruded Boss/Base feature, click Sketch1 from the FeatureManager. Click Extruded Boss/Base to create the feature.

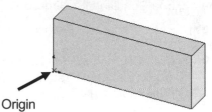

Origin

PLATE Part-Modify Dimensions and Rename

Incorporate design changes into the PLATE. Modify dimension values in the Modify dialog box. Utilize the Rebuild 🔵 tool to update the Extruded Boss/Base feature. Expand Boss-Extrude1 from the FeatureManager to display Sketch1. Rename entries by selecting on the text in the FeatureManager.

Activity: PLATE Part-Modify Dimensions and Rename

Modify the PLATE part.
99) Double-click on the **front face** of Boss-Extrude1 in the Graphics window as illustrated. Boss-Extrude1 is highlighted in the FeatureManager.

Fit the model to the Graphics window.
100) Press the **f** key.

Modify the width dimension.
101) Click the **10**mm dimension in the Graphics window.

102) Enter **5**mm.

Display an Isometric view.
103) Click **Isometric view** 🔲 from the Heads-up View toolbar.

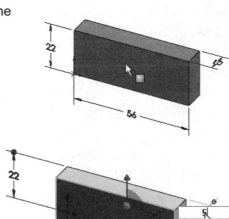

🔆 The Isometric view displays the part in 3D with two equal projection angles.

The Instant3D tool provides the ability to drag geometry and dimension manipulator points to resize or create new features in the Graphics window. Use the on-screen ruler to measure and apply modifications. In this book, you will primarily use the PropertyManager and dialog box to modify model dimensions and to create new features. It is very difficult to apply design intent (End Conditions) using the Instant 3D tool. See SolidWorks Help for additional information.

Incorporate the machining process into the PLATE design. Reference dimensions from the three datum planes with the machined Origin in the lower left hand corner of the PLATE.

Maintain the Origin on the front lower left hand corner of the Extruded Boss/Base feature and co-planar with the Front Plane.

Rename Sketch1 and Boss-Extrude1. The first Extruded Boss/Base feature is named Boss-Extrude1 in the FeatureManager. Sketch1 is the name of the sketch utilized to create the Boss-Extrude1 feature.

The Plus Sign ⊞ icon indicates that additional feature information is available.

Expand the Boss-Extrude1 entry from the FeatureManager.

104) Click the **Plus Sign** ⊞ icon of the Extrude1 feature in the FeatureManager.

Rename Sketch1.

105) Slowly double-click **Sketch1** in the FeatureManager. A white box is displayed around the selected item as illustrated.

106) Enter **Sketch-Base** for the sketch name. The Minus Sign ⊟ icon indicates that the feature information is expanded.

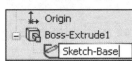

107) Click the **Minus Sign** ⊟ icon to collapse the Boss-Extrude1 feature.

Rename the Boss-Extrude1 feature.

108) Slowly double-click **Boss-Extrude1** in the FeatureManager. A white box is displayed around the selected item.

109) Enter **Base-Extrude** for the feature name as illustrated.

Save the PLATE.

110) Click **Save** 💾.

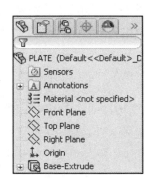

🔍 Additional information on rectangle, dimension, and Extruded feature is available in the SolidWorks Help Topics.

Display Modes, View Modes, View tools and Appearances

Access the display modes, view modes, view tools and appearances from the Standard Views toolbar and the Head-up View toolbar. Apply these tools to display the required modes in your document.

The Apply scene tool adds the likeness of a material to a model in the Graphics window without adding the physical properties of the material. Select a scene from the drop-down menu.

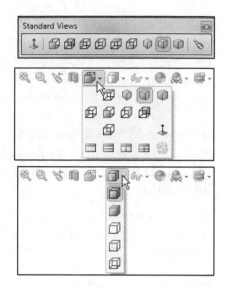

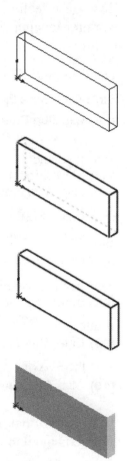

Activity: Display Modes, View Modes, View tools, and Appearances

View the display modes from the Heads-up View toolbar.

111) Click **Wireframe** ⬚.

112) Click **Hidden Lines Visible** ⬚.

113) Click **Hidden Lines Removed** ⬚.

114) Click **Shaded** ⬚.

115) Click **Shaded With Edges** 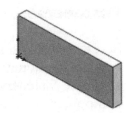.

☀ The Hidden Lines Removed option may take longer to display than the Shaded option depending on your computer configuration and the size of the part and file.

Display modes and View modes remain active until deactivated.

Display the View modes.

116) Click **Previous view** ⬦ to display the previous view of the part in the current window.

117) Click **Zoom to Fit** ⬦ to display the full size of the part in the current window.

118) Click **Zoom to Area** ⬦ .

119) Zoom in on a corner of the PLATE.

120) Click **Zoom to Area** ⬦ to deactivate the tool.

121) Press the **f** key to fit the model to the Graphics window.

☀ Press the lower case z key to zoom out. Press the upper case Z key to zoom in.

122) Right-click **Select** in the Graphics window. View the available view tools.

123) Click **inside** the Graphics window.

Return to a standard Isometric, Shaded With Edges view.

124) Click **Isometric view** ⬦ from the Heads-up View toolbar.

125) Click **Shaded With Edges** ⬦ from the Heads-up View toolbar.

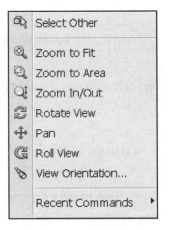

⬦	Select Other
⬦	Zoom to Fit
⬦	Zoom to Area
⬦	Zoom In/Out
⬦	Rotate View
⬦	Pan
⬦	Roll View
⬦	View Orientation...
	Recent Commands ▸

☀ The Normal To view ⬦ tool displays the part ⊥ to the selected plane. The View Orientation tool ⬦ from the Standard Views toolbar creates and displays a custom named view.

☀ To rotate your model, use the middle mouse button. The rotate icon ⬦ is displayed.

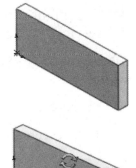

Fasteners

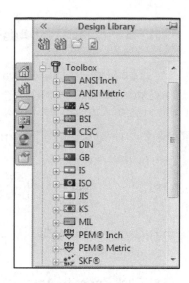

Screws, bolts, nuts are used to join parts. Use standard available fasteners whenever possible. Using standard available fasteners will decrease product cost and reduce component purchase lead time. The SolidWorks Toolbox is a great place to obtain fasteners and hardware. SolidWorks Toolbox is an Add In.

The American National Standard Institute (ANSI) and the International Standardization Organization (ISO) provide standards on various hardware components. The SolidWorks Library contains a variety of standard fasteners to choose from.

Below are a few general selection and design guidelines that are utilized in this text:

- Use standard industry fasteners where applicable.

- Reuse the same fastener types where applicable. Dissimilar screws and bolts require different tools for assembly, additional part numbers and increase inventory storage and cost.

- Decide on the fastener type before creating holes. Dissimilar fastener types require different geometry.

- Create notes on all fasteners. Notes will assist in the development of a Parts list and Bill of Materials.

- Use caution when positioning holes. Do not position holes too close to an edge. Stay one radius head width at a minimum from an edge or between holes. Review manufacturer's recommended specifications.

- Design for service support. Ensure that the model can be serviced in the field and or on the production floor.

Use standard M4x8 Socket Head Cap Screws in this exercise. In SolidWorks:

- M represents a metric screw.

- 4 represents the Nominal size in millimeters.

- 8 represents the overall length of the thread.

Determine the dimensions for the mounting holes from the drawing of the GUIDE CYLINDER assembly.

PLATE Part-Extruded Cut Feature

An Extruded Cut 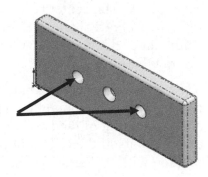 feature removes material, perpendicular, ⊥ to the sketch for a specified depth. The Through All End Condition option creates the holes through the entire depth of the PLATE.

Utilize the Circle Sketch tool to create the hole profile on the front face. The first circle is defined with a center point and a point on its circumference. The second circle utilizes the Ctrl-C (Copy) key to create a copy of the first circle.

Geometric relations are constraints between one or more entities. The holes may appear aligned and equal, but they are not! Add two relations: Insert an Equal relation between the two circumferences and a Horizontal relation between the two center points.

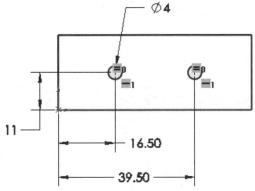

The machinist references all dimensions from the reference datum planes to produce the PLATE holes. Reference all three linear dimensions from the Right datum plane (left vertical edge) and the Top datum plane (bottom horizontal edge). Utilize a dimension to define the diameter of the circle.

The linear dimensions are positioned to the left and below the Extrude1 feature. You will learn more about dimensioning to a drawing standard in Project 3.

☼ Position the dimensions off the sketch profile. Place smaller dimensions to the inside. These dimensioning techniques in the part sketch will save time in dimensioning the drawing.

Design intent is not a static concept that controls changing geometry. As the designer, think about how to maintain the design intent with geometric changes to the model. Dimension and geometric relation schemes vary depending on the design intent of the part. In the next section - there are alternate dimension schemes to define the position of two circles.

☼ Design Intent - Example: If you have a hole in a part that must always be 16.5mm≤ from an edge, you would dimension to the edge rather than to another point on the sketch. As the part size is modified, the hole location remains 16.5mm≤ from the edge.

The following two examples are left as an exercise.

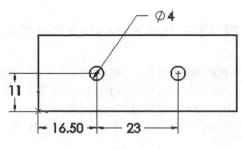

Example A: The center points of the holes reference a horizontal centerline created at the Midpoint of the two vertical lines. A dimension is inserted between the two center points.

Example B: The center points of the circles reference the horizontal line. Insert a vertical line at the Midpoint of the two horizontal lines. The center points of the circles utilize a Symmetric relation with the vertical centerline.

Dimension schemes defined in the part can also be redefined in the future drawing.

Example A

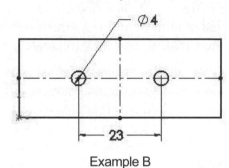

Example B

Activity: PLATE Part-Extruded Cut Feature

View the mouse pointer feedback.

126) Position the mouse pointer on the **front face**. View the Select Face icon . The mouse pointer provides feedback depending on the selected geometry.

127) Position the mouse pointer on the **right edge**. View the Select Edge icon .

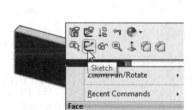

Select the Sketch plane for the holes.

128) Right-click the **front face** of the Base-Extrude feature in the Graphics window as illustrated.

Create a sketch.

129) Click **Sketch** from the Context toolbar. The Sketch toolbar is displayed.

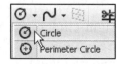

Display the Front view.

130) Click **Front view** . The Origin is located in the lower left corner of the part.

131) Click **Hidden Lines Visible** from the Heads-up View toolbar.

Sketch the first circle.

132) Click the **Circle** Sketch tool. The cursor displays the Circle feedback symbol . The Circle PropertyManager is displayed.

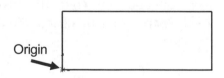

Origin

133) Click a **center point** on the front face diagonally to the right of the Origin as illustrated.

134) Click a **position** to the right of the center point. The circumference and the center point are selected.

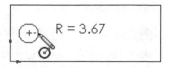

Copy the first circle.
135) Right-click **Select** in the Graphics window to deselect the circle Sketch tool.

136) Hold the **Ctrl** key down.

137) Click and drag the **circumference** of the first circle to the right.

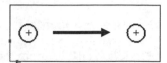

138) Release the **mouse button**.

139) Release the **Ctrl** key. The second circle is displayed.

140) Click a position in the **Graphics window**, off the sketch. The first and second circles are displayed in blue.

Add an Equal relation.
141) Click the **circumference** of the first circle.

142) Hold the **Ctrl** key down.

143) Click the **circumference** of the second circle.

144) Release the **Ctrl** key. Arc1 and Arc2 are displayed in the Selected Entities text box in the Properties PropertyManager.

145) Click **Equal** ☐ from the Add Relations box. The Equal radius/length is displayed in the Existing Relations text box.

146) Click **OK** ✓ from the Properties PropertyManager.

You can right-click the **Make Equal** tool from the Context toolbar.

The SolidWorks default name for curve entities is Arc#. There is an Arc# for each circle. The default name for any point is Point#. The two center points are entities named Point2 and Point4.

To remove unwanted entities from the Selected Entities box, right-click in the **Selected Entities** window. Click **Clear Selections**. The Selected Entities # differs if geometry was deleted and recreated. The Existing Relations box lists Equal radius/length0.

Add a Horizontal relation.

147) Click the first circle **center point**.

148) Hold the **Ctrl** key down.

149) Click the second circle **center point**. The Propeties PropertyManager is displayed. The selected entities are displayed in the Selected Entities box.

150) Release the **Ctrl** key.

151) Click **Horizontal** from the Add Relations box.

152) Click **OK** ✓ from the Properties PropertyManager.

Add a diameter dimension.
153) Click the **Smart Dimension** ✐ Sketch tool.

154) Click the **circumference** of the first circle.

155) Click a **position** diagonally upward.

156) Enter **4**mm.

157) Click the **Green Check mark** ✓ from the Modify dialog box.

☼ If required, click the dimension arrowhead to toggle the dimension arrow as illustrated.

Add a vertical dimension.
158) Click the **bottom horizontal** line of the Base-Extrude feature.

159) Click the **center point** of the first circle.

160) Click a **position** to the left of the profile.

161) Enter **11**mm.

162) Click the **Green Check mark** ✓ from the Modify dialog box.

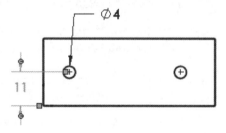

Add a horizontal dimension.
163) Click the **left vertical** line of the Base-Extrude feature.

164) Click the **center point** of the first circle.

165) Click a **position** below the profile.

166) Enter **16.5**mm.

167) Click the **Green Check mark** ✓ from the Modify dialog box.

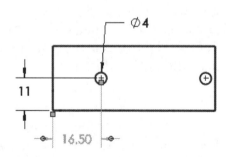

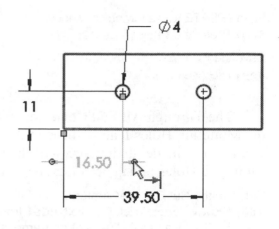

Add a second horizontal dimension.

168) Click the **left vertical** line of the Base-Extrude feature.

169) Click the **center point** of the second circle.

170) Click a **position** below the 16.50 dimension in the Graphics window.

171) Enter **39.5**mm.

172) Click the **Green Check mark** ✓ from the Modify dialog box.

Fit the model to the Graphics window.
173) Press the **f** key.

The drafting standard preference is to place arrows to the inside of the extension lines. Select the arrowhead control blue dots to alternate the arrowhead position.

If required, flip the dimension arrows to the inside.
174) Click the **16.50**mm dimension text in the Graphics window.

175) Click the **arrowhead dot** to display the arrows inside the extension lines.

176) If required, repeat the flip dimension arrow procedure for the **11**mm dimension text.

Fit the PLATE to the Graphics window.
177) Press the **f** key.

178) Click **Isometric view** from the Heads-up View toolbar.

179) Click **Shaded With Edges** from the Heads-up View toolbar.

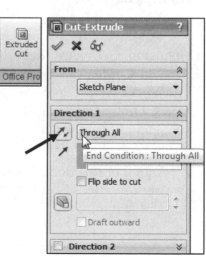

Create an Extruded Cut feature.
180) Click the **Features** tab from the CommandManager.

181) Click the **Extruded Cut** tool from the Features toolbar. The Cut-Extrude PropertyManager is displayed. The direction arrow points into the Extruded feature as illustrated.

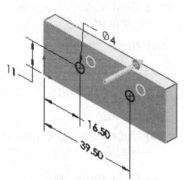

182) Select **Through All** for End Condition in Direction1 as illustrated. Think about the various End Condition options.

183) Click **OK** ✓ from the Cut-Extrude PropertyManager. Cut-Extrude1 is displayed in the FeatureManager.

Fit the PLATE to the Graphics window.
184) Press the **f** key.

Save the PLATE.
185) Click **Save** 💾.

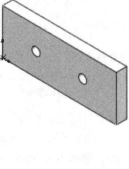

💡 The Through All End Condition option in Direction 1 creates the Mounting Holes feature through the Base-Extrude feature. As you modify the depth dimension of the Base-Extrude, the Mounting Holes update to reflect the change.

Rename the Cut-Extrude1 feature.
186) Slowly double click **Cut-Extrude1** from the FeatureManager. The feature name is highlighted.

187) Enter **Mounting Holes** for new name.

Modify the Base-Extrude depth.
188) Double-click **Base-Extrude** from the FeatureManager.

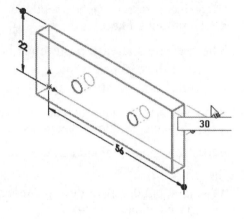

Modify the depth dimension.
189) Click **Hidden Lines Visible** ⬜ from the Heads-up View toolbar.

190) Double-click the **5**mm dimension from the Graphics window.

191) Enter **30**mm.

Return to the original dimension.
192) Click the **Undo** ↺ tool from the Menu bar.

Save the PLATE.
193) Click **Shaded With Edges** ⬜ from the Heads-up View toolbar.

194) Click **Save** 💾.

🔍 Additional details on Circles, Dimension, Geometric Relations and Extruded-Cut feature are available in SolidWorks Help Topics. Index keyword: Circles, Dimension, Add Relations in Sketch and Extruded Cut.

💡 A goal of this book is to expose the new user to various SolidWorks design tools and features.

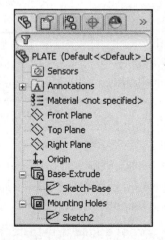

Review the Extruded Boss/Base and Extruded Cut Features

An Extruded Boss/Base feature adds material. An Extruded Cut feature removes material. Extrude features require:

- Sketch Plane

- Sketch

- Geometric relations and dimensions

- Depth and End Condition

PLATE Part-Fillet Feature

The Fillet feature removes sharp edges, strengthens corners and or cosmetically improves appearance. Fillets blend inside and outside surfaces. Fillet features are applied features. Applied features require edges or faces from existing features.

There are many fillet options. The Fillet PropertyManager provides the ability to select either the Manual or FilletXpert tab. Each tab has a separate menu. The Fillet PropertyManager displays the appropriate selections based on the type of fillet you create.

In the next activity, select the Manual tab and single edges for the first Fillet. Select the Tangent propagation option for the second Fillet.

On castings, heat-treated machined parts and plastic molded parts, implement Fillets into the initial design. If you are uncertain of the exact radius value, input a small test radius of 1mm. It takes less time for the manufacturing supplier to modify an existing Fillet dimension than to create a new one.

Activity: PLATE Part-Fillet Feature

Insert a Constant radius edge Fillet.

195) Click **Hidden Lines Visible** from the Heads-up View toolbar.

196) Click the **Fillet** Features tool. The Fillet PropertyManager is displayed.

197) Click the **Manual** tab from the Fillet PropertyManager. The Constant radius Fillet type is selected by default.

198) Enter **1.0**mm in the Fillet Radius box.

199) Click the **4 small corner edges** in the Graphics window of the PLATE as illustrated. The selected entities are displayed in the Items to Fillet box. Accept the default settings.

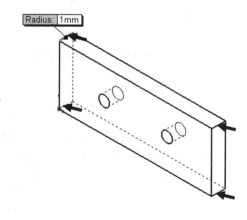

200) Click **OK** ✔ from the Fillet PropertyManager. Fillet1 is displayed in the FeatureManager.

Rename Fillet1.
201) Rename **Fillet1** to **Small Edge Fillet**.

Save the PLATE.
202) Click **Save** 💾.

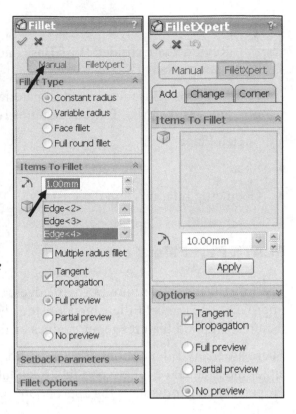

🔅 Click the FilletXpert tab in the Fillet PropertyManager to display the FilletXpert PropertyManager. The FilletXpert can only create and edit a Constant radius fillet type. The FilletXpert provides the ability to: *Create multiple fillets, automatically reorder fillets when required,* and *manage the desired type of fillet corner.* See SolidWorks Help for additional information.

Insert a Constant radius tangent edge Fillet.
203) Click the **front top horizontal edge** as illustrated.

204) Click the **Fillet** ⬙ Features tool. The Fillet PropertyManager is displayed. The Tangent propagation check box is checked.

205) Click the **back top horizontal edge**. The selected edges; Edge<1>, Edge<2> are displayed in the Items To Fillet box.

206) Repeat the process for the **bottom edges**.

207) Enter **1.0**mm in the Fillet Radius box.

208) Click **OK** ✔ from the Fillet PropertyManager. Fillet2 is displayed in the FeatureManager.

Rename Fillet2.
209) Rename **Fillet2** to **Front-Back Edge Fillet**.

Save the PLATE.
210) Click **Save** 💾.

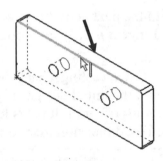

🔅 Save selection time. Utilize Ctrl-Select and choose all edges to fillet. Select the Fillet feature. All Ctrl-Select edges are displayed in the Items to Fillet box or apply the FilletXpert and use the connected to start loop option.

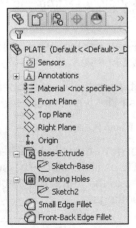

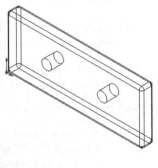

🔅 Minimize the number of Fillet radius sizes created in the FeatureManager. Combine Fillets and Rounds that have a common radius. Select all Fillet/Round edges. Add edges to the Items to Fillet list in the Fillet PropertyManager. Select the small edges, then fillet the larger edges.

PLATE Part-Hole Wizard Feature

The PLATE part requires a Countersink hole. Apply the Hole Wizard 🔲 feature. The Hole Wizard creates simple and complex Hole features by stepping through a series of options (Wizard) to define the hole type and hole placement. The Hole Wizard requires a face or Sketch plane to position the Hole feature. Select the back face.

Activity: PLATE Part-Hole Wizard Feature

Rotate the view.

211) Click and drag the **mouse pointer** using the middle mouse key to the left to rotate the part as illustrated. The Rotate icon ♻ is displayed.

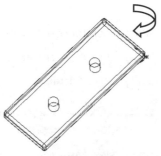

Display the Back view.

212) Click **Back view** 🔲 from the Heads-up View toolbar. The Origin is displayed at the lower right corner.

Insert the Countersink hole.

213) Click the **Hole Wizard** 🔲 Feature tool. The Hole Specification PropertyManager is displayed.

214) Click the **Type** tab.

215) Click the **Countersink** icon.

216) Select **Ansi Metric** for Standard.

217) Select **Flat Head Screw - ANSI B18.6.7M** for Type.

218) Select **M4** for Size.

219) Select **Through All** End Condition. Accept the default settings.

220) Click the **Positions** tab.

Select the Sketch plane for the Hole Wizard.

221) Click the **middle back face** of Base-Extrude as illustrated. The Point ✏ tool is selected.

222) Click **again to place** the center point of the hole.

223) Right-click **Select** to de-select the Point tool.

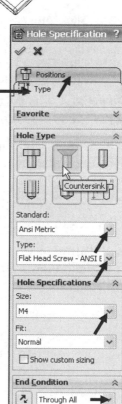

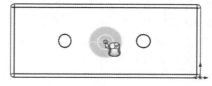

Add a horizontal dimension.

224) Click the **Smart Dimension** ✐ Sketch tool. The Point tool is de-selected and the Smart Dimension icon is displayed.

225) Click the **center point** of the Countersink hole.

226) Click the far **right vertical line**.

227) Click a **position** below the bottom horizontal edge.

228) Enter **56/2**. The dimension value 28mm is calculated.

229) Click the **Green Check mark** ✓ from the Modify dialog box.

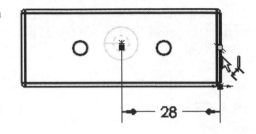

Add a vertical dimension.

230) Click the **center point** of the Countersink hole.

231) Click the **bottom horizontal line**.

232) Click a **position** to the right of the profile.

233) Enter **11**mm.

234) Click the **Green Check mark** ✓ from the Modify dialog box.

235) Click **OK** ✓ from the Dimension PropertyManager.

236) Click **OK** ✓ from the Hole Position PropertyManager. View the results

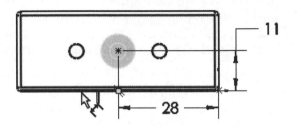

Fit the PLATE to the Graphics window.

237) Press the **f** key.

The Countersink hole is named CSK for M4 Flat Head in the FeatureManager.

☼ Enter dimensions as a formula in the Modify Dialog Box. Example 56/2 calculates 28.

Enter dimensions for automatic unit conversion. Example 2.0in calculates 50.8mm when primary units are set to millimeters (1in. = 25.4mm).

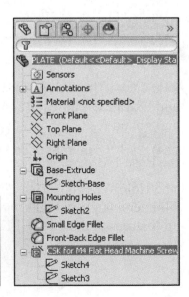

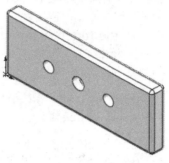

View the PLATE.

238) Click **Isometric view** from the Heads-up View toolbar.

239) Click **Shaded With Edges** from the Heads-up View toolbar.

Save the PLATE.
240) Click **Save** .

Close the PLATE.
241) Click **File**, **Close** from the Menu bar.

Display in the Shaded Isometric view before you save and close the part. A bitmap image of the model is saved. To preview the image in the SolidWorks Open dialog box, select View menu, Thumbnails option.

Additional details on the Hole Wizard feature are available in SolidWorks Help. Keywords: Hole Wizard, holes, simple and complex holes.

Review the PLATE Part

The PLATE part utilized an Extruded Boss/Base feature. An Extruded Boss/Base feature consisted of a rectangular profile sketched on the Front Plane. You added linear dimensions to correspond to the overall size of the PLATE based on the GUIDE-CYLINDER assembly, Piston Plate.

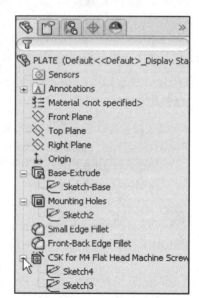

The two holes utilized an Extruded Cut feature. You sketched two circles on the front face and added Equal and Horizontal relations. You utilized Geometric relations in the sketch to reduce dimensions and to maintain the geometric requirements of the part. Linear dimension defined the diameter and position of the circles.

You created fully defined sketches in both the Extruded Base and Extruded Cut feature to prevent future rebuild problems and obtain faster rebuild times. The Fillet feature inserted the edge Fillets and tangent edge Fillets to round the corners of the Base-Extrude feature.

The Hole Wizard feature created the M4 Flat Head Countersink on the back face of the PLATE. You renamed all features in the PLATE FeatureManager.

ROD Part Overview

Recall the functional requirements of the customer:

- The ROD is part of a sub-assembly that positions materials onto a conveyor belt.

- The back end of the ROD fastens to the PLATE.

- The front end of the ROD mounts to the customer's components.

- The customer supplies the geometric requirements for the keyway cut and hole.

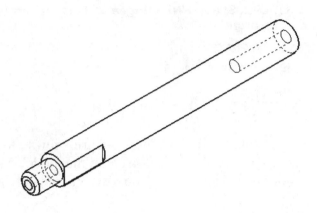

The ROD utilizes an Extruded Boss/Base feature with a circular profile sketched on the Front Plane. The ROD also utilizes the Hole Wizard feature for a simple hole and the Extruded Cut feature. Explore new features and techniques with the ROD.

Utilize the following features to create and modify the ROD part:

- **Extruded Boss/Base** : Create the Extruded Boss/Base feature (Boss-Extrude1) on the Front Plane with a circular sketched profile. Use the Circle Sketch tool.

The plane or face you select for the Base sketch determines the orientation of the part.

- **Hole Wizard** : Utilize the Hole Wizard feature to create the first hole on the front face of the ROD. Later, copy the first hole to the back face of the ROD.

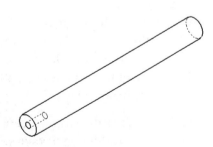

A goal of this text is to expose the new user to various SolidWorks design tools and features. For manufacturing, always apply the Hole Wizard feature to create a simple or complex hole.

- **Chamfer** 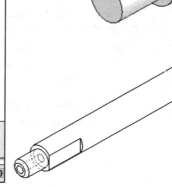: Create the Chamfer feature from the circular edge of the front face. The Chamfer feature removes material along an edge or face. The Chamfer feature assists the ROD by creating beveled edges for ease of movement in the GUIDE-ROD assembly.

- **Extruded Cut** : Create the Extruded Cut feature from a 2D sketched. Convert the edge of the Boss-Extrude1 (Base Extrude) feature to form a Keyway. Utilize the Trim Entities sketch tool to delete sketched geometry.

- **Hole Wizard** : Create the back hole with the Copy/Paste tool. Copy the front hole to the back face. Modify the Hole Wizard hole diameter from 3mm to 4mm.

- **Extruded Cut** : Apply the Rollback bar and Edit Feature function to implement a customer design change. Add a new Extruded Cut feature to the front face.

- **Redefine** the Chamfer feature and Keyway cut feature. Modify the Sketch Plane.

Rollback bar

ROD Part-Extruded Boss/Base Feature

The geometry of the Boss-Extrude1 (Base) feature is a cylindrical extrusion. The Extruded Boss/Base feature is the foundation for the ROD. What is the Sketch plane? Answer: The Front Plane is the Sketch plane for this model. What is the shape of the sketched 2D profile? Answer: A circle.

Activity: ROD Part-Extruded Base Feature

Create a New part.

242) Click **New** ⬚ from the Menu bar.

243) Click the **MY-TEMPLATES** tab. The New SolidWorks Document dialog displays the PART-MM-ANSI Part Template.

244) Double-click **PART-MM-ANSI**.

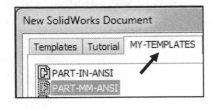

If the MY-TEMPLATES tab is not displayed, click **Options**, **File Locations** from the Menu bar. Add the full pathname to the **MY-TEMPLATES** folder to the Document Templates box.

Save and name the part.

245) Click **Save** 💾.

246) Select **ENGDESIGN-W-SOLIDWORKS\PROJECTS**.

247) Enter **ROD** for File name.

248) Enter **ROD 10MM DIA x 100MM** for Description.

249) Click **Save**.

Select the Sketch plane and create Sketch1.

250) Right-click **Front Plane** from the FeatureManager. Front Plane is your Sketch plane. Think about why the Front Plane was chosen.

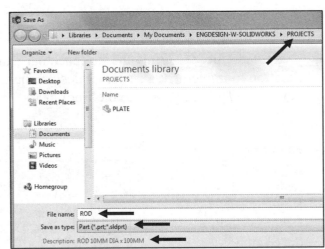

251) Click **Sketch** ✎ from the Context toolbar. The Sketch toolbar is displayed.

To de-activate the display Reference planes, right-click on the **selected plane** in the FeatureManager, click **Hide** 👓 from the Context toolbar or click **View**; uncheck **Planes** from the Menu bar.

Sketch a circle Coincident at the Origin. Only a single dimension is required; to fully define the Base Sketch.

252) Click the **Circle** ⊘ Sketch tool. The Circle PropertyManager is displayed. Use the Origin in the sketch.

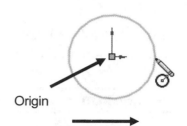

253) Click the **Origin** ↳. The Origin is Coincident with the center point of the circle.

254) Click a **position** to the right of the Origin.

Origin

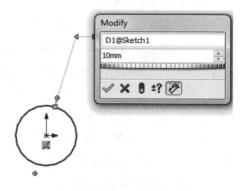

Add a dimension.

255) Right-click the **Smart Dimension** ⟨⟩ Sketch tool.

The Smart Dimension icon is displayed.

256) Click the **circumference** of the circle.

257) Click a **position** diagonally off the profile.

258) Enter **10**mm.

259) Click the **Green Check mark** ✅ from the Modify dialog box.

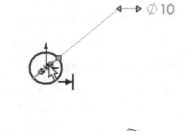

☼ If required, click the arrowhead dot to toggle the direction of the dimension arrow.

Insert an Extruded Boss/Base feature.

260) Click the **Extruded Boss/Base** 🗔 feature tool from the Features tab in the CommandManager. Blind is the default End Condition in Direction 1.

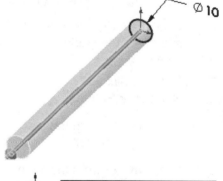

261) Enter **100**mm for Depth. Accept the default settings. Note the direction of the extrude feature.

262) Click **OK** ✅ from the Boss-Extrude PropertyManager. Boss-Extrude1 is displayed in the FeatureManager. Note the location of the Origin.

Fit the model to the Graphics window.

263) Press the **f** key.

264) Rename **Boss-Extrude1** to **Base Extrude** as illustrated.

Save the ROD.

265) Click **Save** 💾.

ROD Part-HOLE Wizard Feature

Utilize the Hole Wizard 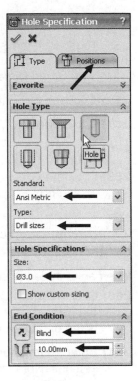 feature to create the first hole on the front face of the ROD. Later, copy the first hole to the back face of the ROD. What is the Sketch Plane for the first Hole Wizard feature? Answer: Front face of the ROD.

> **Activity: ROD Part-Hole Wizard Feature**

Select the Front view.

266) Click **Front view** from the Heads-up view toolbar.

Create a simple ANSI Metric hole using the Hole Wizard Feature.

267) Click the **Hole Wizard** feature tool. The Hole Specification PropertyManager is displayed. Type is the default tab.

268) Click **Hole** for Hole Type.

269) Select **Ansi Metric** for Standard.

270) Select **Drill sizes** for Type.

271) Select **3.0** for Hole Size.

272) Select **Blind** for End Condition.

273) Enter **10mm** for Depth.

274) Click the **Positions** tab.

275) Click a position to the right of the **Origin** as illustrated. Do NOT click the Origin. The

Point tool icon is displayed.

276) Click the **Origin** to place the center point of the hole at the Origin. Note the Geometric relation.

De-select the Point tool.
277) Right-click **Select**.

278) Click **OK** from the Dimension PropertyManager.

279) Click **OK** from the Hole Position PropertyManager. View the results.

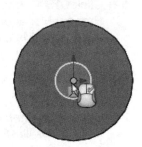

280) Expand Diameter Hole1 in the FeatureManager. View the two sketches. Both sketches are fully defined.

281) Click **Isometric view**  from the Heads-up View toolbar.

282) Click **Hidden Lines Visible** from the Heads-up View toolbar.

Rename the feature.
283) Rename **Diameter Hole1** to **Front Hole**.

Save the ROD.
284) Click **Save** .

The Hole Wizard feature creates either a 2D or 3D sketch for the placement of the hole in the FeatureManager. You can consecutively place multiple holes of the same type. The Hole Wizard creates 2D sketches for holes unless you select a nonplanar face or click the 3D Sketch button in the Hole Position PropertyManager.

Hole Wizard creates two sketches. The first sketch is the revolved cut profile of the selected hole type and the second sketch is the location of the profile. Both sketches should be fully defined.

ROD Part-Chamfer Feature

The Chamfer feature removes material along an edge or face. The Chamfer feature assists the ROD by creating beveled edges for ease of movement in the GUIDE-ROD assembly. There are three Chamfer feature types:

- *Angle distance*. Selected by default. Sets the Distance and Angle in the Chamfer PropertyManager or in the Graphics window.

- *Distance distance*. Enter values for both distances on either side of the selected chamfer edges or Equal Distance and specify a single value.

- *Vertex*. Enter values for the three distances on each side of the selected vertex or click Equal Distance and specify a single value.

Chamfer the front circular edge and utilize the Angle distance option in the next activity.

Selection order does not matter to create a Chamfer or Fillet feature. Example 1: Select edges and faces. Select the feature tool. Example 2: Select the feature tool. Select edges and faces.

Activity: ROD Part-Chamfer Feature

Insert a Chamfer feature.

285) Click the **front outer circular edge**  of the ROD. Note the mouse icon feedback symbol for an edge. The edge turns blue.

286) Click the **Chamfer**  feature tool from the Consolidated toolbar. The Chamfer PropertyManager is displayed. The arrow indicates the chamfer direction.

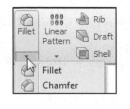

Angle distance is selected by default.

287) Enter **1**mm for Distance. Accept the default 45deg Angle.

288) Click **OK** ✓ from the Chamfer PropertyManager. Chamfer1 is displayed in the FeatureManager.

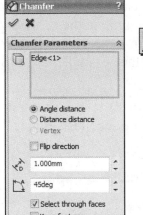

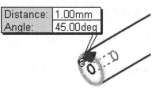

Save the ROD.

289) Click **Save** 💾.

💡 To deselect geometry inside the Chamfer Parameters box, right-click **Clear Selections**. Enter values for Angle distance in the Chamfer Property Manager or inside the Angle distance Pop-up box.

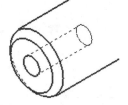

💡 Design Intent is how your part reacts as parameters are modified. Example: If you have a hole in a part that must always go through, apply the Through All End Condition. As the part thickness changes, the hole will always go through the part.

ROD Part-Extruded Cut Feature & Convert Entities Sketch tool

The Extruded Cut 🔲 feature removes material from the front of the ROD to create a keyway. A keyway locates the orientation of the ROD into a mating part in the assembly.

Utilize the Convert Entities Sketch tool in the Sketch toolbar to extract existing geometry to the selected Sketch plane. The Trim Entities Sketch tool deletes sketched geometry. Utilize the Power Trim option from the Trim PropertyManager.

Activity: ROD Part-Extruded Cut Feature and Convert Entities Sketch tool

Select the Sketch plane for the keyway. The Sketch plane is the front circular face of the ROD.

290) Click **Shaded With Edges**  from the Heads-up View toolbar.

291) Right-click the **front circular face** of the ROD. Base Extrude is highlighted in the FeatureManager.

Create a sketch.

292) Click **Sketch** ⤶ from the Context toolbar.

Zoom in on the front face.

293) Click **Zoom to Area** 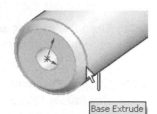 from the Heads-up View toolbar.

294) Zoom in on the front face.

295) Click **Zoom to Area** to deactivate from the Heads-up View toolbar.

Convert the outside edge to the Sketch plane.

296) Click the **outside front circular edge** of the Base Extrude as illustrated.

297) Click the **Convert Entities** Sketch tool from the Sketch toolbar. The system extracts the outside edge and positions it on the Sketch plane. Note: The Convert Entities PropertyManager provides the ability to select more than one entity.

298) Click **OK** ✔ from the Convert Entities PropertyManager.

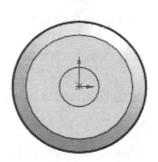

299) Click **Front view** from the Heads-up View toolbar. The front face of the ROD is displayed.

Sketch a vertical line.

300) Click the **Line** ＼ Sketch tool. The Insert Line PropertyManager is displayed.

301) Sketch a **vertical line** as illustrated. The end points of the line extend above and below the diameter of the circle. Note: Insert a

Vertical relation if needed.

De-select the Line Sketch tool.

302) Right-click **Select** in the Graphics window.

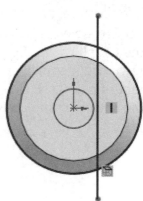

💡 Right-click Select to deselect the existing tool.

Trim the sketch.

303) Click the **Trim Entities** 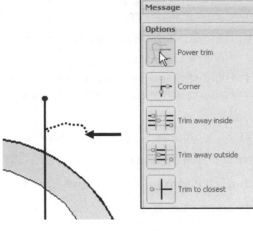 Sketch tool. The Trim PropertyManager is displayed. Note the available options. Click the question mark ? for additional information on each option.

304) Click **Power trim** from the Trim PropertyManager Options box.

305) Click a **position** to the right of the vertical line, above the circular edge.

306) Drag the **mouse pointer** to intersect the vertical line above the circular edge.

307) Release the **mouse button** when you intersect the vertical line. Note: SolidWorks displays a light gray line as you drag. The line is removed.

The Power trim option trims the vertical line to the converted circular edge and adds a Coincident relation between the endpoint of the vertical line and the circle.

308) Click a **position** to the left of the circular edge.

309) Click and drag the **mouse pointer** to intersect the left circular edge.

310) Release the **mouse button**. The left circular edge displays the original profile.

311) Click a **position** to the right of the vertical line, below the circular edge.

312) Drag the **mouse pointer** to intersect the vertical line below the circular edge.

313) Release the **mouse pointer**.

314) Click **OK** from the Trim PropertyManager.

A selection box allows selection of multiple entities at one time.

To Box-select, click the upper left corner of the box in the Graphics window. Drag the mouse pointer to the lower right corner. A blue box is displayed. Release the mouse pointer. If you create the box from left to right, the entities completely inside the box are selected.

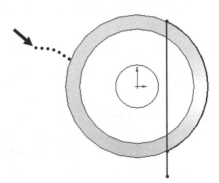

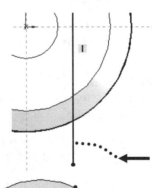

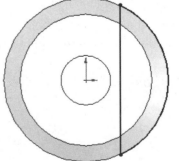

Box-select from left to right.

315) Click a **position** to the upper left corner of the profile as illustrated.

316) Drag the **mouse pointer** diagonally to the lower right corner.

317) Release the **mouse pointer**. The Properties PropertyManager is displayed. The Arc and Line sketched entities are displayed in the Selected Entities dialog box.

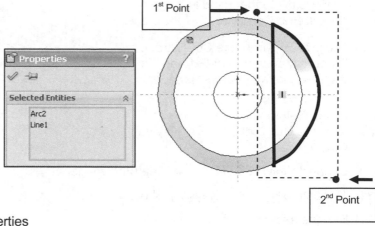

318) Click **OK** from the Properties PropertyManager.

Insert a vertical dimension.

319) Click the **Smart Dimension** Sketch tool. The Smart Dimension icon is displayed.

320) Click the **vertical line** as illustrated.

321) Click a **position** to the right of the profile.

322) Enter **6.0**mm.

323) Click the **Green Check mark** from the Modify dialog box. The vertical line endpoints are coincident with the circle outside edge. The profile is a single closed loop.

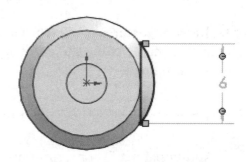

Fit the sketch to the Graphics window.

324) Press the **f** key.

Although dimensions are not required to create features in SolidWorks, dimensions provide location and size information. Models require dimensions for manufacturing.

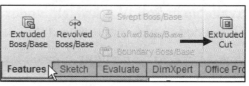

Insert an Extruded Cut feature.

325) Click the **Extruded Cut** feature tool. The Cut-Extrude PropertyManager is displayed.

326) Click **Isometric view** from the Heads-up View toolbar. View the arrow direction.

327) Enter **15**mm for the Depth in Direction 1. Accept the default settings. Note that Blind is the default End Condition.

328) Click **OK** ✔ from the Cut-Extrude PropertyManager. Cut-Extrude1 is displayed in the FeatureManager.

Rename the feature and sketch.

329) Rename **Cut-Extrude1** to **Keyway Cut**.

330) Expand Keyway Cut from the FeatureManager.

331) Rename **Sketch4** to **Keyway Sketch**.

332) Exit the sketch.

Fit the Keyway Sketch to the Graphics window.

333) Press the **f** key.

Save the ROD.

334) Click **Save** 💾.

Rename features or sketches in the FeatureManager by slowly clicking the name twice and entering the new name when the old one is highlighted.

Additional details on Trim Entities, Convert Entities, Extruded Cut, Relations are available in SolidWorks Help. Index keyword: Trim (Sketch), Hole Wizard, Cut (Extruded) and Coincident relations.

The Extruded Cut feature created the Keyway Cut by converting existing geometry. The Hole Wizard feature created the Front Simple Hole on the front face.

The Convert Entities tool extracts feature geometry. Copy features utilizing the Copy/Paste Edit option.

To remove Tangent edges, click **Display/Selections** from the Options menu, check the **Removed** box.

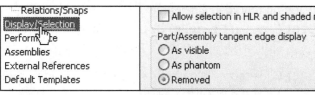

ROD Part-View Orientation, Named View, & Viewport Option

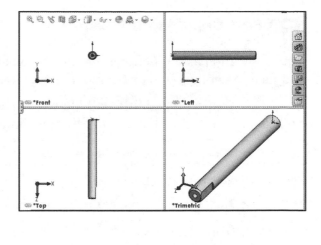

The View Orientation defines the preset position of the ROD in the Graphics window. It is helpful to display various views when creating and editing features. The View Orientation options provide the ability to perform the following tasks:

- Create a named view

- Select a standard view

- Combine views by using the Viewport option

Pressing the **Shift key + Up arrow** rotates the model in 90deg increments. Start in an Isometric view. Hold the Shift key down and press the Up arrow key two times to create a Back Isometric view.

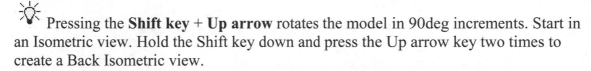

Activity: ROD Part-View Orientation and Named View

Rotate the ROD.
335) Hold the **Shift** key down.

336) Press the **Up arrow** key twice.

337) Release the **Shift** key.

Insert a new view.
338) Press the **Space Bar** on your keyboard or click View Orientation

from the Standard Views toolbar. The Orientation dialog box is displayed.

339) Click the **Push Pin** icon from the Orientation dialog box to maintain the displayed menu.

340) Click the **New View** icon in the Orientation dialog box.

341) Enter **back iso** in the View name text box.

342) Click **OK** from the Named View dialog box.

Close the Orientation menu.
343) **Close** the dialog box. Save the ROD in the new view.

344) Click **Save** .

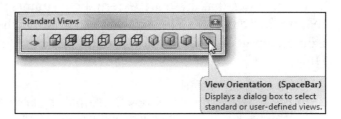

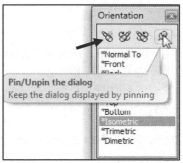

ROD Part-Copy/Paste Feature

The Copy and paste functionality allows you to copy selected sketches and features from one face to another face or from one model to different models.

The ROD requires an additional hole on the back face. Copy the front hole feature to the back face. Edit the sketch and modify the dimensions and geometric relations.

Activity: ROD Part-Copy/Paste Feature

Copy the Front Hole to the Back Face.

345) Click **Back view**  from the Heads up View toolbar.

346) Click **Front Hole** ⊟ ⬛ Ø3.0 (3) Front Hole from the FeatureManager.

347) Click **Edit**, **Copy** from the Menu bar.

348) Click the **back face** of the ROD as illustrated. Note: Do not click the center point.

349) Click **Edit**, **Paste** from the Menu bar.

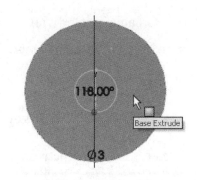

The Copy Confirmation dialog box is displayed. The box states that there are external constraints in the feature being copied. The external constraint is the Coincident geometric relation used to place the Front Hole.

Delete the old constraint.

350) Click **Delete** from the Copy Confirmation dialog box. The back face of the ROD contains a copy of the Front Hole feature.

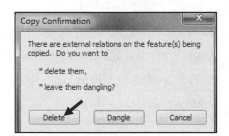

351) Rename **Diameter Hole2** to **Back Hole**.

Edit the Sketch. Locate the Back Hole. Insert a Coincident relation to the Origin.

352) **Expand** Back Hole for the FeatureManager.

353) Right-click **(-) Sketch5**.

354) Click **Edit Sketch**  from the Context toolbar.

355) Click the **center point** of the 3mm circle.

356) Hold the **Ctrl** key down.

357) Click the **Origin** ↳ of the ROD as illustrated.

358) Release the **Ctrl** key. The Properties PropertyManager is displayed. The selected sketch entities are displayed in the Selected Entities box.

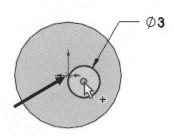

359) Click **Coincident** ⟨ from the Add Relations box. The center point of the back hole and the origin are Coincident.

360) Click **OK** ✓ from the Properties PropertyManager.

Exit the Sketch.

361) Click **Exit Sketch** ⟨ from the Context toolbar.

Modify the diameter dimension of the hole. Edit the Hole Wizard feature.

362) Click **Isometric view** ⟨ from the Heads-up View toolbar.

363) Right-click **Back Hole** from the FeatureManager.

364) Click **Edit Feature** ⟨ from the Context toolbar. The Hole Specification PropertyManager is displayed.

365) Select **4.0** for Hole Size.

366) Click **OK** ✓ from the Hole Specification PropertyManager.

Rename the Back Hole feature.
367) Rename **3.0 (3) Back Hole** to **4.0 (4) Back Hole**.

Save the ROD.
368) Click **Save** 🖫.

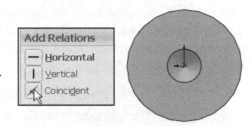

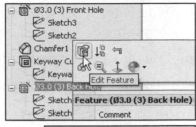

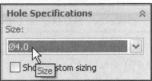

☼ The Confirmation Corner in the sketch Graphics window indicates that a sketch is active.

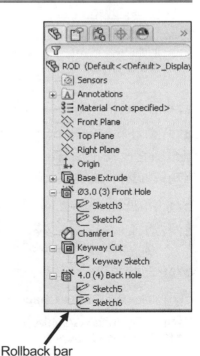

ROD Part-Design Changes with Rollback Bar

You are finished for the day. The phone rings. The customer voices concern with the GUIDE-ROD assembly. The customer provided incorrect dimensions for the mating assembly to the ROD. The ROD must fit into a 7mm hole. The ROD fastens to the customer's assembly with a 4mm Socket head cap screw.

You agree to make the changes at no additional cost since the ROD has not been machined. You confirm the customer's change in writing.

The customer agrees but wants to view a copy of the GUIDE-ROD assembly design by tomorrow! You are required to implement the design change and to incorporate it into the existing part.

Rollback bar

Use the Rollback bar and Edit Feature function to implement a design change. The Rollback bar provides the ability to redefine a feature in any state or order. Reposition the Rollback bar in the FeatureManager.

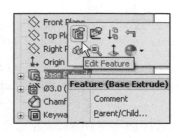

The Edit Feature 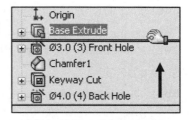 function provides the ability to redefine feature parameters. Implement the design change.

- Modify the ROD with the Extruded Cut feature to address the customer's new requirements.

- Edit the Chamfer feature to include the new edge.

- Redefine a new Sketch plane for the Keyway Cut feature.

In this procedure, you develop rebuild errors and correct rebuild errors.

Activity: ROD Part-Design Changes with Rollback Bar in the FeatureManager

Position the Rollback bar in the FeatureManager.

369) Place the **mouse pointer** over the Rollback bar at the bottom of the FeatureManager. The mouse pointer displays a symbol of a hand.

370) Drag the **Rollback** bar upward below the Base Extrude feature. The Base Extrude feature is displayed.

Select the Sketch plane and display the Sketch toolbar.

371) Right-click the **Front face** of the Base Extrude feature. This is your Sketch plane. Base Extrude is highlighted in the FeatureManager.

372) Click **Sketch** ✏ from the Context toolbar. The Sketch toolbar is displayed.

Sketch a circle Coincident about the Origin.

373) Click the **Circle** ⊙ Sketch tool. The Circle PropertyManager is displayed.

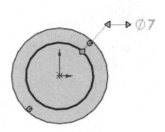

374) Click the **Origin** ↳ of the Base Extrude feature.

375) Click a **position** to the right of the Origin as illustrated.

Add a diameter dimension.

376) Click **Front view** from the Heads-up View toolbar.

377) Right-click the **Smart Dimension** ✐ tool.

378) Click the **circumference** of the circle. Click a **position** diagonally off the profile.

379) Enter **7**mm.

380) Click the **Green Check mark** ✓ from the Modify dialog box.

Fit the ROD to the Graphics window.
381) Press the **f** key.

Insert an Extruded Cut feature.

382) Click the **Extruded Cut** 🔲 feature tool. The Cut-Extrude PropertyManager is displayed.

383) Click the **Flip side to cut** box. The direction arrow points outward. Blind is the default End Condition.

384) Enter **10**mm for Depth. Accept the default settings.

385) Click **OK** ✅ from the Cut-Extrude PropertyManager.

386) Click **Isometric view** 🔳 from the Heads-up View toolbar.

Rename the feature.
387) Rename the **Cut-Extrude2** feature to **Front Cut**.

Save the ROD.
388) Click **Save** 💾.

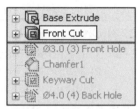

ROD Part-Recover from Rebuild Errors

Rebuild errors can occur when using the Rollback function. A common error occurs when an edge or face is missing. Redefine the edge for the Chamfer feature. When you delete sketch references, Dangling dimensions, sketches and plane errors occur. Redefine the face for the Keyway Sketch plane and the Keyway Sketch.

☀️ Dangling indicates sketch geometry that cannot be resolved. For example, deleting an entity that was used to define another sketch entity

Activity: ROD Part-Recover from Rebuild Errors

Redefine the Chamfer feature.
389) Drag the **Rollback** bar downward below the Chamfer1 feature. A red x 🔶❌ Chamfer1 is displayed next to the name of the Chamfer1 feature. A red arrow is displayed 🔹🔴 ROD next to the ROD part icon. Red indicates a model rebuild error. When you created the Front Cut feature, you deleted the original edge from the Base Extrude feature.

Insert the Chamfer feature on the Front Cut edge.

390) Right-click **Chamfer1** from the FeatureManager.

391) Click **Edit Feature** . The Chamfer1 PropertyManager is displayed. **Missing**Edge<1> is displayed in the Chamfer Parameters box.

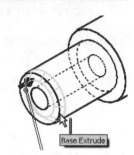

392) Right-click **inside** the Edges and Faces or Vectors box.

393) Click **Clear Selections**.

394) Click the **front edge** of the Base Extrude feature as illustrated. Edge<2> is displayed in the Items to Chamfer box.

395) Enter **1.0**mm for Distance. Accept the default settings.

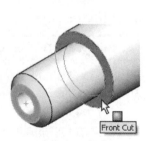

Update the Chamfer feature.

396) Click **OK** from the Chamfer1 PropertyManager.

Redefine the face for the Keyway Sketch plane.

397) Drag the **Rollback** bar downward below the Keyway Sketch.

The Rod Rebuild Error box is displayed. The ⚠ Keyway Sketch indicates a dangling Sketch plane. The Front Cut deleted the original Sketch plane.

Redefine the Sketch plane and sketch relations to correct the rebuild errors.

398) Right-click **Keyway Sketch** from the FeatureManager.

399) Click **Edit Sketch Plane** from the Context toolbar. The Sketch Plane PropertyManager is displayed.

400) Click the **large circular face** of the Front Cut as illustrated. Face<2> is displayed in the Sketch Plane/Face box.

401) Click **OK** from the Sketch Plane PropertyManager. The SolidWorks dialog box is displayed.

402) Click the **Stop and Repair** button. The What's Wrong dialog box is displayed. View the description.

403) Click **Close** from the What's Wrong dialog box.

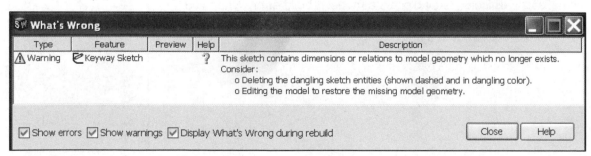

Type	Feature	Preview	Help	Description
⚠ Warning	Keyway Sketch		?	This sketch contains dimensions or relations to model geometry which no longer exists. Consider: o Deleting the dangling sketch entities (shown dashed and in dangling color). o Editing the model to restore the missing model geometry.

☑ Show errors ☑ Show warnings ☑ Display What's Wrong during rebuild   Close   Help

The Keyway Sketch moved to the new Sketch plane. To access the Edit Sketch Plane option, right-click the Sketch entry in the FeatureManager.

Redefine the sketch relations and dimensions to correct the two build errors. Dangling sketch entities result from a dimension or relation that is unresolved. An entity becomes dangling when you delete previously defined geometry.

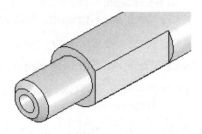

The Display/Delete Relations sketch tool lists dangling geometry.

Edit the Keyway Cut sketch.

404) Right-click **Keyway Cut** from the FeatureManager.

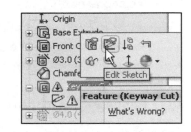

405) Click **Edit Sketch** 🖉 from the Context toolbar.

Display the Sketch relations.

406) Click the **Display/Delete Relations** ⬖ Sketch tool. The Display/Delete Relations PropertyManager is displayed.

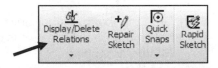

407) Right-click the **On Edge 0** relation.

408) Click **Delete**. The Dangling status is replaced by Satisfied. The sketch is under defined and is displayed in blue.

409) Click **OK** ✔ from the Display/Delete Relations PropertyManager.

When you select a relation from the Relations list box, the appropriate sketch entities are highlighted in the Graphics window. If Sketch Relations, (**View**, **Sketch Relations**) is selected, all icons are displayed, but the icons for the highlighted relations are displayed in a different color.

Display the Front view.

410) Click **Front view** 🗗 from the Heads-up View toolbar.

Add a Coradial relation.

411) Click the **right blue arc** as illustrated.

412) Hold the **Ctrl** key down.

413) Click the **outside circular edge**.

414) Release the **Ctrl** key. The Properties PropertyManager is displayed.

415) Click **Coradial** from the Add Relations box.

416) Click **OK** ✔ from the Properties PropertyManager. The sketch is fully defined and is displayed in black.

Exit the sketch.

417) Rebuild 🔧 the model.

418) Click **Isometric view** 🔲 from the Heads-up View toolbar.

419) Drag the **Rollback** bar downward below the Back Hole feature in the FeatureManager.

Save the ROD.

420) Click **Save** 💾.

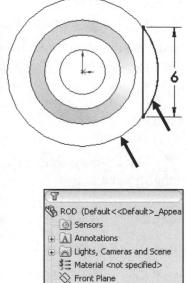

🔅 There are three ways to add geometric relations. Utilize the Add Relations Sketch tool ⏄ use the Properties PropertyManager, or use the Context toolbar.

Utilize the Properties PropertyManager
📋 **Properties** or the Context toolbar when you select multiple sketch entities in the Graphics window. Select the first entity. Hold the Ctrl key down and select the remaining geometry to create Sketch relations. The Properties and Context toolbar technique requires fewer steps.

🔅 View the mouse pointer icon to confirm selection of the correct point, edge, face, etc.

🔅 Rename features or sketches in the FeatureManager by slowly clicking the name twice and entering the new name when the old one is highlighted.

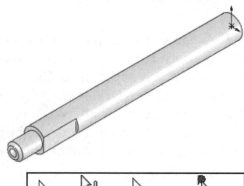

| Face | Edge | Dimension | Vertex |

Modify the depth of the Back Hole.
421) Right-click **Back Hole** from the FeatureManager.

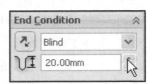

422) Click **Edit Feature** 🗈 from the Context toolbar. The Hole Specification PropertyManager is displayed.

423) Enter **20**mm for End Condition as illustrated.

424) Click **OK** ✅ from the Hole Specification PropertyManager.

Save the ROD part.
425) Click **Save** 🖫.

💡 A machinist rule of thumb states that the depth of a hole does not exceed 10 times the diameter of the hole. The purpose of this book is to expose various SolidWorks design tools and features.

A Countersunk screw fastens the PLATE to the ROD. The depth of the PLATE and the depth of the Back Hole determine the length of the Countersunk screw. Utilize standard length screws.

ROD Part-Edit Part Appearance

Parts are shaded gray by system default. The Edit Part Color option provides the ability to modify part, feature and face color. In the next activity, modify the part color and the machined faces.

Activity: ROD Part-Edit Part Appearance

Edit the part color.
426) Click **Shaded with Edges** ⬜ from the Heads-up View toolbar.

427) Right-click the **ROD Part** 🦴 ROD icon at the top of the FeatureManager.

428) Click the **Appearances** drop-down arrow as illustrated.

429) Click inside the **Edit color** box as illustrated. The Color PropertyManager is displayed.

430) Right-click **ROD** inside the Selection Geometry box.

431) Click **Clear Selections**.

Modify machined surfaces on the ROD part.

432) Click **Select Features** 🗈 in the Selection Geometry box.

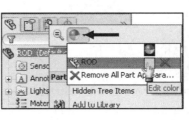

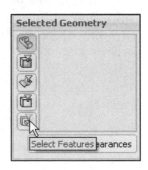

433) Click **Keyway Cut** from the fly-out FeatureManager.

434) Click **Front Hole** from the fly-out FeatureManager.

435) Click **Back Hole** from the fly-out FeatureManager. The selected entities are displayed in the Selected Entities box.

436) Click a **color swatch** for the selected faces. Example: Blue. The ROD faces are displayed in the selected color.

437) Click **OK** ✓ from the Color PropertyManager.

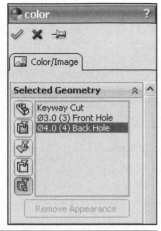

Expand the Show Display Pane.
438) Click the **Show Display Pane arrow**. View the results.

Collapse the Hide Display Plane.
439) Click the **Hide Display Pane arrow** to collapse the FeatureManager.

440) Click **Isometric view** ⬚ from the Heads-up View toolbar.

The ROD part is complete. Save the ROD.
441) Click **Save** 💾.

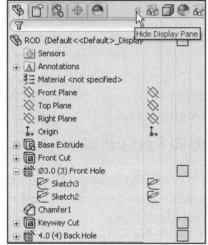

Close the ROD.
442) Click **File**, **Close** from the Menu bar.

☀ DFMXpress is an analysis tool that validates the manufacturability of SolidWorks parts. Use DFMXpress to identify design areas that may cause problems in fabrication or increase the costs of production. As an exercise, run DFMXpress from the Evaluate tab on the ROD. View the results.

☀ Standardize the company part colors. Example: Color red indicates a cast surface. Color blue indicates a machined surface. Green is the selection color in SolidWorks. Minimize the use of the color green to avoid confusion.

Review of the ROD Part

The ROD part utilized an Extruded Boss/Base feature with a circular profile sketched on the Front Plane. Linear and diameter dimensions corresponded to the customer's requirements.

The Chamfer feature removed the front circular edge. The Chamfer feature required an edge, distance and angle parameters. The Extruded Cut feature utilized a converted edge of the Base Extrude feature to form the Keyway Cut. Reuse geometry with the Convert Entity Sketch tool.

You utilized the Hole Wizard feature for a simple hole on the front face to create the Front Hole. You reused geometry. The Front Hole was copied and pasted to the back face to create the Back Hole. The Back Hole dimensions were modified. The Display\Delete Sketch Tool deleted dangling dimensions of the copied feature. You inserted an Extruded Cut feature by moving the Rollback bar to a new position in the FeatureManager. You recovered from the errors that occur with Edit Feature, Edit Sketch, and Edit Sketch plane tools by redefining geometry.

GUIDE Part Overview

The GUIDE part supports the ROD. The ROD moves linearly through the GUIDE. Two slot cuts provide flexibility in locating the GUIDE in the final assembly.

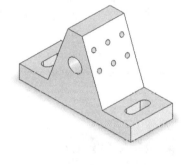

The GUIDE supports a small sensor mounted on the angled right side. You do not have the information on the exact location of the sensor. During the field installation, you receive additional instructions.

Create a pattern of holes on the angled right side of the GUIDE to address the functional requirements.

Address the geometric requirements with the GUIDE part:

- **Extruded Boss/Base** : Use the Extruded Boss/Base feature to create a symmetrical sketched profile for the GUIDE part.

Utilize a centerline and the Dynamic Mirror Sketch tool. The centerline acts as the mirror axis. The copied entity becomes a mirror image of the original across the centerline.

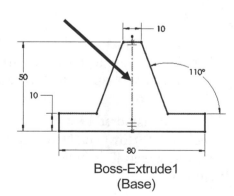

Boss-Extrude1
(Base)

A centerline is a line with a property that makes it exempt from the normal rules that govern sketches. When the system validates a sketch, it does not include centerlines when determining if the profile is disjoint or self-intersecting.

- **Extruded Cut** : Use the Extruded Cut feature with the Through All End Condition to create the first slot. The Extruded Cut feature utilizes the Straight Slot Sketch tool on the top face to form the slot.

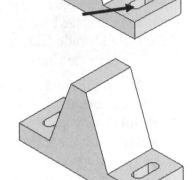

- **Mirror** Use the Mirror feature to create a second slot about the Right Plane. Reuse geometry with the Mirror feature.

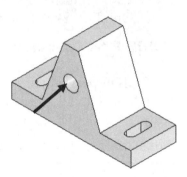

- **Extruded Cut** : Use the Extruded Cut feature with the Through All End Condition to create the Guide Hole. The ROD glides through the Guide Hole. Note: The purpose of this book is to expose various SolidWorks design tools and features.

- **Hole Wizard** Use the Hole Wizard feature to create a M3 tapped hole; seed feature for the Linear Pattern.

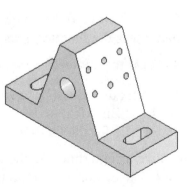

- **Linear Pattern** : Use the Linear Pattern feature to create multiple instances of the M3 tapped hole. Reuse geometry with the Linear Pattern feature.

It is considered best practice to fully define all sketches in the model. However, there are times when this is not practical, generally when using the spline tool to create a freeform shape.

GUIDE Part-Extruded Boss/Base Feature and Dynamic Mirror

The GUIDE utilizes an Extruded Boss/Base feature. What is the Sketch plane? Answer: The Front Plane is the Sketch plane. The Mid Plane End Condition extrudes the sketch symmetric about the Front Plane.

How do you sketch a symmetrical 2D profile? Answer: Use a sketched centerline and the Dynamic Mirror Sketch tool.

Sketching lines and centerlines produces different behavior depending on how you sketch. Review sketching line techniques before you begin the GUIDE part.

There are two methods to create a line:

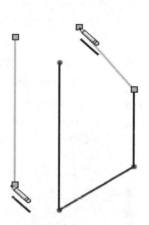

- *Method 1*: Click the first point of the line. Drag the mouse pointer to the end of the line and release. Utilize this technique to create an individual line segment.

- *Method 2*: Click the first point of the line. Release the left mouse button. Move to the end of the line and click again. The endpoint of the first line segment becomes the start point of the second line segment. Move to the end of the second line and click again.

Utilize these methods to create a chain of line segments. To end a chain of line segments, double-click on the endpoint or right-click and select End chain.

Utilize Method 1 to create the individual centerline.

Utilize Method 2 to create the chained profile lines.

Activity: GUIDE Part-Extruded Base Feature and Dynamic Mirror

Create a New part.

443) Click **New** ⬜ from the Menu bar.

444) Click the **MY-TEMPLATES** tab.

445) Double-click **PART-MM-ANSI** in the New SolidWorks Document Template dialog box.

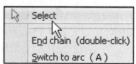

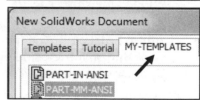

Save and name the part.

446) Click **Save** 🖫.

447) Select **ENGDESIGN-W-SOLIDWORKS\PROJECTS**.

448) Enter **GUIDE** for the File name.

449) Enter **GUIDE SUPPORT** for the Description.

450) Click **Save**. The GUIDE FeatureManager is displayed.

Create the first sketch. Select the Front Plane.

451) Right-click **Front Plane** from the GUIDE FeatureManager. This is your Sketch Plane.

452) Click **Sketch** ✏ from the Context toolbar. The Sketch toolbar is displayed.

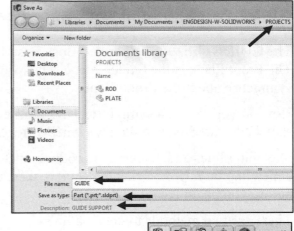

Sketch a centerline.

453) Click the **Centerline** ┊ Sketch tool from the Consolidate toolbar. The Insert Line PropertyManager is displayed.

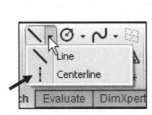

🔆 A centerline is displayed with a series of long and short dashes.

454) Click the **Origin** ↥. The Origin is Coincident with the first point of the sketch. Drag and click a **point** above the Origin as illustrated.

Activate the Mirror Sketch tool.

455) Click **Tools, Sketch Tools, Dynamic Mirror** from the Menu bar.

456) Click the **vertical centerline**. The Line Properties PropertyManager is displayed.

Sketch the profile.

457) Click the **Line** ╲ Sketch tool. The Insert Line PropertyManager is displayed.

458) Sketch the first horizontal line from the **Origin** to the right as illustrated. The mirror of the sketch is displayed on the left side of the centerline. Note: Click the endpoint of the line.

459) Sketch the **four additional line segments** to complete the sketch as illustrated. The last point is coincident with the centerline. The profile is continuous; there should be no gaps or overlaps.

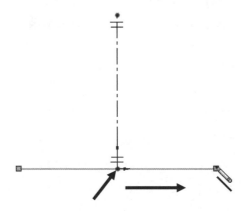

460) Double-click the **end of the line** to end the Sketch.

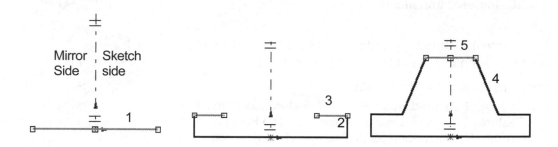

Add angular, vertical and horizontal dimensions.
461) Click the **Smart Dimension** ✧ Sketch tool.

Create an angular dimension.
462) Click the **right horizontal line**.

463) Click the **right-angled line**.

464) Click a **position** between the two lines.

465) Enter **110**.

466) Click the **Green Check mark** ✓.

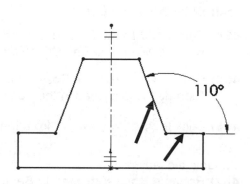

Create two vertical dimensions.
467) Click the **left small vertical line**.

468) Click a **position** to the left of the profile.

469) Enter **10**mm. Click the **Green Check mark** ✓.

470) Click the **bottom horizontal line**.

471) Click the **top horizontal line**.

472) Click a **position** to the left of the profile.

473) Enter **50**mm.

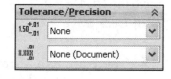

474) Click the **Green Check mark** ✓.

Create two linear horizontal dimensions.
475) Click the **top horizontal line**.

476) Click a **position** above the profile.

477) Enter **10**mm.

478) Click the **Green Check mark** ✓.

479) Click the **bottom horizontal line**.

480) Click a **position** below the profile.

481) Enter **80**mm. Click the **Green Check mark** ✓.

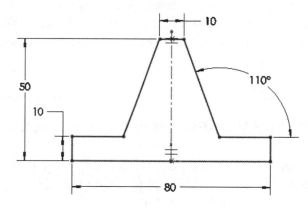

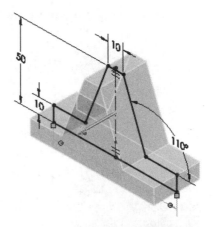

Select edges instead of points to create linear dimensions. Points are removed when Fillet and Chamfer features are added.

Dimension angles, smaller line segments, and then larger line segments to maintain the profile shape.

Insert an Extruded Boss/Base feature.

482) Click the **Extruded Boss/Base** feature tool from the Features tab in the CommandManager. The Boss-Extrude PropertyManager is displayed. The direction arrow points to the front.

483) Enter **30**mm for Depth.

484) Select **Mid Plane** for End Condition in Direction1. Think Symmetry with the Mid Plane End Condition option.

485) Click **OK** from the Boss-Extrude PropertyManager. Boss-Extrude1 is displayed in the FeatureManager.

486) Click **Isometric view** from the Heads-up View toolbar.

Rename the feature.
487) Rename **Boss-Extrude1** to **Base Extrude**.

Save the GUIDE.
488) Click **Save**.

There are two Mirror Sketch tools in SolidWorks: *Mirror* and *Dynamic Mirror*. Mirror requires a predefined profile and a centerline. Select **Mirror**, select the **profile** and the **centerline**. Dynamic Mirror requires a centerline. As you sketch, entities are created about the selected centerline.

Think design intent. When do you use the various End Conditions and Geometric sketch relations? What are you trying to do with the design? How does the component fit into an Assembly?

GUIDE Part-Extruded Cut Slot Profile

Create an Extruded Cut feature. Utilize the Straight Slot Sketch tool to create the 2D sketch profile for the slot.

Activity: GUIDE Part-Extruded Cut Slot Profile

Create the Slot sketch on the top right face of the model.

489) Click **Top view** ⬚ from the Heads-up View toolbar

490) Right-click the **top right face** of the GUIDE in the Graphics window as illustrated.

491) Click **Sketch** ✏ from the Context toolbar. The Sketch toolbar is displayed.

Apply the Straight Slot Sketch tool. The Straight Slot Sketch tool requires three points,

492) Click the **Straight Slot Sketch** ⬚ tool from the Consolidated Slot toolbar as illustrated. The Slot PropertyManager is displayed.

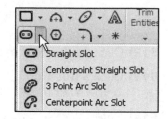

493) Click the **first point** as illustrated.

494) Click the **second point** directly above the first point.

495) Click the **third point** directly to the right of the second point. The Slot Sketch is created.

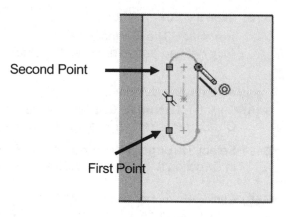

Second Point

First Point

🔅 Create a slot without using the Slot Sketch tool by sketching two lines, two 180-degree tangent arcs and apply the needed geometric relations.

Add dimensions.

496) Click the **Smart Dimension** ✏ Sketch tool.

497) Click the **Origin** ⤬ as illustrated.

498) Click the **center point** of the top arc. Note: When the mouse pointer is positioned over the center point, the center point is displayed in red.

499) Click a **position** above the profile.

500) Enter **30**mm.

501) Click the **Green Check mark** ✓ .

Create a vertical dimension.

502) Click the **top horizontal edge** of the GUIDE.

503) Click the **center point** of the top arc.

504) Click a **position** to the right of the profile.

505) Enter **10**mm.

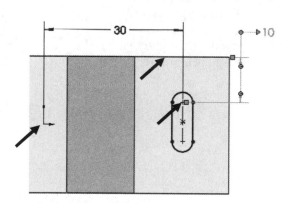

506) Click the **Green Check mark** ✓ .

Create a dimension between the two arc center points.

507) Click the top arc **center point**. When the mouse pointer is positioned over the center point, the center point is displayed in red

508) Click the second arc **center point**.

509) Click a **position** to the right of the profile.

510) Enter **10**mm.

511) Click the **Green Check mark** ✓ .

Create a radial dimension.

512) Click the **bottom arc** radius.

513) Click a **position** diagonally below the profile.

514) Enter **3**mm.

515) Click the **Green Check mark** ✓ . The Sketch is fully defined and is displayed in black.

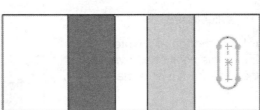

Insert an Extruded Cut feature.

516) Click the **Extruded Cut** 📄 feature tool. The Cut-Extrude PropertyManager is displayed.

517) Select **Through All** (Design Intent) for End Condition in Direction1. Accept all default settings.

518) Click **OK** ✓ from the Cut-Extrude PropertyManager. Cut-Extrude1 is displayed in the FeatureManager.

519) Display an **Isometric view**  from the Heads-up View toolbar.

Rename the feature.
520) Rename **Cut-Extrude1** to **Slot Cut**.

Save the GUIDE.
521) Click **Save** 💾.

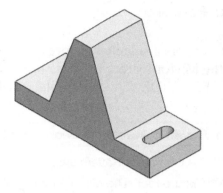

🔆 To remove Tangent edges, click **Display/Selections** from the Options menu, check the **Removed** box.

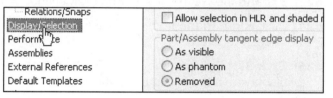

SolidWorks contains additional methods to create sketch slots. Here are three examples that you can explore as an exercise:

- Sketch a rectangle. Delete two opposite line segments. Insert two arcs. Extrude the profile.

- Sketch a single vertical centerline. Utilize Offset Entities, Bidirectional option and Cap Ends option. Extrude the profile.

- Utilize a Design Library Feature.

The machined, metric, straight slot feature provides both the Slot sketch and the Extruded Cut feature. You define the references for dimensions, overall length and width of the Slot. Explore the SolidWorks Design Library later in the book.

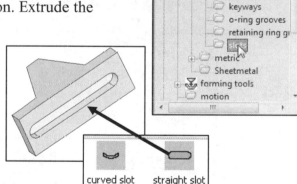

curved slot   straight slot

GUIDE Part-Mirror Feature

The Mirror feature  mirrors a selected feature about a mirror plane. Utilize the Mirror feature to create the Slot Cut on the left side of the GUIDE. The mirror plane is the Right Plane.

Activity: GUIDE Part-Mirror Feature with Geometric Pattern Option

Insert a Mirror feature with the Geometric Pattern options.

522) Click the **Mirror** feature tool. The Mirror PropertyManager is displayed.

523) Expand the GUIDE fly-out FeatureManager in the Graphics window.

524) Click **Right Plane** from the fly-out FeatureManager. Right Plane is displayed in the Mirror Faces/Plane box.

525) Click **Slot Cut** from the fly-out FeatureManager for the Features to Mirror. Slot Cut is displayed in the Features to Mirror box.

526) Check the **Geometry Pattern** box.

527) Click **OK** from the Mirror PropertyManager. Mirror1 is displayed in the FeatureManager.

Save the GUIDE part.

528) Click **Save**.

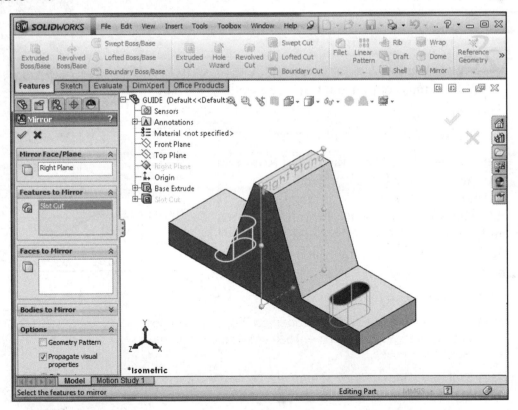

Fit the GUIDE part to the Graphics window.
529) Press the **f** key.

💡 Save rebuild time; check the Geometry pattern option in the Mirror Feature PropertyManager. Each instance is an exact copy of the faces and edges of the original feature.

🔍 Additional details on Geometry Pattern, Mirror Feature and Sketch Mirror are available in SolidWorks Help. Index keywords: Geometry, Mirror Features and Mirror Entities.

GUIDE Part-Holes

The Extruded Cut 🔲 feature removes material from the front face of the GUIDE to create a Guide Hole. The ROD moves through the Guide Hole.

Create the tapped holes with the Hole Wizard feature and the Linear Pattern feature. Apply dimensions to the tapped holes relative to the Guide Hole.

Activity: GUIDE Part-Holes

Create a sketch on the front face.
530) Right-click the **front face** of the GUIDE in the Graphics window. This is your Sketch Plane.

531) Click **Sketch** ✏ from the Context toolbar.

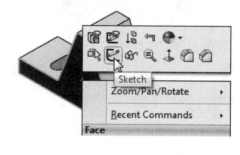

532) Click **Normal To view** ⬍ from the Heads-up View toolbar. The front face of the GUIDE is displayed.

533) Click the **Circle** ⊘ Sketch tool. The Circle PropertyManager is displayed.

534) Sketch a **circle** in the middle of the GUIDE as illustrated.

Add a Vertical relation between the Origin and the center point of the circle.
535) Right-click **Select** to deselect the Circle sketch tool.

536) Click the **Origin** ⤱. Hold the **Ctrl** key down.

537) Click the **center point** of the circle.

538) Release the **Ctrl** key. The Properties PropertyManager is displayed. The selected entities are displayed in the Selected Entities box.

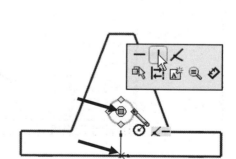

539) Right-click **Make Vertical** from the Context toolbar in the Graphics window.

540) Click **OK** ✔ from the Properties PropertyManager.

Add dimensions.

541) Click the **Smart Dimension** ✎ Sketch tool.

542) Click the circle **circumference**.

543) Click a **position** below the horizontal line.

544) Enter **10**mm.

545) Click the **Green Check mark** ✅.

546) Click the **center point** of the circle.

547) Click the **Origin**.

548) Click a **position** to the right of the profile.

549) Enter **29**mm.

550) Click the **Green Check mark** ✅.

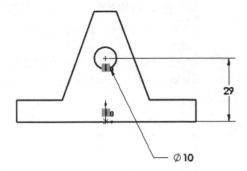

Create an Extruded Cut feature.

551) Click the **Extruded Cut** 🔲 feature tool. The Cut-Extrude PropertyManager is displayed.

552) Select **Through All** for End Condition in Direction1. Accept the default conditions.

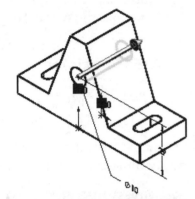

553) Click **OK** ✅ from the Cut-Extrude PropertyManager. Cut-Extrude2 is displayed in the FeatureManager.

554) Click **Isometric view** 🔲 from the Heads-up View toolbar.

555) Click **Hidden Lines Removed** 🔲 from the Heads-up View toolbar.

Rename the feature.
556) Rename **Cut-Extrude2** to **Guide Hole**.

Display the Temporary Axes.
557) Click **View**; check **Temporary Axes** from the Menu bar. The Temporary Axes are displayed in the Graphics window.

Create the Tapped Hole.

558) Click the **Hole Wizard** 🔲 feature tool. The Hole Specification PropertyManager is displayed.

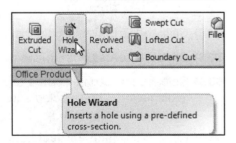

559) Click the **Type** tab.

560) Click **Straight Tap** for Hole Specification.

561) Select **Ansi Metric** for Standard.

562) Select **Bottoming Tapped hole** for Type.

563) Select **M3x0.5** for Size.

564) Select **Blind** for End Condition. Accept the default conditions.

565) Click the **Positions** tab.

566) Click the **right angled face** below the Guide Hole Temporary Axis as illustrated. The Point icon is displayed 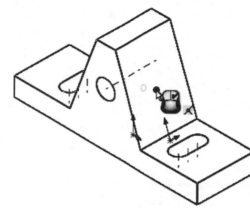.

567) Click **again** to place the center of the hole.

568) Click **Normal To view** ⊥ from the Heads-up View toolbar to view the Sketch plane.

Dimension the tapped hole relative to the Guide Hole axis. Add the horizontal dimension.

569) Click the **Smart Dimension** ✧ Sketch tool. The Point tool is de-selected. The Smart Dimension icon is displayed.

570) Click the **center point** of the tapped hole. Note: When the mouse pointer is positioned over the center point, the center point is displayed in red.

571) Click the **right vertical edge**.

572) Click a **position** above the profile.

573) Enter **25**mm.

574) Click the **Green Check mark** ✅.

Add a vertical dimension.
575) Click the **center point** of the tapped hole.

576) Click the horizontal **Temporary Axis**.

577) Click a **position** to the right of the profile.

578) Enter **4**mm.

579) Click the **Green Check mark** ✅.

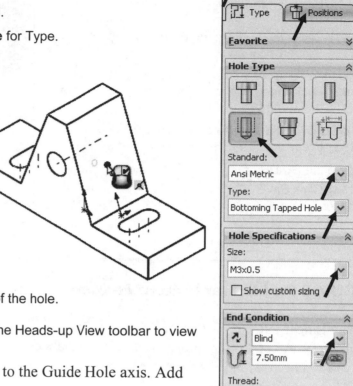

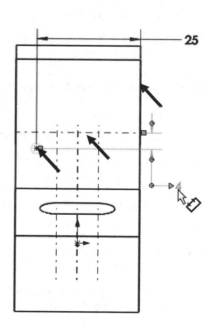

Display the tapped hole.

580) Click **OK** ✓ from the Dimension PropertyManager

581) Click **OK** ✓ from the Hole Position PropertyManager.

Hide the Temporary Axes.
582) Click **View**; uncheck **Temporary Axes** from the Menu bar.

Fit GUIDE to the Graphics window.
583) Press the **f** key.

Save the GUIDE.
584) Click **Save** 🖫.

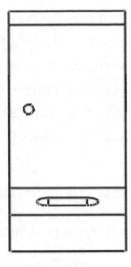

GUIDE Part-Linear Pattern Feature

The Linear Pattern ⁛ feature creates multiple copies of a feature in an array. A copy of a feature is called an instance. A Linear Pattern feature requires:

- One or two directions

- A Feature, Face or Body to pattern

- Distance between instances

- Number of instances

Insert a Linear Pattern feature of tapped holes on the right-angled face of the GUIDE.

Activity: GUIDE Part-Linear Pattern Feature with Geometry Pattern option

Display the Right view.
585) Click **Right view** ⬚ from the Heads-up View toolbar.

Insert a Linear Pattern feature with the Geometry Pattern option.
586) Click the **Linear Pattern** ⁛ feature tool. The Linear Pattern PropertyManager is displayed.

587) **Expand** GUIDE from the fly-out FeatureManager.

588) Click inside the **Features to Pattern** box.

589) Click **M3X0.5 Tapped Hole1** from the fly-out FeatureManager. Click inside the **Pattern Direction** box for Direction 1.

590) Click the **top horizontal edge** for Direction 1. Edge<1> is displayed in the Pattern Direction box. The direction arrow points to the right.

591) Enter **10**mm in the Direction 1 Spacing box. The direction arrow points to the right. If required, click the Reverse Direction button.

592) Enter **3** in the Number of Instances box for Direction 1.

593) Click inside the **Pattern Direction** box for Direction 2.

594) Click the **left vertical edge** for Direction 2. The direction arrow points upward. If required, click the Reverse Direction button.

595) Enter **12**mm in the Direction 2 Spacing box.

596) Enter **2** in the Number of Instances box.

597) Check **Geometry Pattern** in the Options box.

Display the Linear Pattern feature.

598) Click **OK** from the Linear Pattern PropertyManager. LPattern1 is displayed in the FeatureManager.

599) Click **Isometric view** from the Heads-up View toolbar.

Save the GUIDE.

600) Click **Save** .

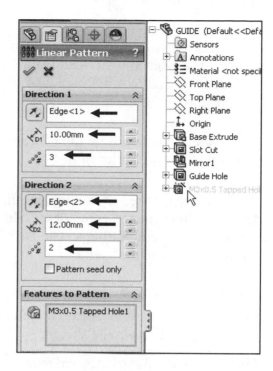

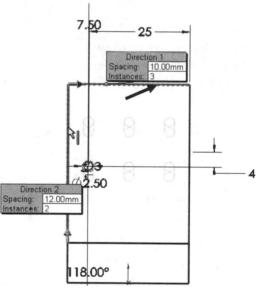

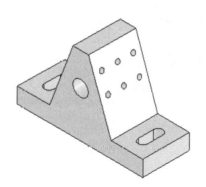

Reuse geometry. The Linear Pattern feature with the Geometry Pattern option saves rebuild time. Rename the seed feature in the part pattern. The seed feature is the first feature for the pattern, Example: M3x0.5 Tapped Hole1. Assemble additional parts to the seed feature in the assembly.

Additional details on the Linear Pattern feature are available in SolidWorks Help.

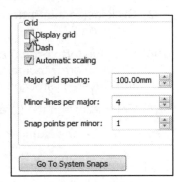

Save time when sketching by utilizing the Grid. Click **Options**, **Document Properties**, **Grid/Snap** to control display and spacing settings. The Display grid check box turns on/off the Grid in the sketch.

Hide geometric relations to clearly display profile lines in the sketch. Check **View**, **Sketch Relations** to hide/display geometric relations in the sketch.

GUIDE Part-Materials Editor and Mass Properties

Apply Materials such as steel, aluminum and iron to a SolidWorks part through the Material entry in the FeatureManager. Visual Properties and Physical Properties are dispalyed in the Materials Editor. Visual Properties are displayed in the Graphics window and section views of the part. The Physical Property value for Density, 8000 kg/m^3 propogates to the Mass Properties tool.

Activity: GUIDE Part-Materials Editor and Mass Properties

Apply AISI 304 material to the model.

601) Right-click the **Materials** folder in the FeatureManager.

602) Click **Edit Material**. The Materials dialog box is displayed.

Expand the Steel folder.

603) **Expand** the Steel folder from the SolidWorks Materials box.

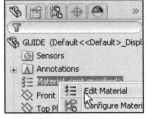

604) Select **AISI 304**. View the available materials and information provided in the Materials dialog box. The Materials dialog box is updated from previous years.

605) Click **Apply**.

606) Click **Close** from the Materials dialog box. AISI 304 is displayed in the GUIDE FeatureManager.

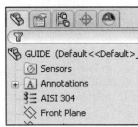

Calculate the Mass Properties of the GUIDE.

607) Click the **Evaluate** tab from the CommandManager.

608) Click the **Mass Properties** ⚖ tool. The Mass Properties dialog box is displayed.

Set the units to display.

609) Click the **Options** button.

610) Click the **Use custom settings** button.

611) Enter **4** in the Decimal box.

612) Click **OK** from the Mass/Section Property Options dialog box. Review the Mass properties of the GUIDE.

613) Click **Close** from the Mass Properties dialog box.

Save the GUIDE.

614) Click **Save** 💾.

Close the GUIDE.

615) Click **File**, **Close** from the Menu bar.

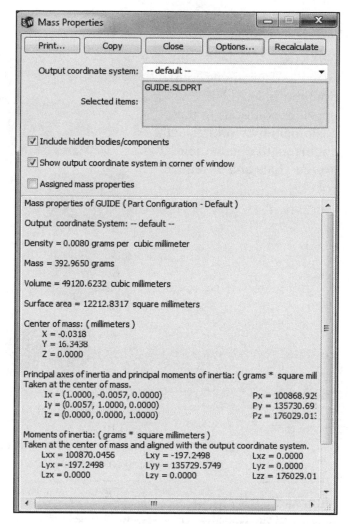

A material that is selected in the Material dialog box propagates to other SolidWorks applications: Mass Properties, SolidWorks SimulationXpress, PhotoWorks and PhotoView to name a few. Specify a material early in the design process. Define a material for the SW library or define your own material.

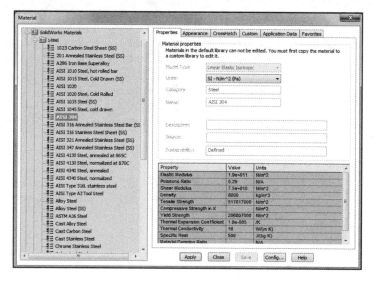

Manufacturing Considerations

CSI Mfg.
(www.compsources.com)
Southboro, MA, USA
provides solutions to their
customers by working with
engineers to obtain their
design goals and to reduce
cost.

Customer Focus
Courtesy of CSI Mfg (www.compsources.com)
Southboro, MA

High precision turned
parts utilize a variety of
equipment such as single
and multi-spindle lathes
for maximum output.

MultiDECO 20/8b, 8 spindle, full CNC 23 axis automatic
Courtesy of Tornos USA (www.tornosusa.com)

Obtain knowledge about the manufacturing process from your colleagues before you design the part.

Ask questions on material selection, part tolerances, machinability and cost. A good engineer learns from past mistakes of others.

Review the following design guidelines for Turned Parts.

Guideline 1: Select the correct material.

Examples of Cost Effective Materials
- AISI 303
- BRASS
- AL 2011

Consider part functionality, operating environment, cost, "material and machine time", machinability and surface finish. Utilize standard stock sizes and material that is readily available. Materials are given a machinability rating. Brass (100), AISI 303(65) and AISI 304(30).

Example: Using AISI 304 vs. AISI 303. AISI 303 is easier to machine due to its higher sulfur content.

DFMXpress is an analysis tool that validates the manufacturability of SolidWorks parts. Use DFMXpress to identify design areas that may cause problems in fabrication or increase the costs of production of a new tool.

Guideline 2: Minimize tool changes. Maximize feature manufacturability.

The Extruded Cut, Hole, Chamfer and Fillet features require different tooling operations. Will changes in form or design reduce the number of operations and the cost of machining?

Example:

- Avoid Fillets on inside holes to remove sharp edges. Utilize the Chamfer feature to remove sharp edges.

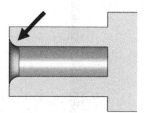 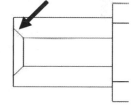

Fillet Edge Expensive   Chamfer Edge Preferred

- Utilize common size holes for fewer tool changes.

- Leave space between the end of an external thread and the shoulder of the rod. Rule of thumb: 1-2 pitch of thread, minimum.

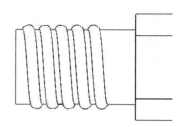

- Leave space at the end of an internal thread for the tool tip endpoint.

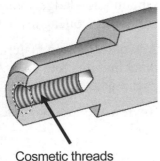

Cosmetic threads

To display a cosmetic thread, right-click the Annotations folder in the FeatureManager, click Details. Check the Cosmetic threads box and the Shaded cosmetic thread box. Click OK. Select Shaded With Edges for display modes. View the results.

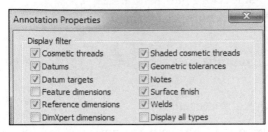

Guideline 3: Optimize the design for production. Provide dimension, tolerance and precision values that can be manufactured and inspected.

- Check precision on linear, diameter and angular dimensions. Additional non-needed decimal places increase cost to a part.

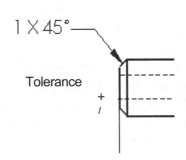

Example:

The dimension 45 at +/- .5° indicates the dimension is machined between 44.5° and 45.5.

The dimension 45.00 at +/- .05° indicates the dimension is machined between 44.95 and 45.05.

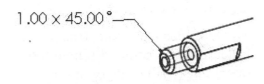

This range of values adds additional machining time and cost to the final product.

Modify the dimension decimal display in the Tolerance/Precision box in the part, assembly or drawing.

Set Document Properties, Dimensions, Tolerance and Precision for the entire document.

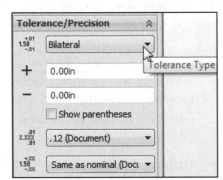

In Project 3, create drawings for the GUIDE part. Modify the dimension values and tolerance on the ROD and GUIDE to create a clearance fit.

Sketch Entities and Sketch Tools

The Sketch Entities and Sketch Tools entries in the tools menu contain additional options not displayed in the default Sketch toolbar. Each option contains information in Help, SolidWorks Help Topics.

If a sketch produces errors during the feature creation process, utilize the **Tools**, **Sketch Tools**, **Check Sketch for Feature** option.

Example: Select Base Extrude for Feature usage. Click the Check button. The error message, "The sketch has more than one open contour" is displayed.

The most common errors are open contours (gaps) or overlapping contours. Edit the sketch and correct the error. Click the Check button. The message, "No problems found" is displayed for a valid sketch.

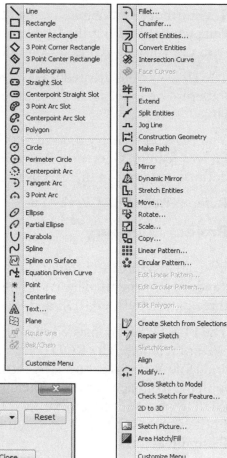

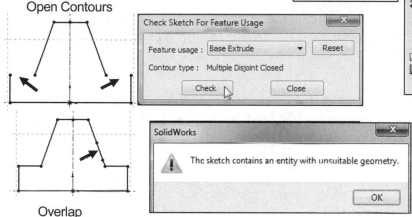

Review of the GUIDE Part

The GUIDE part utilized an Extruded Boss/Base feature. The 2D sketched profile utilized the Dynamic Mirror Sketch tool to build symmetry into the part.

The linear and diameter dimensions corresponded to the ROD and GUIDE-CYLINDER assembly.

The Extruded Cut feature created the Slot-Cut. The Mirror feature created a second Slot-Cut symmetric about the Right Plane. The Guide Hole utilized the Extruded Cut feature with the Through All End Condition.

The Hole Wizard created the M3 x0.5 Tapped Hole, the seed feature in the Linear Pattern. The Linear Pattern feature created multiple instances of the M3 x0.5 Tapped Hole.

You reused geometry to save design time and rebuild time. The Mirror feature, Linear Pattern feature, Geometry Pattern option and Sketch Mirror are time saving tools. You defined materials in the part used by other software applications.

Refer to Help, SolidWorks Tutorials, and Lesson 1-Parts for additional information.

Project Summary

You created three file folders to manage your project models, document templates and vendor components. The PART-ANSI-MM Part Template was developed and utilized for three parts: ROD, GUIDE and PLATE.

The PLATE part consisted of an Extruded Bose/Base, Extruded Cut, Fillet and Hole Wizard features. The ROD consisted of an Extruded Boss/Base, Extruded Cut, Chamfer, and Hole Wizard feature. The ROD illustrated the process to manipulate design changes with the Rollback bar, Edit Sketch and Edit Feature / Edit Sketch commands. The GUIDE part utilized an Extruded Boss/Base feature sketch with the Dynamic Mirror Entities tool. The GUIDE also utilized the Extruded Cut, Mirror, Hole Wizard and Linear Pattern features.

All sketches for the Extruded Boss/Base and Extruded Cut features utilized geometric relationships. Incorporate geometric relations such as Horizontal, Vertical, Symmetric, Equal, Collinear and Coradial into your sketches.

In Project 2, insert the ROD, GUIDE and PLATE parts into a new assembly. In Project 3, create an assembly drawing and a detailed drawing of the GUIDE part.

Are you ready to start Project 2? Stop! Examine and create additional parts. Perform design changes. Take chances, make mistakes and have fun with the various features and commands in the project exercises.

Think design intent. When do you use the various End Conditions and geometric sketch relations? What are you trying to do with the design? How does the component fit into an assembly?

Project Terminology

Add Relations: A Sketch tool that provides a way to connect related geometry. Some common relations in this book are Concentric, Tangent, Coincident and Collinear.

Applied Feature: A classification of a feature type that utilizes a model's existing edges or faces. Fillets and chamfers are examples of this type of feature.

Base Feature: The Base feature is the first feature that is created in a part. The Base feature is the foundation of the part. Keep the Base feature simple. The Base feature geometry for the PLATE part is an extrusion. An Extruded Boss/Base feature requires a sketch plane, sketch and extruded depth.

Centerline: A centerline is a construction line utilized for symmetry and revolved features.

Chamfer feature: An applied feature used to create a beveled edge. The front face of the ROD utilized the Chamfer.

CommandManager: The CommandManager is a context-sensitive toolbar that dynamically updates based on the toolbar you want to access. By default, it has toolbars embedded in it based on the document type.

Copy and Paste: Copy and paste simple sketched features and some applied features onto a planar face. Multi-sketch features such as sweeps and lofts cannot be copied. Utilize the Ctrl C / Ctrl V keys to Copy/Paste.

Dimensions: Dimensions measure geometry. A dimension consists of dimension text, arrows, dimensions lines, extension lines and or leader lines.

Document: In the SolidWorks application, each part, assembly, and drawing is referred to as a document, and each document is displayed in a separate window.

Extruded Boss/Base: Extends a profile along a path normal to the profile plane for some depth. The movement along that path becomes the solid model. An Extruded Boss/Base adds material. The Boss-Extrude1 feature is the first feature in a part. The PLATE utilized an Extruded Boss/Base feature. The profile was a rectangle sketched on the Front Plane.

Extruded Cut: Removes material from a solid. The Extruded Cut extends a profile along a path for some depth. The PLATE utilized an Extruded Cut feature to create two holes on the front face. The Through All End condition in Direction 1 links the depth of the Extruded Cut to the depth of the Extruded Base.

Features: Features are geometry building blocks. Features add or remove material. Create features from sketched profiles or from edges and faces of existing geometry. Features are classified as either sketched or applied.

Fillet: Removes sharp edges. Fillets are generally added to the solids, not to the sketch. By the nature of the faces adjacent to the selected edge, the system knows whether to create a round (removing materials) or a fillet (adding material). Some general filleting design rules:

- Leave cosmetic fillets until the end.

- Create multiple fillets that will have the same radius in the same command.
- Fillet order is important.
- Fillets create faces and edges that are used to generate more fillets.
- In the PLATE part you created the small edge fillets first.

Geometric Relations: A relation is a geometric constraint between sketch entities or between a sketch entity and a plane, axis, edge, or vertex. Relations force a behavior on a sketch element to capture the design intent.

Hole Wizard: The Hole Wizard feature is used to create specialized holes in a solid. Create simple, tapped, Counterbore and Countersink holes using a systematic procedure. Use the Hole Wizard feature to create a Countersink hole at the center of the PLATE. Use the Hole Wizard to create tapped holes in the GUIDE.

Linear Pattern: Repeats features or geometry in an array. The Linear Pattern feature requires the number of instances and the spacing between instances.

Mass Properties: A tool used to report the part density, mass, volume, moment of inertia and other physical properties.

Materials Editor: The Materials Editor provides visual properties and physical properties for a part. AISI 304 Stainless Steel was applied to the GUIDE. The cross hatching display propagate to the drawing. The physical properties propagate to SolidWorks Simulation and other applications.

Menus: Menus provide access to the SolidWorks commands. There are Pop-up, fly-out, and context menus.

Mirror Entities: Involves creating a centerline and using the Dynamic Mirror option. The centerline acts as the mirror axis. Sketch geometry on one side of the mirror axis. The copied entity becomes a mirror image of the original across the centerline. Mirror Entities requires a predefined centerline and a sketched profile.

Mirror Feature: Creates an instance (copy) of features about a plane. The GUIDE Slot Cut utilized the Mirror feature.

Mouse Buttons: The left and right mouse buttons have distinct meanings in SolidWorks. The left button is used to select geometry and commands. The right mouse button displays pop-up menus. The middle mouse button rotates the model in the Graphics window.

Origin: The model origin appears in blue and represents the (0,0,0) coordinate of the model. When a sketch is active, a sketch origin appears in red and represents the (0,0,0) coordinate of the sketch. Dimensions and relations can be added to the model origin, but not to a sketch origin.

Plane: A plane is an infinite and flat surface. They display on the screen with visible edges. The PLATE, ROD and GUIDE all utilize the Front Plane for the Extruded Base feature.

Rebuild: Regenerates the model from the first feature to the last feature. After a change to the dimensions, rebuild the model to cause those changes to take effect.

Sketch: The name to describe a 2D profile is called a sketch. 2D Sketches are created on flat faces and planes within the model. Typical geometry types are lines, arcs, rectangles, circles and ellipses.

Sketched Features: The Sketched feature is based on a 2-D sketch. The sketch is transformed into a solid by extrusion, rotation, sweeping or lofting.

Sketch Plane: When you open a new part document, first you create a sketch. The sketch is the basis for a 3D model. You can create a sketch on any of the default planes (Front Plane, Top Plane, and Right Plane), or a created plane.

Status of a Sketch: Four states are utilized in this Project:

- Under Defined: There is inadequate definition of the sketch, (Blue).

- Fully Defined: Has complete information, (Black).

- Over Defined: Has duplicate dimensions, (Red).

- Dangling: Geometry or Sketch Plane has been deleted, (Mustard Yellow-Brown).

System Feedback: System feedback is provided by a symbol attached to the cursor arrow indicating your selection. As the cursor floats across the model, feedback is provided in the form of symbols, riding next to the cursor.

Template: Templates are part, drawing, and assembly documents that include user-defined parameters and are the basis for new documents.

Toolbars: The toolbar menus provide shortcuts enabling you to quickly access the most frequently used commands.

Unit System: SolidWorks provides the following unit systems: MKS (meter, kilogram, second), CGS (centimeter, gram, second), MMGS (millimeter, gram, second), IPS (inch, pound, second), and Custom. Custom lets you set the Length units, Density units, and Force.

A goal of this book is to expose various SolidWorks design tools and features. For manufacturing, use the Hole Wizard feature to create a simple or complex hole.

During the initial SolidWorks installation, you are requested to select either the ISO or ANSI drafting standard. ISO is typically an European drafting standard and uses First Angle Projection. The book is written using the ANSI (US) overall drafting standard and Third Angle Projection for drawings.

Questions

1. Identify at least four key design areas that you should investigate before starting a project.

2. Why is file management important?

3. Identify the steps in starting a SolidWorks session.

4. Identify the three default Reference planes in SolidWorks.

5. Describe a Base sketch. Provide an example.

6. True or False: The plane or face you select for the Base sketch determines the orientation of the part.

7. True or False: In a fully defined sketch, all sketch entities are displayed in black.

8. True or False: In an over defined sketch, all sketch entities are displayed in blue.

9. True or False: It is considered very poor design practice not to fully define a sketch.

10. Describe a Base feature. Provide an example.

11. Identify the Drafting Standard which is used in the United States.

12. SolidWorks provides system feedback by attaching a symbol to the mouse pointer cursor. The system feedback symbol indicates what you are selecting or what the system is expecting you to select. What is the feedback symbol for a face, edge and vertex?

13. Describe the difference between First Angle Projection and Third Angle Projection.

14. How do you recover from a Rebuild error?

15. When should you apply the Hole Wizard feature vs. the Extruded Cut feature?

16. Describe the difference between an Extruded Boss/Base feature and an Extruded Cut feature.

17. Describe a Fillet feature. Provide an example.

18. Name the command keys used to Copy sketched geometry.

19. Describe the difference between the Edit Feature and Edit Sketch tool.

20. Describe a Chamfer feature. Provide an example.

21. Describe the Rollback bar function. Provide an example.

22. Identify three types of Geometric Relations that you can apply to a sketch.

23. Describe the procedure in creating a Part Document template.

24. Describe the procedure to modify material in an existing part.

25. Describe the procedure to set the Overall drafting standard, units and precision for a Part document.

26. Describe the procedure to sketch an Elliptical profile.

27. Identify the name of the following feature tool icons.

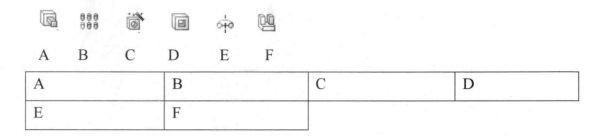

A   B   C   D   E   F

A	B	C	D
E	F		

28. Identify the name of the following Sketch tool icons.

A   B   C   D   E   F   G   H   I   J

A	B	C	D
E	F	G	H
I	J		

29. Identify the surfaces with the appropriate letter that will appear in the FRONT view, TOP view and RIGHT view.

FRONT view surfaces: _____

TOP view surfaces: _____

RIGHT view surfaces:_____

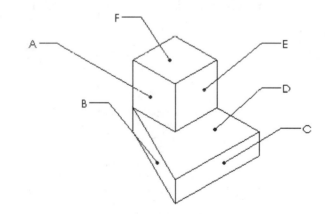

25. Identify the Angle of Projection for the provided illustration.

Angle of Projection_____.

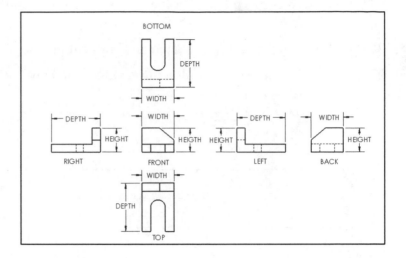

Exercises

Exercise 1.1 and Exercise 1.2

A fast and economical way to manufacture parts is to utilize Steel Stock, Aluminum Stock, Cast Sections, Plastic Rod and Bars and other stock materials.

Stock materials are available in different profiles and various dimensions from tool supply companies.

- Create a new L-BRACKET part and a T-SECTION part. Each part contains an Extruded Base feature. Apply units from the below tables.

- Utilize the Front Plane for the Sketch plane.

L-BRACKET   T-SECTION

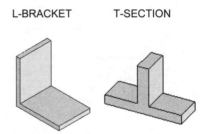

Exercise 1.1: L-BRACKET

Ex.:	A:	B:	C:	LENGTH:	Units:	
1.1a	3	3	5/8	10	IN	
1.1b	4	4	3/4	10		
1.1c	6	4	1/2	25		
1.1d	8	6	3/4	25		
1.1e	12	12	3/2	25		
1.1f	7	4	3/4	30		
1.1g	75	75	10	250	MM	
1.1h	75	100	10	250		
1.1i	200	100	25	500		

Exercise 1.2: T-SECTION

Ex.:	A:	B:	C:	LENGTH:	Units:	
1.2a	3	3	5/8	30	IN	
1.2b	4	4	3/4	30		
1.2c	6	6	1	30		
1.2d	75	75	10	250	MM	
1.2e	75	100	10	250		
1.2f	200	100	25	500		

Exercise 1.2a: PART DOCUMENT TEMPLATES

Create a Metric (MMGS) Part document template.

- Use the ISO drafting standard and the MMGS unit system.
- Set the Drawing/Units and precision to the appropriate values.
- Name the Template: PART-MM-ISO.

Exercise 1.2b: PART DOCUMENT TEMPLATES

Create an English (IPS) Part document template.

- Use the ANSI drafting standard and the IPS unit system.
- Set Drawing/Units and precision to the appropriate values.
- Name the Template: PART-IN-ANSI.

Exercise 1.3: Identify the Sketch plane for the Boss-Extrude1 feature. View the Origin location. Simplify the number of features.

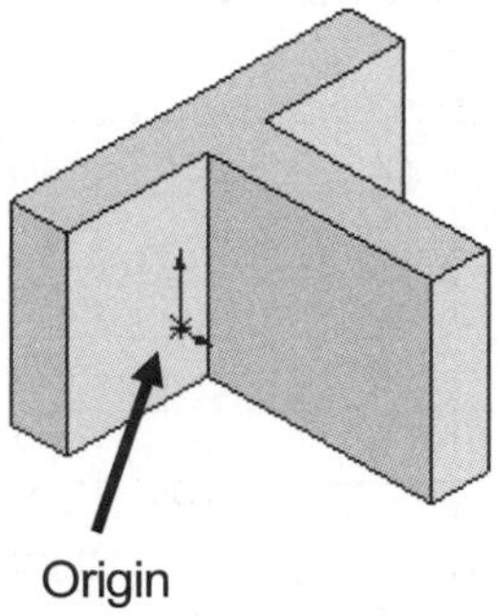

A: Top Plane

B: Front Plane

C: Right Plane

D: Left Plane

Correct answer _____.

Origin

Exercise 1.4: Identify the Sketch plane for the Boss-Extrude1 feature. View the Origin location. Simplify the number of features.

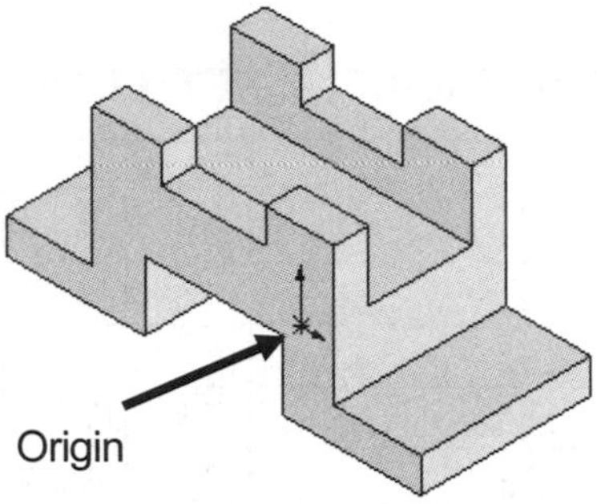

A: Top Plane

B: Front Plane

C: Right Plane

D: Left Plane

Correct answer _____.

Origin

Exercise 1.5: Identify the Sketch plane for the Boss-Extrude1 feature. View the Origin location. Simplify the number of features.

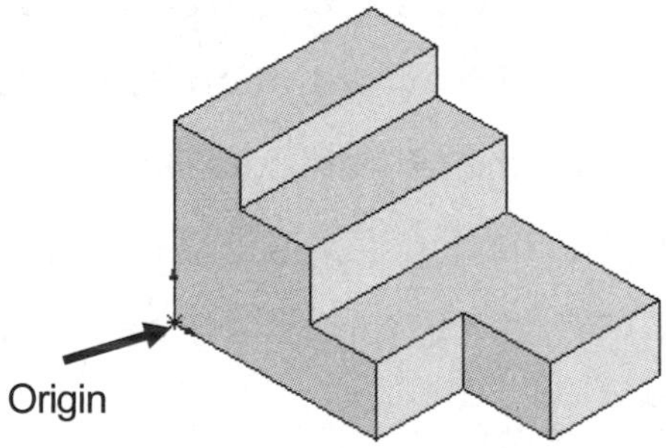

A: Top Plane

B: Front Plane

C: Right Plane

D: Left Plane

Correct answer _____.

Origin

Exercise 1.6: Identify the Sketch plane for the Boss-Extrude1 feature. View the Origin location. Simplify the number of features.

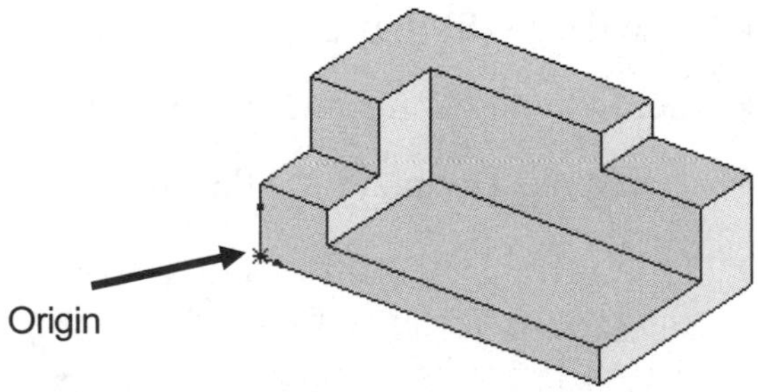

A: Top Plane

B: Front Plane

C: Right Plane

D: Left Plane

Correct answer _____.

Origin

Exercise 1.7: Identify the Material category for 6061 Alloy.

A: Steel

B: Iron

C: Aluminum Alloys

D: Other Alloys

E: None of the provided

Correct answer _____ .

Exercise 1.8: AXLE

Create an AXLE part as illustrated with dual units. Overall drafting standard – ANSI. IPS is the primary unit system.

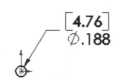
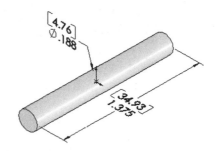

- Utilize the Front Plane for the Sketch plane.

- Utilize the Mid Plane End Condition. The AXLE is symmetric about the Front Plane. Note the location of the Origin.

Exercise 1.9: SHAFT COLLAR

Create a SHAFT COLLAR part as illustrated with dual system units; MMGS (millimeter, gram, second) and IPS (inch, pound, second). Overall drafting standard - ANSI.

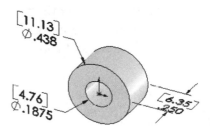

- Utilize the Front Plane for the Sketch plane. Note the location of the Origin.

Exercise 1.10: FLAT BAR - 3 HOLE

Create the FLAT BAR - 3 HOLE part as illustrated with dual system units; MMGS (millimeter, gram, second) and IPS (inch, pound, second). Overall drafting standard – ANSI.

- Utilize the Front Plane for the Sketch plane.

- Utilize the Center point Straight Slot Sketch tool.

- Utilize a Linear Pattern feature for the three holes. The FLAT BAR – 3 HOLE part is stamped from 0.060in, [1.5mm] Stainless Steel.

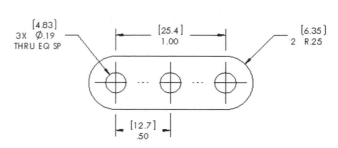

Exercise 1.11: FLAT BAR - 9 HOLE

Create the the FLAT BAR - 9 HOLE part. Overall drafting standard - ANSI.

- The dimensions for hole spacing, height and end arcs are the same as the FLAT BAR - 3 HOLE part.

- Utilize the Front Plane for the Sketch plane.

- Utilize the Centerpoint Straight Slot Sketch tool.

- Utilize the Linear Pattern feature to create the hole pattern. The FLAT BAR - 9 HOLE part is stamped from 0.060in, [1.5mm] 1060 Alloy.

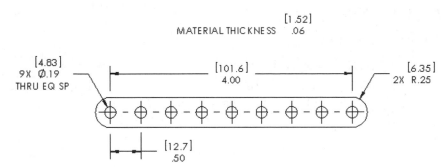

Exercise 1.12: SIMPLE BLOCK

Create the illustrated part. Note the location of the Origin. Overall drafting standard - ANSI.

- Calculate the overall mass of the illustrated model. Apply the Mass Properties tool.

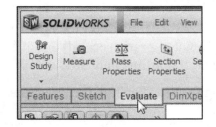

- Think about the steps that you would take to build the model.

- Review the provided information carefully.

- Units are represented in the IPS, (inch, pound, second) system.

- A = 3.50in, B = .70in

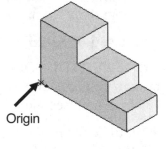

Origin

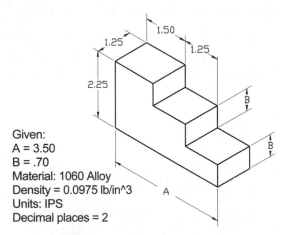

Given:
A = 3.50
B = .70
Material: 1060 Alloy
Density = 0.0975 lb/in^3
Units: IPS
Decimal places = 2

Exercise 1.13: SIMPLE T

Create the illustrated part. Note the location of the Origin. Overall drafting standard - ANSI.

- Calculate the overall mass of the illustrated model. Apply the Mass Properties tool.

- Think about the steps that you would take to build the model.

- Review the provided information carefully. Units are represented in the IPS, (inch, pound, second) system.

- A = 3.00in, B = .75in

Given:
A = 3.00
B = .75
Material: Copper
Density = 0.321 lb/in^3
Units: IPS
Decimal places = 2

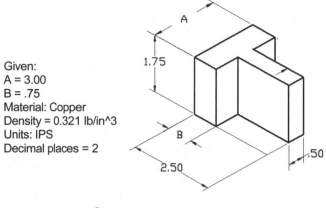

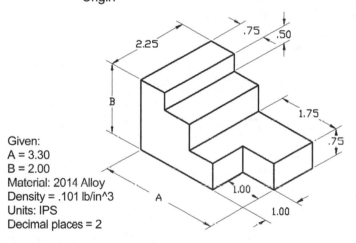

Origin

Exercise 1.14: SIMPLE BLOCK STEP

Create the illustrated part. Note the location of the Origin. Overall drafting standard - ANSI.

- Calculate the volume of the part and locate the Center of mass relative to the origin with the provided information.

- Apply the Mass Properties tool.

- Think about the steps that you would take to build the model.

- Review the provided information carefully.

Given:
A = 3.30
B = 2.00
Material: 2014 Alloy
Density = .101 lb/in^3
Units: IPS
Decimal places = 2

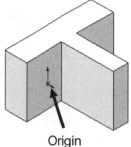

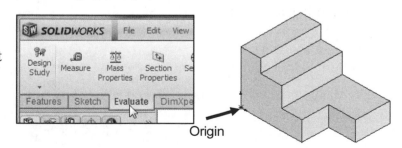

Origin

Exercise 1.15: DRAWING 1

Create the part from the illustrated ANSI - MMGS Third Angle Projection drawing: Front, Top, Right and Isometric views.

Note: The location of the Origin.

- Apply 1060 Alloy for material.

- Calculate the Volume of the part and locate the Center of mass relative to the origin. Use the Mass Properties tool.

- Think about the steps that you would take to build the model. The part is symmetric about the Front Plane.

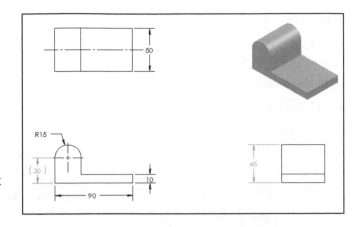

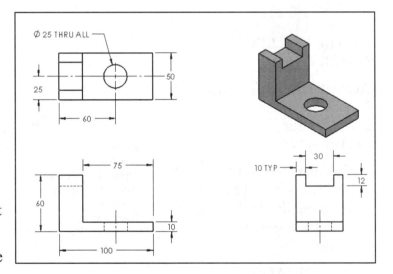

Exercise 1.16: DRAWING 2

Create the part from the illustrated ANSI - MMGS Third Angle Projection drawing: Front, Top, Right and Isometric views.

Note the location of the Origin. Assume Symmetry.

- Apply the Hole Wizard feature.

- Apply 1060 Alloy for material.

- The part is symmetric about the Front Plane.

- Calculate the Volume of the part and locate the Center of mass relative to the origin, Use the Mass Properties tool.

- Think about the steps that you would take to build the model.

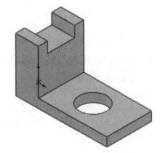

Exercise 1.17: BASIC PART

Create the part from the illustrated ANSI - MMGS Third Angle Projection drawing: Front, Top, Right and Isometric views.

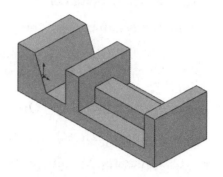

Note: The location of the Origin.

- Apply Cast Alloy steel for material.

- The part is symmetric about the Front Plane.

- Calculate the Volume of the part and locate the Center of mass relative to the origin. Use the Mass Properties tool.

Think about the steps that you would take to build the model. Do you need the Right view for manufacturing? Does it add any valuable information?

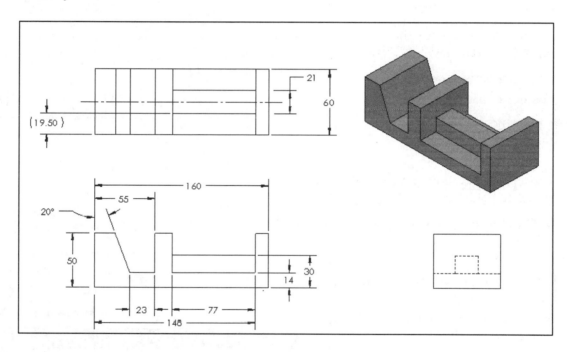

Exercise 1.18: COSMETIC THREAD

Apply a Cosmetic thread: 1/4-20x2 UNC. A Cosmetic thread represents the inner diameter of a thread on a boss or the outer diameter of a thread.

Copy and open the Cosmetic Thread part from the Chapter 1 Homework folder on the DVD. Note: Copy the model from the DVD to your hard drive. Do not work directly from the read-only DVD.

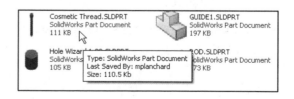

Create a Cosmetic thread.

Click the bottom edge of the part as illustrated.

Click Insert, Annotations, Cosmetic Thread from the Menu bar menu. View the Cosmetic Thread PropertyManager. Edge<1> is displayed.

Select Blind for End Condition.

Enter 1.00 for depth.

Enter .200 for min diameter.

Enter ¼-20-2 UNC 2A in the Thread Callout box.

Click OK ✓ from the Cosmetic Thread FeatureManager.

Expand the FeatureManager. View the Cosmetic Thread feature.

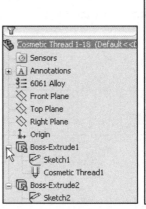

If needed, right-click the Annotations folder, click Details.

Check the Cosmetic threads, and Shaded cosmetic threads boxes.

Click OK. View the cosmetic thread on the model.

🔆 The Thread Callout: ¼-20-2UNC 2A is automatically inserted into a drawing document, if the drawing document is in the ANSI drafting standard.

🔆 ¼-20-2 UNC 2A - ¼ inch major diameter - 20 threads / inch, Unified National Coarse thread series, Class 2 (General Thread), A -External threads.

Exercise 1.19: HOLE BLOCK

Create the Hole-Block using the Hole Wizard feature as illustrated. Create the Hole-Block part on the Front Plane.

The Hole Wizard 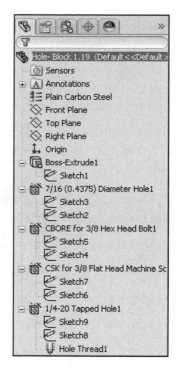 feature creates either a 2D or 3D sketch for the placement of the hole in the FeatureManager. You can consecutively place multiple holes of the same type. The Hole Wizard creates 2D sketches for holes unless you select a non-planar face or click the 3D Sketch button in the Hole Position PropertyManager.

Hole Wizard creates two sketches. The first sketch is the revolved cut profile of the selected hole type and the second sketch - center placement of the profile. Both sketches should be fully defined.

Create a rectangular prism 2 inches wide by 5 inches long by 2 inches high. On the top surface of the prism, place four holes, 1 inch apart

- Hole #1: Simple Hole Type: Fractional Drill Size, 7/16 diameter, End Condition: Blind, 0.75 inch deep.

- Hole #2: Counterbore hole Type: for 3/8 inch diameter Hex bolt, End Condition: Through All.

- Hole #3: Countersink hole Type: for 3/8 inch diameter Flat head screw, 1.5 inch deep.

- Hole #4: Tapped hole Type, Size ¼-20, 1 inch Blind hole deep.

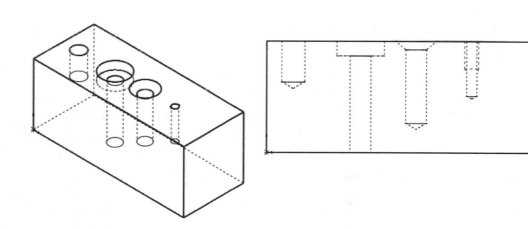

Exercise 1.20: 3D SKETCH / HOLE WIZARD

Create the Part using the Hole Wizard feature. Apply the 3D sketch placement method as illustrated in the FeatureManager. Insert and dimension a hole on a cylindrical face.

Copy and open Hole Wizard 1.20 from the Chapter 1 Homework folder on the DVD. Copy the model to your folder on the computer.

Click the Hole Wizard 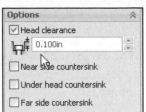 Features tool. The Hole Specification PropertyManager is displayed.

Select the Counterbore Hole Type.

Select Ansi Inch for Standard.

Select Socket Head Cap Screw for fastener Type.

Select 1/4 for Size. Select Normal for Fit.

Select Through All for End Condition.

Enter .100 for Head clearance in the Options box.

Click the Positions Tab. The Hole Position PropertyManager is displayed.

Click the 3D Sketch button. SolidWorks

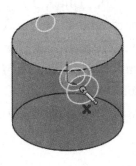 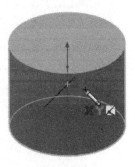

displays a 3D interface with the Point tool active.

When the Point tool is active, wherever you click, you will create a point

Click the cylindrical face of the model as illustrated. The selected face is displayed in orange. This indicates that an OnSurface sketch relations will be created between the sketch point and the cylindrical face. The hole is displayed in the model.

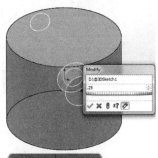

Insert a dimension between the top face and the Sketch point.

Click the Smart Dimension Sketch tool.

Click the top flat face of the model and the sketch point.

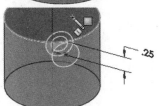

Enter .25in.

Locate the point angularly around the cylinder. Apply construction geometry.

Activate the Temporary Axes. Click View; check the Temporary Axes box from the Menu bar toolbar.

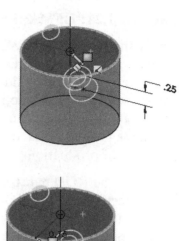

Click the Line ＼ Sketch tool. Note: 3D sketch is still activated.

Ctrl+click the top flat face of the model. This moves the red space handle origin to the selected face. This also constrains any new sketch entities to the top flat face. Note the mouse

pointer ✎✏ ⋌ icon.

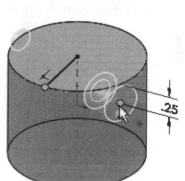

Move the mouse pointer near the center of the activated top flat face as illustrated. View the small black circle. The circle indicates that the end point of the line will pick up a Coincident relation.

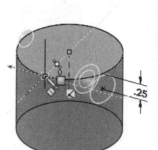

Click the center point of the circle.

Sketch a line so it picks up the AlongZ sketch relation. The cursor displays the relation to be applied. This is a very important step!

Create an AlongY sketch relation between the center point of the hole on the cylindrical face and the endpoint of the sketched line as illustrated.

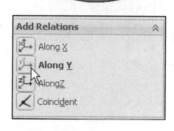

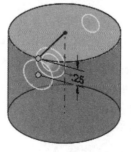

Click OK ✔ from the Properties PropertyManager.

Click OK ✔ from the Hole Position PropertyManager.

Expand the FeatureManager and view the results. The two sketches are fully defined.

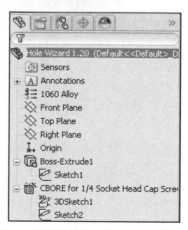

Close the model.

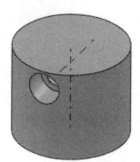

Exercise 1.21: DFMXpress

Apply the DFMXpress Wizard. DFMXpress is an analysis tool that validates the manufacturability of SolidWorks parts. Use DFMXpress to identify design areas that may cause problems in fabrication or increase the costs of production.

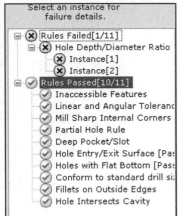

- Open the ROD part from this Project. Note: The ROD part is also located in the Chapter 1 - Homework folder on the DVD. Copy the part to your local folder if needed.

- Click the Evaluate tab in the CommandManager.

- Click DFMXpress Analysis Wizard.

- Click the RUN button.

- Expand each folder. View the results.

- Make any needed changes and save the ROD in a DIFFERENT file folder.

Project 2

Fundamentals of Assembly Modeling

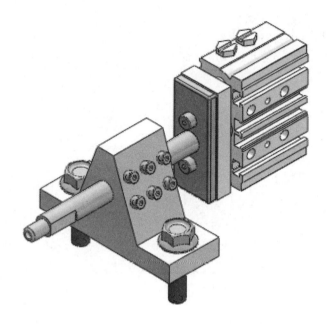

Below are the desired outcomes and usage competencies based on the completion of Project 2.

Project Desired Outcomes:	Usage Competencies:

• Two assemblies:
   o GUIDE-ROD assembly
   o CUSTOMER assembly
	• Insert and edit components, mates and SmartMates in an assembly using the Bottom-up assembly design approach.

• Use an assembly from SMC USA. o GUIDE-CYLINDER	• Use and assemble components from 3D ContentCentral.
• Flange bolt part from the SolidWorks Design Library.	• Assemble components from the SolidWorks Design Library.

• Two parts:
   o 4MMCAPSCREW
   o 3MMCAPSCREW
	• Apply the Revolved Base, Extruded Cut, Chamfer, Component Pattern features along with the Copy, Edit and suppress tools.

Notes:

Project 2 - Fundamentals of Assembly Modeling

Project Objective

Provide an understanding of the Bottom-up assembly design approach. Insert existing parts into an assembly. Orient and position the components in the assembly using Standard Mates.

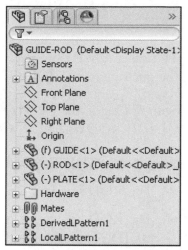

Create the GUIDE-ROD assembly. Utilize the ROD, GUIDE and PLATE parts. The ROD, GUIDE and PLATE parts were created in Project 1.

Create the CUSTOMER assembly. The CUSTOMER assembly consists of two sub-assemblies: GUIDE-ROD and GUIDE-CYLINDER. Use a downloaded assembly from 3D ContentCentral.

Insert the Flange bolt from the SolidWorks Design Library into the GUIDE-ROD assembly. Utilize a Revolved Base feature to create the 4MMCAPSCREW. Copy the 4MMCAPSCREW part and modify dimensions to create the 3MMCAPSCREW part.

On the completion of this project, you will be able to:

- Understand the Assembly FeatureManager Syntax.

- Insert parts / sub-components into an assembly.

- Insert and edit Standard mates and SmartMates into an assembly.

- Rename parts and copy assemblies with internal references.

- Incorporate design changes into an assembly.

- Insert a sub-component from the SolidWorks Design Library.

- Modify, Edit and Suppress features in an assembly.

- Recover from Mate errors in an assembly.

- Suppress/Un-suppress component features.

- Create an Isometric Exploded view of an assembly.

- Create a Section view of an assembly.

- Apply and edit a material in a component.

- Apply the Revolved Base, Extruded Cut, Chamfer and Component Pattern features.

Project Situation

The PLATE, ROD and GUIDE parts were created in Project 1. Perform the following steps:

Step 1: Insert the ROD, GUIDE and PLATE into a GUIDE-ROD assembly.

Step 2: Obtain the customer's GUIDE-CYLINDER assembly using 3D ContentCentral. The assembly is obtained to guarantee proper fit between the GUIDE-ROD assembly and the customer's GUIDE-CYLINDER assembly.

Step 3: Create the CUSTOMER assembly. The CUSTOMER assembly combines the GUIDE-ROD assembly with the GUIDE-CYLINDER assembly.

Review the GUIDE-ROD assembly design constraints:

- The ROD requires the ability to travel through the GUIDE.

- The ROD keyway is parallel to the right surface of the GUIDE. The top surface of the GUIDE is parallel to the work area.

- The ROD mounts to the PLATE. The GUIDE mounts to a flat work surface.

The PISTON PLATE is the front plate of the GUIDE-CYLINDER assembly. The PLATE from the GUIDE-ROD assembly mounts to the PISTON PLATE. Create a rough sketch of the conceptual assembly.

Rough Sketch of Design Situation: CUSTOMER assembly

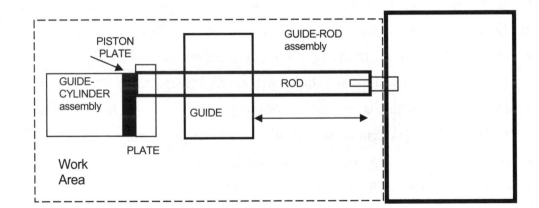

An assembly combines two or more parts. In an assembly, parts are referred to as components. Design constraints directly influence the assembly design process. Other considerations indirectly affect the assembly design, namely: cost, manufacturability and serviceability.

Project Overview

Translate the rough conceptual sketch into a SolidWorks assembly.

The GUIDE-ROD is the first assembly.

Determine the first component of the assembly.

The first component is the GUIDE. The GUIDE remains stationary. The GUIDE is a fixed component.

The action of assembling components in SolidWorks is defined as mates.

Mates are relationships between components that simulate the construction of the assembly in a manufacturing environment. Mates effect/restrict the *six degrees* of freedom in an assembly.

The CUSTOMER assembly combines the GUIDE-CYLINDER assembly with the GUIDE-ROD assembly.

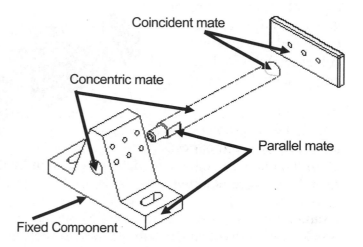

GUIDE-ROD assembly

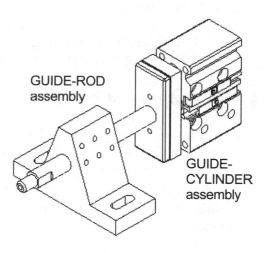

CUSTOMER assembly

Assembly Modeling Approach

In SolidWorks, components and their assemblies are directly related through a common file structure. Changes in the components directly affect the assembly and vise a versa. You can create assemblies using the Bottom-up assembly approach, Top-down assembly approach or a combination of both methods. This chapter focuses on the Bottom-up assembly approach.

The Bottom-up approach is the traditional method that combines individual components. Based on design criteria, the components are developed independently. The three major steps in a Bottom-up assembly approach are: Create each component independent of any other component in the assembly, insert the components into the assembly, and mate the components in the assembly as they relate to the physical constraints of your design.

In the Top-down assembly approach, major design requirements are translated into assemblies, sub-assemblies and components.

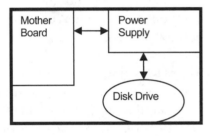

In the Top-down approach, you do not need all of the required component design details. Individual relationships are required.

Example: A computer. The inside of a computer can be divided into individual key sub-assemblies such as a: motherboard, disk drive, power supply, etc. Relationships between these sub-assemblies must be maintained for proper fit.

Use the Bottom-up design approach for the GUIDE-ROD assembly and the CUSTOMER assembly.

Linear Motion and Rotational Motion

In dynamics, motion of an object is described in linear and rotational terms. Components possess linear motion along the x, y and z-axes and rotational motion around the x, y and z-axes. In an assembly, each component has *six degrees* of freedom: three translational (linear) and three rotational. Mates remove degrees of freedom. All components are rigid bodies. The components do not flex or deform.

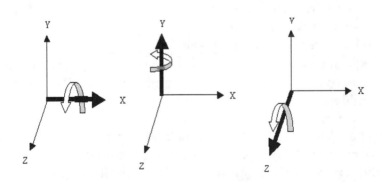

GUIDE-ROD assembly

The GUIDE-ROD assembly consists of six components.

- GUIDE
- ROD
- PLATE
- Flange bolt
- 4MMCAPSCREW
- 3MMCAPSCREW

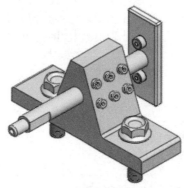

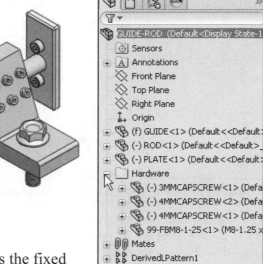

The first component is the GUIDE. The GUIDE is the fixed component in the GUIDE-ROD assembly.

The second component is the ROD. The ROD translates linearly through the GUIDE.

The third component is the PLATE. The PLATE is assembled to the ROD and the GUIDE-CYLINDER assembly.

The forth component is the Flange bolt. Obtain the Flange bolt from the SolidWorks Design Library.

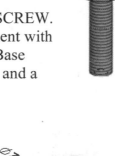

The fifth component is the 4MM CAPSCREW. Create the 4MM CAPSCREW component with a Revolved Base feature. A Revolved Base feature requires a centerline for an axis and a sketched profile. The profile is rotated about the axis to create the feature.

The sixth component is the 3MMCAPSCREW. Create the 3MMCAPSCREW from the 4MMCAPSCREW. Reuse geometry to save time.

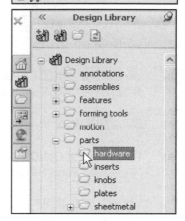

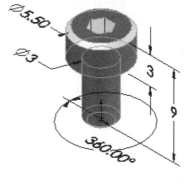

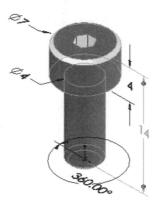

Insert a Component Pattern feature of the 3MMCAPSCREW part. A Component Pattern is created in the assembly. Reference the GUIDE Linear Pattern of Tapped Holes to locate the 3MMCAPSCREWs.

Activity: Create the GUIDE-ROD Assembly

Close all SolidWorks documents.

1) Click **Windows**, **Close All** from the Menu bar before you begin this project.

Open the GUIDE part.

2) Click **Open** from the Menu bar.

3) Select the **ENGDESIGN-W-SOLIDWORKS\PROJECTS** folder.

4) Select **Part** for the drop-down menu.

5) Double-click **GUIDE**. The GUIDE FeatureManager is displayed.

Create the GUIDE-ROD assembly.

6) Click **New** from the Menu bar. The New SolidWorks Documents dialog box is displayed.

7) Double-click **Assembly** from the default Templates tab. The Begin Assembly PropertyManager is displayed.

The Begin Assembly PropertyManager and the Insert Component PropertyManager is displayed when a new or existing assembly is opened, if the Start command when creating new assembly box is checked.

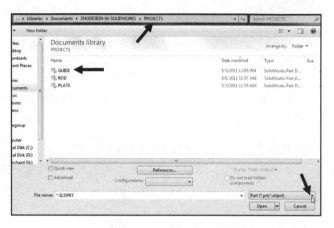

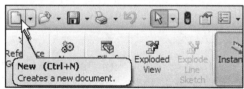

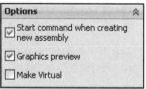

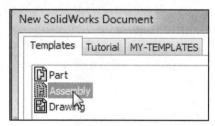

When a part is inserted into an assembly it is called a component. The Begin Assembly PropertyManager is displayed to the left of the Graphics window. GUIDE is listed in the Part/Assembly to Insert box.

Insert the GUIDE and fix it to the assembly Origin.

8) Double-click **GUIDE** in the Open documents box.

9) Click **OK** ✔ from the Begin Assembly PropertyManager. The GUIDE part icon ⊞ 🐚 (f) GUIDE<1> is displayed in the assembly FeatureManager

The GUIDE name is added to the assembly FeatureManager with the symbol (f).

The symbol (f) represents a fixed component. A fixed component cannot move and is locked to the assembly Origin. The first component in an assembly should be fixed or fully defined.

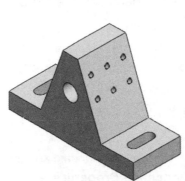

☀ To fix the first component to the Origin, you can click OK ✔ from the Begin Assembly PropertyManager or click the Origin in the Graphics window.

☀ To remove the fixed state (f), right-click the **fixed component name** in the FeatureManager. Click **Float**. The component is free to move.

☀ Your default system document templates may be different if you are a new user of SolidWorks vs. an existing user who has upgraded from a previous version.

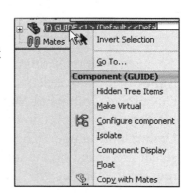

Save and name the assembly.

10) Click **Save As** from the Menu bar.

 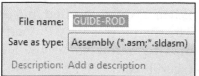

11) Select the **ENGDESIGN-W-SOLIDWORKS\PROJECTS** folder.

12) Enter **GUIDE-ROD** for File name.

13) Click **Save**.

Close the GUIDE part.

14) Click **Window**, **GUIDE** from the Menu bar.

15) Click **File**, **Close** from the Menu bar. The GUIDE-ROD assembly is the open document.

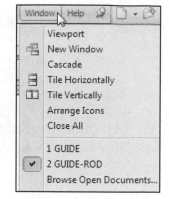

Select the Ctrl-Tab keys to quickly alternate between open SolidWorks documents. Select inside Close 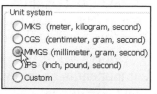 to close the current SolidWorks document. The outside Close 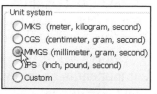 exits SolidWorks.

Set the GUIDE-ROD assembly Drafting Standard and units.

16) Click **Options**, **Document Properties** tab.

17) Select **Ansi** for Drafting Standard.

18) Click the **Units** folder.

19) Click **MMGS** (millimeter, gram, second) for Unit system. Accept the default settings.

20) Click **OK** from the Document Properties - Units dialog box.

Save the GUIDE-ROD assembly.

21) Click **Save**.

To customize the CommandManager, right-click on an existing tab. Click Customize Command Manager. Check the option to display in the CommandManager. The options you select are based on the tool types you require for the design.

GUIDE-ROD Assembly-Insert Component

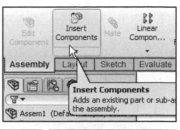

The first component is the foundation of the assembly. The GUIDE is the first component in the GUIDE-ROD assembly. The ROD is the second component in the GUIDE-ROD assembly. Add components to assemblies utilizing the following techniques:

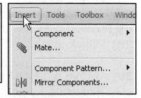

- Utilize the Insert

  Components  Assembly tool.

- Utilize Insert, Component from the Menu bar.

- Drag a component from Windows Explorer (3D ContentCentral) into the Assembly.

- Drag a component from the SolidWorks Design Library into the Assembly.

- Drag a component from an Open part file into the Assembly.

Activity: GUIDE-ROD Assembly-Insert Component

Insert the ROD component into the assembly. Use the Insert Components tool.

22) Click the **Assembly** tab from the CommandManager.

23) Click the **Insert Components** Assembly tool. The Insert Component PropertyManager is displayed.

24) Click the **BROWSE** button from the Open documents box.

25) Browse to the **ENGDESIGN-W-SOLIDWORKS\PROJECTS** folder.

26) Select **Part** for file type.

27) Double-click the **ROD** part. The ROD is displayed in the Graphics window.

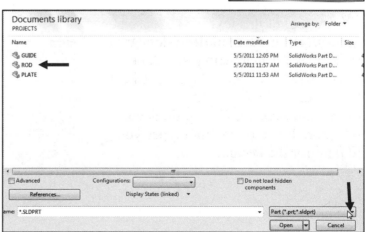

The ROD part icon is displayed in the Open
documents text box. The mouse pointer displays the
ROD component when positioned inside the GUIDE-
ROD Graphics window.

28) Click a **position** to the left of the GUIDE as
illustrated.

Re-position the ROD component in the Graphics window.
29) Click and drag the **ROD** in the Graphics window.

30) Move the **ROD** on the right side of the GUIDE as
illustrated.

Fit the model to the Graphics window.
31) Press the **f** key.

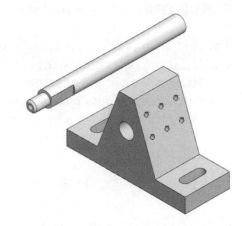

The component movement in the assembly is
determined by its degrees of freedom.

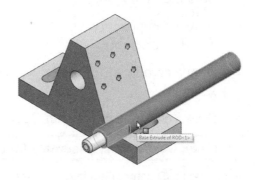

Degree of freedom is geometry that is not
defined by dimensions or relations and is free to
move. In a 2D sketch, there are three degrees of
freedom: *movement along the X axis, Y axis* and
rotation about the Z axis (the axis normal to the
Sketch plane).

Save the GUIDE-ROD assembly.
32) Click **Save** 💾.

33) Click **Yes** to save and rebuild the model. The ROD
component is displayed in the GUIDE-ROD
FeatureManager. The ROD part is free to move
and rotate.

A goal of this text is to expose new
SolidWorks users to various SolidWorks design
tools, features and methods.

Review the FeatureManager Syntax.

34) **Expand** (f) GUIDE<1> in the FeatureManager. The features are displayed in the FeatureManager. Note: Base Extrude, Slot Cut, Guide Hole and M3x0.5 Tapped Hole1 contain additional sketches. Note: Features were renamed in Project 1. Example: Boss-Extrude1 was renamed to Base Extrude.

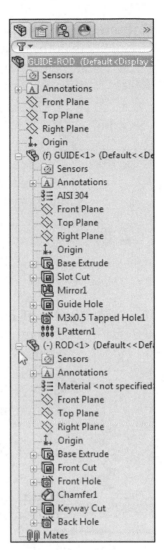

In Project 1, you applied AISI 304 as a material to the GUIDE part.

35) **Expand** (-) ROD<1> in the FeatureManager. The features are displayed in the FeatureManger.

36) Click the **Minus** ⊟ icon to the left of the GUIDE entry and ROD entry to collapse the list.

A Plus ⊞ icon indicates that additional feature information is available. A Minus ⊟ icon indicates that the feature list is fully expanded. Manipulating the FeatureManager is an integral part of the assembly. In the step-by-step instructions, expand and collapse are used as follows:

Expand - Click the Plus ⊞ icon.

Collapse - Click the Minus ⊟ icon.

FeatureManager Syntax

Entries in the FeatureManager design tree have specific definitions. Understanding syntax and states saves time when creating and modifying parts and

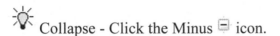

assemblies. Review the six columns of the MGPTube component syntax in the FeatureManager as illustrated.

Column 1: A Resolved component (not in lightweight 🏵 state) displays a plus ⊞ icon. The plus icon indicates that additional feature information is available. A minus ⊟ icon displays the fully expanded feature list.

Column 2: Identifies a component's (part or assembly) relationship with other components in the assembly.

Component or Part States:	
Symbol:	**State:**
⊞ 🗇	Resolved part. A yellow part icon indicates a resolved state. A blue part icon indicates a selected, resolved part. The component is fully loaded into memory and all of its features and mates are editable.
⊞ 🗇	Lightweight part. A blue feather on the part icon indicates a lightweight state. When a component is lightweight, only a subset of its model data is loaded in memory.
⊞ 🗇	Out-of-Date Lightweight. A red feather on the part icon indicates out-of-date references. This option is not available when the Large Assembly Mode is activated.
🗇	Suppressed. A gray icon indicates the part is not resolved in the active configuration.
🗇	Hidden. A clear icon indicates the part is resolved but invisible.
🗇	Hidden Lightweight. A transparent blue feather over a transparent component icon indicates that the component is lightweight and hidden.
🗇	Hidden, Out-of-Date, Lightweight. A red feather over a clear part icon indicates the part is hidden, out-of-date and lightweight.
⊞ 🗇	Hidden Smart Component. A transparent star over a transparent icon indicates that the component is a Smart Component and hidden.
⊞ 🗇	Smart Component. A star overlay is displayed on the icon of a Smart Component.
⊟ 🗇	Rebuild. A rebuild is required for the assembly or component.
🗇	Resolved assembly. Resolved (or unsuppressed) is the normal state for assembly components. A resolved assembly is fully loaded in memory, fully functional, and fully accessible.

Column 3: The MGPTube part is fixed (f). You can fix the position of a component so that it cannot move with respect to the assembly Origin. By default, the first part in an assembly is fixed; however, you can float it at any time.

It is recommended that at least one assembly component is either fixed, or mated to the assembly planes or Origin. This provides a frame of reference for all other mates, and helps prevent unexpected movement of components when mates are added. The Component Properties are:

Component Properties in an assembly:	
Symbol:	**Relationship:**
(-)	A minus sign (–) indicates that the part or assembly is under-defined and requires additional information.
(+)	A plus sign (+) indicates that the part or assembly is over-defined.
None	The Base component is mated to three assembly reference planes. It is fully defined.
(f)	A fixed symbol (f) indicates that the part or assembly does not move.
(?)	A question mark (?) indicates that additional information is required on the part or assembly.

Column 4: MGPTube - Name of the part.

Column 5: The symbol <#> indicates the particular inserted instance of a component. The symbol <1> indicates the first inserted instance of a component, "MGPTube" in the assembly. If you delete a component and reinsert the same component again, the <#> symbol increments by one.

Column 6: The Resolved state displays the MGPTube icon with an External reference symbol, "- >". The state of External references is displayed as follows:

- If a part or feature has an external reference, its name is followed by –>. The name of any feature with external references is also followed by –>.

- If an external reference is currently out of context, the feature name and the part name are followed by ->?

- The suffix ->* means that the reference is locked.

- The suffix ->x means that the reference is broken.

There are modeling situations in which unresolved components create rebuild errors. In these situations, issue the forced rebuild, Ctrl+Q. The Ctrl+Q option rebuilds the model and all its features. If the mates still contain rebuild errors, resolve all the components below the entry in the FeatureManager that contains the first error.

Mate Types

Mates provide the ability to create geometric relationships between assembly components. Mates define the allowable directions of rotational or linear motion of the components in the assembly. Move a component within its degrees of freedom in the Graphics window, to view the behavior of an assembly.

Mates are solved together as a system. The order in which you add mates does not matter. All mates are solved at the same time. You can suppress mates just as you can suppress features.

The Mate PropertyManager provides the ability to select either the *Mates* or *Analysis* tab. Each tab has a separate menu. The *Analysis* tab requires the ability to run SolidWorks Motion. The Analysis tab is not covered in this book. The Mate PropertyManager displays the appropriate selections based on the type of mate you create. The components in the GUIDE-ROD assembly utilize Standard Mate types.

Review the *Standard*, *Advanced* and *Mechanical* Mates types.

Standard Mates:

Components are assembled with various mate types. The Standard Mate types are:

- **Coincident Mate**: Locates the selected faces, edges, or planes so they use the same infinite line. A Coincident mate positions two vertices for contact

- **Parallel Mate**: Locates the selected items to lie in the same direction and to remain a constant distance apart.

- **Perpendicular Mate**: Locates the selected items at a 90° angle to each other.

- **Tangent Mate**: Locates the selected items in a tangent mate. At least one selected item must be either a conical, cylindrical, or spherical face.

- **Concentric Mate**: Locates the selected items so they can share the same center point.

- **Lock Mate**: Maintains the position and orientation between two components.

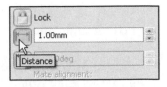

- **Distance Mate**: Locates the selected items with a specified distance between them. Use the drop-down arrow box or enter the distance value directly.

- **Angle Mate**: Locates the selected items at the specified angle to each other. Use the drop-down arrow box or enter the angle value directly.

There are two Mate Alignment options. The Aligned option positions the components so that the normal vectors from the selected faces point in the same direction. The Anti-Aligned option positions the components so that the normal vectors from the selected faces point in opposite directions.

Use for positioning only. When selected, components move to the position defined by the mate, but a mate is not added to the FeatureManager design tree. A mate appears in the Mates box so you can edit and position the components, but nothing appears in the FeatureManager design tree when you close the Mate PropertyManager.

Advanced Mates:

The Advanced Mate types are:

- **Symmetric Mate**: Positions two selected entities to be symmetric about a plane or planar face. A Symmetric Mate does not create a Mirrored Component.

- **Width Mate**: Centers a tab within the width of a groove.

- **Path Mate**: Constrains a selected point on a component to a path.

- **Linear/Linear Coupler Mate**: Establishes a relationship between the translation of one component and the translation of another component.

- **Distance (Limit) Mate**: Locates the selected items with a specified distance between them. Use the drop-down arrow box or enter the distance value directly.

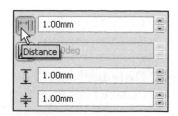

- **Angle Mate**: Locates the selected items at the specified angle to each other. Use the drop-down arrow box or enter the angle value directly.

Mechanical Mates:

The Mechanical Mate types are:

- **Cam Mate**: Forces a plane, cylinder, or point to be tangent or coincident to a series of tangent extruded faces.

- **Hinge**: Limits the movement between two components to one rotational degree of freedom. It has the same effect as adding a Concentric mate plus a Coincident mate.

- **Gear Mate**: Forces two components to rotate relative to one another around selected axes.

- **Rack Pinion Mate**: Provides the ability to have Linear translation of a part, rack causes circular rotation in another part, pinion, and vice versa.

- **Screw Mate**: Constrains two components to be concentric, and also adds a pitch relationship between the rotation of one component and the translation of the other.

- **Universal Joint Mate**: The rotation of one component (the output shaft) about its axis is driven by the rotation of another component (the input shaft) about its axis.

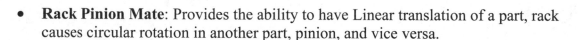 Mates reflect the physical behavior of a component in an assembly. In this project, the two most common Mate types are Concentric and Coincident.

GUIDE-ROD Assembly-Mate the ROD Component

Recall the initial assembly design constraints:

- The ROD requires the ability to travel through the GUIDE.

- The face of the Keyway in the ROD is parallel to the right face of the GUIDE.

Utilize Concentric and Parallel mates between the ROD and GUIDE. The Concentric mate utilizes the cylindrical face of the shaft with the cylindrical face of the GUIDE Hole. The Parallel mate utilizes two planar faces from the ROD Keyway Cut and the right face of the GUIDE.

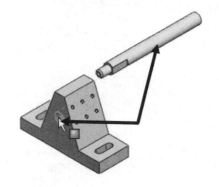

Concentric Mate - 2 Cylindrical faces   Parallel - 2 Planar faces

The Concentric and Parallel mate provides the ability for the ROD to translate linearly through the GUIDE Hole. The ROD does not rotate.

Use the following steps to create a Mate:

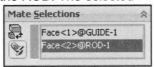

- Click the Mate tool ✎ from the Assembly toolbar.

- Select the geometry from the first component, (usually the part).

- Select the geometry from the second component, (usually the assembly).

- Select the Mate type.

- Click OK ✔ to create the Mate. View the created mate.

Activity: GUIDE-ROD Assembly-Mate the ROD Component

Insert a Concentric mate between two faces.

37) Click the **Mate** ✎ Assembly tool. The Mate PropertyManager is displayed.

38) Click the **inside cylindrical face** of the Guide Hole as illustrated. Note: The icon feedback symbol.

39) Click the **cylindrical face** of the ROD. The selected faces are displayed in the Mate Selections box. Concentric mate is selected by default.

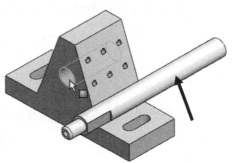

40) Click the **Green Check mark** 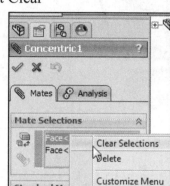 from the Mates Pop-up toolbar to insert the Concentric Mate. The Mate PropertyManager remains open on the left side of the Graphics window. Concentric1 is displayed in the Mates box.

Review the Mate Selections, the cylindrical face of the Guide Hole and the cylindrical face of the ROD. If the Mate Selections are not correct, right-click a position inside the Mate Selections box and select Clear Selections.

The Mate Pop-up toolbar minimizes the time required to create a Standard Mate. Utilize the Mate Pop-up toolbar for this project.

Lock   Flip Mate Alignment

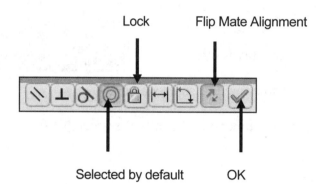

Selected by default   OK

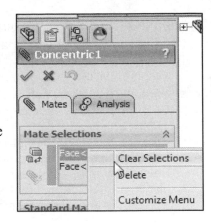

When selecting faces, position the mouse pointer in the middle of the face. Do not position the mouse pointer near the edge of the face. If the wrong face or edge is selected, perform one of the following actions:

- Click the face or edge again to remove it from the Mate Selections box.

- Right-click in the Mate Selections box. Click Clear Selections or delete to remove all geometry or a single entity from the Mate Selections box.

- Utilize the Undo button to begin the Mate command again.

The ROD is Concentric with the GUIDE. The ROD has the ability to move and rotate while remaining concentric to the GUIDE hole.

Move and rotate the ROD in the Graphics window.

41) Click and drag the **ROD** in a horizontal direction. The ROD travels linearly in the GUIDE.

42) Click and drag the **ROD** in a vertical direction. The ROD rotates in the GUIDE.

43) Rotate the **Rod** until the Keyway cut is approximately parallel to the right face of the GUIDE.

44) Display origins in the Assembly. Click **View**; check **Origins** from the Menu bar.

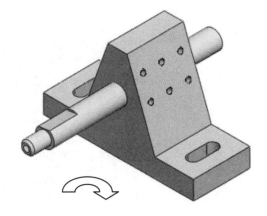
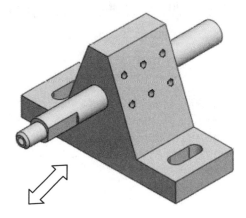

Recall the second assembly design constraint. The flat end of the ROD must remain parallel to the right surface of the GUIDE.

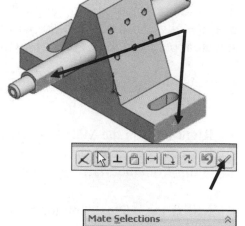

Insert a Parallel mate between two faces.

45) Click the **Keyway face** of the ROD.

46) Click the **flat right face** of the GUIDE. The selected faces are displayed in the Mate Selections box. A Mate error message is displayed. The default mate type (Concentric) would over define the assembly. Insert a Parallel mate.

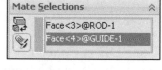

47) Click **Parallel** ⟍ from the Mate Pop-up toolbar.

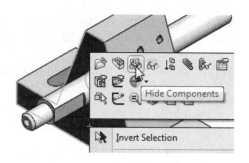

48) Click the **Green Check mark** ✔ to add a Parallel mate. Parallel1 is created.

49) Click **OK** ✔ from the Mate PropertyManager.

Move the ROD in the Graphics window.

50) Click and drag the **ROD** in a horizontal direction and position it approximately in the center of the GUIDE.

Hide the GUIDE component in the Graphics window.

51) Right-click on the front face of the **GUIDE** in the Graphics window.

52) Click **Hide components** from the Context toolbar. The GUIDE component is not displayed in the Graphics window. Note the change in color in the FeatureManager.

Display the Mate types.

53) **Expand** the Mates folder from the FeatureManager.

Display the full Mate names.

54) Drag the **vertical FeatureManager border** to the right. View the two created mates: Concentric1 and Parallel1.

Save the GUIDE-ROD assembly.

55) Click **Save** 💾.

56) Click **Rebuild and Save** the document.

The ROD mates reflect the physical constraints in the GUIDE-ROD assembly. The ROD is under defined - indicated by a minus sign (-).

The Concentric mate allows the ROD to translate freely in the Z direction through the Guide Hole. The Parallel Mate prevents the ROD from rotating in the Guide Hole.

The GUIDE part icon ⊞ ⤵ (f) GUIDE<1> is displayed with no color in the FeatureManager to reflect the Hide state.

GUIDE-ROD Assembly-Mate the PLATE Component

Recall the initial design constraints.

- The ROD is fastened to the PLATE.

- The PLATE part mounts to the GUIDE-CYLINDER PISTON PLATE part.

Apply the Rotate Component ⤷ tool to position the PLATE before applying the Mates.

Activity: GUIDE-ROD Assembly-Mate the PLATE Component

Insert the PLATE component into the assembly from the PROJECTS folder.

57) Click the **Insert Components** 🖑 Assembly tool. The Insert Component PropertyManager is displayed.

58) Click the **BROWSE** button from the Open documents box.

59) Browse to the **PROJECTS** folder.

60) Double-click the **PLATE** part.

61) Click a **position** behind the ROD in the Graphics window as illustrated. The PLATE component is added to the GUIDE-ROD FeatureManager.

Fit the model to the Graphics window.
62) Press the **f** key.

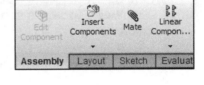

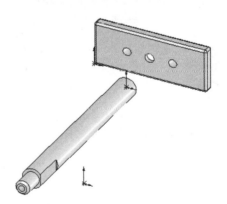

Rotate the PLATE.

63) Click the **Rotate Component** 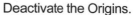 Assembly tool. The Rotate Component PropertyManager is displayed. View your options.

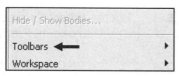

64) Click and drag the **front face** of the PLATE downward until the PLATE rotates approximately 90°.

65) Click **OK** ✔ from the PropertyManager to deactivate the tool.

66) Click a **position** in the Graphics window, to the right of the PLATE, to deselect any faces or edges.

Deactivate the Origins.
67) Click **View**; uncheck **Origins** from the Menu bar.

Use Selection filters to select difficult individual features such as: *faces*, *edges* and *points*. Utilize the Filter Faces tool to select the hidden ROD Back Hole face from the Selection Filter toolbar.

Display the Selection Filter toolbar. Activate the Filter Faces tool.
68) Click **View**, **Toolbars** from the Menu bar. View the available toolbars.

69) Click **Selection Filter**. The Selection Filter toolbar is displayed.

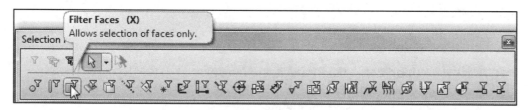

70) Click **Filter Faces** from the Selection Filter toolbar as illustrated. The Selection Filter icon is displayed in your mouse pointer.

To deactivate the Filter Faces tool, click Clear All Filters from the Selection Filter toolbar.

Insert a Concentric mate between two faces.

71) Click **WireFrame** ⬚ from the Heads-up View toolbar.

72) Click the **Mate** ✎ Assembly tool. The Mate PropertyManager is displayed.

73) Click the **center inside cylindrical face** from the PLATE Countersink hole. The center-cylindrical face turns green.

74) Click the **cylindrical face** of the ROD. The selected faces are displayed in the Mate Selections box. Concentric is selected by default.

75) Click the **Green Check mark** ✔ from the Mate Pop-up toolbar. Concentric2 is created.

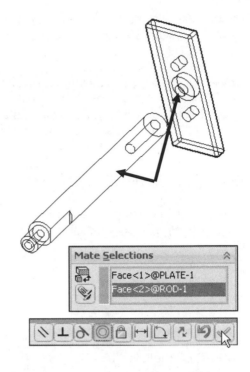

Note: Review the Mate Selections, the cylindrical face of the PLATE and the cylindrical face of the ROD. If the Mate Selections are not correct, right-click a position inside the Mate Selections box and select Clear Selections.

76) Click and drag the **PLATE** behind the ROD.

Insert a Coincident mate between two faces.

77) Press the **left arrow** key to rotate the view until the back face of the ROD is visible.

78) Click the **back circular face** of the ROD.

79) Press the **right arrow** key to rotate the view until the PLATE front face is visible.

80) Click the **front rectangular face** of the PLATE. The selected faces are displayed in the Mate Selections box. Coincident is selected by default.

81) Click the **Green Check mark** ✔ from the Mate Pop-up toolbar. Coincident2 is created.

82) Click **Shaded With Edges** ⬛ from the Heads-up View toolbar.

83) Click **Isometric view** ⬗ from the Heads-up View toolbar.

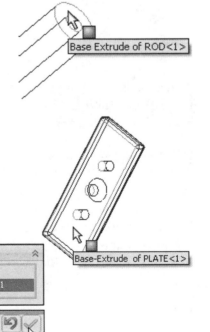

Insert a Parallel mate between two faces.

84) Press the **Shift + z** keys to zoom in on the ROD.

85) Click the ROD **Keyway Cut** flat face.

86) Click the PLATE right **rectangular face** as illustrated. The selected faces are displayed in the Mate Selections box.

87) Click **Parallel** from the Mate Pop-up toolbar.

88) Click the **Green Check mark** ✅ from the Mate Pop-up toolbar. Parallel2 is created.

89) Click **OK** ✅ from the Mate PropertyManager.

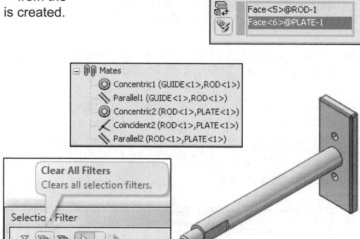

Clear all filters in SolidWorks.

90) Click the **Clear All Filters** 🔖 icon from the Selection Filter toolbar.

Save the GUIDE-ROD assembly.

91) Click **Save** 💾. View the results.

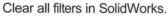

Mates reflect the physical relations bewteen the PLATE and the ROD. The Concentric mate aligns the PLATE Countersink Hole and the ROD cylindrical face. The Coincident mate eliminates translation between the PLATE front face and the ROD back face.

The Parallel mate removes PLATE rotation about the ROD axis. Create the Parallel mate.

☀ A Distance mate of 0 provides additional flexibility over a Coincident mate. A Distance mate value can be modified. Utilize a Coincident mate when mating faces remain coplanar.

The mouse pointer displays the Filter icon when the Selection Filter is activated. Deactivate Selection Filters when not required.

Activate/ Deactivate Filters using the following keys:

Filter for edges	Press e
Filter for faces	Press x
Filter for vertices	Press v
Hide/Show all Filters	F5
Off/On all Selected Filters	F6

Accidentally pressing the e, x or v keys can activate a Filter. If the mouse pointer displays the Filter icon, you cannot select geometry, dimensions or text. Press the F5 key to display the Selection Filter toolbar. Select Clear All Filters.

GUIDE-ROD Assembly-Mate Errors

Mate errors occur when component geometry is over defined. Example: You added a new Concentric Mate between the PLATE bottom Mounting Hole and the ROD cylindrical face.

The ROD back hole cannot physically exist with a Concentric Mate to both the PLATE middle CSK Hole and bottom Mounting Hole.

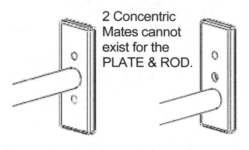

2 Concentric Mates cannot exist for the PLATE & ROD.

- Review the design intent. Know the behavior of the components in the assembly.

- Review the messages and symbols in the FeatureManager.

- Utilize Delete, Edit Feature and Undo commands to recover from Mate errors.

View the Mates for a component. Right-click a component (of the assembly or of a sub-assembly) and click the View Mates tool from the Context toolbar.

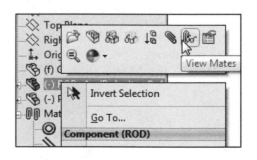

To view the mates for more than one component, hold the Ctrl key down, select the components, then right-click, and click the View Mates tool.

A Mate problem displays the following icons:

⚠ **Warning**. The mate is satisfied, but is involved in over defining the assembly.

⊗ **Error**. The mate is not satisfied.

Insert the second Concentric mate which will create a Mate error in the following steps.

Activity: GUIDE-ROD Assembly-Mate Errors

Insert a Concentric mate between two faces.

92) Click the **Mate** ✎ Assembly tool. The Mate PropertyManager is displayed.

93) Click the **outside cylindrical face** of the ROD as illustrated.

94) Click the **bottom Mounting Hole inside cylindrical face** of the PLATE. The selected faces are displayed in the Mate Selections box. A Mate error message is displayed. Adding a Concentric mate would over-define the assembly.

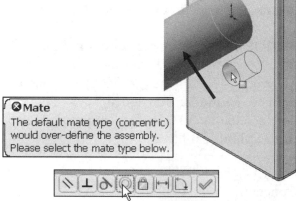

⊗ Mate
The default mate type (concentric) would over-define the assembly. Please select the mate type below.

95) Click **Concentric** from the Mate Pop-up toolbar. A second Mate error message is displayed. The components cannot be moved to a position which satisfies this mate.

96) Click **Close** ✖ from the PropertyManager to return to the Assembly FeatureManager.

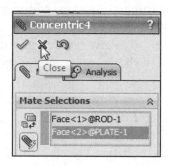

Review the created Mates.

97) **Expand** the Mates folder in the FeatureManager. View the created mates and their icons.

If you delete a Mate and then recreate it, the Mate number will be different. View the mate symbols in the Mates folder.

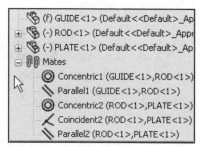

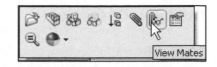

Review the mates for a component.

98) Right-click **ROD** from the Graphics window.

99) Click **View Mates** from the Context toolbar. The Mates dialog box is displayed. Note: The GUIDE is Hidden in the FeatureManager.

Components involved in the mate system for the selected components are vaguely transparent in the Graphics window. Components not involved are hidden.

100) **Close** the pop-up mate dialog box to return to the GUIDE-ROD FeatureManager.

Organize the Mates names. Rename Mate names with descriptive names for clarity.

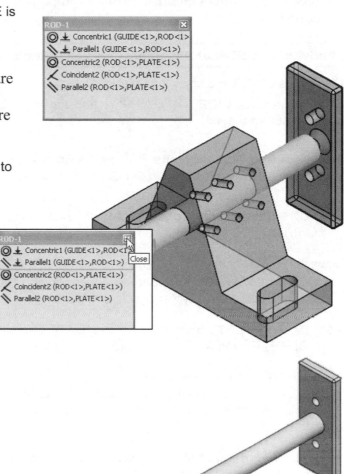

Collision Detection

The Collision Detection assembly function detects collisions between components as they move or rotate. A collision occurs when geometry on one component coincides with geometry on another component. Place components in a non-colliding position; then test for collisions.

Activity: GUIDE-ROD Assembly-Collision Detection

Display the GUIDE and apply the Collision Detection tool.

101) Right-click **GUIDE** (f) GUIDE<1> from the FeatureManager.

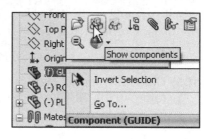

102) Click **Show components** from the Context toolbar. The GUIDE is displayed in the Graphics window.

103) Click **Shaded with Edges** from the Heads-up View toolbar.

Move the PLATE behind the GUIDE.

104) Click the **Move Component** Assembly tool. The Move Component PropertyManager is displayed.

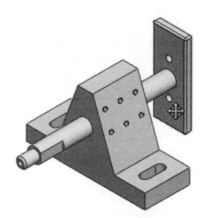

105) Drag the **PLATE** backward until the PLATE clears the GUIDE. The PLATE is free to translate along the Z-axis.

Apply the Collision Detection tool.

106) Click the **Collision Detection** checkbox. The Stop at collision check box is selected by default in the Options box.

107) Check **Highlight faces**, **Sound** and **Ignore complex surfaces** from the Advanced Options box.

108) Drag the **PLATE** forward. The GUIDE back, top and angled right faces turn blue when the PLATE front face collides with the back face of the GUIDE.

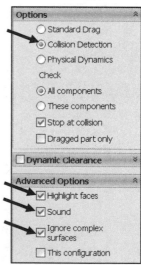

Return the PLATE to the original position.

109) Drag the **PLATE** backward until the ROD is approximately halfway through the GUIDE.

110) Click **OK** from the Move Component PropertyManager.

Save the GUIDE-ROD assembly.
111) Click **Save** 🖫.

112) Click **Rebuild and Save** the document.

Modify Component Dimension

Modify part dimensions in the assembly. Utilize Rebuild to update the part and the assembly. You realize from additional documentation that the Slot in the GUIDE is 4mm. Modify the right Slot Cut feature dimensions in the GUIDE-ROD assembly. Rebuild the assembly. The left Mirror Slot Cut and right Slot Cut update with the new value.

Activity: GUIDE-ROD Assembly-Modify Component Dimension

Modify the Slot of the Guide.
113) Double-click on the right **Slot Cut** of the GUIDE in the Graphics window. The Slot Cut dimensions are displayed in the Graphics window.

Modify the radial dimension.
114) Double-click **R3** in the Graphics window.

115) Enter **4**mm.

116) Click **Rebuild** from the Modify dialog box.

117) Click the **Green Check mark** ✅ from the Modify dialog box. Note: R4 is displayed in blue.

118) Click **OK** ✅ from the Dimension PropertyManager.

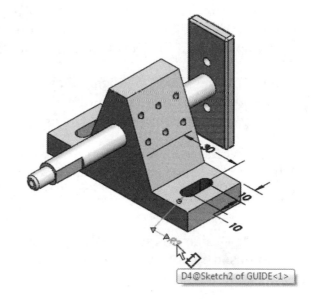

Save the GUIDE-ROD assembly.
119) Click **Save** 🖫.

120) Click **Save All**.

🔍 Additional details on Assembly, Mates, Mate Errors, Collision Detection, Selection Filters are available in SolidWorks Help.

Keywords: Standard Mates, Mate PropertyManager, Mates (Diagnostics), Collision Detection, Design Methods in Assembly and Selection Filters.

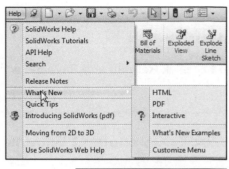

Additional information on Mate Diagnostics is located in the What's New section.

SolidWorks Design Library

A parts library contains components used in a design creation. The SolidWorks Design Library 🗂 tab in the Task Pane provides a central location for reusable elements such as parts, assemblies and sketches. It does not recognize non-reusable elements such as SolidWorks drawings, text files, or other non-SolidWorks files.

The Design Library consists of annotations, assemblies, features, forming tools, motion, parts, routing, smart components, Toolbox (Add-in) and 3D ContentCentral (models from suppliers). SolidWorks Add-ins are software applications.

Your company issued a design policy. The policy states that you are required to only use parts that are presently in the company's parts library. The policy is designed to lower inventory cost, purchasing cost and design time.

In this project, the SW Design Library parts simulate your company's part library. Utilize a hex flange bolt located in the Hardware folder. Specify a new folder location in the Design Library to quickly locate the components utilized in this project.

🔆 The Design Library saves time locating and utilizing components in an assembly. Note: In some network installations, depending on your access rights, additions to the Design Library are only valid in new folders.

Activity: GUIDE-ROD Assembly-SolidWorks Design Library

Open and obtain components from the SolidWorks Design Library.

121) Click the **Design Library** 🗂 tab from the Task Pane as illustrated. The Design Library menu is displayed in the Graphics window.

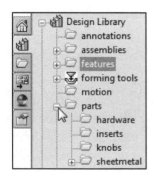

Pin the Design Library (Auto Show) to remain open.
122) Click **Pin** 🗔. **Expand** Design Library.

Select the Flange bolt from the parts folder.
123) Expand the parts folder.

124) Double-click the **hardware** folder. The hardware components are displayed. The flange bolt icon represents a family of similar shaped components in various configurations.

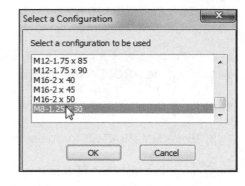

Add the first flange bolt to the assembly.
125) Click and drag the **flange bolt** icon to the right of the GUIDE-ROD assembly in the Graphics window.

126) Release the **mouse button**.

127) Select the 8mm flange bolt, **M8-1.25 x 30** from the drop down menu.

128) Click **OK** from the Select a Configuration dialog box.

Insert the second flange bolt.
129) Click a **position** to the left of the GUIDE-ROD assembly.

130) Click **Cancel** ✖ from the Insert Component PropertyManager. Two flange bolts are displayed in the Graphics window.

Rotate the flange bolts.
131) Click the **shaft** of the first flange bolt.

132) Click the **Rotate Component** ⑤ Assembly tool. The Rotate Component PropertyManager is displayed.

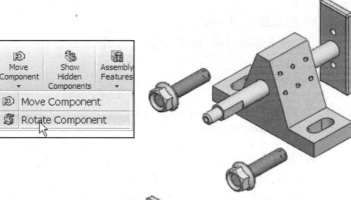

133) Click and drag the **shaft** of the first flange bolt in a vertical direction as illustrated. The flange bolt rotates.

134) Click and drag the **shaft** of the second flange bolt in a vertical direction.

135) Click **OK** ✔ from the Rotate Component PropertyManager.

Add a new file location to the Design Library.

136) Click the **Add File Location** 🗎 tool from the Design Library. The Choose Folder dialog box is displayed.

137) Click the **drop-down arrow** from the Look in box.

138) Select the **PROJECTS** folder.

139) Click **OK** from the Choose Folder dialog box. The PROJECTS folder is added to the Design Library.

140) Click the **PROJECTS** folder to display your parts and assemblies.

Add a new folder to the Design Library Parts.
141) Click the **PROJECTS** folder in the Design Library.

142) Click **Create New Folder**.

143) Enter **MY-PLATES** for folder name.

144) Click the **MY-PLATES** folder. The folder is empty.

Un-pin the Design Library.
145) Click **Pin** 📌.

Save the GUIDE-ROD assembly.
146) Click **Isometric view** 🟦 from the Heads-up View toolbar.

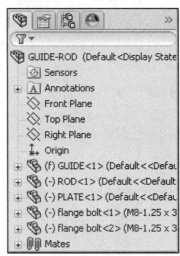

147) Click **Save** 💾.

148) Click **Yes** to rebuild. The two flange bolts are displayed in the FeatureManager.

Caution: Do not drag the PLATE part into the MY-PLATES folder at this time. The GUIDE-ROD assembly references the PLATE in the PROJECTS folder. Insert parts into the Design Library when the assembly is closed. This action is left as an exercise.

💡 There are files of type part (*.sldprt) and Library feature part (*.sldlfp). The flange bolt is a Library feature part. The PLATE is a part. The PLATE cannot be saved as a Library feature part because it contains a Hole Wizard feature. Save the PLATE as a part (*.sldprt) when the assembly is not opened.

GUIDE-ROD Assembly-Insert Mates for Flange bolts

Insert a Concentric mate for the first flange bolt.

149) Click the **Mate** ✎ Assembly tool. The Mate PropertyManager is displayed.

150) Click the first flange bolt **cylindrical face** as illustrated.

151) Click the **right Slot Cut back radial face**. The selected faces are displayed in the Mate Selections box. Concentric is selected by default.

152) Click the **Green Check mark** ✅ from the Mate Pop-up toolbar.

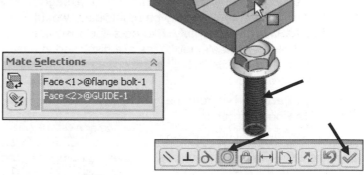

💡 Later, delete the Concentric mate and center the Flange to the GUIDE using Temporary Axes.

Insert a Coincident mate for the first flange bolt.

153) Rotate the view. Press the **up arrow key** until you display the flat bottom face of the bolt.

154) Click the flange bolt **flat bottom face.**

155) Rotate the view.

156) Click the **GUIDE top right face** as illustrated. The selected faces are displayed in the Mate Selections box. Coincident is selected by default in the Mate Pop-up toolbar.

157) Click the **Green Check mark** ✅ from the Mate Pop-up toolbar.

158) Click **Isometric view** 🔲 from the Heads-up View toolbar. The first flange bolt is free to rotate about its centerline in the Slot Cut.

Insert a Parallel mate between two faces.

159) Click the **front face** of the hex head as illustrated.

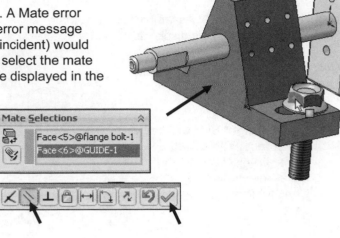

160) Click the **front face** of the GUIDE. A Mate error message is displayed. The Mate error message states, "The default mate type (coincident) would over-define the assembly. Please select the mate type below. The selected faces are displayed in the Mate Selections box.

161) Click **Parallel** ⬉ from the Mate Pop-up toolbar.

162) Click the **Green Check mark** ✔ from the Mate Pop-up toolbar.

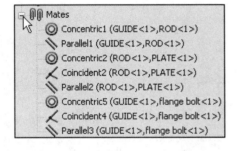

163) Click **OK** ✔ from the Mate PropertyManager. The first FLANGE BOLT is fully defined.

View the created Mates.

164) **Expand** the Mates folder in the FeatureManager. View the created mates.

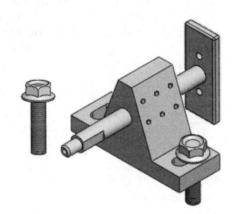

☀ Display the Mates in the FeatureManager to check that the components and the Mate types correspond to your design intent.

☀ If you delete a Mate and then recreate it, the Mate numbers will be different (increase).

Insert a Concentric mate for the second flange bolt.

165) Click **Mate** 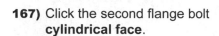 from the Assembly toolbar. The Mate PropertyManager is displayed.

166) Rotate the view. Press the **right arrow key** until you display the left Slot Cut back radial face.

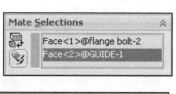

167) Click the second flange bolt **cylindrical face**.

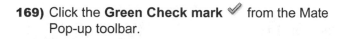

168) Click the **left Slot Cut back radial face**. The selected faces are displayed in the Mate Selections box. Concentric mate is selected by default.

169) Click the **Green Check mark** ✓ from the Mate Pop-up toolbar.

Insert a Coincident mate.
170) Rotate the view. Press the **arrow key** until you display the flat bottom face of the bolt.

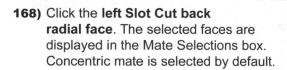

171) Click the flange bolt **flat bottom face**.

172) Rotate the view.

173) Click the **GUIDE top left face**. The selected faces are displayed in the Mate Selections box. Coincident is selected by default.

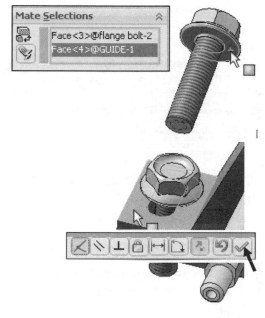

174) Click the **Green Check mark** ✓ from the Mate Pop-up toolbar.

175) Click **Isometric view** from the Heads-up View toolbar. View the results.

Later, delete the Concentric mate and center the first Flange to the GUIDE using Temporary Axes. Then delete the second Flange and apply a Linear Component Pattern tool to create the second Flange from the First Flange.

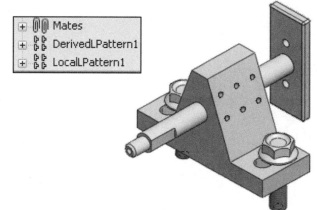

Insert a Parallel mate between two faces.

176) Press the **direction arrows** to display the front face of the left flange bolt.

177) Click the **front face** of the left flange bolt hex head.

178) Click the **front face** of the GUIDE. A Mate error message is displayed. The Mate error message states, "The default mate type (coincident) would over-define the assembly. Please select the mate type below

179) Click **Parallel** ⟍ from the Mate Pop-up toolbar.

180) Click the **Green Check mark** ✔ from the Mate Pop-up toolbar.

181) Click **OK** ✔ from the Mate PropertyManager. The second flange bolt is fully defined.

182) Click **Isometric view** ▽ from the Heads-up View toolbar.

Save the GUIDE-ROD assembly.

183) Click **Save** 🖫.

🔅 Copy components directly in the Graphics window. Hold the Ctrl key down. Click and drag the component from the FeatureManager directly into the Graphic window to create a new instance (copy). Release the Ctrl key. Release the mouse button.

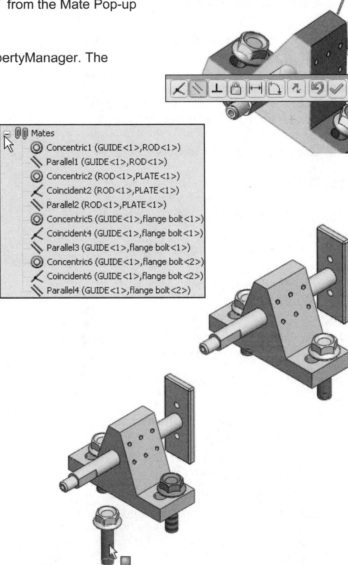

Socket Head Cap Screw Part

The PLATE mounts to the PISTON PLATE of the GUIDE CYLINDER assembly with two M4x0.7 Socket Head Cap Screws. Create a simplified version of the 4MMCAPSCREW based on the ANSI B 18.3.1M-1986 standard.

How do you determine the overall length of the 4MMCAPSCREW? Answer: The depth of the PLATE plus the required blind depth of the PISTON PLATE provided by the manufacturer.

When using fasteners to connect two plates, a design rule of thumb is to use a minimum of 75% to 85% of the second plate's blind depth. Select a common overall length available from your supplier.

The 4MMCAPSSCREW is created from three features. The Base feature is a Revolved feature. The Revolved Base ⊕ feature creates the head and shaft of the 4MMCAPSCREW. The Chamfer 🗋 feature inserts two end cuts. The Extruded Cut 🔲 feature utilizes a hex profile. Utilize the Polygon Sketch ⊕ tool to create the hexagon.

Activity: Socket Head Cap Screw-4MMCAPSCREW Part

Create the 4MMCAPSCREW.
184) Click **New** 🗋 from the Menu bar.

185) Click the **MY-TEMPLATES** tab from the New SolidWorks Document dialog box.

186) Double-click **PART-MM-ANSI**. The Part FeatureManager is displayed.

Save and name the part.
187) Click **Save** 🖫.

188) Select the **ENGDESIGN-W-SOLIDWORKS\PROJECTS** folder.

189) Enter **4MMCAPSCREW** for File name.

190) Enter **CAP SCREW, 4MM** for Description.

191) Click **Save**. The 4MMCAPSCREW FeatureManager is displayed.

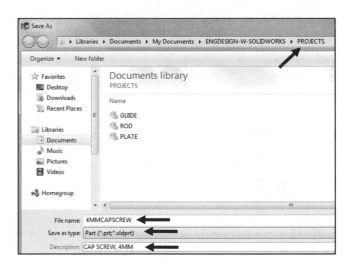

Create the Base sketch on the Front Plane.
192) Right-click **Front Plane** from the FeatureManager.

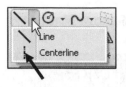

193) Click **Sketch** ✎ from the Context toolbar.

194) Click the **Centerline** ┊ Sketch tool. The Insert Line PropertyManager is displayed.

195) Click **Front view** ⬛ from the Heads-up View toolbar.

Sketch a vertical centerline.

196) Click the **Origin** ↳ in the Graphics window. The Origin is coincident is with the first point.

197) Click a **position** directly above the Origin as illustrated.

Sketch the illustrated profile of the 4MMCAPSCREW using the Line Sketch tool.
198) Click the **Line** ＼ Sketch tool. The Insert Line PropertyManager is displayed.
Note: You can right-click the Line sketch tool from the Sketch Entities box.

199) Click the **Origin** ↳.

200) Click a **position** to the right of the Origin to create a horizontal line.

201) Sketch the first **vertical line** to the right of the centerline.

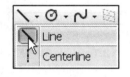

202) Sketch the second **horizontal line**.

203) Sketch the second **vertical line**.

204) Sketch the third **horizontal line**. The endpoint of the line is Coincident with the centerline. The centerline extends above the third horizontal line. The Sketch profile is closed.

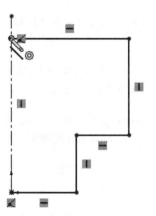

Deselect the Line Sketch tool.
205) Right-click **Select** in the Graphics window to deselect the Line Sketch tool.

A diameter dimension for the revolved sketch requires a centerline, profile line and a dimension position to the left of the centerline. A dimension position directly below the bottom horizontal line creates a radial dimension.

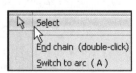

🔆 Insert smaller dimensions first, then larger dimensions to maintain the shape of the sketch profile.

Fit the sketch to the Graphics window.
206) Press the **f** key.

Insert dimensions on the Base Sketch.
207) Click the **Smart Dimension** ✎ from the Sketch tool.

Add a bottom diameter dimension.
208) Click the **centerline**.

209) Click the **first vertical line**.

210) Click a **position** below and to the left of the Origin to create
a diameter dimension.

211) Enter **4**mm.

212) Click the **Green Check mark** ✅ from the Modify dialog box.

Add a vertical dimension.
213) Click the **second vertical line**.

214) Click a **position** to the right of the profile.

215) Enter **4**mm.

216) Click the **Green Check mark** ✅ from the Modify dialog box.

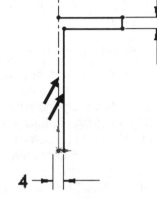

Add a top diameter dimension.
217) Click the **centerline**.

218) Click the **second vertical line**.

219) Click a **position** to the left of the Origin and above the second
horizontal line to create a diameter dimension.

220) Enter **7**mm.

221) Click the **Green Check mark** ✅ from the Modify dialog box.

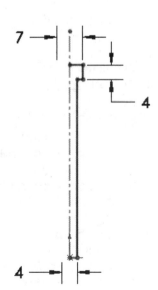

Create an overall vertical dimension.
222) Click the **top horizontal line**.

223) Click the **Origin**.

224) Click a **position** to the right of the profile.

225) Enter **14**mm. Click the **Green Check mark** 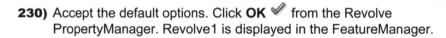.

Deselect the Smart Dimension Sketch tool.
226) Right-click **Select**.

227) Click the **centerline** in the Graphics window for axis for revolution. The Line Properties PropertyManager is displayed.

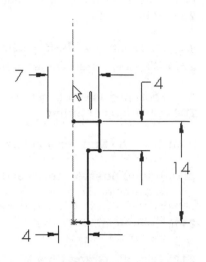

Create the first feature. Insert a Revolved Base feature.

228) Click **Revolved Boss/Base** from the Features toolbar.

229) Click **Yes** to the question, "The sketch is currently open. A non-thin revolution feature requires a closed sketch. Would you like the sketch to be automatically closed?" The Revolve PropertyManager is displayed

Note: The "Yes" button causes a vertical line to be automatically sketched from the top left point to the Origin. The Graphics window displays a preview of the Revolved Base feature.

230) Accept the default options. Click **OK** from the Revolve PropertyManager. Revolve1 is displayed in the FeatureManager.

Fit the model to the Graphics window.
231) Press the **f** key.

Insert a Chamfer feature.
232) Click the **Chamfer** feature tool. The Chamfer PropertyManager is displayed.

233) Click the **top circular edge** of the 4MMCAPSCREW head as illustrated.

234) Click the **bottom circular edge** of the 4MMCAPSCREW shaft as illustrated. The selected entities are displayed in the Chamfer Parameters box.

235) Enter **0.40**mm in the Distance box. Accept the default settings.

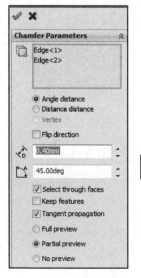

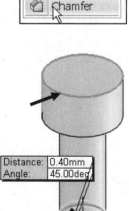

236) Click **OK** ✅ from the Chamfer PropertyManager. Chamfer1 is displayed in the FeatureManager.

Save the 4MMCAPSCREW part.
237) Click **Save** 💾.

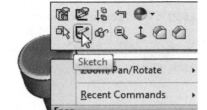

💡 Angle distance values for the Chamfer feature can be entered directly in the Pop-up box | Distance: 0.40mm | Angle: 45.00deg | from the Graphics window.

Create the Hex Extruded Cut feature.
238) Rotate the view. Rotate the **middle mouse button** to display the top circular face of 4MMCAPASCREW.

Insert a sketch.
239) Right-click the **top circular face** of Revolve1 for the Sketch plane as illustrated. Click **Sketch** 💳 from the Context toolbar.

240) Click **Top view** 🔲 from the Heads-up View toolbar.

Sketch a hexagon.
241) Click the **Polygon** ⬡ Sketch tool. The Polygon PropertyManager is displayed.

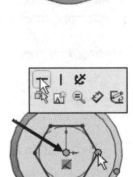

242) Click the **Origin** ↳. Click a **position** to the right as illustrated.

Insert a Horizontal relation.
243) Right-click **Select** to deselect the Polygon Sketch tool. Click the **Origin** ↳. Hold the **Ctrl** key down.

244) Click the **right point** of the hexagon. Release the **Ctrl** key. The Properties PropertyManager is displayed.

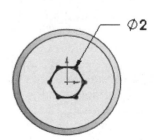

245) Right-click **Make Horizontal** ▬ from the Pop-up Context toolbar.

246) Click **OK** ✅ from the Properties PropertyManager.

Add a dimension.
247) Click the **Smart Dimension** 🖉 Sketch tool.

248) Click the **inscribed circle**. Click a position **diagonally** to the right of the profile.

249) Enter **2mm**. Click the **Green Check mark** ✅.

Insert an Extruded Cut feature.

250) Click the **Extruded Cut** 📦 feature tool. The Cut-Extrude PropertyManager is displayed.

251) Enter **4**mm for the Depth. Accept the default settings. Click **OK** ✓ from the Cut-Extrude PropertyManager. Cut-Extrude1 is displayed in the FeatureManager.

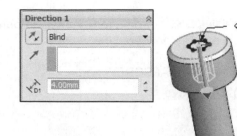

252) Click **Isometric view** ◈ from the Heads-up View toolbar.

Save the 4MMCAPSCREW.
253) Click **Save** 💾.

SmartMates

A SmartMate is a Mate that automatically occurs when a component is placed into an assembly. The mouse pointer displays a SmartMate feedback symbol when common geometry and relationships exist between the component and the assembly.

SmartMates are Concentric or Coincident. A Concentric SmartMate assumes that the geometry on the component has the same center as the geometry on an assembled reference. A Coincident Plane SmartMate assumes that a plane on the component lies along a plane on the assembly. As the component is dragged into place, the mouse pointer provides feedback such as:

Mating Entities:	*Type of Mate*:	*Icon Feedback*:
• Two linear edges	Coincident	
• Two planar faces	Coincident	
• Two vertices	Coincident	
• Two conical faces	Concentric	
• Two circular edges	Coincident / Concentric	

Coincident/Concentric SmartMate

The most common SmartMate between a screw/bolt and a hole is the Coincident/Concentric SmartMate. The following technique utilizes two windows. The first window contains the 4MMCAPSCREW part and the second window contains the GUIDE-ROD assembly. Zoom in on both windows to view the Mate reference geometry. Drag the part by the shoulder edge into the assembly window. View the mouse pointer for Coincident/Concentric feedback . Release the mouse pointer on the circular edge of the PLATE Mounting Hole.

Activity: Coincident/Concentric SmartMate

Display the 4MMCAPSCREW part and the GUIDE-ROD assembly.
254) Click **Window**, **Tile Horizontally** from the Menu bar.

255) **Zoom in** on the PLATE to view the Mounting Holes.

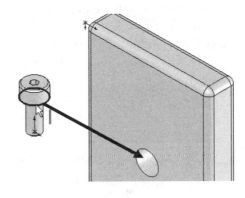

Insert the first 4MMCAPSCREW.
256) Click and drag the **circular edge** of the 4MMCAPSCREW part into the GUIDE-ROD assembly Graphics window.

257) Release the mouse pointer on the **top circular edge** of the PLATE. The mouse pointer displays the Coincident/ Concentric circular edges feedback icon.

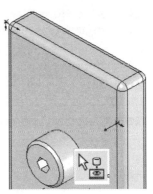

Insert the second 4MMCAPSCREW.
258) Click and drag the **circular edge** of the 4MMCAPSCREW part into the GUIDE-ROD assembly Graphics window.

259) Release the mouse pointer on the **bottom circular edge** of the PLATE. The mouse pointer displays the Coincident/ Concentric circular edges feedback icon .

260) **Maximize** the GUIDE-ROD assembly.

Fit the GUIDE-ROD assembly to the Graphics window.
261) Press the **f** key.

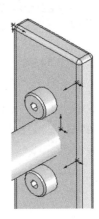

Zoom in before dragging a component into the assembly to select the correct circular edge for a Coincident/Concentric Mate.

The circular edge of the 4MMCAPSCREW produces both the Coincident/Concentric Mate. The cylindrical face of the 4MMCAPSCREW produces only a Concentric Mate. View the mouse pointer for the correct feedback.

Insert a Parallel mate for the first 4MMCAPSCREW.

262) Click the **Mate** Assembly tool. The Mate PropertyManager is displayed.

263) Expand the GUIDE-ROD GUIDE-ROD icon from the fly-out FeatureManager.

264) Click **Right Plane** of the 4MMCAPSCREW<1> in the fly-out FeatureManager as illustrated.

265) Click **Right Plane** of the 4MMCAPSCREW<2> in the fly-out FeatureManager. The selected planes are displayed in the Mate Selections box.

266) Click **Parallel** from the Mate Pop-up toolbar.

267) Click the **Green Check mark** from the Mate dialog box.

268) Click **OK** from the Mate PropertyManager.

Fit the assembly to the Graphics window.
269) Press the **f** key.

270) Click **Isometric view** from the Heads-up View toolbar

Save GUIDE-ROD assembly.
271) Click **Save** .

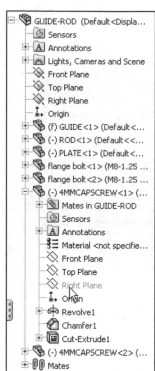

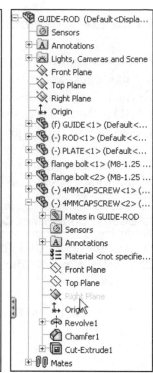

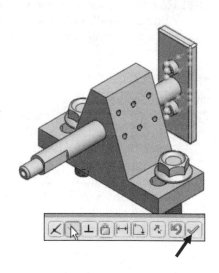

The ROD, PLATE and 4MMCAPSCREWS are free to translate along the z-axis. Their component status remains under defined, (-) in the GUIDE-ROD FeatureManager.

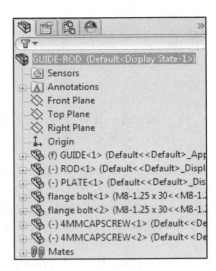

The GUIDE is fixed (f) to the assembly Origin. The flange bolts are fully defined since they are mated only to the GUIDE.

☀ Understand how Mates reflect the physical behavior in an assembly. Move and rotate components to test the mate behavior. The correct mate selection minimizes rebuild time and errors in the assembly.

Tolerance and Fit

The ROD travels through the GUIDE in the GUIDE-ROD assembly. The shaft diameter of the ROD is 10mm. The hole diameter in the GUIDE is 10mm. A 10mm ROD cannot be inserted into a 10mm GUIDE hole without great difficulty! (Press Fit)

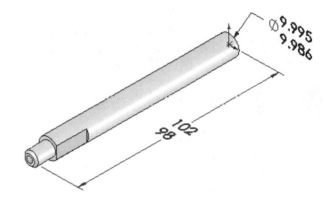

Note: The 10mm dimension is the nominal dimension. The nominal dimension is approximately the size of a feature that corresponds to a common fraction or whole number.

Tolerance is the difference between the maximum and minimum variation (Limits) of a nominal dimension and the actual manufactured dimension.

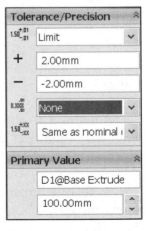

Example: A ROD has a nominal dimension of 100mm with a tolerance of ± 2mm, (100mm ± 2mm). This translates to a part with a possible manufactured dimension range between 98mm (lower limit) to 102mm (Upper limit). The total ROD tolerance is 4mm.

Note: Design rule of thumb: Design with the maximum permissible tolerance. Tolerance flexibility saves in manufacturing time and cost.

The assembled relationship between the ROD and the GUIDE is called the fit. The fit is defined as the tightness or looseness between two components. This project discusses three major types of fits:

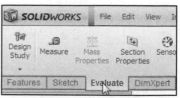

- **Clearance fit** - The shaft diameter is less than the hole diameter. See SolidWorks Help for additional information.

- **Interference fit** - The shaft diameter is larger than the hole diameter. The difference between the shaft diameter and the hole diameter is called interference. See SolidWorks Help for additional information.

- **Transition fit** - Clearance or interference can exist between the shaft and the hole. See SolidWorks Help for additional information.

You require a Clearance fit between the shaft of the ROD and the Guide Hole of the GUIDE. There are multiple categories for Clearance fits. Dimension the GUIDE hole and ROD shaft for a Sliding Clearance fit. All below dimensions are in millimeters.

Use the following values:

Hole	Maximum	10.015mm.
	Minimum	10.000mm.
Shaft	Maximum	9.995mm.
	Minimum	9.986mm.
Fit	Maximum	10.015 - 9.986 = .029 Max. Hole - Min. Shaft.
	Minimum	10.000 - 9.995 = .005 Min. Hole - Max. Shaft.

Calculate the maximum variation:

Hole Max: 10.015 - Hole Min: 10.000 = .015 Hole Max. Variation.

Select features from the FeatureManager and the Graphics window. In the next activity, locate feature dimensions with the FeatureManager for the GUIDE.

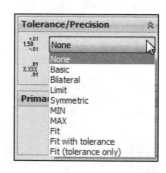

Locate feature dimensions in the Graphics window for the ROD. Select the dimension text; then apply the Tolerance/Precision through options in the Dimension PropertyManager.

Activity: Tolerance and Fit

Locate the dimension in the FeatureManager.
272) Expand GUIDE from the GUIDE-ROD FeatureManager.

273) Double-click **Guide Hole** from the FeatureManager to display the dimensions.

Add the maximum and minimum Guide Hole dimensions.
274) Click the Ø**10** dimension. The Dimension PropertyManager is displayed.

275) Select **Limit** from the Tolerance/Precision box.

276) Select **.123** for three place Precision

277) Enter **0.015**mm for Maximum Variation.

278) Enter **0.000**mm for Minimum Variation.

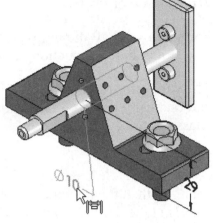

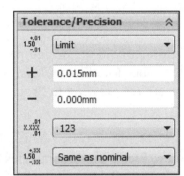

279) Click **OK** from the Dimension PropertyManager.

Add maximum and minimum Shaft dimensions.
280) Double-click the **Base-Extrude** (**ROD**) component from the Graphics window.

281) Click the Ø**10** diameter dimension.

282) Select **Limit** from the Tolerance/Precision box.

283) Select **.123** for three place Precision

284) Enter **-0.005**mm for Maximum Variation.

285) Enter **-0.014**mm for Minimum Variation.

286) Click **OK** from the Dimension PropertyManager.

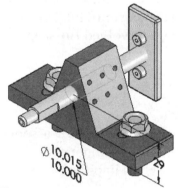

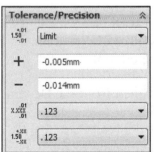

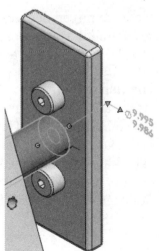

Save the GUIDE-ROD assembly.
287) Click **Save** .

288) Click **Save All**.

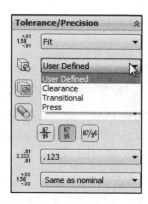

ISO symbol Hole/Shaft Classification is applied to an individual dimension for Fit, Fit with tolerance, or Fit (tolerance only) types. Classification can be: *User Defined*, *Clearance*, *Transitional* or *Press*.

For a hole or shaft dimension, select a classification from the list. The Hole/Shaft designation for a Sliding Fit is H7/g6.

The values for Maximum and Minimum tolerances are calculated automatically based on the diameter of the Hole/Shaft and the Fit Classification.

Utilize Hole/Shaft Classification early in the design process. If the dimension changes, then the tolerance updates. The Hole/Shaft Classification propagates to the details in the drawing. You will create the drawing in Project 3. See SolidWorks Help for additional information.

Additional details on Tolerance, Precision, SmartMates, Revolved Feature, and Design Library are available in Help, SolidWorks Help Topics. Keywords: Tolerances (Dimension), Fit Tolerance, SmartMates, Feature Based Mates, Geometry Based Mates and Revolved Boss/Base.

Review of the GUIDE-ROD Assembly

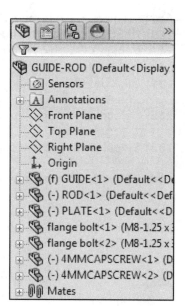

The GUIDE-ROD assembly combined the GUIDE, ROD, and PLATE components. The GUIDE was the first component inserted into the GUIDE-ROD assembly.

Mates removed degrees of freedom. Concentric, Coincident and Parallel Mates were utilized to position the ROD and PLATE with respect to the GUIDE.

The flange bolts were obtained from the Design Library. You utilize a Revolved Base feature to create the 4MMCAPSCREW. The Revolved Base feature contained an axis, sketched profile and an angle of revolution. The Polygon Sketch tool was utilized to create the hexagon Extruded Cut feature. The 4MMCAPSCREW utilized the Concentric/Coincident SmartMate option.

Exploded View

The Exploded View illustrates how to assemble the components in an assembly. Create an Exploded View with four steps in the ROD-GUIDE assembly. Click and drag components in the Graphics window. The Manipulator icon

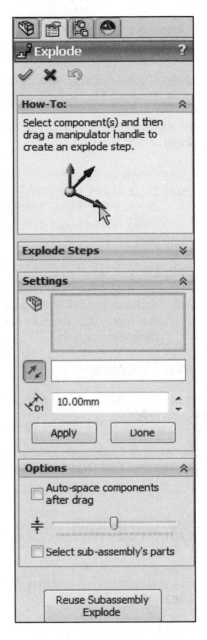

indicates the direction to explode. Select an alternate component edge for the Explode direction. Drag the component in the Graphics window or enter an exact value in the Explode distance box. In this activity, manipulate the top-level components in the assembly.

In the project exercises, create exploded views for each sub-assembly and utilize the Re-use sub-assembly explode option in the top level assembly.

Access the Explode view option as follows:

- Right-click the configuration name in the ConfigurationManager.

- Select the Exploded View tool in the Assembly toolbar.

- Select Insert, Exploded View from the Menu bar.

The Exploded View feature uses the Explode PropertyManager as illustrated.

Activity: GUIDE-ROD Assembly-Exploded View

Insert an Exploded view in the Assembly

289) Click the **Exploded View** 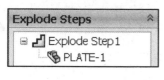 Assembly tool. The Explode PropertyManager is displayed.

Fit the model to the Graphics window.
290) Press the **f** key.

Create Explode Step 1.
291) Click the **PLATE** component in the Graphics window. The selected entity is displayed in the Settings box.

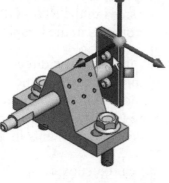

292) Enter **100**mm in the Explode distance box. The direction of the explode view is towards the back. If required, click the **Reverse direction** button.

293) Click **Apply**.

294) Click **Done**. Explode Step1 is created.

If the exact Explode distance value is not required, click and drag the Manipulator handle as illustrated in the next step.

Create Explode Step2.
295) Click the **ROD** component from the Graphics window.

Fit the model to the Graphics window.
296) Press the **f** key.

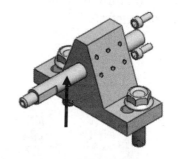

297) Drag the **blue/yellow manipulator handle** backward to position between the PLATE and the GUIDE as illustrated.

298) Click **Done** from the Settings box. Explode Step2 is created.

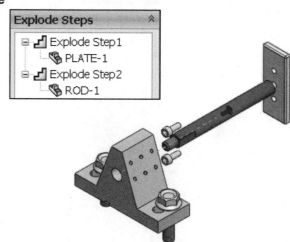

Create Explode Step3.
299) Click the **left flange bolt** from the Graphics window.

300) Drag the **vertical manipulator**

handle  upward above the GUIDE.

301) Click **Done** from the Settings box. Explode Step3 is created.

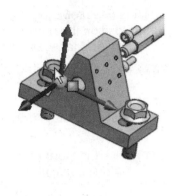

Create Explode Step4.
302) Click the **right flange bolt** from the Graphics window.

303) Drag the **vertical green manipulator**

handle 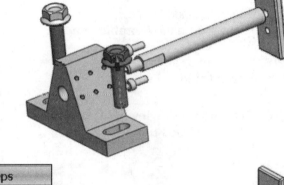 upward above the GUIDE.

304) Click **Done** from the Settings box. Explode Step4 is created.

305) Click **OK** ✔ from the Explode PropertyManager.

306) Click **Isometric view** ⬛ from the Heads-up View toolbar.

307) Click **Save** 💾.

Explode Steps
- Explode Step 1
  - PLATE-1
- Explode Step 2
  - ROD-1
- Explode Step 3
  - flange bolt-2
- Explode Step 4
  - flange bolt-1

Split the FeatureManager to view the Exploded Steps.

308) Position the **mouse pointer** at the top of the FeatureManager. The mouse pointer displays the Split bar ⊹.

309) Drag the **Split bar** half way down to display two FeatureManager windows.

310) Click the **ConfigurationManager** 🔠 tab to display the Default configuration in the lower window.

311) **Expand** Default <Display State-1>.

312) **Expand** ExplView1 to display the four Exploded Steps.

Fit the Exploded view to the Graphics window.
313) Press the **f** key.

Remove the Exploded state.
314) Right-click in the **Graphic window**.

315) Click **Collapse** from the Pop-up menu.

316) Click **Isometric view** 🔲 from the Heads-up View toolbar.

Animate the Exploded view.
317) Right-click **ExplView1** from the ConfigurationManager.

318) Click **Animate explode**. The Animation Controller dialog box is displayed. Play ▷ is selected by default. View the animation.

319) Click **Stop** ■ to end the animation.

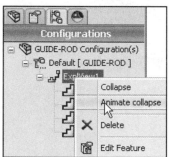

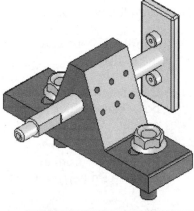

Close the Animation Controller.

320) Close ⊠ the Animation Controller. The GUIDE-ROD is in the Exploded state.

321) Right-click **Collapse** in the GUIDE-ROD Graphics window.

Display the FeatureManager.

322) Return to a single FeatureManager. Drag the **Split bar** upward to display one FeatureManager window.

323) Click the **Assembly FeatureManager** 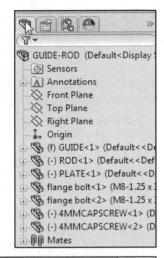 tab.

Save the GUIDE-ROD assembly.

324) Click **Save** 🖫.

Click the Motion Study tab at the bottom of the Graphics window. Select Basic Motion from the MotionManager. Click the Animation Wizard to create a simple animation. Click Play to view the Animation Wizard. Click the Model tab at the bottom of the Graphics window to return to the SolidWorks screen.

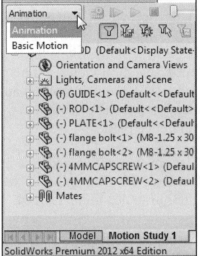

Note: The SolidWorks Animator application is required to record an AVI file through the Animation Controller. Play the animation files through the Windows Media Player. The time required to create the animation file depends on the number of components in the assembly and the options selected.

Stop the animation before closing the Animation Controller toolbar to avoid issues. Utilize SolidWorks Animator for additional control over the Explode/Collapse motion in the assembly.

Create the Exploded steps in the order that you would disassemble the assembly. Collapse the assembly to return to the original assembled position. Your animations will appear more realistic based on the order of the Explode steps.

Review of the GUIDE-ROD assembly Exploded View

You created an Exploded View in the GUIDE-ROD assembly. The Exploded View displayed the assembly with its components separated from one another.

The Exploded View animation illustrated how to assemble and disassemble the GUIDE-ROD assembly through the collapse and explode states.

Section View

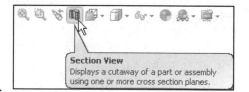

A Section view  displays the internal cross section of a component or assembly. The Section view dissects a model like a knife slicing through a stick of butter. Section views can be performed anywhere in a model. The location of the cut corresponds to the Section plane. A Section plane is a planar face or reference plane.

Use a Section view to determine the interference between the 10mm Guide Hole and the Linear Pattern of 3mm holes in the GUIDE.

Activity: GUIDE-Section View

Open the GUIDE part.
325) Right-click the **GUIDE** component in the Graphics window.

326) Click **Open Part** ➘ from the shortcut toolbar. The GUIDE part is displayed in the Graphics window.

Insert a Section View on the Front Plane of the GUIDE.
327) Click **Front Plane** from the GUIDE FeatureManager.

328) Click **Section View** from the Heads-up View toolbar in the Graphics window. The Section View PropertyManager is displayed.

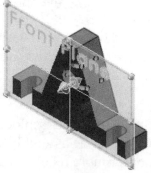

Save the Section View.

329) Click **Save** from the Section View PropertyManager. The Save As dialog box is displayed. SectionView is the default view name.

330) Click **Save**.

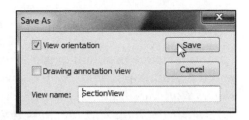

Click the drop-down arrow from the View Orientation tool as illustrated. The SectionView1 is saved as a custom view.

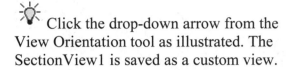
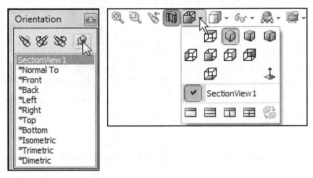

Display and Pin the View Orientation dialog box.

331) Press the **Space** bar. View the Orientation box with the new view.

332) Close ✕ the Orientation view dialog box.

333) Return to a full standard view. Click **Front view** from the Heads-up View toolbar. If needed, click Section view from the Heads-up View toolbar to return to a full view.

334) Click **Hidden Lines Visible** from the Heads-up View toolbar.

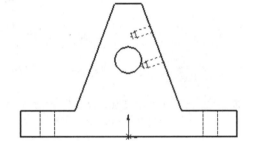

Save the GUIDE part.

335) Click **Save**.

The Section view tool detects potential problems before manufacturing. The Section view tool determines the interference between the 10mm Guide hole and the Linear Pattern of 3mm holes in the GUIDE.

The GUIDE, ROD and PLATE components all share a common file structure with the GUIDE-ROD assembly. What component do you modify? How will changes affect other components? Let's analyze the interference problem and determine a solution.

Analyze an Interference Problem

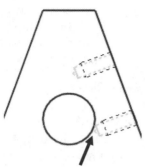

An interference problem exists between the 10mm Guide Hole and the 3mm tapped holes. Review your design options: 1.) *Reposition the Guide Hole*, 2.) *Modify the size of the Guide Hole*, 3.) *Adjust the length of the 3mm holes* or 4.) *Reposition the Linear Pattern feature*.

The first three options affect other components in the assembly. The GUIDE-CYLINDER assembly and PLATE determine the Guide Hole location. The ROD diameter determines the size of the Guide Hole. The position sensor requires the current depth of the 3mm Holes. Reposition the Linear Pattern feature by modifying the first 3mm Thru Hole.

Activity: GUIDE-Analyze an Interference Problem

Modify the Thru Hole dimensions.
336) Expand the M3x0.5 Tapped Hole1 feature from the FeatureManager.

337) Click **Isometric view** from the Heads-up View toolbar.

338) Click **Shaded With Edges** from the Heads-up View toolbar.

339) Double-click **Sketch5** from the FeatureManager to display the position dimensions.

340) Double-click the **4**mm dimension created from the temporary axis of the Guide Hole.

341) Enter **6**mm.

342) Rebuild the model.

343) Click the **Green Check mark** from the Modfiy dialog box.

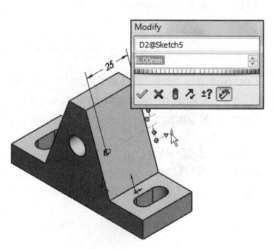

Close the GUIDE part.
344) Click **File**, **Close** from the Menu bar.

345) Click **Yes** to Save changes to the GUIDE.

346) Click **Yes** to rebuild.

Display the Front view.
347) Click **Wireframe** from the Heads-up View toolbar.

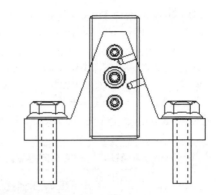

348) Click **Front view** . The GUIDE-ROD assembly
updates to display the changes to the GUIDE part.

Return to an Isometric view.

349) Click **Isometric view** .

350) Click **Shaded With Edges** .

Save the GUIDE-ROD assembly.

351) Click **Save** .

SolidWorks provides an
Interference Detection tool and a
Clearance Verification tool to analyze
assemblies. See SolidWorks help for
additional information.

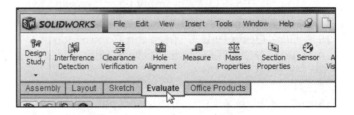

Analyze issues at the part level first. Working at the part level reduces rebuild time
and complexity. Return to the assembly and review the modifications.

Additional details on Explode, Collapse and Section View are available in
SolidWorks Help. Keywords: Exploded View (Collapse, Assemblies) and Section View.

The last component to insert to the GUIDE-CYLINDER is the 3MMCAPSCREW.
Utilize the Save As Copy option to copy the 4MMCAPSCREW to the
3MMCAPSCREW part.

Save As Copy Option

Conserve design time and cost. Modify existing parts and
assemblies to create new parts and assemblies. Utilize the Save
as copy option to avoid updating the existing assemblies with
new file names.

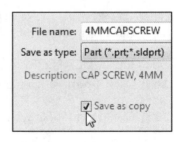

The 4MMCAPSCREW was created earlier. The GUIDE requires
3MMCAPSCREWs to fasten the sensor to the 3mm Tapped Hole Linear
Pattern.

Start with the 4MMCAPSCREW. Utilize the Save as copy option. Enter the
3MMCAPSCREW for the new file name.

The Save as copy option prevents the 3MMCAPSCREWS from replacing the 4MMCAPSCREWS in the GUIDE-CYLINDER assembly.

Important: Check the Save as copy check box. The Save as copy box check box creates a copy of the current part with no references to existing assemblies that utilize the part. The 3MMCAPSCREW is the new part name. Modify the dimensions of the Revolved Base feature to create the 3MMCAPSCREW.

Activity: 4MMCAPSCREW-Save as Copy Option

Open the 4MMCAPSCREW part.

352) Right-click the **4MMCAPSCREW front face** component in the GUIDE-ROD assembly Graphics window.

353) Click **Open Part** from the Context toolbar. The 4MMCAPSCREW part is displayed.

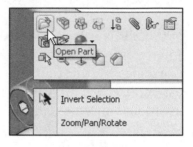

Apply the Save As Copy option.

354) Click **Save As** from the Menu bar.

355) Click **OK** to the warning message that the 4MMCAPSCREW is referenced by open documents.

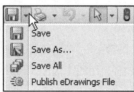

356) Select the **ENGDESIGN-W-SOLIDWORKS\PROJECTS** folder.

357) Check the **Save as copy** box.

358) Enter **3MMCAPSCREW** for File name.

359) Enter **CAP SCREW, 3MM** for Description.

360) Click **Save**. The 4MMCAPSCREW part remains open.

Close the 4MMCAPSCREW part.

361) Click **File**, **Close** from the Menu bar.

Open the 3MMCAPSCREW part.

362) Click **Open** from the Menu bar.

363) Click **Part** for file type in the Projects folder.

364) Double-click **3MMCAPSCREW**. The 3MMCAPSCREW FeatureManager is displayed.

365) Click **Revolve1** in the FeatureManager. The dimensions are displayed.

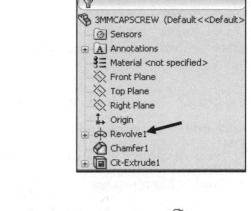

Fit the model to the Graphics window.
366) Press the **f** key.

Modify the Revolve dimensions.
367) Click the vertical dimension **4**mm.

368) Enter **3**mm.

369) Click the depth dimension **14**mm.

370) Enter **9**mm.

371) Click the diameter dimension **4**mm.

372) Enter **3**mm.

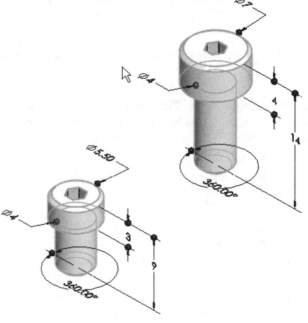

373) Click the diameter dimension **7**mm.

374) Enter **5.5**mm.

Save the 3MMCAPSCREW part.
375) Click **SAVE** 💾.

With the Save as Copy option, the current document remains open. Close the open document. Select the copied document to open and modify.

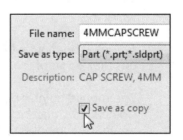

The GUIDE Linear Pattern of 3mm tapped holes requires six 3MMCAPSCREWs. Do you remember the seed feature in the Linear Pattern?

The first 3mm tapped hole is the seed feature. The seed feature is required for a Component Pattern in the GUIDE-ROD assembly.

GUIDE-ROD Assembly-Feature Driven Component Pattern

There are various major methods to define a pattern in an assembly:

- Linear Component Pattern tool

- Circular Component Pattern tool

- Feature Driven Component Pattern tool

- Mirror Components tool

Utilize the Feature Driven Component Pattern tool to create instances of the 3MMCAPSCREW. Insert the 3MMCAPSCREW part into the GUIDE-ROD assembly.

Activity: GUIDE-ROD Assembly-Feature Driven Component Pattern

---

Insert and mate the 3MMCAPSCREW.

376) Click **Window**, **Tile Horizontally** from the Menu bar. The 3MMCAPSCREW and the GUIDE-ROD assembly are displayed.

The 3MMCAPSCREW and the GUIDE-ROD assembly are the open documents. Close all other documents. Select the Close ⊠ icon. Click Window, Tile Horizontally again to display the two open documents.

377) **Zoom in** on the bottom left circular edge of the GUIDE left Tapped Hole.

378) Click and drag the **bottom circular edge** of the 3MMCAPSCREW into the GUIDE-ROD assembly.

379) Release the mouse button on the **bottom left circular edge** of the GUIDE left Tapped Hole.

The mouse pointer displays the Coincident/Concentric Circular edges feedback symbol. The 3MMCAPSCREW part is positioned in the bottom left Tapped Hole.

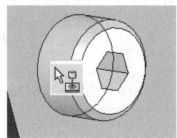

Insert a Feature Derived Component Pattern.
380) Maximize the GUIDE-ROD assembly.

Fit the GUIDE-ROD assembly to the Graphics window.
381) Press the **f** key.

382) Click the **Feature Driven Component Pattern**  tool from the Consolidated Assembly toolbar. The Feature Driven PropertyManager is displayed. 3MMCAPSCREW<1> is displayed in the Components to Pattern box.

383) Click inside the **Driving Feature** box.

384) Expand GUIDE in the GUIDE-ROD fly-out FeatureManager.

385) Click **LPattern1** under GUIDE<1> from the fly-out FeatureManager. LPattern1@GUIDE-1 is displayed in the Driving Feature box.

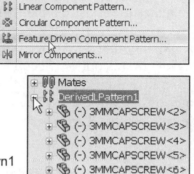

386) Click **OK** ✔ from the Feature Driven PropertyManager. The DerivedLPattern1 is displayed in the FeatureManager.

387) Click **Isometric view** ⬚ from the Heads-up View toolbar.

Save the GUIDE-ROD assembly.
388) Click **Save** 🖫.

Close all models.
389) Click **Window**, **Close All** from the Menu bar.

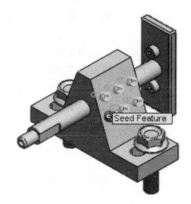

Reuse geometry. Utilize patterns early in the design process. A Linear Pattern feature in the part is utilized as a Feature Driven Component Pattern in the assembly.

Review the Component Pattern feature for the 3MMCAPSCREWs

You utilized the Save as copy option to copy the 4MMCAPSCREW to the 3MMCAPSCREW. The 3MMCAPSCREW was mated to the GUIDE 3MM Tapped Hole. You utilized a Feature Derived Component Pattern tool to create an array of 3MMCAPSCREWs. The Feature Derived Component Pattern feature was derived from the GUIDE Linear Pattern feature of Tapped Holes.

Redefining Mates and Linear Component Pattern Feature

In the modeling process, you modify and edit sketches and features. You are also required to redefine mates. The two flange bolts are not centered with the Slot Cut of the GUIDE. Redefine the mates to reposition the flange bolt at the center of the Slot Cut.

How do you redefine existing mates? Answer: First, review the mates for an individual component with the View Mates option. Second, determine the mates to redefine and the mates to delete. Third, utilize the Edit Feature tool to redefine the mate. No cylindrical face exists at the center of the Slot Cut. Delete the existing Concentric mate and insert a new Coincident mate between the Temporary Axis of the flange bolt and a sketched point in the right Slot Cut.

The original Coincident mate referenced an edge on the GUIDE. When you delete the Concentric mate, this Coincident mate acts like a hinge. Redefine the Coincident mate between the bottom face of the flange bolt and the top face of the GUIDE.

Delete the second flange bolt. Utilize the Linear Component Pattern feature to create an instance of the flange bolt. Additionally, the Linear Component Pattern feature developed in the assembly is called LocalLPattern1. Note: If you delete a mate and then recreate the mate, your instance number will be different in the next activity.

Activity: Redefining Mates and Linear Component Pattern

View the created mates.
390) Open the GUIDE-ROD assembly.

391) Right-click the **flange bolt<1>** component from the Graphics window.

392) Click **View Mates** from the Context toolbar. The pop-up mate dialog box is diplayed. There are three mates. Review their mate references in the Graphics window.

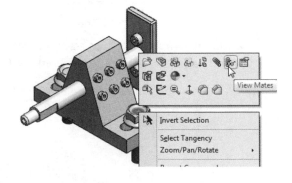

Delete the Concentric mate.

393) Right-click the **Concentric** mate from the dialog box.

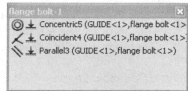

394) Click **Delete**.

395) Click **Yes** to the question: Do you really want to delete this? The flange bolt is free to translate along the right top face of the GUIDE.

396) **Close** the Mate dialog box.

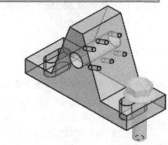

Open the GUIDE part.

397) Right-click the **GUIDE** component from the Graphics window.

398) Click **Open Part** . The GUIDE FeatureManager is displayed.

Display the sketch.

399) **Expand** Slot Cut in the FeatureManager.

400) Right-click **Sketch2**.

401) Click **Show**. View the sketch in the Graphics window. Note: The Show command displays the sketch in the GUIDE part, in order to select the point in the GUIDE-ROD assembly.

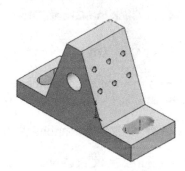

Save the GUIDE part.

402) Click **Isometric view** from the Heads-up View toolbar.

403) Click **Save** .

Open the GUIDE-ROD assembly.

404) Open the **GUIDE-ROD** assembly.

405) Click and drag the **first flange bolt** in front of the right Slot Cut as illustrated.

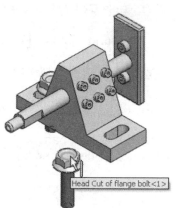

Display the Temporary Axes and Sketches.

406) Click **View**; check **Temporary Axes** from the Menu bar. The Temporary Axes are displayed. If required, click **View**; check **Sketches** from the Menu bar.

Insert a Coincident mate.

407) Click the **Mate** 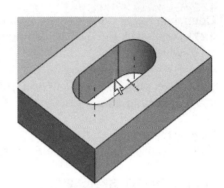 Assembly tool. The Mate PropertyManager is displayed.

408) Clear all Mate Selections.

409) Click the **flange bolt Temporary Axis** as illustrated. Note: Move the mouse pointer over the center of the flange bolt to view the Temporary Axis. The mouse icon displays  .

410) Click the center **Temporary Axis** in the midpoint of the Slot Cut sketch as illustrated. The selected entities are displayed in the Mate Selections box. Coincident mate is selected by default.

411) Click the **Green Check mark** ✓ from the Mate dialog box.

412) Click **OK** ✓ from the Mate PropertyManager.

Hide the reference geometry.

413) Click **View**; uncheck **Temporary Axes** from the Menu bar.

414) Click **View**; uncheck **Sketches** from the Menu bar.

Hide the Slot Cut sketch.

415) Press **Ctrl Tab** to display the GUIDE part.

416) Expand Slot Cut in the FeatureManager.

417) Right-click **Sketch2** in the FeatureManager as illustrated.

418) Click **Hide** 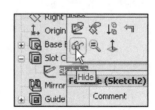 from the Context toolbar.

Close the GUIDE part.

419) Click **File**, **Close** from the Menu bar.

420) Click **Yes** to Save changes to GUIDE.

Delete the second flange bolt.

421) Right-click the **flange bolt <2>** component in the FeatureManager.

422) Click **Delete**. Click **Yes** to confirm.

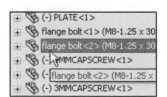

Insert a Linear Component Pattern feature.

423) Click the **right flange bolt** in the Graphics window as illustrated.

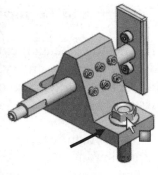

424) Click the **Linear Component Pattern** tool from the Consolidated toolbar. The Linear Pattern PropertyManager is displayed. Flange bolt<1> is displayed in the Components to Pattern box.

425) Click the **front horizontal edge** for Direction 1. The Direction arrow points to the left.

426) Enter **60**mm for Spacing.

427) Enter **2** for Number of Instances

428) Click **OK** from the Linear Pattern PropertyManager. LocalLPattern1 is displayed in the FeatureManager.

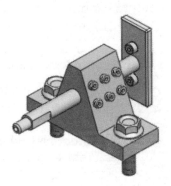

Save the GUIDE-ROD assembly.

429) Click **Isometric view** from the Heads-up View toolbar.

430) Click **Save** .

431) Click **Save All**.

A goal of this book is to expose new users to various SolidWorks design tools, features and methods.

Before you redefine or edit a Mate, save the assembly. One Mate modification can lead to issues in multiple components that are directly related. Understand the Mate Selections syntax. The Mate PropertyManager lists the Mate Type, geometry selected, and component reference (part/assembly and instance number).

The 3mm Tapped Holes utilized the Feature Driven Component Pattern feature. The flange bolts utilized the Linear Component feature. Utilize the Feature Driven Component Pattern feature when a part contains a pattern to reference. Utilize the Linear Component feature when a part does not contain a pattern reference.

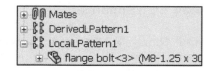

Folders and Suppressed Components

The FeatureManager entries increase as components are inserted into the Graphics window of an assembly. Folders reduce the length of the FeatureManager in the part and assembly. Folders also organize groups of similar components. Organize hardware in the assembly into folders.

Suppress features, parts and assemblies that are not displayed. During model rebuilding, suppressed features and components are not calculated. This saves rebuilding time for complex models. The names of the suppressed features and components are displayed in light gray.

Create a folder in the FeatureManager named Hardware. Drag all individual bolts and screws into the Hardware Folder. The DerivedLPattern1 feature and the LocalLPattern1 feature cannot be dragged into a folder. Suppress the Hardware folder. Suppress the DerivedLPattern1 and LocalLPattern1 feature.

Activity: Folders and Suppressed Components

Create a new folder in the FeatureManager.
432) Right-click **flange bolt<1>** in the GUIDE-ROD FeatureManager.

433) Click **Create New Folder**. Folder1 is displayed in the FeatureManager

434) Enter **Hardware** for folder name.

435) Click and drag **flange bolt<1>** in the FeatureManager into the Hardware folder. The mouse pointer displays the Move Component ⇦ icon.

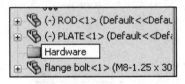

436) Repeat the above process for the **4MMCAPSCREW<1>**, **4MMCAPSCREW<2>** and **3MMCAPSCREW<1>** components. All screws and bolts are located in the Hardware folder.

437) **Expand** the Hardware folder. View the components.

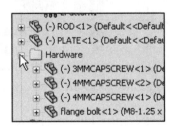

Suppress the Folder and Patterns.
438) Right-click the **Hardware** folder in the FeatureManager.

439) Click **Suppress**.

440) Right-click **DerivedLPattern1** in the FeatureManager.

441) Click **Suppress**.

442) Right-click **LocalLPattern1**.

443) Click **Suppress**.

Display the GUIDE-ROD assembly.

444) Click **Isometric view**  from the Heads-up View toolbar.

Save the GUIDE-ROD assembly.

445) Click **Save** 💾. View the FeatureManager.

Close all parts and assemblies.

446) Click **Window**, **Close All** from the Menu bar.

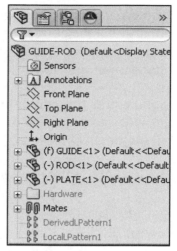

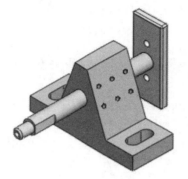

💡 Standardize on folder names such as Hardware or Fillets so colleagues recognize folder names. Place a set of continuous features or components into an individual folder.

Make-Buy Decision-3D ContentCentral

In a make-buy decision process, a decision is made on which parts to manufacture and which parts to purchase.

In assembly modeling, a decision is also made on which parts to design and which parts to obtain from libraries and the World Wide Web. SolidWorks contains a variety of designed parts in their Design Library.

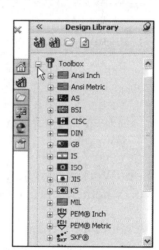

The SolidWorks Toolbox is a library of feature based design automation tools for SolidWorks. The Toolbox uses Window's drag and drop functionality with SmartMates. Fasteners are displayed with full thread details. Un-suppress the correct feature to obtain full thread display in the Graphics window.

💡 Activate the SolidWorks Toolbox. Click **Options**, **Add-Ins** from the Menu bar tool. Check the **SolidWorks Toolbox** box and the **SolidWorks Toolbox Browser** box. Click **OK**.

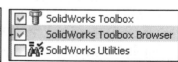

SolidWorks SmartFastener uses the Hole Wizard tool to automatically SmartMate the corresponding Toolbox fasteners in an assembly. The fastener is sized and inserted into the assembly based on the Dimensioning Standard of the Hole.

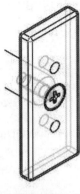

SolidWorks provides a tool know as 3D ContentCentral. 3D ContentCentral provides access to 3D models from component suppliers and individuals in all major CAD formats.

3D ContentCentrol provides the following access:

- 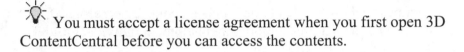 **Supplier Content** : Links to supplier Web sites with certified 3D models. You can search the 3D ContentCentral site, view, configure, and evaluate models online, and download models for your documents.

- **User Library** : Links to models from individuals using 3D PartStream.NET.

You must accept a license agreement when you first open 3D ContentCentral before you can access the contents.

To submit your model to the 3D ContentCentral user library:

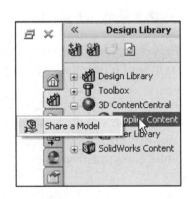

- Right-click the 3D ContentCentral ● icon.

- Select Share a Model . Follow the directions on the screen to enter your profile information and to upload your model.

SolidWorks Corporation will review your model and inform you whether it is accepted for sharing.

Vendors utilize this service to share model information with their customers. SolidWorks users share information through 3D ContentCentral.

In the next section, create the CUSTOMER Assembly from the MGPM12-1010 assembly. The MGPM12-1010 assembly was downloaded from 3D ContentCentral. The manufacturer of the MGPM12-1010 assembly is SMC of USA. Use the assembly that is located on the DVD in the book in the **ENGDESIGN-W-SOLIDWORKS\VENDOR COMPONENTS** file folder.

CUSTOMER Assembly

Three documents for this project are contained in the
**ENGDESIGN-W-SOLIDWORKS\VENDOR
COMPONENTS** folder on the DVD in the book. Copy
these files directly from the DVD to your hard drive to the
correct folder. Work directly from these files.

◈ MGPM12-1010
◈ MGPM12-1010_MGPRod(12M)
◈ MGPM12-1010_MGPTube(12M)

The MGPM12-1010_MGPTube(12M) part and MGPM12-
1010_MGPRod(12M) part contain references to the
MGPM12-1010 assembly.

☀ When parts reference an assembly, open the assembly first. Then open the individual
parts from within the Assembly FeatureManager.

Create the CUSTOMER assembly. The CUSTOMER assembly combines the GUIDE-
ROD assembly and the GUIDE-CYLINDER (MGPM12-10) assembly. The GUIDE-
ROD assembly is fixed to the CUSTOMER assembly Origin.

Activity: Create the CUSTOMER Assembly

Copy the three needed files from the DVD in the book to your ENGDESIGN-W-SOLIDWORKS\
VENDOR COMPONENTS folder.
447) Insert the **DVD** into your computer.

448) Copy the **three** needed files from the DVD in the
ENGDESIGN-W-SOLIDWORKS\VENDOR
COMPONENTS folder to your ENGDESIGN-W-
SOLIDWORKS\VENDOR COMPONENTS folder on your
hard drive. Work from you hard drive.

◈ MGPM12-1010
◈ MGPM12-1010_MGPRod(12M)
◈ MGPM12-1010_MGPTube(12M)

Open the MGPM12-1010 assembly from your hard drive.
449) Double-click the **MGPM12-1010** assembly from
the ENGDESIGN-W-SOLIDWORKS\VENDOR
COMPONENTS folder on your hard dirve. The
MGPM12-1010 FeatureManager is displayed.

450) If needed, click **Yes** to rebuild.

Fit the model to the Graphics window.
451) Press the **f** key.

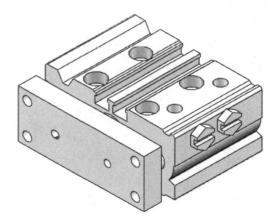

Open the GUIDE-ROD assembly.

452) Double-click the **GUIDE-ROD** assembly from the PROJECTS folder.

Create the CUSTOMER assembly.

453) Click **New** ☐ from the Menu bar.

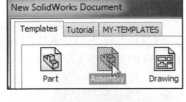

454) Double-click **Assembly** from the Templates tab. The Begin Assembly PropertyManager is displayed.

Display the Origin.

455) If required, click **View**; check **Origins** from the Menu bar. The MGPM12-1010 assembly and the GUIDE-ROD assembly are displayed in the Part/Assembly to Input box.

456) Double-click **GUIDE-ROD** from the Open documents box.

457) Click **OK** ✔ from the Begin Assembly PropertyManager to fix the GUIDE-ROD to the Origin.

Save the assembly.

458) Click **Save** 🖫.

459) Select **ENGDESIGN-W-SOLIDWORKS\PROJECTS** for folder.

460) Enter **CUSTOMER** for File name.

The Pack and Go functionality gathers all related files for a model design and copies them into a folder or zip file. Use Pack and Go for an Assembly or Drawing document to save the document and the reference documents.

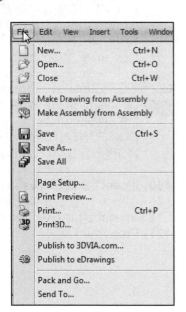

461) Enter **GUIDE-ROD AND GUIDE-CYLINDER ASSEMBLY** for Description.

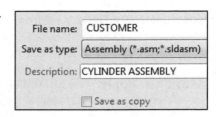

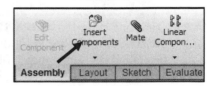

462) Click **Save**.

GUIDE-ROD is the first component in the CUSTOMER assembly. Insert the second component.

Activity: CUSTOMER Assembly-Insert Component

Insert the MGPM12-1010 assembly.

463) Click the **Insert Components** Assembly tool. The Insert Component PropertyManager is displayed.

464) Click **MGPM12-1010** from the Part/Assembly box.

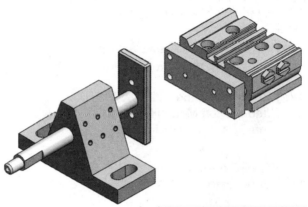

465) Click a **position** in the Graphics window behind the GUIDE-ROD assembly as illustrated. The MGPM12-1010<1> component is added to the CUSTOMER FeatureManager.

Fit the model to the Graphics window.
466) Press the **f** key.

Deactivate the Origins for clarity.
467) Click **View**; uncheck **Origins** from the Menu bar.

Deactivate the Planes for clarity if required.
468) Click **View**; uncheck **Planes** from the Menu bar.

Re-position the MGPM12-1010<1> component.
469) Click the **MGPM12-1010<1>** component in the FeatureManager. It is displayed in blue in the Graphics window.

470) Click the **Rotate Component** Assembly tool. The Rotate Component PropertyManager is displayed.

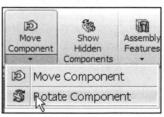

471) Select **By Delta XYZ** from the Rotate box.

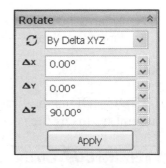

472) Enter **90** in the delta Z box.

473) Click **Apply**. The MGPM12-1010<1> component rotated 90 degrees in the Graphics window.

474) Click **OK** ✔ from the Rotate Component PropertyManager.

Fit the model to the Graphics window.
475) Press the **f** key.

476) Click and drag the **MGPM12-1010 assembly** behind the GUIDE-ROD assembly. Note: The PISTON PLATE is behind the GUIDE-ROD assembly.

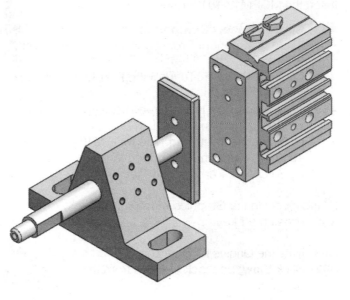

Zoom in on the PISTON PLATE component.

477) Click **Zoom to Area** 🔍 from the Heads-up View toolbar.

478) **Zoom in** on the PISTON PLATE and the top MountHoles face of the PLATE as illustrated.

479) Click **Zoom to Area** 🔍 to deactivate.

Insert a Concentric mate between to faces.
480) Click the **Mate** ✎ Assembly tool. The Mate PropertyManager is displayed.

481) Click the **inside MountHoles** face of the PLATE.

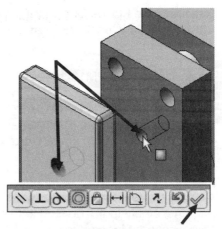

482) Click the **inside Mounting Hole** face of the PISTON PLATE. Both faces are selected. The selected faces are displayed in the Mate Selections box. Concentric mate is selected by default.

483) Click the **Green Check mark** ✔ from the Mate dialog box.

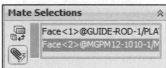

☀ The bottom hole feature contains the Slot Cut and a Mounting Hole. The Mounting Hole is hidden. Utilize the Select Other option to select the hidden Mounting Hole.

Insert a Concentric mate between two faces. Use the Select Other tool.

484) Click **View**; check **Temporary Axes** from the Menu bar.

Select the hidden Mounting Hole.

485) Click **Zoom to Area** 🔍 .

486) Zoom in on the bottom Slot and the MountHoles. Do not select the Slot.

487) Click **Zoom to Area** 🔍 to deactivate.

488) Right-click a **position** behind the Slot Cut on the Piston Plate.

489) Click **Select Other**. The Select Other dialog box is displayed. The Select Other box lists geometry in the selected region of the Graphics window. There are edges, faces and axes. The list order and geometry entries depend on the selection location of the mouse pointer.

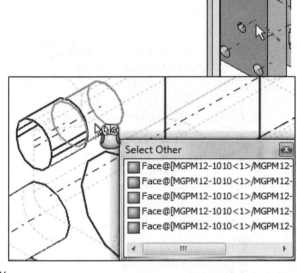

490) Position the **mouse pointer** over the Mounting Hole in the Piston Plate.

491) Click the inside **Mounting Hole** face.

492) Click the **bottom inside Mounting Hole face** of the PLATE. The selected faces are displayed in the Mate Selections box. Concentric is selected by default.

493) Click the **Green Check mark** ✓ from the Mate dialog box.

Fit the model to the Graphics window.
494) Press the **f** key.

Insert a Coincident mate between two faces.
495) Click and drag the **PISTON PLATE** backward to create a gap.

496) Click the **front face** of the PISTON PLATE. Press the **left arrow key** to rotate the view until you can see the back face of the PLATE.

497) Click the **back face** of the PLATE. The two selected faces are displayed in the Mate Selections box. Coincident is selected by default.

498) Click the **Green Check mark** from the Mate dialog box.

499) Click **OK** from the Mate PropertyManager.

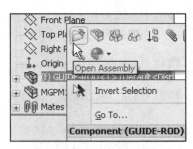

Un-suppress all components.
500) Right-click the **GUIDE-ROD** component in the CUSTOMER FeatureManager.

501) Click **Open Assembly** from the Context toolbar. The GUIDE-ROD Assembly FeatureManager is displayed.

502) Right-click the **Hardware** folder in the FeatureManager.

503) Click **UnSuppress**.

504) Right-click **DerivedLPattern1** in the FeatureManager.

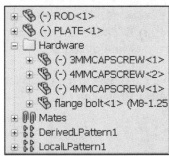

505) Click **UnSupress**.

506) Right-click **LocalLPattern1** in the FeatureManager.

507) Click **UnSuppress**.

Return to the CUSTOMER assembly.
508) Press **Ctrl+Tab**. Select **CUSTOMER assembly**.

509) Click **Isometric view** from the Heads-up View toolbar.

Deactivate the Temporary Axes.
510) Click **View**; uncheck **Temporary Axes** from the
Menu bar.

Save the CUSTOMER assembly in the ENGDESIGN-W-
SOLIDWORKS/PROJECTS file folder.
511) Click **Save** 💾.

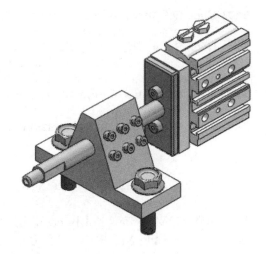

Copy the CUSTOMER Assembly - Apply Pack and Go

Copying an assembly in SolidWorks is not the same as copying a document in Microsoft
Word. The CUSTOMER assembly contains references to its parts and other sub-
assemblies. Your task is to provide a copy of the CUSTOMER assembly to a colleague
for review. Copy the CUSTOMER assembly and all of the components (references) into
a different file folder named CopiedModels. Reference all component file locations to the
new file folder. Apply the SolidWorks Pack and Go tool.

Activity: CUSTOMER Assembly - SolidWorks Pack and Go tool

Apply the SolidWorks Pack and Go tool to save the CUSTOMER assembly to a new file folder.
512) Click **File**, **Pack and Go** from the Menu bar. Note your save
options: Save to folder or Save to Zip file.

513) Click the **Browse** button.

514) Select the **Documents** folder.

515) Click the **Make New Folder** button.

516) Enter **CopiedModels** for Folder
Name.

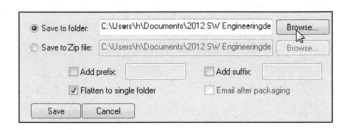

517) Click **OK** from the Browse For Folder dialog box.

518) Click **Save**. The CUSTOMER assembly remains open.

Close all files.
519) Click **Windows**, **Close All** from the Menu bar.

520) Click **Yes**. View the saved assembly in the CopiedModels folder.

 Review of the CUSTOMER Assembly

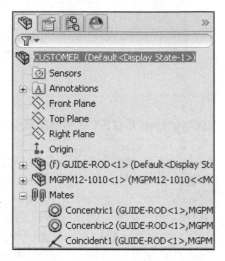

The CUSTOMER assembly combined the GUIDE-ROD assembly and the MGPM12-1010 assembly. In the design process, you decided to obtain the MGPM12-1010 assembly in SolidWorks format from 3D ContentCentral.

The GUIDE-ROD assembly is the first component inserted into the CUSTOMER assembly. The GUIDE-ROD assembly is fixed (f) to the CUSTOMER assembly Origin.

You inserted and mated the MGPM12-1010 assembly to the GUIDE-ROD assembly. The flange bolts and cap screws utilized SmartMates to create Concentric and Coincident mates.

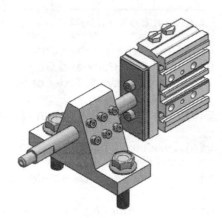

In the GUIDE-ROD assembly, you utilized a Feature Driven Component Pattern and a Linear Component Pattern for the flange bolt and 3MMCAPSCREW. You modified and redefined the flange bolt mates.

The Save As option with the Reference option copied the CUSTOMER assembly and all references to a new folder location.

Project Summary

You created the GUIDE-ROD assembly from the following parts: GUIDE, ROD, PLATE, Flange Bolt, 3MMCAPSCREW and 4MMCAPSCREW. The components were oriented and positioned in the assembly using Concentric, Coincident, Parallel and Concentric/Coincident SmartMates.

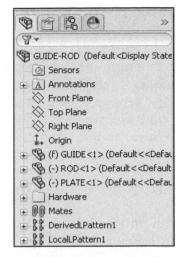

Mates were redefined in the GUIDE-ROD assembly to center the Flange Bolts in the Slot Cut of the GUIDE. In the GUIDE-ROD assembly, you utilized the Feature Driven Component Pattern feature for the 3MMCAPSCREW and the Local Component Pattern feature for the Flange Bolt.

The CUSTOMER assembly contained the MGPM12-1010 assembly and the GUIDE-ROD assembly. The assemblies utilized a Bottom-up design approach. In a Bottom-up design approach, you possess all the required design information for the individual components.

Project 2 is completed. In Project 3, you will create an assembly drawing and a detailed drawing of the GUIDE.

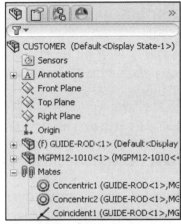

Think design intent. When do you use the various End Conditions and Geometric sketch relations? What are you trying to do with the design? How does the component fit into an Assembly?

Use the Defeature tool to remove details from a part or assembly and save the results to a new file in which the details are replaced by dumb solids (that is, solids without feature definition or history). You can then share the new file without revealing all the design details of the model.

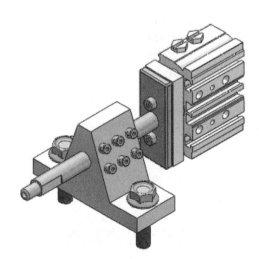

Project Terminology

Add Relations: Constraints utilized to connect related geometry. Some common relations are Horizontal, Concentric, Equal, Coincident and Collinear. The Add Relations tool is invoked through the Sketch toolbar. Relations are also added by selecting the Ctrl key, geometric entities and then the relation through the Properties PropertyManager.

Appearances: Used to add visual properties to a model, such as color or realistic material textures, without affecting the physical properties of the model.

Arc: A single portion of a circle equaling less than 360 degrees.

Assembly: An assembly combines two or more parts. In an assembly, parts are referred to as components. The file extension for an assembly is *.SLDASM.

Assembly Configuration: A single variation of an assembly with multiple versions.

Axis: A straight line that can be used to create model geometry, features, or patterns. An axis can be made in a number of different ways, including using the intersection of two planes.

Base Feature: The first feature in your model; and listed in the FeatureManager.

Base Sketch: The first sketch of the Base feature and listed in the FeatureManager.

Bottom-up assembly design approach: An assembly modeling technique where you create parts and then insert them into an assembly. In this approach, you possess all of the required design information for the individual components.

Box-select: A selection method used in parts, assemblies, and drawings to select entities by dragging a selection box with the mouse pointer. Dragging from left to right will select all items that fall within the area of the box.

Centerpoint Arc: A sketch tool that creates an arc from a centerpoint, a start point, and an endpoint.

Chamfer: Removes material with a beveled cut from an edge or face. The Chamfer requires a distance and an angle or two distances.

Closed Profile: Also called a closed contour, it is a sketch or sketch entity with no exposed endpoints: for example, a circle or polygon.

Coincident: A geometric condition where two or more entities share the same point in space.

Coincident Mate: A mate tool used to mate two selected entities in components of an assembly share the same point.

CommandManager: A context-sensitive toolbar that is docked about the Graphics window by default and dynamically updates to the appropriate toolbar based on the modeling environment and selected tab.

Component Pattern, Derived: A pattern of components in an assembly based on a feature pattern of an existing component. The 3MMCAPSCREW pattern in the assembly was derived from the GUIDE Linear Pattern of 3mm Tapped Holes.

Component Pattern, Linear: A pattern of components in an assembly based on reference geometry in the assembly. A Linear Component Pattern is called a LocalLPattern. The two flange bolts were redefined with a Linear Component Pattern.

Component: A part or sub-assembly inserted into an assembly. Changes in the components directly affect the assembly and vise a versa.

Concentric: A geometric condition where two or more circles or arcs share the same centerpoint.

Concentric Mate: A mate tool used to specify that two selected circular edges or cylindrical faces are to share the same centerpoint or axis.

Confirmation Corner: A way to accept or cancel features or sketches in the upper-right corner of the Graphics window.

Context Toolbar: A toolbar that is displayed when selecting items in the Graphics window or FeatureManager that provides access to the most common set of tools.

Construction Line: A sketch entity that aids in the creation of other geometry without actually being part of the created geometry.

Convert Entities: A sketch tool that projects one or more curves onto the current sketch plane. Select an edge, loop, face, curve, or external sketch contour, set of edges, or set of sketch curves.

Coordinate System: A system of planes used to assign Cartesian coordinates to features, parts, and assemblies. Part and assembly documents contain default coordinate systems; other coordinate systems can be defined with reference geometry. Coordinate systems can be used with measurement tools and for exporting documents to other file formats.

Cosmetic Thread: An annotation type that is used to represent the inner diameter of a threaded boss and the other diameter of a threaded hole and can include a hole callout in a drawing document.

Cursor Feedback: Feedback is provided by a symbol attached to the cursor arrow indicating your selection.

Degrees of Freedom: The directions a component can freely move in 3D space. Three of the degrees consist of the translations in the directions of X, Y, Z. The next three degrees consist of the rotation around the x-, y-, and z axes.

Dimension: Values that define the overall size and shape of a feature or the relationship between components.

Dimension Line: A portion of a dimension used to indicate the extent and direction of a dimension. The dimension line, in mechanical applications is terminated at both ends with an arrowhead and contains the value of the dimension.

Dimensioning Standard: A set of drawing and detailing options developed by national and international organizations. A few key dimensioning standard options are: ANSI, ISO, DIN, JIS, BSI, GOST and GB.

Design Intent: How a model reacts to change and updates.

Display Pane: A portion of the FeatureManager that can be expanded to view the various display settings in a part, assembly or drawing document.

Document Templates: A part, assembly, or drawing document that includes user-defined parameters or standard default parameters that is used in the creation of a new SolidWorks document.

Document Structure: The relationship between a part document, assembly document, and a drawing document in SolidWorks.

Document Types: Documents that are used by a specific program. Types are determined by the extension followed by a period in the filename. Examples: Part (*SLDPRT), Assembly (*SLDASM), and Drawing (*SLDDRW).

Drag Handle: A point on a sketch or feature that can be selected and dragged to create or edit an extrusion.

Edit Feature: A tool utilized to modify existing feature parameters. Right-click the feature in the FeatureManager. Click Edit Feature.

Edit Sketch: A tool utilized to modify existing sketch geometry. Right-click the feature in the FeatureManager. Click Edit Sketch.

Elastic Modulus, E: The stress required to cause one unit of strain. Utilized in Material Properties.

End Condition: An option located in the PropertyManager. End Condition determines how a feature extends and the design intent of the model.

Extension Lines: Also called projection lines – extension lines are the lines of a dimension that extend from a part that are used to indicate the locations on the part where the dimension applies.

Exploded view: Displays an assembly with its components separated from one another. An Exploded view is used to show how to assemble or disassemble the components. With SolidWorks Animator, exploded view collapse and explode animations are recorded to an avi file.

Extruded Cut: Cut features are used to remove material from a solid. This is the opposite of the Extruded Boss/Base features.

Feature: A single shape that is used along with other features to create a part or assembly.

FeatureManager: The Pane by default located to the left of the Graphics window that lists the features and sketches used to create the part, assembly or drawing document.

Fillet: A rounded corner on a sketch or a rounded edge of a model.

Fits: There are three major types of fits addressed in this Project:

- Clearance fit - The shaft diameter is less than the hole diameter.

- Interference fit - The shaft diameter is larger than the hole diameter. The difference between the shaft diameter and the hole diameter is called interference.

- Transition fit - Clearance or interference can exist between the shaft and the hole.

Fully Defined: A sketch where all lines and curves in the sketch, and their positions, are described by dimensions or relations, or both, and cannot be moved. Fully defined sketch entities are displayed in black.

Geometric Relations: A relation is a geometric constraint between sketch entities or between a sketch entity and a plane, axis, edge, or vertex. Relations force a behavior on a sketch element to capture the design intent.

Graphics Window: The largest working window within SolidWorks were parts, assemblies and drawings are displayed.

Heads-up View Toolbar: A transparent toolbar located in the upper portion of the Graphics window by default.

Hidden Geometry: Geometry that is not displayed. Utilize Hide/Show to control display of components in an assembly.

Hidden Lines: A line type used in 2D drafting to represent edges that can't be seen normally in the view - which are hidden by other geometry.

Hole Wizard: The Hole Wizard feature is used to create specialized holes in a solid. The HoleWizard creates simple, tapped, counterbore and countersunk holes using a step-by-step procedure.

Instant3D: A tool that provides the ability to quickly create and modify feature in part and assembly documents. If does not offer End Condition options.

Isometric View: A graphical projection of a model in a drawing that displays the model with its x-, y-, and z- axes spaced 120° apart and with the vertical z-axis.

Line types: A term describing the display of a line in a drawing including its type and thickness.

Modify Dimension: The act of changing a dimension. Double-click on a dimension. Enter the new value in the Modify dialog box. Click Rebuild.

Move Component/Rotate Component: Components in an assembly are translated by selecting the middle mouse button. To rotate a component in an assembly, utilize Rotate Component from the Assembly toolbar. Additional options for Collision Detection are available from the Move Component option on the Assembly toolbar.

Material: Specified substance from which a part is intended to be made from.

Model: 3D geometry within a part or assembly document.

Mass Properties. A tool that evaluates the characteristics of a part or an assembly such as volume, surface area, centroid, etc.

Mates: The action of assembling components in SolidWorks is defined as Mates. Mates are geometric relationships that align and fit components in an assembly. Mates remove degrees of freedom from a component. Mates require geometry from two different components.

Open Profile: Also called an open contour, it is a sketch or sketch entity with endpoints exposed. For example, a U-shaped profile is open.

Origin: The point where the three axes intersect at the 0,0,0 coordinate of the model or sketch. Always know the location of the origin when modeling.

Orthographic Projection: A single view that is meant to represent a 3D object as a flat 2D object.

Over Defined: A sketch is over defined when dimensions or relations are either in conflict or redundant.

Part: A single 3D object made up of features. A part can become a component in an assembly, and it can be represented in 2D in a drawing. Examples of parts are bolt, pin, plate, and so on. The extension for a SolidWorks part file name is .SLDPRT.

Plane: Flat construction geometry. Planes can be used for a 2D sketch, section view of a model, a neutral plane in a draft feature, and others.

Rebuild: After changes are made to the dimensions, rebuild the model to cause those changes to take affect.

Relation: A geometric constraint between sketch entities or between a sketch entity and a plane, axis, edge, or vertex. Relations can be added automatically or manually.

Revolved Boss/Base: Creates a feature from an axis, a profile sketch and an angle of revolution. The Revolve Boss/Base feature adds material to the part. The 4MMCAPSCREW utilized a Revolve Base feature.

Save As: Utilize the References button to save all parts and subassemblies in a new assembly to a different file folder. Utilize the Save as copy option to create a copy of an existing part.

Section View: Displays the internal cross section of a component or assembly. Section views can be performed anywhere in a model. The location of the cut corresponds to the Section plane. A Section plane is a planar face or reference plane.

SimulationXpress: A Finite Element Analysis (FEA) tool. SimulationXpress calculates the displacement and stress in a part based on material, restraints and static loads. Results are displayed for von Mises stress and deflection.

Sketch: A collection of (lines, arcs, circles, rectangles, etc.) on a plane or face that forms the basis for a feature such as a base or a boss.

SmartMates: A SmartMate is a Mate that automatically occurs when a component is placed into an assembly. The mouse pointer displays a SmartMate feedback symbol when common geometry and relationships exist between the component and the assembly. SmartMates are Coincident, Concentric or Concentric / Coincident.

- Coincident SmartMate: The mating entities are either two linear edges or two planar faces.

- Concentric SmartMate: The mating entities are either two conical faces, or two temporary axes, or one conical face and 1 temporary axis.

- Concentric / Coincident SmartMate: The mating entities are either two circular edges, "the edges do not have to be complete circles, or two circular patterns on flanges.

Sub-assemblies: Sub-assemblies are components inserted into an assembly. They behave as a single piece of geometry. When an assembly file is added to an existing assembly, it is referred to as a sub-assembly, (*.SLDASM) or component.

Suppressed features and components: Suppressed features, parts and assemblies are not displayed. During model rebuilding, suppressed features and components are not calculated. Features and components are suppressed at the component or assembly level in the FeatureManager. The names of the suppressed features and components are displayed in light gray.

Template: A document (part, assembly, or drawing) that forms the basis of a new document. It can include user-defined parameters, annotations, predefined views, geometry, etc.

Toolbars: The toolbars provide shortcuts enabling you to access the most frequently used commands.

Trim Entities: A sketch tool used to delete selected sketched geometry.

Units: Used in the measurement of physical quantities. Decimal inch dimensioning and Millimeter dimensioning are the two types of common units specified for engineering parts and drawings.

Questions

1. Describe an assembly or sub-assembly.

2. What are Mates and why are they important in assembling components?

3. Name and describe the three major types of Fits.

4. Name and describe the two assembly modeling techniques in SolidWorks.

5. Describe Dynamic motion.

6. In an assembly, each component has_____# degrees of freedom? Name them.

7. True or False. A fixed component cannot move and is locked to the Origin.

8. Identify the procedure to create a Revolved Base feature.

9. Describe the different types of SmartMates. Utilize on line help to view the mouse pointer feedback icons for different SmartMate types.

10. How are SmartMates used?

11. Identify the process to insert a component from the Design Library.

12. Describe a Section view.

13. What are Suppressed features and components? Provide an example.

14. True or False. If you receive a Mate Error you should always delete the component from the assembly.

15. List the names of the following icons from the Assembly toolbar.

A B C D E F G H I

A	B	C
D	E	F
G	H	I

Exercises

Exercise 2.1: Weight-Hook Assembly

Create the Weight-Hook assembly. The Weight-Hook assembly has two components: WEIGHT and HOOK.

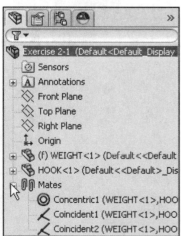

- Create a new assembly document. Insert the WEIGHT part from the Chapter2 - Homework folder in the book DVD.

- Fix the WEIGHT to the Origin as illustrated in the Assem1 FeatureManager.

- Insert the HOOK part from the Chapter2 - Homework folder into the assembly.

- Insert a Concentric mate between the inside top cylindrical face of the WEIGHT and the cylindrical face of the thread. Concentric is the default mate.

- Insert the first Coincident mate between the top edge of the circular hole of the WEIGHT and the top circular edge of Sweep1, above the thread. Coincident is the default mate. The HOOK can rotate in the WEIGHT.

- Fix the position of the HOOK. Insert the second Coincident mate between the Right Plane of the WEIGHT and the Right Plane of the HOOK. Coincident is the default mate.

- Expand the Mates folder and view the created mates.

Exercise 2.2: Weight-Link Assembly

Create the Weight-Link assembly. The Weight-Link assembly has two components and a sub-assembly: Axle component, FLATBAR component, and the Weight-Hook sub-assembly that you created in Exercise 2.1.

- Create a new assembly document. Insert the Axle part from the Chapter2 - Homework folder in the book DVD.

- Fix the Axle component to the Origin of the assembly.

- Insert the FLATBAR part from the Chapter2 - Homework folder in the book DVD.

- Insert a Concentric mate between the Axle cylindrical face and the FLATBAR inside face of the top circle.

- Insert a Coincident mate between the Front Plane of the Axle and the Front Plane of the FLATBAR.

- Insert a Coincident mate between the Right Plane of the Axle and the Top Plane of the FLATBAR. Position the FLATBAR as illustrated.

- Insert the Weight-Hook sub-assembly that you created in exercise 2.1.

- Insert a Tangent mate between the inside bottom cylindrical face of the FLATBAR and the top circular face of the HOOK, in the Weight-Hook assembly. Tangent mate is selected by default.

- Insert a Coincident mate between the Front Plane of the FLATBAR and the Front Plane of the Weight-Hook sub-assembly. Coincident mate is selected by default. The Weight-Hook sub-assembly is free to move in the bottom circular hole of the FLATBAR.

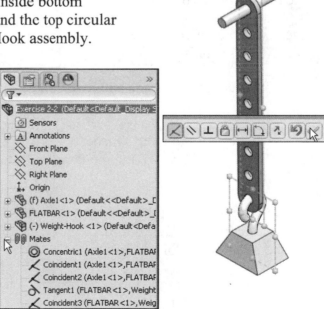

Exercise 2.3: Counter Weight Assembly

Create the Counter Weight assembly as illustrated using SmartMates and Standard mates. Components are supplied in the Chapter 2 - Homework/Counter-Weight folder on the DVD. Copy all components to your working folder. The Counter Weight consists of the following items:

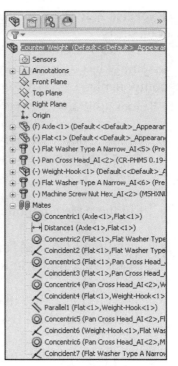

- Weight-Hook sub-assembly

- Weight

- Eye Hook

- Axle component

- Flat component

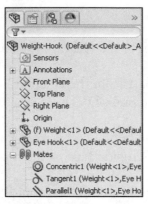

- Flat Washer Type A from the SolidWorks Toolbox

- Pan Cross Head Screw from the SolidWorks Toolbox

- Flat Washer Type A from the SolidWorks toolbox

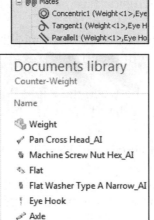

- Machine Screw Nut Hex from the SolidWorks Toolbox

Fix the Axle component to the Origin of the assembly.

Bring in other components and sub-assemblies. Apply all needed mates.

Use SmartMates with the Flat Washer Type A Narrow_AI, Machine Screw Nut Hex_AI and the Pan Cross Head_AI components.

Use a Distance mate to fit the Axle in the middle of the Flat. Note a Symmetric mate could replace the Distance mate. Think about the design of the assembly.

The symbol (f) represents a fixed component. A fixed component cannot move and is fixed to the assembly Origin.

Exercise 2.4: Binder Clip Assembly

- Create a simple Gem binder clip. You see this common item every day.

- Create an ANSI - IPS model.

- Create two components – BASE and HANDLE.

- Apply material to each component and address all needed mates. Think about where you would start. Think about how the binder clip assembly would move.

- What would be the Base Sketch for each component?

- What are the dimensions? Approximate the dimensions from a small or large Gem binder clip.

- You are the designer. View the sample Assembly FeatureManager. Your Assembly FeatureManager can (should) be different. This is just ONE way of many to create the assembly.

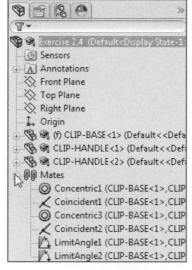

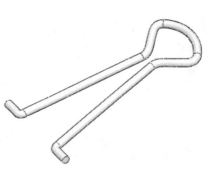

Below is a sample assembly of the Binder Clip from my Freshman Engineering class. You are the designer. Be creative.

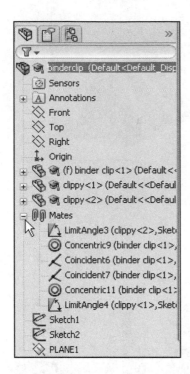

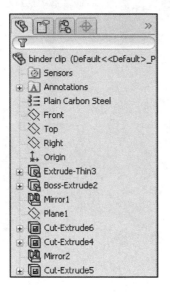

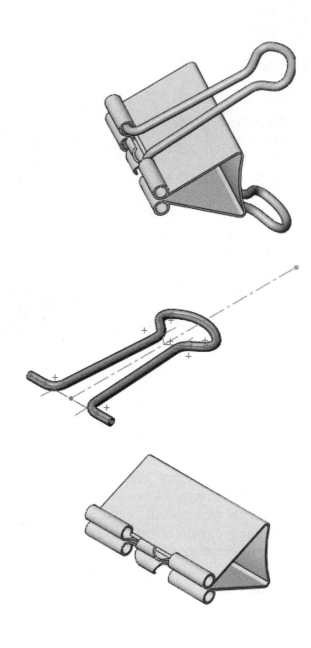

Exercise 2.5: Limit Mate Assembly (Advanced Mate Type)

- Open the assembly (Limit Mate) from the Chapter 2 - Homework/Limit Mate folder located on the DVD in the book.

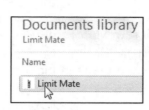

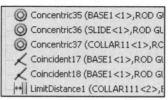

- Copy the assembly (components and sub-assemblies) to your working folder. Do not work directly from the DVD.

- Insert a Limit Mate to restrict the movement of the Slide Component - lower and upper movement.

- Use the Measure tool to obtain max and min distances.

- Use SolidWorks Help for additional information.

A Limit Mate is an Advanced Mate type. Limit mates allow components to move within a range of values for distance and angle. You specify a starting distance or angle as well as a maximum and minimum value.

- Save the model and move the slide to view the results in the Graphics window. Think about how you would use this mate type in other assemblies.

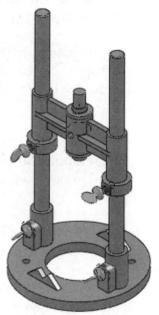

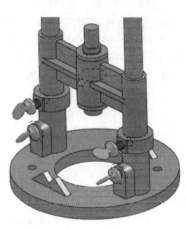

**Exercise 2.6: Screw Mate Assembly
(Mechanical Mate Type)**

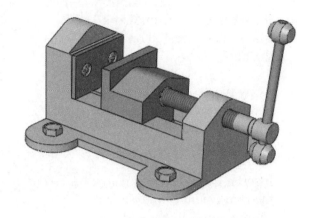

- Open the assembly (Screw Mate) from the Chapter 2 - Homework/Screw Mate folder located on the DVD in the book.

- Copy the assembly (components and sub-assemblies) to your working folder. Do not work directly from the DVD.

- Insert a Screw mate between the inside Face of the Base and the Face of the vice. A Screw is a Mechanical Mate type. It is not a Standard mate type.

A Screw mate constrains two components to be concentric, and also adds a pitch relationship between the rotation of one component and the translation of the other. Translation of one component along the axis causes rotation of the other component according to the pitch relationship. Likewise, rotation of one component causes translation of the other component. Use SolidWorks Help if needed.

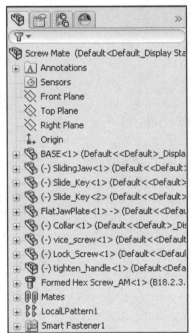

Use the Select Other tool (See SolidWorks Help if needed) to select the proper inside faces and to create the Screw mate for the assembly.

- Rotate the handle and view the results. Think about how you would use this mate type in other assemblies.

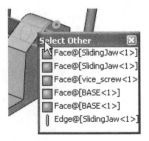

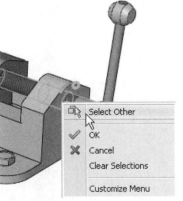

Exercise 2.7: Angle Mate Assembly

- Open the assembly (Angle Mate) from the Chapter 2 - Homework/ Angle Mate folder located on the DVD in the book.

- Copy the assembly (components and sub-assemblies) to your working folder. Do not work directly from the DVD.

- Move the Handle in the assembly. The Handle is free to rotate. Set the angle of the Handle.

- Insert an Angle mate (165 degrees) between the Handle and the Side of the valve using Planes. An Angle mate places the selected items at the specified angle to each other.

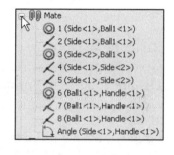

- The Handle has a 165-degree Angle mate to restrict flow through the valve. Think about how you would use this mate type in other assemblies.

Exercise 2.8: Angle Mate Assembly (Cont:)

Open the SolidWorks FlowXpress Tutorial under the Design Analysis folder. Follow the directions.

Create two end caps (lids) for the ball value using the Top-down Assembly method. Note the Reference - In-Content symbols in the FeatureManager.

- Modify the Appearance of the body to observe the change - enhance visualization.

- Apply the Select-other tool to obtain access to hidden faces and edges.

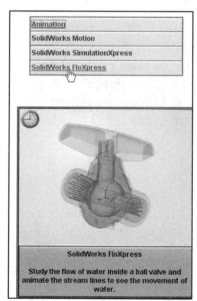

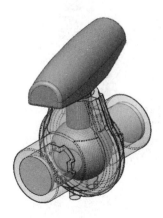

Exercise 2.8: 4Bar Linkage Assembly

Create the 4Bar linkage assembly as illustrated. The four bar linkage assembly has five simple components. Create the five simple components. Assume dimensions. You are the designer.

View the avi file from the Chapter 2 - Homework folder located on the DVD in the book for required movement. Insert all needed mates.

In an assembly, fix (f) the first component to the origin or fully define it to the three default sketch planes of the Assembly.

Insert additional components and insert needed mates to simulate the movement of a 4 bar linkage assembly.

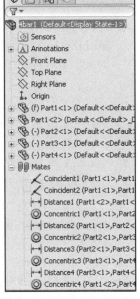

Read the section on Coincident, Concentric and Distance mates in SolidWorks Help.

Create a base with text for extra credit.

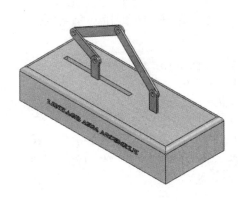

Below are sample models from my Freshman Engineering class. Note the different designs to maintain the proper movement of the 4 bar linkage.

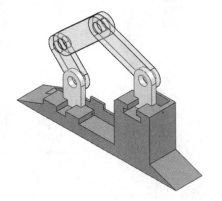

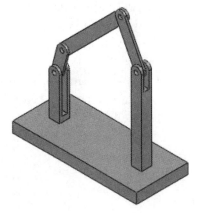

Exercise 2.9: BASE PLATE

- Design a BASE PLATE part to fasten three CUSTOMER assemblies. The BASE PLATE is the first component in the BASE-CUSTOMER assembly:

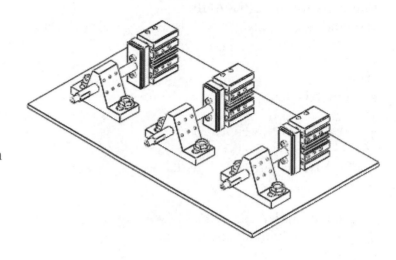

- Create six holes. Determine the hole location based on the CUSTOMER assembly.

- The BASE PLATE thickness is 10mm. The material is 1060 Alloy.

Exercise 2.10: PLATE4H-Part & Assembly

Create a PLATE4H part. Utilizes four front outside holes of the MGPM12-1010 assembly. The MGPM12-1010 zip file is located in the Chapter2 - Homework folder in the book DVD.

- Manually sketch the dimensions of the PLATE4H part.

- Dimension the four outside holes.

- The Material thickness is 10mm.

- Create the new assembly, PLATE4H-GUIDECYLINDER.

- Mate the PLATE4H part to the MGPM12-1010 assembly.

MGPM12-1010 assembly

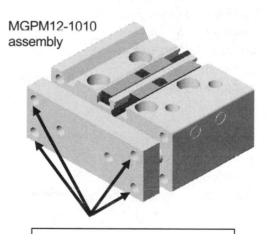

Fasten PLATE4H to the front face of the MGPM12-1010 assembly

Exercise 2.11: LINKAGE Assembly

In the Project 1 exercises, you created four parts for the LINKAGE assembly.

- Axle part

- SHAFT COLLAR part

- FLAT BAR - 3 HOLE part

- FLAT BAR - 9 HOLE part

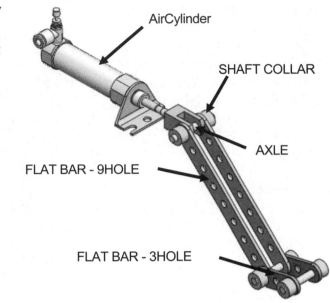

AirCylinder

SHAFT COLLAR

AXLE

FLAT BAR - 9HOLE

FLAT BAR - 3HOLE

Create the LINKAGE assembly with the SMC AirCylinder and the four parts from Project1.

The AirCylinder is the first sub-assembly in the LINKAGE assembly. When compressed air goes in to the air inlet, the Piston Rod is pushed out.

Without compressed air, the Piston Rod is returned by the force of the spring.

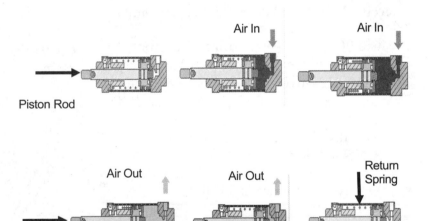

Air In

Air In

Piston Rod

Air Out

Air Out

Return Spring

Piston Rod

The Piston Rod linearly translates the ROD CLEVIS in the LINKAGE assembly.

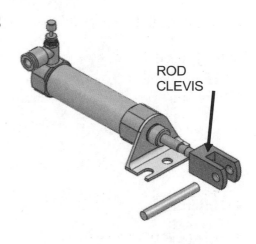

ROD CLEVIS

- Create a new assembly. Insert the AirCylinder assembly. The AirCylinder assembly is fixed to the Origin.

The AirCylinder assembly is available on the Multimedia DVD in the Chapter2 - Homework folder.

- Insert the Axle.

- Insert a Concentric mate between the cylindrical face of the Axle and the inside circular face of the ROD CLEVIS hole.

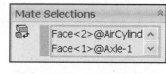

- Insert a Coincident mate to align the Axle in the ROD CLEVIS holes. The Axle is symmetric about its Front Plane.

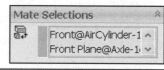

- Insert the first FLAT BAR - 9 HOLE component as illustrated.

- Utilize a Concentric mate and Coincident mate. The FLAT BAR - 9 HOLE rotates about the Axle.

- Insert the second FLAT BAR - 9 HOLE component on the other side. Repeat the Concentric mate and Coincident mate for the second FLAT BAR - 9 HOLE. The second FLAT BAR - 9 HOLE is free to rotate about the AXLE, independent from the first FLAT BAR.

- Insert a Parallel mate between the two top narrow faces of the FLAT BAR - 9 HOLE. The two FLAT BAR - 9 HOLEs rotate together.

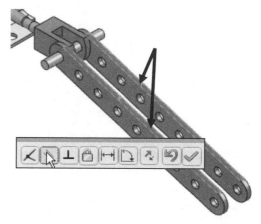

- Insert the first SHAFT-COLLAR component. Insert a Concentric and Coincident mate.

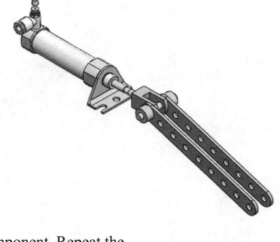

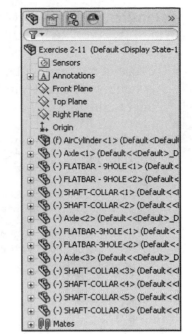

- Insert the second SHAFT-COLLAR component. Repeat the Concentric and Coincident mate.

- Insert the second instance of the Axle. Mate the AXLE to the bottom hole of the FLAT BAR - 9 HOLE. Insert a Concentric and Coincident mate.

- Insert the two FLAT-BAR - 3 HOLE components, the third Axle, four SHAFT-COLLAR components with the needed mates to complete the assembly.

Exercise 2.12: MOTION LINKAGE Assembly Motion Study

The Motion Study tool provides the ability to apply a Physical simulation to an assembly.

* Open the MOTION-LINKAGE assembly. The MOTION LINKAGE assembly is available on the Multimedia DVD in the Chapter2 - Homework folder.

* Click the Motion Study1 tab in the lower left corner of the Graphics window.

* Select Basic Motion from the Motion Study Manager.

Apply a Rotary Motor to a face of a component in the assembly.

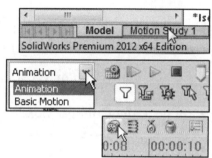

* Click the Motor icon. The Motor PropertyManager is displayed.

* Click Rotary Motor.

* Click the front face of the FLATBAR as illustrated. A red Rotary Motor icon is displayed. The red direction arrow points counterclockwise.

* Accept the default settings. Click OK.

* Click the Play from Start icon. View the model in the Graphics window.

* Click Save.

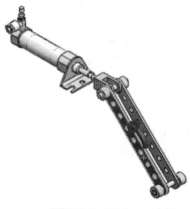

- Click OK. View the results. Click NO.

- Click the Model tab in the bottom left corner of the Graphics window to exit Motion Study.

Exercise 2.13: WIZARD-MOTION-LINKAGE Assembly

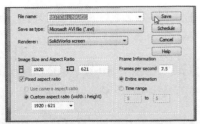

Create a Motion Study using the Wizard.
- Open WIZARD-MOTION-LINKAGE assembly. The WIZARD-MOTION-LINKAGE assembly is available on the Multi-media DVD in the Chapter2 - Homework folder.

- Select Basic Motion.

- Click the Motion Study1 tab in the lower left corner of the Graphics window.

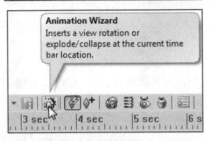

- Click the Animator Wizard icon.

- Check the Collapse box.

- Click Next.

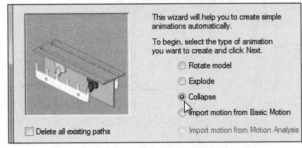

- Enter 10 seconds for Duration.

- Enter 0 for Start Time.

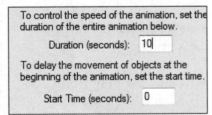

- Click Finish. View the model in the Graphics window

- Click Play from Start.

- Click Yes. View the model in the Graphics window.

- Click the Model tab to return to the SolidWorks model.

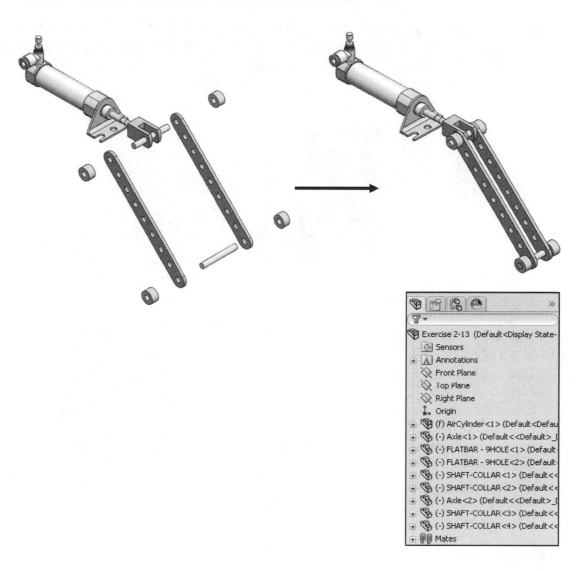

Exercise 2.14: U-BLOCK Assembly

Create the illusteated assembly. The assembly contains the following components: two U-Bracket parts, four Pin parts, and one Square block part. The required parts are located on the DVD in the Chapter2 - Homework folder.

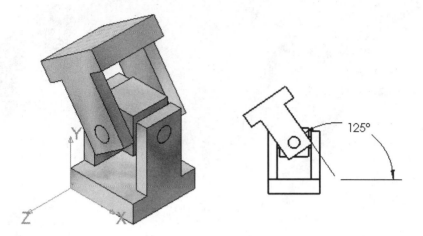

- U-Bracket, (Item 1): Material: AISI 304. Two U-Bracket components are combined together Concentric to opposite holes of the Square block component. The second U-Bracket component is positioned with an Angle mate, to the right face of the first U-Bracket and a Parallel mate between the top face of the first U-Bracket and the top face of the Square block component. Angle A = 125deg.

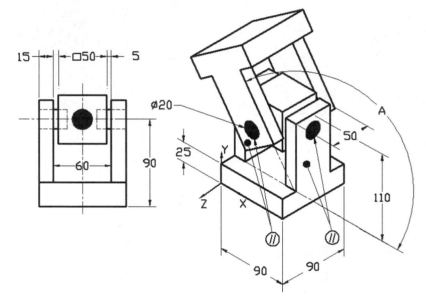

- Square block, (Item 2): Material: AISI 304. The Pin components are mated Concentric and Coincident to the 4 holes in the Square block, (no clearance). The depth of each hole = 10mm.

- Pin, (Item 3): Material: AISI 304. The Pin components are mated Concentric to the hole, (no clearance). The end face of the Pin components are Coincident to the outer face of the U-Bracket components. The Pin component has a 5mm spacing between the Square block component and the two U-Bracket components.

- Locate the Center of mass of the model with respect to the illustrated coordinate system. In this example, start with the U-Bracket part.

Create the coordinate location for the assembly.

- Select the front bottom left vertex of the first U-Bracket component as illustrated.

- Click Insert, Reference Geometry, Coordinate System from the Menu bar. The Coordinate System PropertyManager is displayed.

- Click OK from the Coordinate System PropertyManager. Coordinate System1 is displayed.

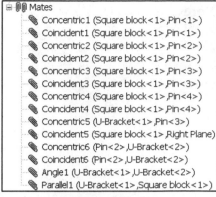

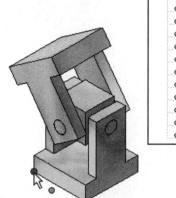

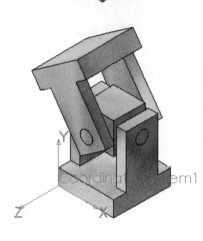

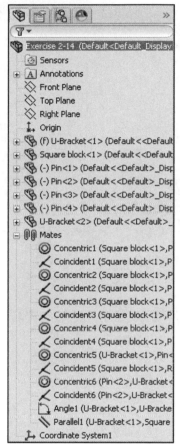

Notes:

Project 3

Fundamentals of Drawing

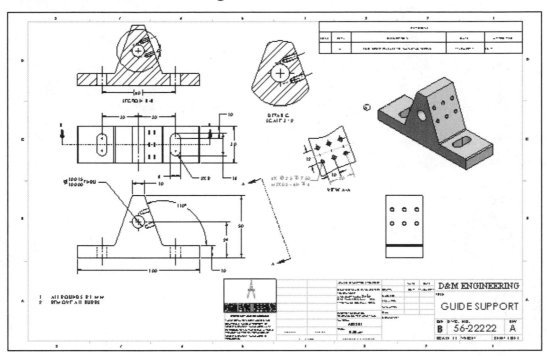

Below are the desired outcomes and usage competencies based on the completion of Project 3.

Project Desired Outcomes:	Usage Competencies:
• B-ANSI-MM Drawing Template.	• Generate a Drawing Template with Document Properties and Sheet Properties.
• CUSTOM-B Sheet Format.	• Produce a Sheet Format with Custom Sheet Properties, Title block, Company logo and more.
• GUIDE Drawing.	• Create Standard Orthographic, Auxiliary, Detail and Section views. • Insert, create, and modify dimensions and annotations.
• GUIDE-ROD Drawing with a Bill of Materials.	• Knowledge to develop and incorporate a Bill of Materials with Custom Properties.

Notes:

Project 3 - Fundamentals of Drawing

Project Objective

Provide an understanding of Drawing Templates, Part drawings, Assembly drawings, details and annotations.

Create a B-ANSI-MM Drawing Template. Create a CUSTOM-B Sheet Format. The Drawing Template contains Document Property settings. The Sheet Format contains a Company logo, Title block, Revision table and Sheet information.

Create the GUIDE drawing. Display Standard Orthographic, Section, Auxiliary and Detail drawing views. Insert, create and modify part and component dimensions.

Create a GUIDE-ROD assembly drawing with a Bill of Materials. Obtain knowledge to develop and incorporate a Bill of Materials with Custom Properties.

On the completion of this project, you will be able to:

- Create a new Drawing Template.

- Generate a customized Sheet Format with Custom Properties.

- Open, Save and Close Drawing documents.

- Produce a Bill of Materials with Custom Properties.

- Insert and position views on a Multi Sheet drawing.

- Set the Dimension Layers.

- Insert, move, and modify dimensions from a part into a drawing view.

- Insert Annotations: Center Mark, Centerline, Notes, Hole Callouts and Balloons.

- Use Edit Sheet Format and Edit Sheet mode.

- Insert a Revision table.

- Modify dimensioning scheme.

- Create Parametric drawing notes.

- Link notes in the Title block to SolidWorks properties.

- Rename parts and drawings.

Project Situation

The individual parts and assembly are completed. What is the next step? You are required to create drawings for various internal departments, namely; production, purchasing, engineering, inspection and manufacturing.

Each drawing contains unique information and specific footnotes. Example: A manufacturing drawing would require information on assembly, Bill of Materials, fabrication techniques and references to other relative documents.

Project Overview

Generate two drawings in this project:

- GUIDE drawing with a customized Sheet Format.

- GUIDE-ROD assembly drawing with a Bill of Materials.

The GUIDE drawing contains three Standard Orthographic views and an Isometric view.

Do you remember what the three Principle Orthographic Standard views are? They are: Top, Front and Right side (Third Angle Projection).

Three new views are introduced in this project: Detailed view, Section view and Auxiliary view. Orient the views to fit the drawing sheet. Incorporate the GUIDE dimensions into the drawing with inserted Annotations.

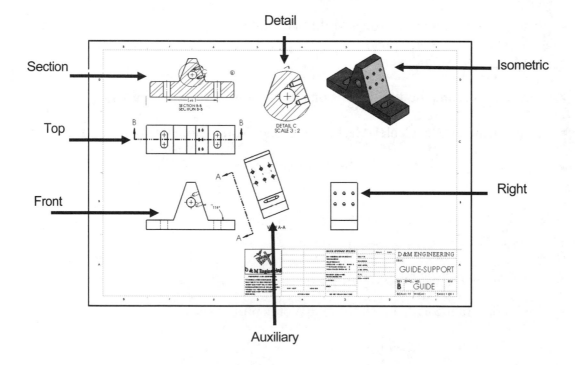

The GUIDE-ROD assembly drawing contains an Isometric Exploded view.

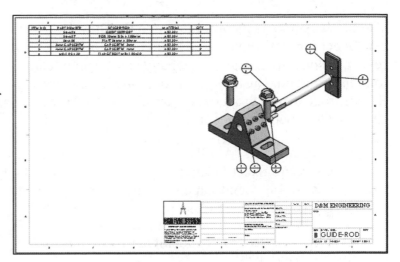

The drawing contains a Bill of Materials with balloon text with bent leaders and magnetic lines.

Both drawings utilize a custom Sheet Format containing a Company logo, Title block and Sheet information.

There are two major design modes used to develop a SolidWorks drawing:

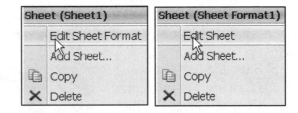

- *Edit Sheet Format*

- *Edit Sheet*

The Edit Sheet Format mode provides the ability to:

- Change the Title block size and text headings.

- Incorporate a company logo.

- Add a drawing, design, or company text.

The Edit Sheet mode provides the ability to:

- Add or modify drawing views.

- Add or modify drawing view dimensions.

- Add or modify text and more.

Drawing Template and Sheet Format

The foundation of a SolidWorks drawing document is the Drawing Template. Drawing size, drawing standards, company information, manufacturing, and or assembly requirements, units and other properties are defined in the Drawing Template.

The Sheet Format is incorporated into the Drawing Template. The Sheet Format contains the border, Title block information, revision block information, company name, and or company logo information, Custom Properties, and SolidWorks Properties. Custom Properties and SolidWorks Properties are shared values between documents.

Utilize the standard B (ANSI) Landscape size Drawing Template with no Sheet Format. Set the Units, Font and Layers. Modify a B (ANSI) Landscape size Sheet Format to create a Custom Sheet Format and Custom Drawing Template.

1. Set Sheet Properties and Document Properties for the Drawing Template.

2. Insert Custom Properties: CompanyName, Revision, Number, DrawnBy, DrawnDate, Company Logo, Third Angle Projection Logo, etc. for the Sheet Format.

3. Save the Custom Drawing Template and Custom Sheet Format in the MY-TEMPLATE file folder.

	Property Name	Type		Value / Text Expression
1	CompanyName	Text		D&M ENGINEERING
2	Revision	Text		A
3	Number	Text		56-22222
4	DrawnBy	Text		DCP
5	DrawnDate	Text		1/24/2011
6	<Type a new property>			

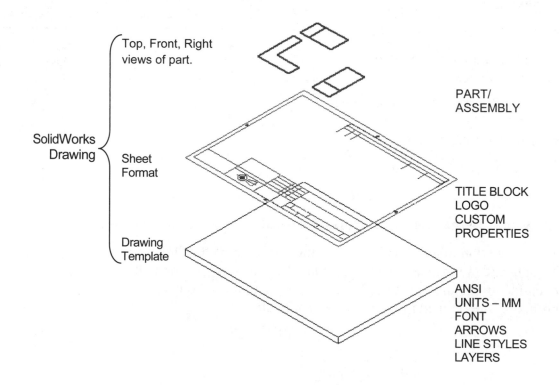

Top, Front, Right views of part.

SolidWorks Drawing

Sheet Format

Drawing Template

PART/ ASSEMBLY

TITLE BLOCK
LOGO
CUSTOM
PROPERTIES

ANSI
UNITS – MM
FONT
ARROWS
LINE STYLES
LAYERS

Views from the part or assembly are inserted into the SolidWorks Drawing.

A Third Angle Projection scheme is illustrated in this project. For non-ANSI dimension standards, the dimensioning techniques are the same, even if the displayed arrows and text size are different.

For printers supporting millimeter paper sizes, select A3 (ANSI) Landscape (420mm x 297mm).

The default Drawing Templates with Sheet Format displayed contain predefined Title block Notes linked to Custom Properties and SolidWorks Properties.

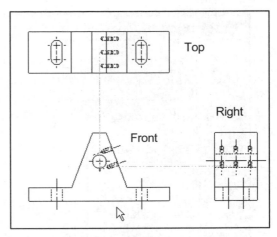

Third Angle Projection

Activity: Drawing Template

Close all documents.
1) Click **Window**, **Close All** from the Menu bar.

Create a B (ANSI) Landscape, Third Angle Projection drawing document.

2) Click **New** from the Menu bar.

3) Double-click **Drawing** from the Templates tab.

4) If needed, **uncheck** the **Only show standard formats** box.

5) Click **OK** from the Sheet Format/ Size dialog. At this time accept the default sheet properties.

6) Click **Cancel** ✖ from the Model View PropertyManager. The Draw1 FeatureManager is displayed.

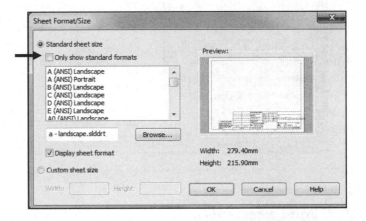

If the Start command when creating new drawing option is checked, the Model View PropertyManager is selected by default.

Draw1 is the default drawing name. Sheet1 is the default first sheet name. The CommandManager alternates between the View Layout, Annotation, Sketch and Evaluate tabs.

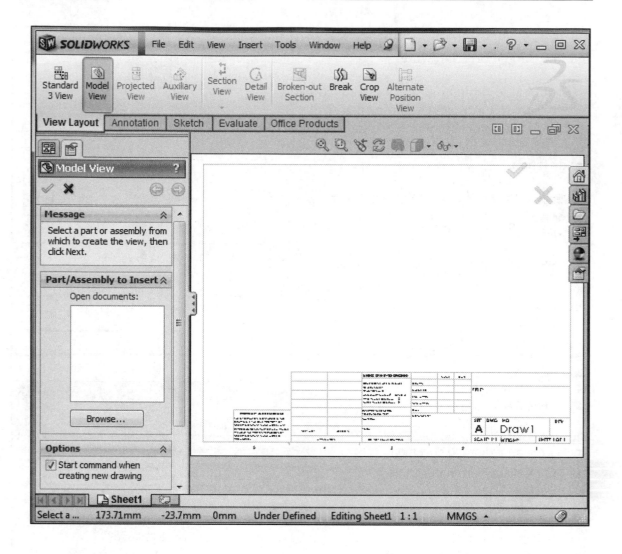

Most commands in this project are accessed through the CommandManager. To display individual toolbars, right-click in the gray area and check the required toolbar.

The B (ANSI) Landscape Standard Sheet border defines the drawing size, 17″ x 11″ (431.8mm x 279.4mm).

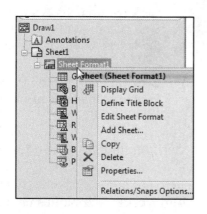

Expand Sheet1 in the FeatureManager. Right-click **Sheet Format1** and view your options. The purpose of this book is to expose the new user to various tools and methods.

Set Drawing Size, Sheet Properties and Document Properties for the Drawing Template. Sheet Properties control Sheet Size, Sheet Scale and Type of Projection. Document Properties control the display of dimensions, annotations and symbols in the drawing.

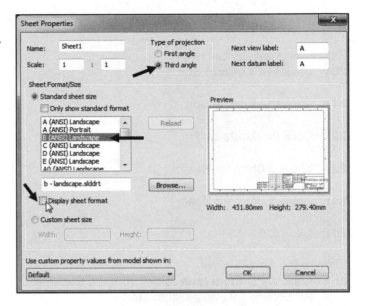

Set Sheet Properties. View your options in the Sheet Properties dialog box.

7) Right-click in the **Graphics window**.

8) Click **Properties** . The Sheet Properties dialog box is displayed.

9) Select Sheet Scale **1:1**.

10) Select **B (ANSI) Landscape** from the Standard sheet size box.

11) Uncheck the **Display sheet format** box.

12) Select **Third Angle** for Type of projection.

13) Click **OK** from the Sheet Properties dialog box.

Set Document Properties.

14) Click **Options** , **Document Properties** tab from the Menu bar.

15) Select **ANSI** for Overall drafting standard.

16) Click the **Units** folder.

17) Select **MMGS** for Unit system.

18) Select **.12** for basic unit length decimal place.

19) Select **None** for basic unit angle decimal place.

Detailing options provide the ability to address: dimensioning standards, text style, center marks, extension lines, arrow styles, tolerance and precision.

There are numerous text styles and sizes available in SolidWorks. Companies develop drawing format standards and use specific text height for Metric and English drawings.

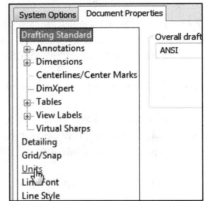

Type	Unit	Decimals
Basic Units		
Length	millimeters	.12
Dual Dimension Length	inches	.12
Angle	degrees	None
Mass/Section Properties		None
Length	millimeters	.1
		.12
Mass	grams	.123
		.1234
Per Unit Volume	millimeters^3	.12345
		123456

Numerous engineering drawings use the following format:

- Font: Century Gothic - All capital letters.

- Text height: .125in. or 3mm for drawings up to B (ANSI) Size, 17in. x 22in.

- Text height: .156in. or 5mm for drawings larger than B (ANSI) Size, 17in x 22in.

- Arrow heads: Solid filled with a 1:3 ratio of arrow width to arrow height.

Set the Annotations font height.
20) Click the **Annotations** folder.

21) Click the **Font** button.

22) Click the **Units** button.

23) Enter **3.0**mm for Height.

24) Enter **1.0**mm for Space.

25) Click **OK**.

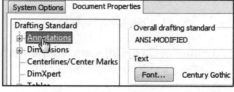

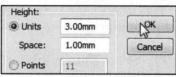

Change the Arrow Height.
26) Click the **Dimensions** folder from the Document Properties column as illustrated.

27) Enter **1**mm for arrow Height.

28) Enter **3**mm for arrow Width.

29) Enter **6**mm for arrow Length.

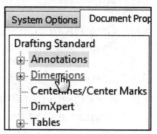

Set Section/View size.
30) **Expand** the View Labels folder from the Document Properties column.

31) Click the **Section** folder.

32) Enter **2**mm for arrow Height.

33) Enter **6**mm for arrow Width.

34) Enter **12**mm for arrow Length.

35) Click **OK** from the Document Properties - Section dialog box.

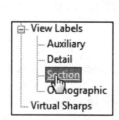

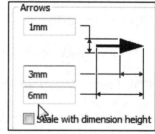

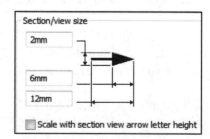

Drawing Layers organize dimensions, annotations, and geometry. Create a new drawing layer to contain dimensions and notes. Create a second drawing layer to contain hidden feature dimensions.

Select the Light Bulb ♀ to turn On/Off Layers. Dimensions placed on the hidden are turned on and off for clarity and can be recalled for parametric annotations.

Display the Layer toolbar.

36) **Right-click** in the gray area to the right of the word Help in the Menu bar. Check **Layer** if the Layer toolbar is not active. The Layer toolbar is displayed.

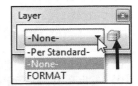

37) Click the **Layer Properties** ▣ file folder from the Layer toolbar. The Layers dialog box is displayed.

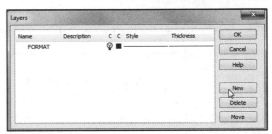

Create the Dimension Layer.

38) Click the **New** button. Enter **Dim** in the Name column.

39) **Double-click** under the Description column. Enter **Dimensions** in the Description column.

Create the Notes Layer.

40) Click the **New** button. Enter **Notes** for Name.

41) **Double-click** under the Description column.

42) Enter **General Notes** for Description.

Create the Hidden Dims Layer.

43) Click the **New** button. Enter **Hidden Dims** for Name. **Double-click** under the Description column. Enter **Hidden Insert Dimensions** for Description.

Dimensions placed on the Hidden Dims Layer are not displayed on the drawing until the Hidden Dims Layer status is On. Set the Layer Color to locate dimensions on this layer easily.

Turn the Hidden Insert Dimension Layer Off.

44) Click **On/Off** ♀. The light bulb is displayed in light gray or white.

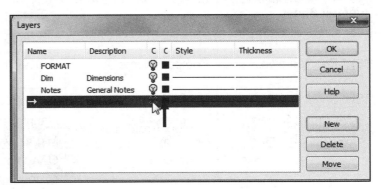

Set the Layer Color.

45) Click the **small black square** ♀ ■ in the Hidden Dims row.

Select a Color Swatch from the Color dialog box.
46) Select **Dark Blue**.

47) Click **OK**.

48) Click **OK** from the Layers dialog box.

The current Layer is Hidden Dims. Set the current Layer to None before saving the Drawing Template.

Set None for Layer.
49) Click the **Layer drop-down arrow**.

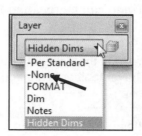
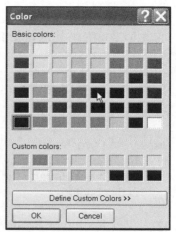

50) Click **None**. None is displayed in the Layer toolbar.

The Drawing Template contains the drawing Size, Document Properties and Layers. The Overall drafting standard is ANSI and the Units are in millimeters. The current Layer is set to None. The Drawing Template requires a Sheet Format. The Sheet Format contains Title block information. The Title block contains vital part or assembly information. Each company may have a unique version of a Title block.

Sheet Format and Title block

The Sheet Format contains the Title block, Revision block, Company logo and Custom Properties. The Title block contains text fields linked to System Properties and Custom Properties.

System Properties are determined from the SolidWorks documents. Custom Property values are assigned to named variables. Save time. Utilize System Properties and define Custom Properties in your Sheet Formats.

System Properties Linked to fields in default Sheet Formats:	Custom Properties of drawings linked to fields in default Sheet Formats:		Custom Properties of parts and assemblies linked to fields in default Sheet Formats:
SW-File Name (in DWG. NO. field):	CompanyName:	EngineeringApproval:	Description (in TITLE field):
SW-Sheet Scale:	CheckedBy:	EngAppDate:	Weight:
SW-Current Sheet:	CheckedDate:	ManufacturingApproval:	Material:
SW-Total Sheets:	DrawnBy:	MfgAppDate:	Finish:
	DrawnDate:	QAApproval:	Revision:
	EngineeringApproval:	QAAppDate:	

Utilize the standard landscape B (ANSI) Sheet Format (17in. x 11in.) or the standard-A (ANSI) Sheet Format (420mm x 297mm) to create a Custom Sheet Format.

Activity: Sheet Format and Title block

Display the standard B (ANSI) Landscape Sheet Format.

51) Right-click in the **Graphics window**.

52) Click **Properties**. The Sheet Properties dialog box is displayed.

53) Click the **Standard sheet size** box.

54) Select **B (ANSI) Landscape**.

55) Check the **Display sheet format** box. The default Sheet Format, b - landscape.slddrt is displayed.

56) Click **OK** from the Sheet Properties dialog box.

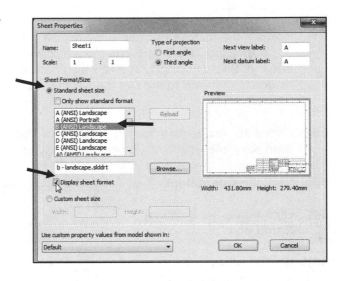

The default Sheet Format is displayed in the Graphics window. The FeatureManager displays Draw1, Sheet 1.

A SolidWorks drawing contains two edit modes:

1. *Edit Sheet*

2. *Edit Sheet Format*

Insert views and dimensions in the Edit Sheet mode.

Modify the Sheet Format text, lines or Title block information in the Edit Sheet Format mode.

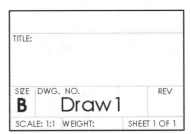

The CompanyName Custom Property is located in the Title block above the TITLE box. There is no value defined for CompanyName. A small text box indicates an empty field. Define a value for the Custom Property CompanyName. Example: D&M ENGINEERING.

A goal of this book is to expose the new user to various tools and methods. An option is the Define Title Block command as illustrated. See SolidWorks Help for additional information.

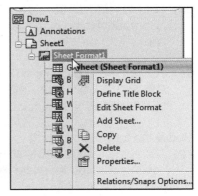

Activate the Edit Sheet Format mode.

57) Right-click in the **Graphics window**.

58) Click **Edit Sheet Format**. The Title block lines turn blue.

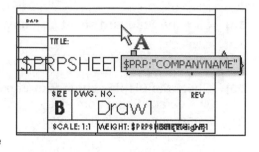

View the right side of the Title block.

59) Click **Zoom to Area** 🔍. **Zoom in** on the Sheet Format Title block.

60) Click **Zoom to Area** 🔍 to deactivate.

Define CompanyName Custom Property.

61) Position the **mouse pointer** in the middle of the box above the TITLE box. The mouse pointer displays Sheet Format1. The box also contains the hidden text, linked to the CompanyName Custom Property.

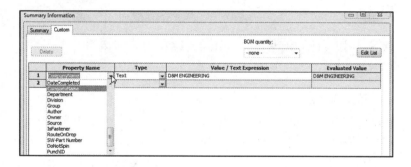

62) Click **File**, **Properties** from the Menu bar. The Summary Information dialog box is displayed.

63) Click the **Custom** tab.

64) Click inside the **Property Name box**.

65) Click the **drop-down arrow** 🔽 in the Property Name box.

66) Select **CompanyName** from the Property List.

67) Enter **D&M ENGINEERING** (or your company name) in the Value/Text Expression box.

68) Click inside the **Evaluated Value** box. The CompanyName is displayed in the Evaluated Value box.

69) Click **OK**. The Custom Property, "$PRP:COMPANYNAME", Value "D&M ENGINEERING" is displayed in the Title block.

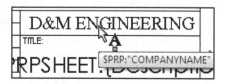

Modify the font size.

70) Double-click **D&M ENGINEERING**. The Formatting dialog box and the Note PropertyManager is displayed.

71) Click the **drop-down arrows** to set the Text Font and Height.

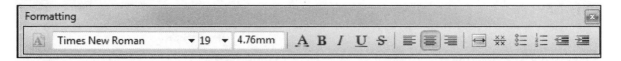

72) Click the **Style buttons** and **Justification buttons** to modify the selected text.

73) Click **OK** ✔ from the Note PropertyManager. View the results.

☼ Click a position outside the selected text box to save and exit the text.

The Tolerance block is located in the Title block. The Tolerance block provides information to the manufacturer on the minimum and maximum variation for each dimension on the drawing. If a specific tolerance or note is provided on the drawing, the specific tolerance or note will override the information in the Tolerance block.

General tolerance values are based on the design requirements and the manufacturing process.

☼ Create Sheet Formats for different parts types. Example: sheet metal parts, plastic parts and high precision machined parts. Create Sheet Formats for each category of parts that are manufactured with unique sets of Title block notes.

Modify the Tolerance block in the Sheet Format for ASME Y14.5 machined, millimeter parts. Delete unnecessary text. The FRACTIONAL text refers to inches. The BEND text refers to sheet metal parts. The Three Decimal Place text is not required for this millimeter part.

Modify the Tolerance Note.
74) Double-click the text **INTERPRET GEOMETRIC TOLERANCING PER:**

75) Enter **ASME Y14.5** as illustrated.

76) Click **OK** ✔ from the Note PropertyManager.

77) Right-click the **Tolerance block** text.

78) Click **Edit Text**.

79) Delete the text **INCHES**.

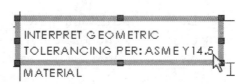

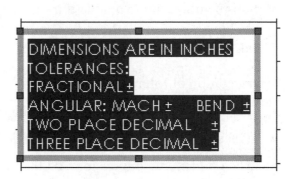

80) Enter **MILLIMETERS**.

81) Delete the line **FRACTIONAL +-**. Delete the text **BEND +-**.

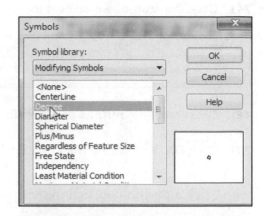

Enter ANGULAR tolerance.
82) Click a **position** at the end of the line.

83) Enter **0**. Click **Add Symbol** 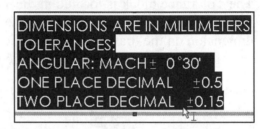 from the Text Format box.

84) Select **Degree** from the Modifying Symbols library.

85) Click **OK**. Enter **30′** for minutes of a degree.

Modify the TWO and THREE PLACE DECIMAL LINES.
86) Delete the **TWO** and **THREE PLACE DECIMAL lines**.

87) Enter **ONE PLACE DECIMAL +- 0.5**.

88) Enter **TWO PLACE DECIMAL +- 0.15**.

89) Click **OK** ✓ from the Note PropertyManager.

90) Right-click **Edit Sheet** in the Graphics window.

Save Draw1. Fit the drawing to the Graphics window.
91) Press the **f** key.

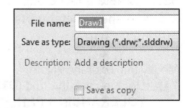

92) Click **Save** 💾. Accept the default name. View the results.

💡 Draw1 is the default drawing file name. This name is temporary. In the next activity, invoke Microsoft Word. Always save before selecting another software application.

Various symbols are available through the Symbol button in the Text dialog box. The ± symbol is located in the Modify Symbols list. The ± symbol is displayed as <MOD-PM>. The degree symbol ° is displayed as <MOD-DEG>.

Interpretation of tolerances is as follows:

- The angular dimension 110 is machined between 109.5 and 110.5.

- The dimension 2.5 is machined between 2.0 and 3.0.

- The Guide Hole dimension 10.000/10.015 is machined according to the specific tolerance on the drawing.

Company Logo

A Company logo is normally located in the Title block of the drawing. You can create your own Company logo or copy and paste an existing picture.

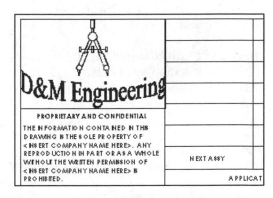

☀ The COMPASS.jpeg file is enclosed on the DVD in the book. Copy all folders and files from the DVD to your hard drive. Do not work directly from the DVD. Insert the provided Company logo in the Edit Sheet Format mode.

☀ If you have your own Logo, use it for the drawing and skip the process of copying it from the DVD.

Activity: Company Logo

Insert a Company Logo.

93) Copy the folder **LOGO** from the Multi-media DVD (**ENGDESIGN-W-SOLIDWORKS\LOGO**) to your hard drive. Note: If you have your own Logo, use it for the drawing and skip the process of copying it from the DVD.

94) Right-click **Edit Sheet Format** in the Graphics window.

95) Click **Insert**, **Picture** from the Menu bar. The Open dialog box is displayed.

96) Select the **Logo.jpg file** from your hard drive.

97) Click **Open**. The Sketch Picture PropertyManager is displayed.

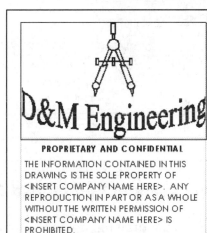

98) Drag the picture handles to size the **picture** to the left side of the Title block. Note: Text was added to the picture.

99) Click **OK** ✅ from the Sketch Picture PropertyManager.

☀ Text can be added to create a custom logo. You can insert a picture or an object.

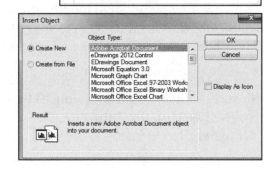

Return to the Edit Sheet mode.

100) Right-click in the **Graphics window**.

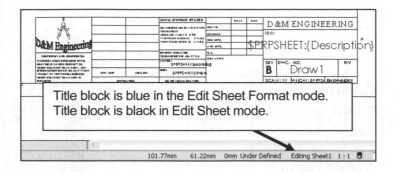

Title block is blue in the Edit Sheet Format mode.
Title block is black in Edit Sheet mode.

101) Click **Edit Sheet**. The Title block is displayed in black.

Return to Edit Sheet mode.

102) Right-click in the **Graphics window**.

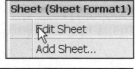

103) Click **Edit Sheet**. The Title block is displayed in black/gray.

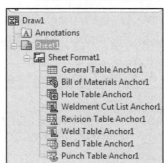

Fit the Sheet Format to the Graphics window.

104) Press the **f** key.

Draw1 displays Editing Sheet1 in the Status bar. The Title block is displayed in black when in Edit Sheet mode.

Save Sheet Format and Save As Drawing Template

Save the drawing document in the Graphics window in two forms: Sheet Format and Drawing Template. Save the Sheet Format as a custom Sheet Format named CUSTOM-B. Use the CUSTOM-B (ANSI) Sheet Format for the drawings in this project. The Sheet format file extension is .slddrt.

The Drawing Template can be displayed with or without the Sheet Format. Combine the Sheet Format with the Drawing Template to create a custom Drawing Template named B-ANSI-MM. Utilize the File, Save As option to save a Drawing Template. The Drawing Template file extension is .drwdot.

🔆 Select the Save as type option first, then select the Save in folder to avoid saving in default SolidWorks installation directories.

The System Options, File Locations, Document Templates option is only valid for the current session of SolidWorks in some *network locations*. Set the File Locations option in order to view the MY-TEMPLATES tab in the New Document dialog box.

Activity: Save Sheet Format and Save As Drawing Template

Save the Sheet Format.

105) Click **File**, **Save Sheet Format** from the Menu bar. The Save Sheet Format dialog box is displayed. The file extension for Sheet Format is .slddrt.

106) Select **ENGDESIGN-W-SOLIDWORKS\MY-TEMPLATES** for Save In File Folder.

107) Enter **CUSTOM-B** for File name.

108) Click **Save** from the Save Sheet Format dialog box.

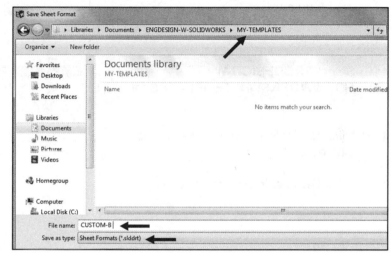

🔆 The book is designed to expose the new user to many tools, techniques and procedures. It may not always use the most direct tool or process.

Save the Drawing Template.

109) Click **Save As** from the Menu bar.

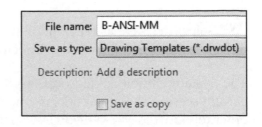

110) Click **Drawing Templates(*.drwdot)** from the Save as type box.

111) Select **ENGDESIGN-W-SOLIDWORKS\MY-TEMPLATES** for Save In File Folder.

112) Enter **B-ANSI-MM** for File name.

113) Click **Save**.

Set System Options - File Locations.

114) Click **Options** , **File Locations** from the Menu bar.

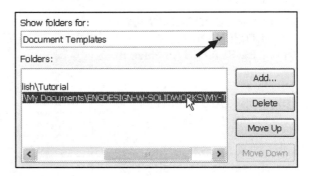

115) Click **Add**.

116) Select **ENGDESIGN-W-SOLIDWORKS\MY-TEMPLATES** for folder.

117) Click **OK** from the Browse For Folder dialog box.

118) Click **OK** to exit System Options.

119) Click **Yes**.

Close all files.

120) Click **Window**, **Close All** from the Menu bar. SolidWorks remains open, no documents are displayed.

Utilize Drawing Template descriptive filenames that contain the size, dimension standard and units.

Combine customize Drawing Templates and Sheet Formats to match your company's drawing standards. Save the empty Drawing Template and Sheet Format separately to reuse information.

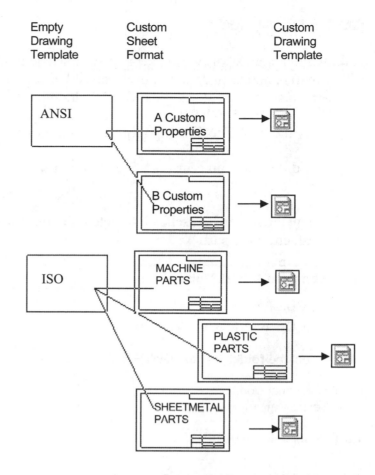

🔍 Additional details on Drawing Templates, Sheet Format and Custom Properties are available in SolidWorks Help Topics.

Keywords: Documents (templates, properties) Sheet Formats (new, new drawings, note text), Properties (drawing sheets), Customized Drawing Sheet Formats.

⚙️ Review Drawing Templates

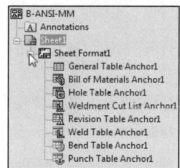

A custom Drawing Template was created from the default Drawing Template. You modified Sheet Properties and Document Properties to control the Sheet size, scale, annotations, dimensions and layers.

The Sheet Format contained Title block and Custom Property information. You inserted a Company Logo and modified the Title block.

The Save Sheet Format option was utilized to save the CUSTOM-B.slddrt Sheet Format. The File, Save As option was utilized to save the B-ANSI-MM.drwdot Template. The Sheet Format and Drawing Template were saved in the MY-TEMPLATES folder.

💡 As an exercise, explore the sub-folders and their options under Sheet Format1.

GUIDE Part-Modify

A drawing contains part views, geometric dimensioning and tolerances, centerlines, center marks, notes, custom properties and other related information. Perform the following tasks before starting the GUIDE drawing:

- Verify the part. The drawing requires the associated part.

- View dimensions in each part. Step through each feature of the part and review all dimensions.

- Review the dimension scheme to determine the required dimensions and notes to manufacture the part.

Activity: GUIDE Part-Modify

Open the GUIDE part.
121) Click **Open** ✏ from the Menu bar.

122) Select the **folder** that the GUIDE document is in.

Modify the dimensions.
123) Select **Part** for file type.

124) Double-click **GUIDE**.

125) Click **Hidden Lines Visible** ◰ from the Heads-up View toolbar.

126) Double-click **Base-Extrude** from the GUIDE FeatureManager.

127) Click the **80** dimension.

128) Enter **100**mm.

129) Click **inside** the Graphics window.

Save the GUIDE part.
130) Click **Shaded With Edges** ◰ from the Heads-up View toolbar.

131) Click **Save** 💾.

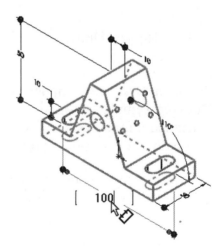

🔅 Review part history with the Rollback bar to understand how the part was created. Position the Rollback bar at the top of the FeatureManager. Drag the Rollback bar below each feature. When working between features, right-click in the FeatureManager. Select Roll to Previous\Roll to End.

GUIDE Part Drawing

The GUIDE drawing consists of multiple views, dimensions and annotations. The GUIDE part was designed for symmetry. Add or redefine dimensions in the drawing to adhere to a Drawing Standard. Add dimensions and notes to the drawing in order to correctly manufacture the part.

Address the dimensions for three features:

- The right Slot Cut is not dimensions to an ASME Y14 standard.

- No dimensions exist for the left Mirror Slot Cut.

- The Guide Hole and Linear Pattern of Tapped Holes require notes.

The GUIDE part remains open. Create a new GUIDE drawing. Utilize the B-ANSI-MM Drawing Template. Utilize the Model View tool to insert the Front view into Sheet1. Utilize the Auto-start Projected View option to project the Top, Right and Isometric views from the Front view.

Activity: GUIDE Part Drawing

Create the GUIDE Drawing.

132) Click **New** from the Menu bar.

133) Double-click **B-ANSI-MM** from the MY-TEMPLATES tab. The Model View PropertyManager is displayed.

134) Click **Cancel** from the Model View PropertyManager. The Draw2 FeatureManager is displayed.

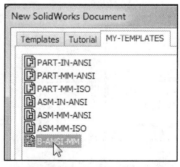

Review the Draw2 FeatureManager.
135) Expand Sheet1 from the FeatureManager.

The current drawing name is Draw2 if the second new drawing is created in the same session of SolidWorks. The current sheet name is Sheet1. Sheet1 is the current Sheet.

Insert four Drawing Views.

136) Click **Model View**  from the View Layout tab in the CommandManager. The Model View PropertyManager is displayed.

137) Double-click **GUIDE** from the Part/Assembly to Insert box.

Insert four views.

138) Check the **Create multiple views** box.

139) Click *Front, *Top and *Right from the Orientation box. Note: All four views are selected. *Isometric is selected by default.

140) Click **OK** ✓ from the Model View PropertyManager. Click **Yes**. The four views are displayed on Sheet1.

141) Click inside the **Isometric view boundary**. The Drawing View4 PropertyManager is displayed.

142) Click **Shaded With Edges** ⬚ from the Display Style box.

143) Click **OK** ✓ from the Drawing View4 PropertyManager. If required, hide any annotations, origins or dimensions as illustrated.

By default, the Center marks-holes box is checked under; **Options**, **Document Properties**, **Detailing** from the Menu bar.

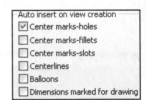

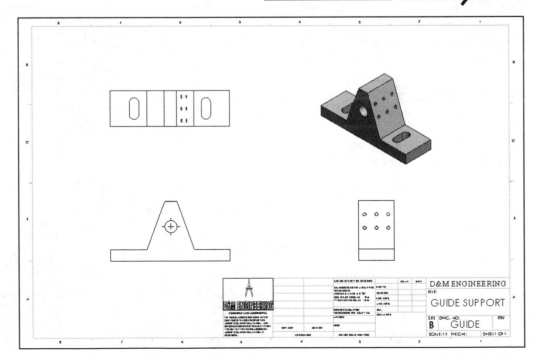

Save the GUIDE drawing.

144) Click **Save As** from the Menu bar.

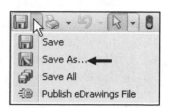

145) Select **PROJECTS** for Save in folder. GUIDE is the default filename. Drawing is the default Save as type.

146) Click **Save**. Note: The correct display modes need to be selected, dimensions to be added, along with Center Marks, Centerlines, additional views, Custom Properties, etc.

The DWG. NO. box in the Title block displays the part File name, GUIDE. The TITLE: box in the Title block displays GUIDE SUPPORT.

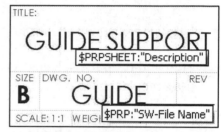

Predefined text in the CUSTOM-B Sheet Format links the Properties: $PRPSHEET:"Description" and $PRP:"SW-FileName". The Properties were defined in the GUIDE part utilizing File, Save As. Properties in the Title block are passed from the part to the drawing.

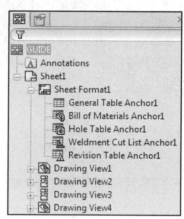

Always confirm your File name and Save in folder. Projects deal with multiple File names and folders. Select Save as type from the drop down list. Do not enter the extension. The file extension is entered automatically.

Each drawing has a unique file name. Drawing file names end with a .SLDDRW suffix. Part file names end with a .SLDPRT suffix. A drawing or part file can have the same prefix. A drawing or part file **can't have the same suffix**.

Example: Drawing file name: GUIDE.SLDDRW. Part file name: GUIDE.SLDPRT. The current file name is GUIDE.SLDDRW.

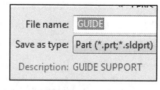

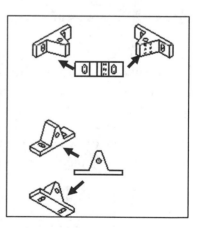

The GUIDE drawing contains three Principle Views (Standard Orthographic Views): Front, Top, Right and an Isometric View. You

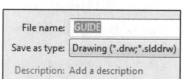

created the views with the Model View option.

Drawing views are inserted as follows:

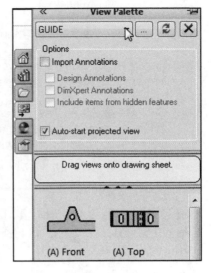

- Utilize the Model View tool from the View Layout tab in the CommandManager.

- Drag and drop a part into the drawing to create three Standard Views.

- Predefine views in a custom Drawing Template.

- Drag a hyperlink through Internet Explorer.

- Drag and drop an active part view from the View Palette located in the Task Pane. With an open part, drag and drop the selected view into the active drawing sheet.

The Top view and Right view is projected off the view you place in the Front view location! Any view can be dragged and dropped into the Front view location of a drawing.

The View Palette from the Task Pane populates when you:

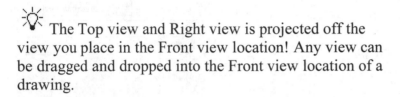

- Click Make Drawing from Part/Assembly

- Browse to a document from the View Palette

- Select from a list of open documents in the View Palette

Move Views and Properties of the Sheet

The GUIDE drawing contains four views. Reposition the view on a drawing. Provide approximately 1in. - 2in., (25mm - 50mm) between each view for dimension placement.

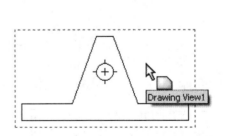

Move Views on Sheet1 to create space for additional Drawing View placement. The mouse pointer provides

feedback in both the Drawing Sheet and Drawing View modes. The mouse

pointer displays the Drawing Sheet icon when the Sheet properties and commands are executed.

The mouse pointer displays the Drawing View icon when the View properties and commands are executed.

View the mouse pointer for feedback to select Sheet, View, and Component and Edge properties in the Drawing.

Sheet Properties

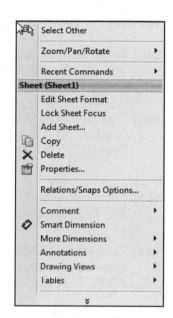

- Sheet Properties display properties of the selected sheet.

  Right-click in the sheet boundary to view the available commands.

View Properties

- View Properties display properties of the selected view.

  Right-click inside the view boundary. Modify the View Properties in the Display Style box or the View Toolbar.

Component Properties

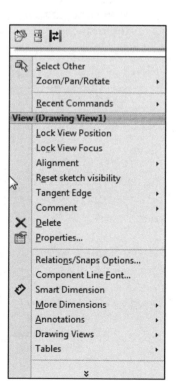

- Component Properties display properties of the selected

  component. Right-click on the face of the component. View the available options.

Edge Properties

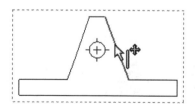

- Edge Properties display properties of the selected geometry. Right-click on an edge inside the view boundary. View the available options.

Activity: Move Views and Properties of the Sheet

Modify and move the front view.

147) Click inside the **Drawing View1** (Front) view boundary. The mouse pointer displays the Drawing View 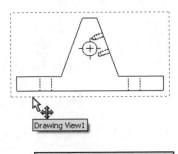 icon. The view boundary is displayed in blue.

148) Click **Hidden Lines Visible** ⬚ from the Display Style box.

149) Position the **mouse pointer** on the edge of the Front view until the Drawing Move View ✛ icon is displayed.

150) Click and drag **Drawing View1** in an upward vertical direction. The Top and Right view move aligned to Drawing View1 (Front).

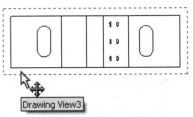

Modify the top view.

151) Click inside the **Top view boundary**, (Drawing View3). The Drawing View3 PropertyManager is displayed.

152) Click **Hidden Lines Removed** ⬚ from the Display Style box. Various display styles provides the ability to select and view features of a part.

153) Click **OK** ✓ from the Drawing View3 PropertyManager. Later, address Centerlines, Center Marks, Tangent Edges Removed, Display styles, Customer properties, etc. to finish the drawing.

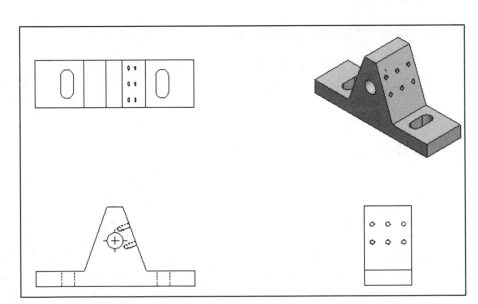

🔆 Select the dashed view boundary to move a view in the drawing.

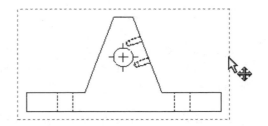

Auxiliary View, Section View and Detail View

The GUIDE drawing requires additional views to document the part. Insert an Auxiliary view, Section view and Detail view from the View Layout tab in the CommandManager. Review the following view terminology before you begin the next activity.

Auxiliary View

The Auxiliary view ✍ drawing tool provides the ability to display a plane parallel to an angled plane with true dimensions.

A primary Auxiliary view is hinged to one of the six Principle Orthographic views. Create a Primary Auxiliary view that references the angled edge in the Front view.

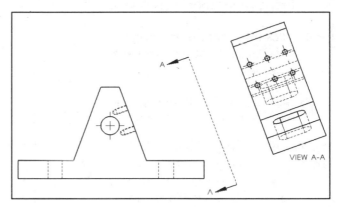

Section View

The Section view ⇵ drawing tool provides the ability to display the interior features of a part. Define a cutting plane with a sketched line in a view perpendicular to the Section view.

Create a full Section view by sketching a section line in the Top view.

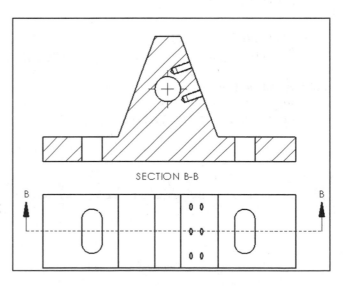

Detail View

The Detailed view 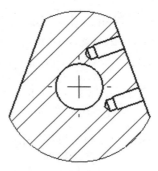 drawing tool provides the ability to enlarge an area of an existing view. Specify location, shape and scale.

Create a Detail view from a Section view with a 3:2 scale.

The book is designed to expose the new user to many tools, techniques and procedures. It does not always use the most direct tool or process.

DETAIL C
SCALE 3 : 2

Activity: Auxiliary Drawing View

Insert an Auxiliary view.

154) Click the **View Layout** tab from the CommandManager.

155) Click the **Auxiliary View** drawing tool. The Auxiliary View PropertyManager is displayed.

156) Click the **right angled edge** of the GUIDE in the Front view as illustrated.

157) Click a **position** to the right of the Front view as illustrated.

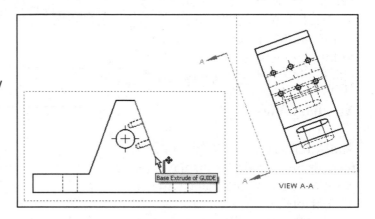

VIEW A-A

Position the Auxiliary View.

158) Click and drag the **section line A-A midpoint** toward Drawing View1. The default label, A is displayed in the Arrow box.

159) Click the **OK** from the PropertyManager.

160) **Rename** the new view (Drawing View5) to Auxiliary in the FeatureManager as illustrated.

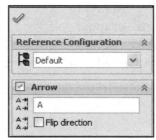

Activity: Section View

Insert a Section view.

161) Click the **Section View** ⇵ drawing tool. The Section View PropertyManager is displayed. Locate the midpoint of the left vertical line for a reference as illustrated.

162) Position the **mouse pointer** over the center of the left vertical line. *Do not click the midpoint.*

163) Drag the **mouse pointer** to the left.

164) Click a **position** to the left of the profile.

165) Click a **position** horizontally to the right of the profile as illustrated. The Section line extends beyond the left and right profile lines.

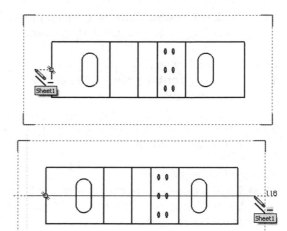

Position the Section View.

166) Click a **position** above the Top view. The section arrows point downward.

167) Check **Flip direction** from the Section Line box. The section arrows point upward. If required, enter **B** for Section View Name in the Label box.

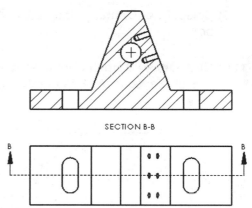

SECTION B-B

168) Click **OK** ✅ from the Section View B-B PropertyManager. Section View B-B is displayed in the FeatureManager. If required, hide any annotations.

Save the drawing.

169) Click **Save** 💾.

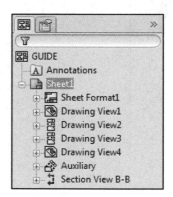

🔅 The material in the GUIDE part determines the hatch pattern in the GUIDE drawing.

Activity: Detail View

Insert a Detail view.

170) Click the **Detail View** 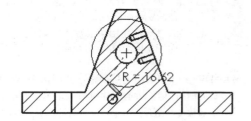 drawing tool. The Detail View PropertyManager is displayed.

Sketch a Detail circle.

171) Click the **center point** of the Guide Hole in the Section view.

172) Click a **position** to the lower left of the Guide Hole to complete the circle.

Position the Detail View.

173) Click a **position** to the right of the Section View. If required, enter **C** for Detail View Name in the Label box.

Change the scale.

174) Check **Use custom scale** option.

175) Select **User Defined**. Enter **3:2** in the Scale box.

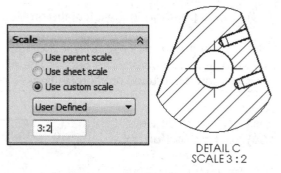

176) Click **OK** ✔ from the Detail View C PropertyManager. Detail View C is displayed in the FeatureManager.

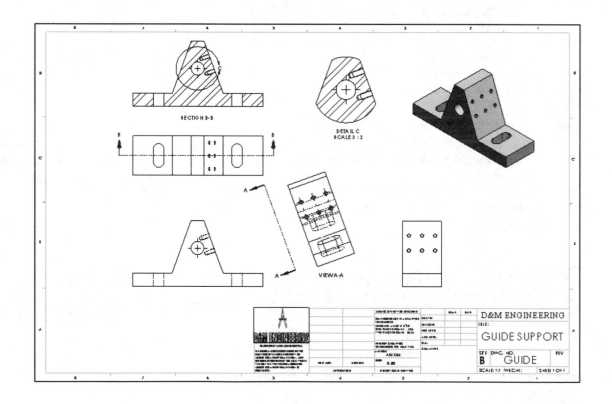

Save the GUIDE drawing.

177) Click **Save** .

Partial Auxiliary View - Crop View

Create a Partial Auxiliary view from the Full Auxiliary view. Sketch a closed profile in the active Auxiliary view. Create the Profile with a closed Spline. Create a Partial Auxiliary view.

Crop the view. The 6mm dimension references the centerline from the Guide Hole. For Quality Assurance and Inspection of the GUIDE part, add a dimension that references the Temporary Axis of the Guide Hole. Sketch a centerline collinear with the Temporary Axis.

Activity: Partial Auxiliary View-Crop View

Select the view.

178) Click **Zoom to Area** from the Heads-up View toolbar.

179) Zoom in on the Auxiliary view.

180) Click **Zoom to Area** to deactivate.

181) Click inside the **Auxiliary view boundary**. The Auxiliary PropertyManager is displayed.

182) Click **Hidden Lines Removed** .

Sketch a closeD Spline profile.
183) Click the **Sketch** tab from the CommandManager.

184) Click the **Spline** Sketch tool.

185) Click seven or more **positions** clockwise to create the closed Spline as illustrated. The first point is Coincident with the last point. The Sketch profile is closed.

Insert a Partial Auxiliary View.

186) Click the **Crop View** drawing tool from the View Layout tab in the CommandManager.

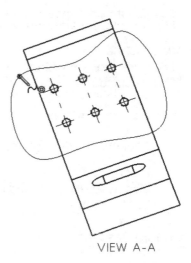

VIEW A-A

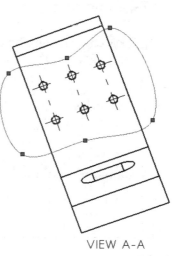

VIEW A-A

187) Click **OK** ✅ from the Spline PropertyManager. The Crop View is displayed on Sheet1.

Insert a sketched centerline.
188) Click inside the **Auxiliary view boundary**. The Auxiliary PropertyManager is displayed.

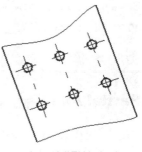

VIEW A-A

189) Click **View**; check **Temporary Axes** from the Menu bar. The Temporary Axis for the Guide Hole is displayed.

190) Click **Sketch** tab from the CommandManager.

191) Click the **Centerline** ⫶ Sketch tool. The Insert Line PropertyManager is displayed.

192) Sketch a **centerline** parallel, above the Temporary axis. The centerline extends approximately 5mm to the left and right of the profile lines. Note: If needed insert a Parallel relation to the temporary axis.

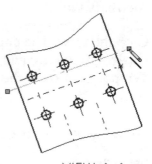

VIEW A-A

193) Right-click **Select** to deselect the Centerline Sketch tool.

Add a Collinear relation between the centerline and temporary axis.
194) Click the **centerline**. The Line Properties PropertyManager is displayed.

195) Hold the **Ctrl** key down.

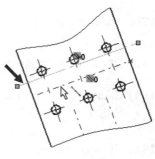

196) Click the **Temporary Axis**. The mouse pointer displays the Axis feedback ⬉ icon. The Properties PropertyManager is displayed. Axis<1> and Line1 is displayed in the Selected Entities box.

197) Release the **Ctrl** key.

198) Click **Collinear** from the Add Relations box.

199) Click **OK** ✅ from the Properties PropertyManager.

Hide the Temporary Axis.
200) Click **View**; uncheck **Temporary Axes** from the Menu bar.

Move the views to allow for ample spacing for dimensions and notes.

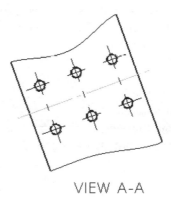

VIEW A-A

🔍 Additional information on creating a New Drawing, Model View, Move View, Auxiliary View, Section View, and Detail View are location in the SolidWorks Help Topics section. Keywords: New (drawing document), Auxiliary View, Detail View, Section View and Crop View.

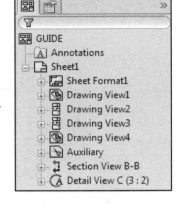

Review the GUIDE Drawing

You created a new drawing, GUIDE with the B-ANSI-MM Drawing Template. The GUIDE drawing utilized the GUIDE part in the Model View PropertyManager. The Model View PropertyManager allowed new views to be inserted with a View Orientation. You selected Front, Top, Right and Isometric to position the GUIDE views.

Additional views were required to fully detail the GUIDE. You inserted the Auxiliary Section, Detail, Partial Auxiliary and Crop view. You moved the views by dragging the view boundary. The next step is to insert the dimensions and annotations to detail the GUIDE drawing.

Display Modes and Performance

Display modes for a Drawing view are similar to a part. When applying Shaded With Edges, select either Tangent edges removed or Tangent edges As phantom from the System Options section.

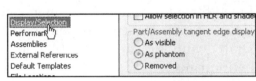

Mechanical details require either the Hidden Lines Visible mode or the Hidden Lines Removed display mode. Select Shaded/Hidden Lines Removed to display Auxiliary views to avoid confusion.

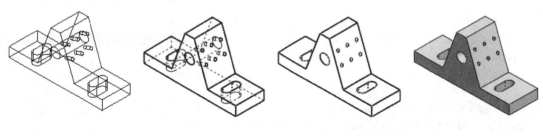

Wireframe   Hidden Line Visible   Hidden Line Removed   Shaded With Edges

Tangent Edges Visible provides clarity for
feature edges. To address the ASME Y14.5
standard, use Tangent Edges With Font
(Phantom lines) or Tangent Edges
Removed. Right-click in the view boundary
to access the Tangent Edge options.

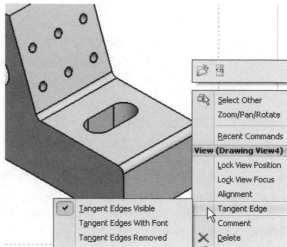

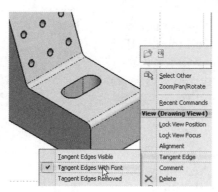

Drawing views can be displayed in High
quality and Draft quality. In High quality, all
model information is loaded into memory. By
default, drawing views are displayed in High
quality.

In Draft quality, only minimum model
information is loaded into memory. Utilize
Draft quality for large assemblies to increase
performance.

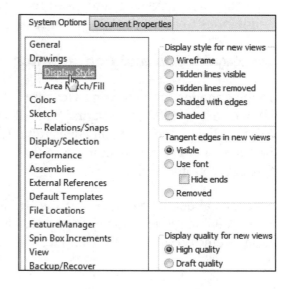

Utilize **Options**, **System Options**,
Drawings, **Display Style** to control the quality
of a view.

By default, SolidWorks will populate Section
and Detail views before other views on your
drawing.

Use the Pack and Go tool to save an assembly
or drawing with references. The Pack and Go tool
saves either to a folder or creates a zip file to e-mail.
View SolidWorks help for additional information.

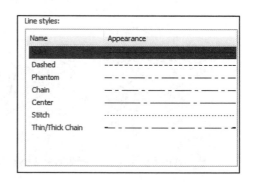

Detail Drawing

The design intent of this project is to work with dimensions inserted from parts and to incorporate them into the drawings. Explore methods to move, hide and recreate dimensions to adhere to a drawing standard.

There are other solutions to the dimensioning schemes illustrated in this project. Detail drawings require dimensions, annotations, tolerance, materials, Engineering Change Orders, authorization, etc. to release the part to manufacturing and other notes prior to production.

Review a hypothetical "worse case" drawing situation. You just inserted dimensions from a part into a drawing. The dimensions, extensions lines and arrows are not in the correct locations. How can you address the position of these details? Answer: Dimension to an ASME Y14.5M standard.

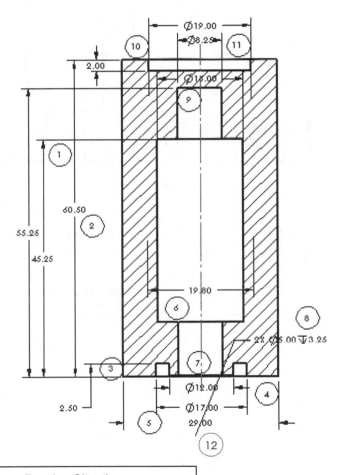

No.	Situation:
1	Extension line crosses dimension line. Dimensions not evenly spaced.
2	Largest dimension placed closest to profile.
3	Leader lines overlapping.
4	Extension line crossing arrowhead.
5	Arrow gap too large.
6	Dimension pointing to feature in another view. Missing dimension – inserted into Detail view (not shown).
7	Dimension text over centerline, too close to profile.
8	Dimension from other view – leader line too long.
9	Dimension inside section lines.
10	No visible gap.
11	Arrows overlapping text.
12	Incorrect decimal display with whole number (millimeter), no specified tolerance.

Worse Case Drawing Situation

The ASME Y14.5M 2009 standard defines an engineering drawing standard. Review the twelve changes made to the drawing to meet the standard.

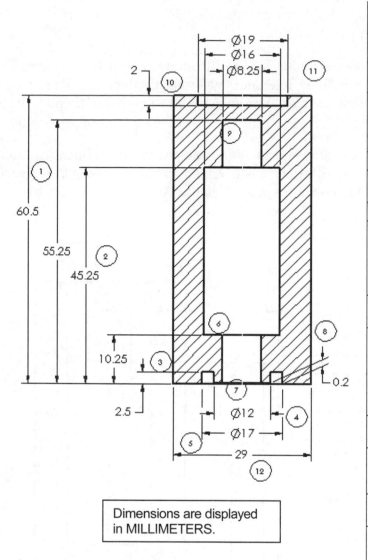

No.	Preferred Application of the Dimensions:
1	Extension lines do not cross unless situation is unavoidable. Stagger dimension text.
2	Largest dimension placed farthest from profile. Dimensions are evenly spaced and grouped.
3	Arrow heads do not overlap.
4	Break extension lines that cross close to arrowhead.
5	Flip arrows to the inside.
6	Move dimensions to the view that displays the outline of the feature. Ensure that all dimensions are accounted for.
7	Move text off of reference geometry (centerline).
8	Drag dimensions into their correct view boundary. Create reference dimensions if required. Slant extension lines to clearly illustrate feature.
9	Locate dimensions outside off section lines.
10	Create a visible gap between extension lines and profile lines.
11	Arrows do not overlap the text.
12	Whole numbers displayed with no zero and no decimal point (millimeter).

Apply these dimension practices to the GUIDE drawing. Manufacturing utilizes detailed drawings. A mistake on a drawing can cost your company substantial loss in revenue. The mistake could result in a customer liability lawsuit.

As the designer, dimension and annotate your parts clearly to avoid common problems and mistakes.

Insert Dimensions from the part.
Dimensions you created for each part feature are inserted into the drawing.

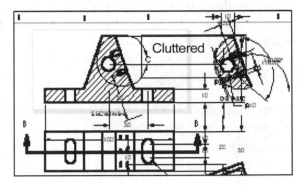

Select the first dimensions to display for the Front view. Do not select the Import Items into all Views option for complex drawings. Dimension text is cluttered and difficult to locate.

Follow a systematic, "one view at a time" approach for complex drawings. Insert part feature dimensions onto the Dims Layer for this project.

Activity: Detail Drawing-Insert Model Items

Set the Dimension Layer.
201) Click the **drop-down arrow** from the Layer dialog box. If needed, click View, Toolbars, Layer from the Main menu to display the Layer dialog box.

202) Click **Dim**.

Insert dimensions into Drawing View1 (Front).
203) Click inside the **Drawing View1** boundary. The Drawing View1 PropertyManager is displayed.

204) Click the **Model Items** tool from the Annotation tab in the CommandManager. The Model Items PropertyManager is displayed.

205) Select **Entire model** from the Source box. Drawing View1 is displayed in the Destination box. Note: At this time, we will not click the Hole callout button to import the Hole Wizard information into the drawing. Click the Accept the default settings as illustrated.

206) Click **OK** ✓ from the Model Items PropertyManager. Dimensions are displayed in the Front view.

A goal of this text is to expose the new user to various tools, techniques and procedures. It may not always use the most direct tool or process.

Drawing dimension location is dependent on:

- Feature dimension creation

- Selected drawing views

Note: The Import items into all views option, first inserts dimensions into Section Views and Detail Views. The remaining dimensions are distributed among the visible views on the drawing.

Move Dimensions in the Same View

Move dimensions within the same view. Use the mouse pointer to drag dimensions and leader lines to a new location.

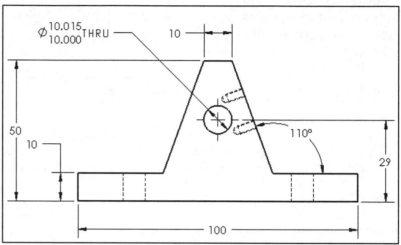

Leader lines reference the size of the profile. A gap must exist between the profile lines and the leader lines. Shorten the leader lines to maintain a drawing standard. Use the blue Arrow control buttons to flip the dimension arrows.

Insert part dimensions into the Top view. The Top view displays crowded dimensions. Move the overall dimensions. Move the Slot Cut dimensions.

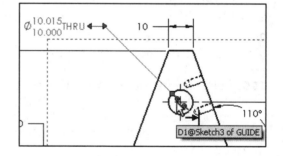

Place dimensions in the view where they display the most detail. Move dimensions to the Auxiliary View. Hide the diameter dimensions and add Hole Callouts. Display the view with Hidden Lines Removed.

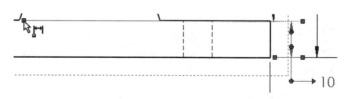

Illustrations may vary depending on your SolidWorks release version.

Activity: Detail Drawing-Move Dimensions

Move the linear dimensions in Drawing View1.

207) Zoom to area on Drawing View1.

208) Click and drag the vertical dimension text **10, 29,** and **50** to the right as illustrated.

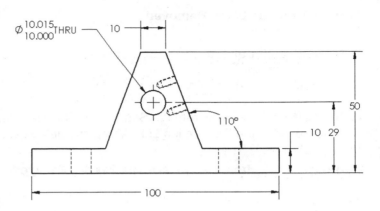

Create a gap between the extension lines and the profile lines for the 10mm vertical dimension.

209) Click the vertical dimension text **10**. The vertical dimension text, extension lines and the profile lines are displayed in blue.

210) Click and drag the **square blue endpoints** approximately 10mm's from the right vertex as illustrated. A gap is created between the extension line and the profile.

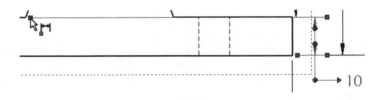

A gap exists between the profile line and the leader lines. Drag the blue endpoints to a vertex, to create a gap

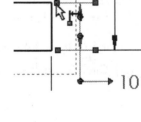

☼ The smallest linear dimensions are closest to the profile.

☼ Click the dimension Palette rollover ⬦ button to display the dimension palette. Use the dimension palette in the Graphics window to save mouse travel to the Dimension PropertyManager. Click on a dimension in a drawing view, and modify it directly from the dimension palette..

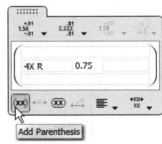

Fit the drawing to the Graphics window.
211) Press the **f** key.

Insert dimensions into Drawing View3.

212) Click inside the **Drawing View3** boundary. The Drawing View3 PropertyManager is displayed.

213) Click **Hidden Lines Removed**.

214) Click the **Model Items** tool from the Annotation tab in the CommandManager. The Model Items PropertyManager is displayed.

215) Select **Entire model** from the Source box. Drawing View3 is displayed in the Destination box. Accept the default settings.

216) Click **OK** from the Model Items PropertyManager.

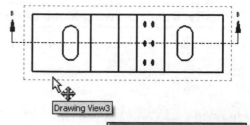

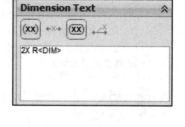

Move the vertical dimensions.

217) Click and drag the two vertical Slot Cut dimensions, **10** to the right of the Section arrow as illustrated.

218) Flip the **arrows** to the inside.

219) Click the **dimension text** and drag the text outside the leader lines. Hide all other dimensions and annotations as illustrated.

220) Enter **2X** for the Radius as illustrated.

Fit the drawing to the Graphics window.

221) Press the **f** key.

Save the GUIDE drawing.

222) Click **Save**.

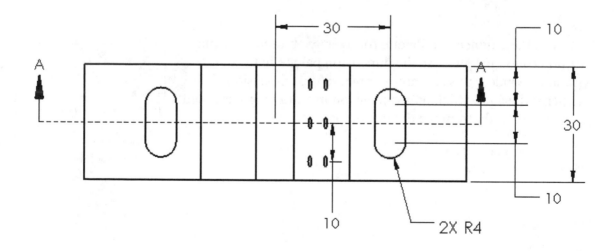

Insert dimensions into the Auxiliary View.
223) Click inside the Auxiliary view boundary.

224) Click the **Model Items** 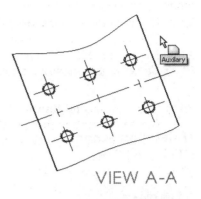 Drawing tool. The Model Items PropertyManager is displayed.

225) Select **Entire Model** from the Source box. Auxiliary is displayed in the Destination box.

226) Click **Hole Wizard Locations** from the Dimensions box. Note: Do not click the Hole Callout button at this time. You will manual create the Hole Wizard Callout in the drawing.

227) Click **OK** ✔ from the Model Items PropertyManager.

228) Move the dimensions off the view as illustrated. If needed, zoom in on the dimensions to move them.

Note: The dimensions for the Linear Pattern of Holes are determined from the initial Hole Wizard position dimensions and the Linear Pattern dimensions. Your dimensioning standard requires the distance between the holes in a pattern.

Do not over dimension. In the next steps, hide the existing dimensions and add a new dimension.

In the next section if required, click **Options, Document Properties, Dimensions** from the Menu bar. Uncheck the **Add parentheses by default** box. Click **OK**.

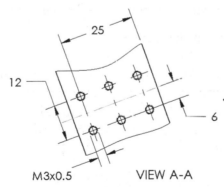

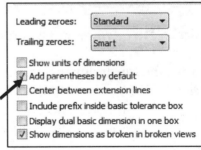

VIEW A-A

M3x0.5   VIEW A-A

Hide the dimensions.
229) Right-click **25**. Click **Hide**. Right-click **6**. Click **Hide**.

230) Right-click **M3x0.5**. Click **Hide**. Note: if required, hide any other dimensions or annotations.

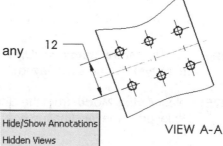

VIEW A-A

To show a hidden dimension, click **View**, **Hide/Show Annotations** from the Menu bar. Click the Hidden dimension.

Add a dimension.
231) Click **Smart Dimension** from the Annotation toolbar. The Dimension PropertyManager is displayed.

232) Click **Smart dimensioning** in the Dimension Assist Tools box.

233) Click the **center point** of the bottom left hole. Click the **center point** of the bottom right hole.

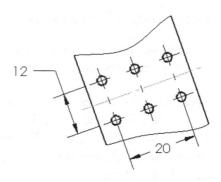

234) Click a **position** below the profile as illustrated.

235) Check **OK** from the Dimension PropertyManager.

Save the drawing.
236) Click **Save** .

Move Dimensions to a Different View

Move the linear dimension 10 that defines the Linear Hole Pattern feature from Drawing View3 (Top) to the Auxiliary view. When moving dimensions from one view to another, utilize the Shift key and only drag the dimension text. Release the dimension text inside the view boundary. The text will not switch views if positioned outside the view boundary.

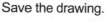
Activity: Move Dimensions to a Different View

Move dimensions from the Top view to the Auxiliary view.
237) Press the **z key** approximately 4 times to view the dimensions in the Top view.

238) Hold the **Shift** key down. Click and drag the vertical dimension **10**, between the 2 holes from the Top view to the Auxiliary view.

239) Release the **mouse button** and the **Shift** key when the mouse pointer is inside the Auxiliary view boundary.

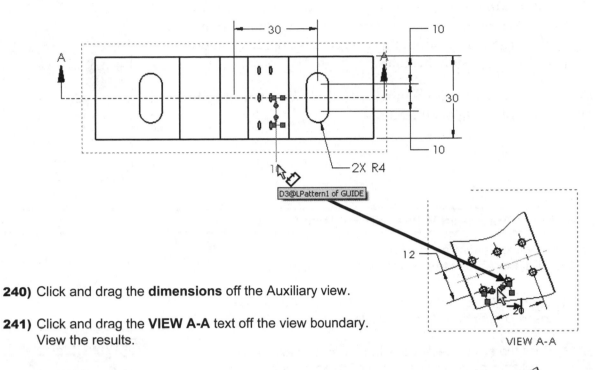

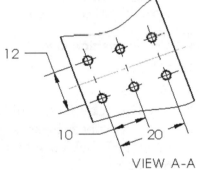

240) Click and drag the **dimensions** off the Auxiliary view.

241) Click and drag the **VIEW A-A** text off the view boundary. View the results.

Save the drawing.

242) Click **Save** 💾. View the results.

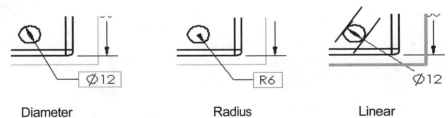

VIEW A-A

Dimension Holes and the Hole Callout

Simple holes and other circular geometry are dimensioned in three ways: *Diameter*, *Radius*, *Linear* (between two straight lines).

Diameter   Radius   Linear

The holes in the Auxiliary view require a diameter dimension and a note to represent the six holes. Use the Hole Callout to dimension the holes. The Hole Callout function creates additional notes required to dimension the holes.

The dimension standard symbols are displayed automatically when you use the Hole Wizard feature.

Activity: Dimension Holes and the Hole Callout

Dimension the Linear Pattern of Holes.

243) Click the **Annotation** tab from the CommandManager.

244) Click the **Hole Callout** ⊔⌀ tool. The Hole Callout tool inserts information from the Hole Wizard.

245) Click the **circumference** of the lower left circle in the Auxiliary view as illustrated. The tool tip, M3x0.5 Tapped Hole1 of GUIDE is displayed.

246) Click a **position** to the bottom left of the Auxiliary view.

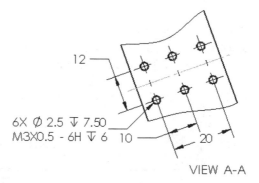

VIEW A-A

The Hole Callout text displayed in the Dimension Text box depends on the options utilized in the Hole Wizard feature and the Linear Pattern feature.

Remove the trailing zeros for ASME Y14 millimeter display.

247) Select **.1** from the Primary Unit Tolerance/Precision box.

248) Click **OK** ✔ from the Dimension PropertyManager. The Hole Callout is deactivated.

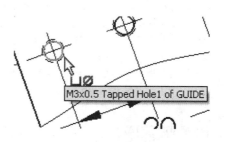

Display the Hole Wizard Callout information automatically by selecting the Hole callout button in the Model Items PropertyManager.

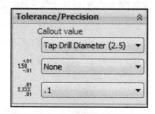

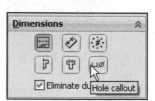

Know inch/mm decimal display. The ASME Y14.5 standard states:

- For millimeter dimensions <1, display the leading zero. Remove trailing zeros.

- For inch dimensions <1, delete the leading zero. The dimension is displayed with the same number of decimal places as its tolerance.

Symbols are located on the bottom of the Dimension Text dialog box. The current text is displayed in the text box. Example:

- <NUM_INST>: Number of Instances in a Pattern.

- <MOD-DIAM>: Diameter symbol ⌀.

- <HOLE-DEPTH>: Deep symbol ⏊.

- <HOLE-SPOT>: Counterbore symbol ⊔.

- <DIM>: Dimension value 3.

Fit the Drawing to the Graphics window.
249) Press the **f** key.

If needed, insert dimension text.
250) Click inside the **Drawing View1** boundary. The Drawing View1 PropertyManager is displayed.

251) Click the Guide Hole dimension ⌀**10.015/10.000** in Drawing View1.

252) Enter text **THRU** in the Dimension box. THRU is displayed on the drawing in blue.

253) Click **OK** ✔ from the Dimension PropertyManager.

Save the drawing.
254) Click **Save** 💾.

 Access Notes, Hole Callouts and other Dimension Annotations through the Annotations toolbar menu. Access Annotations with the right mouse button, click Annotations.

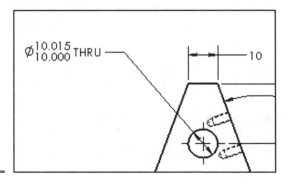

Center Marks and Centerlines

Hole centerlines are composed of alternating long and short dash lines. Centerlines indicate symmetry. Centerlines also identify the center of a circle, axes or cylindrical geometry.

Center Marks represents two perpendicular intersecting centerlines. The default Size

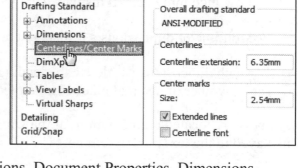

and display of the Center Mark is set in Options, Document Properties, Dimensions, Centerlines/Center Marks as illustrated

Center Marks are inserted on view creation by default. Modify the position and size of the Center Marks. The following three steps illustrate how to create a new Center Mark.

Activity: Center Marks and Centerlines

Insert a Center mark into the Front view. If a Center mark exists in Drawing View1, skip the next few steps.

255) Click inside the **Drawing View1** boundary.

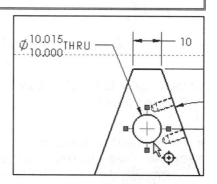

256) Click the **Center Mark** ⊕ tool from the Annotation tab in the CommandManager. The Center Mark PropertyManager is displayed.

257) Click the **Guide Hole circumference**. The Center mark is inserted.

258) Click **OK** ✓ from the Center Mark PropertyManager.

Dimension standards require a gap between the Center Mark and the end points of the leader line. Currently, the leader line overlaps the Center Mark.

Insert Centerlines into the drawing views.

259) Click the **Centerline** tool from the Annotation tab in the CommandManager. The Centerline PropertyManager is displayed.

260) Click inside the Front view, **Drawing View1** boundary.

261) Insert the needed **Centerlines**.

262) Check the **Select View** box in the Auto Insert dialog box.

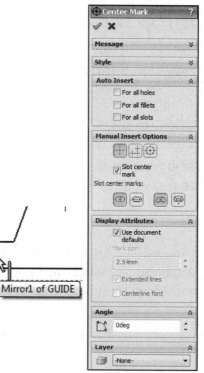

263) Click inside the **Detail view** boundary. Centerlines are displayed.

264) Click inside the **Section view** boundary. Centerlines are displayed. Click **OK** ✔ from the Centerline PropertyManager.

Save the drawing.
265) Click **Save** 🖫.

Activate the Top view and insert Center marks.
266) Click inside the **Drawing View3** boundary. Click the **Center Mark** ⊕ tool from the Annotation tab in the CommandManager. The Center Mark PropertyManager is displayed.

267) Uncheck the **Slot center mark** box. Click the **four arcs**. The center marks are displayed.

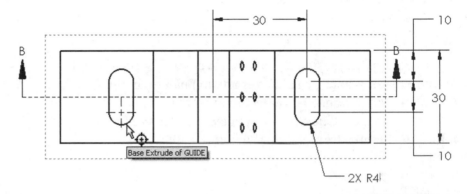

268) Uncheck the **Use document defaults** box.

269) Enter **1**mm for Mark size.

270) Un-check the **Extended lines** box. Accept the default settings.

271) Click **OK** ✔ from the Center Mark PropertyManager. View the results.

Save the drawing.
272) Click **Save** 🖫. View the results.

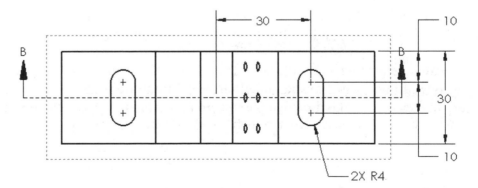

Insert an Annotation Centerline.

273) Click the **Centerline** 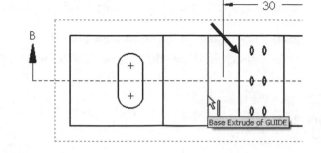 tool from the Annotation tab. The Centerline PropertyManager is displayed.

274) Click the **left edge** of the GUIDE hole as illustrated.

275) Click the **right edge** of the GUIDE hole. The centerline is displayed.

276) Click **OK** ✔ from the Centerline PropertyManager.

Save the drawing.

277) Click **Save** 💾. View the results.

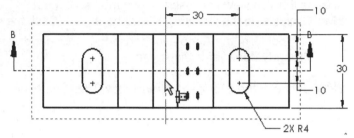

💡 Click the dimension Palette rollover 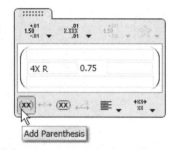 button to display the dimension palette. Use the dimension palette in the Graphics window to save mouse travel to the Dimension PropertyManager. Click on a dimension in a drawing view, and modify it directly from the dimension palette.

Modify the Dimension Scheme

The current feature dimension scheme represents the design intent of the GUIDE part. The Mirror Entities Sketch tool built symmetry into the Extruded Base sketch. The Mirror feature built symmetry into the Slot Cuts.

The current dimension scheme for the Slot Cut differs from the ASME 14.5M Dimension Standard for a slot. Redefine the dimensions for the Slot Cut according to the ASME 14.5M Standard.

The ASME 14.5M Standard requires an outside dimension of a slot. The radius value is not dimensioned. The left Slot Cut was created with the Mirror feature.

Create a centerline and dimension to complete the detailing of the Slot Cut. Sketch the vertical dimension, 10. The default arc conditions are measured from arc center point to arc center point. The dimension extension lines are tangent to the top arc and bottom arc.

Activity: Modify the Dimension Scheme

Modify the Slot Cut dimension scheme.

278) Click **Layer Properties** from the Layer toolbar. The Layers dialog box is displayed.

Set the On/Off icon to modify the status of the Layers.

279) Click the **Dims** Layer On. Dims is the current layer.

280) Click the **Notes** Layer On.

281) Click the **Hidden Dims** Layer Off.

282) Click the **Format** Layer On.

283) Click **OK** from the Layers dialog box.

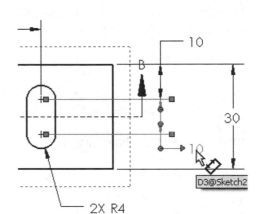

284) Click **Zoom to Area** from the Heads-up View toolbar.

285) **Zoom in** on the right slot of Drawing View3 as illustrated.

Hide the dimension between the two arc center points.

286) Click the **10** dimension on the right bottom side of Drawing View3.

287) Select **Hidden Dims** from the Layer box in the Dimension PropertyManager.

288) Click **OK** ✔ from the Dimension PropertyManager. The 10 dimension is not displayed.

Create a vertical dimension.

289) Click the **Smart Dimension** ✎ Sketch tool. The Dimension PropertyManager is displayed.

290) Click **Smart dimensioning** from the Dimension Assist Tools box.

291) Click the top of the **top right arc**. Do not select the Center Mark of the Slot Cut.

292) Click the bottom of the **bottom right arc**. Do not select the Center Mark of the Slot Cut.

293) Click a **position** to the right of Drawing View3.

294) Click the **Leaders** tab in the Dimension PropertyManager.

295) Click **Max** for First arc condition. Click **Max** for Second arc condition.

296) Click **OK** ✔ from the Dimension PropertyManager.

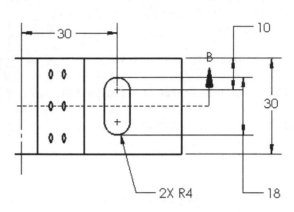

If needed, modify the Radius text to 2X R.
297) Click **2X R4** in Drawing View2.

298) Delete **R<DIM>** in the Dimension text box.

299) Click **Yes** to the Confirm dimension value text override message.

300) Enter **RR** for Dimension text.

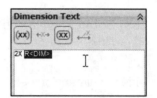

301) Click **OK** ✔ from the Dimension PropertyManager.

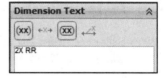

Add a dimension.
302) Click the **Smart Dimension** ⟨ tool.

303) Click **Smart dimensioning** from the Dimension Assist Tools box.

304) Click the **left vertical line** of the right Slot Cut.

305) Click the **right vertical line** of the right Slot Cut.

306) Click a **position** below the horizontal profile line.

Drawing dimensions are added in the drawing document. Model dimensions are inserted from the part. Utilize Smart Dimension in the GUIDE drawing to create drawing dimensions.

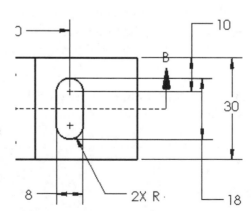

307) Click the **centerline**.

308) Click the left **Slot Cut arc top center point**.

309) Click a **position** above the top horizontal line as illustrated. 30 is the dimension.

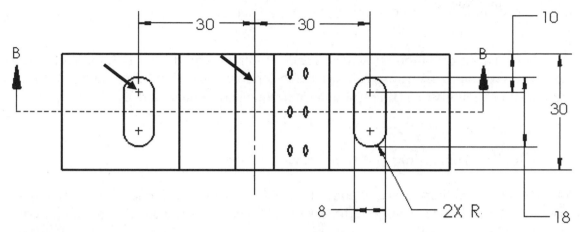

Insert a Reference dimension in the Section view.
310) Click the **left centerline** in the Section view.

311) Click the **right centerline** in the Section view.

312) Click a **position** below the bottom horizontal line.

313) Click the **Add Parenthesis** box in the Dimension Text.

314) Click **OK** ✓ from the Dimension PropertyManager.

Save the drawing.
315) Click **Save** 💾.

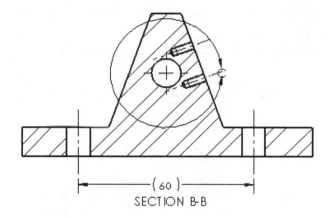

SECTION B-B

Design parts to maximize symmetry and geometric relations to fully define sketches. Minimize dimensions in the part. Insert dimensions in the drawing to adhere to machining requirements and Dimension Standards.

Examples: Insert reference dimensions by adding parentheses and redefine a slot dimension scheme according to the ASME Y14.5 standard.

Additional Information Dimensions and Annotations are found in SolidWorks Help Topics. Keywords: Dimensions (circles, extension lines, inserting into drawings, move, parenthesis), Annotations (Hole Callout, Centerline and Center Mark).

Review Dimensions and Annotations

You inserted part dimensions and annotations into the drawing. Dimensions were moved to new positions. Leader lines and dimension text were repositioned. Annotations were edited to reflect the drawing standard.

Centerlines and Center Marks were inserted into each view and were modified in the PropertyManager. You modified Hole Callouts, dimensions, annotations and referenced dimensions to conform to the drawing standard.

GUIDE Part-Insert an Additional Feature

The design process is dynamic. We do not live in a static world. Create and add an edge Fillet feature to the GUIDE part. Insert dimensions into Drawing View1; (Front).

Activity: GUIDE Part-Insert an Additional Feature

Open the Guide part.
316) Right-click inside the **Drawing View1** boundary.

317) Click **Open Part** from the Content toolbar.

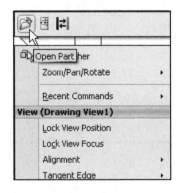

 If the View menu is not displaycd, right-click Select. Tools in the Annotation toolbar remain active until they are deactivated.

Display Hidden Lines to select edges to fillet.
318) Click **Hidden Lines Visible** from the Heads-up View toolbar.

Display the Isometric view.
319) Click **Isometric view** 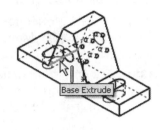 from the Heads-up View toolbar.

Create a Fillet feature.
320) Click the **hidden edge** as illustrated.

321) Click the **Fillet** feature tool. The Fillet PropertyManager is displayed.

322) Enter **1**mm in the Radius box.

323) Click the other **3 edges** as illustrated. Each edge is added to the Items to Fillet list. Accept the default settings.

324) Click **OK** from the Fillet PropertyManager. Fillet1 is displayed in the FeatureManager.

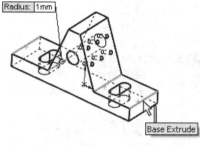

Save the GUIDE part.
325) Click **Save**.

Open the Drawing.
326) Right-click the **GUIDE Part** GUIDE icon in the FeatureManager.

327) Click **Open Drawing**. The GUIDE drawing is displayed.

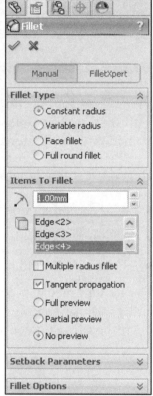

Fit the drawing to the Graphics window.
328) Press the **f** key. Note: Hide any unwanted annotations.

Display the GUIDE features.
329) **Expand** Drawing View1 from the FeatureManager.

330) **Expand** the GUIDE part from the FeatureManager.

331) Click **Fillet1** from the Drawing FeatureManager. Insert the dimensions for the Fillet1 feature on the Dims Layer. The current layer should be Dims. If required, select Dims.

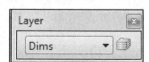

The Model Items, Import into drawing view, Import items into all views option insert dimensions into the Section view and Detail view before other types of views on the drawing. The Fillet feature dimension in the part is displayed in the Section view.

Insert dimensions into Drawing View1.

332) Click the **Model Items** 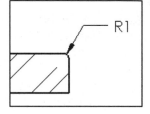 tool from the Annotation tab. The Model Items PropertyManager is displayed. Accept the default conditions.

333) Click **OK** ✅ from the Model Items PropertyManager. Note: Hide any un-wanted annotation or dimension in the Detail view.

Position the Fillet text.

334) Use the Shift key to **move the R1 text** from Drawing View1 to the Section view.

335) Click and drag the **R1** text off the profile in the Section view. The R1 text is displayed either on the left or right side of the Section view. Note: It may come in directly into the Section view.

Save the GUIDE drawing.

336) Click **Save** 💾.

General Notes and Parametric Notes

Plan ahead for general drawing notes. Notes provide relative part or assembly information. Example: Material type, material finish, special manufacturing procedure or considerations, preferred supplier, etc.

Below are a few helpful guidelines to create general drawing notes:

- Use capital letters

- Use left text justification

- Font size should be the same size as the dimension text

Create Parametric notes by selecting dimensions in the drawing. Example: Specify the Fillet radius of the GUIDE as a note in the drawing. If the radius is modified, the corresponding note is also modified.

Hide superfluous feature dimensions. Do not delete feature dimensions. Recall a hidden dimension by using the View, Hide/Show Annotations from the Menu bar. Utilize a Layer to Hide/Show superfluous feature dimensions with the Layer On/Off icon.

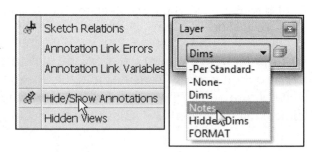

Activity: General Notes and Parametric Notes

Locate the R1 text used to create the edge Fillet.

337) Position the **mouse pointer** over the R1 text in the Section view. The text displays the dimension name: "D1@Fillet1 of Guide".

Insert a Parametric drawing note.

338) Click the **Notes** layer from the Layer toolbar.

339) Click the **Note A** tool from the Annotation tab in the CommandManager. The Note PropertyManager is displayed.

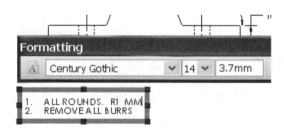

340) Click a **start point** in the lower left corner of the Graphics window, to the left of the Title block, below Drawing View1.

Type two lines of notes in the Note text box.

341) Line 1: Enter **1. ALL ROUNDS**.

342) Press the **Space** key.

343) Click **R1**. The radius value R1 is added to the text box.

344) Press the **Space** key.

345) Enter **MM**.

346) Press the **Enter** key.

347) Line 2: Enter **2. REMOVE ALL BURRS**.

348) Click **OK** ✓ from the Note PropertyManager.

Do not double dimension a drawing with a note and the corresponding dimension. Hide the radial dimension.

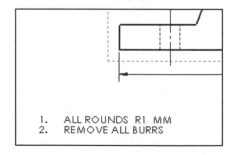

Hide the R1 dimension in the drawing.
349) Click the **R1** text in the Section view.

350) Click **Hidden Dims** from the Layer drop-down menu.

351) Click **OK** ✓ from the PropertyManager. R1 is hidden.

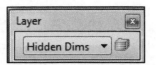

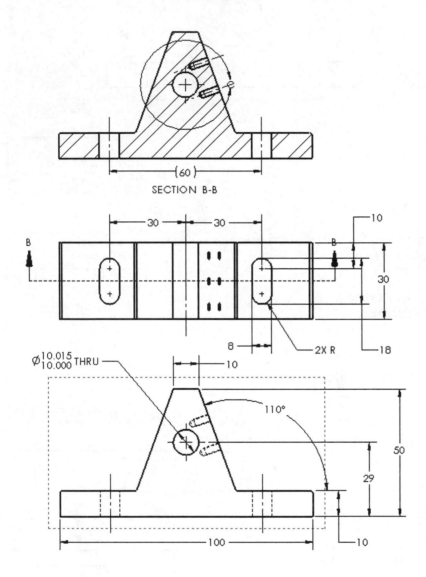

SECTION B-B

1.   ALL ROUNDS R1 MM
2.   REMOVE ALL BURRS

An Engineering Change Order (ECO) or Engineering Change Notice (ECN) is written for each modification in the drawing. Modifications to the drawing are listed in the Revision Table.

 Obtain the ECO number before you institute the change to manufacturing.

Revision Table

The Revision Table provides a drawing history. Changes to the document are recorded systematically in the Revision Table. Insert a Revision Table into the GUIDE drawing. The default columns are as follows: Zone, Rev, Description, Date and Approved.

Zone utilizes the row letter and column number contained in the drawing boarder. Position the Rev letter in the Zone area. Enter the Zone letter/number.

Enter a Description that corresponds to the ECO number. Modify the date if required. Enter the initials/name of the engineering manager who approved the revision.

Activity: Revision Table

Insert a Revision table.

352) Click **None** from the Layer drop-down menu.

353) Click the **Revision Table** tool from the Consolidated toolbar in the CommandManager. The Revision Table PropertyManager is displayed. Accept the defaults.

354) Click **OK** from the Revision Table PropertyManager. The Revision table is displayed in the top right corner of the Sheet.

355) Zoom in on the first row in the table.

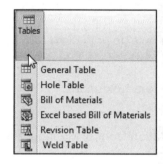

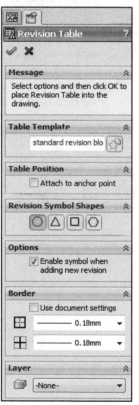

Insert a row. Create the first revision.

356) Right-click the **Revision Table**.

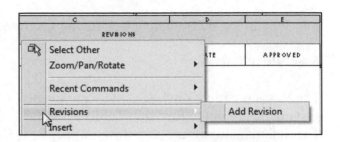

357) Click **Revisions**, **Add Revision**. The Revision letter A and the current date are displayed. The Revision Symbol is displayed on the mouse pointer.

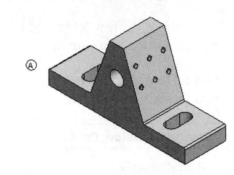

358) Click a **position** in the Isometric view to place the icon.

359) Click **OK** ✓ from the Revision Symbol PropertyManager.

360) Double-click the **text box** under the Description column.

361) Enter **ECO 32510 RELEASE TO MANUFACTURING** for DESCRIPTION.

362) Click a **position** outside of the text box.

363) Double-click the **text box** under APPROVED.

364) Enter Documentation Control Manager's Initials, **DCP**.

365) Click a **position** outside of the text box.

		REVISIONS		
ZONE	REV.	DESCRIPTION	DATE	APPROVED
	A	ECO 32510 RELEASE TO MANUFACTURING	11/24/2011	DCP

The Revision Property, $PRP:"Revision", in the Title block is linked to the REV. Property in the REVISIONS table. The latest REV. letter is displayed in the Title block.

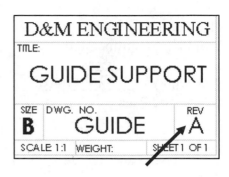

Revision Table Zone Numbers, Dates and Approved names are inserted as an exercise. The Revision Table and Engineering Change Order documentation are integrated in the drawing and the design process.

To save time, incorporate the Revision Table into your custom Sheet Formats.

Part Number and Document Properties

Engineers manage the parts they create and modify. Each part requires a Part Number and Part Name. A part number is a numeric representation of the part. Each part has a unique number. Each drawing has a unique number. Drawings incorporate numerous part numbers or assembly numbers.

There are software applications that incorporate unique part numbers to create and perform:

- Bill of Materials

- Manufacturing procedures

- Cost analysis

- Inventory control / Just in Time (JIT)

You are required to procure the part and drawing numbers from the documentation control manager.

Utilize the following prefix codes to categories created parts and drawings. The part name, part number and drawing numbers are as follows:

Category:	Prefix:	Part Name:	Part Number:	Drawing Number:
Machined Parts	56-	GUIDE	56-A26	56-22222
		ROD	56-A27	56-22223
		PLATE	56-A28	56-22224
Purchased Parts	99-	FLANGE BOLT	99-FBM8x1.25	999-551-8
Assemblies	10-	GUIDE-ROD	10-A123	10-50123

Link notes in the Title block to SolidWorks Properties. The title of the drawing is linked to the GUIDE Part Description, GUIDE-SUPPORT. Create additional notes in the Title block that complete the drawing.

Additional notes are required in the Title block. The text box headings: SIZE B, DWG. NO., REV., SCALE, WEIGHT and SHEET OF are entered in the SolidWorks default Sheet Format. Properties are variables shared between documents and applications. Define the Document Properties in the GUIDE drawing. Link the Document Properties to the notes in the Title block.

Activity: Part Number and Document Properties

Enter Summary information for the GUIDE drawing.

366) Click **File**, **Properties** from the Menu bar. The Summary Information dialog box is displayed.

367) Click the **Summary** tab.

368) Enter your **initials** for Author. Example: DCP.

369) Enter **GUIDE, ROD** for Keywords.

370) Enter **GUIDE for customer ABC** for Comments.

Select the Custom tab.

371) Click the **Custom** tab. Row 1 contains the Company Name. Row 2 contains the latest revision. The revision letter is displayed in the Revisions Table.

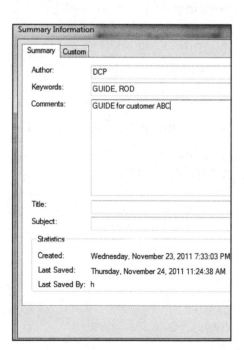

372) Click the **third row** under the Property Name.

373) Select **Number** from the Name drop-down menu.

374) Click in the **Value/Text Expression box**.

375) Enter **56-22222** for Value. Click the **Evaluated Value** box.

Add DrawnBy and DrawnDate in Custom Properties.
376) Click the **forth row** under the Property Name.

377) Select **DrawnBy** from the Name drop-down menu.

378) Click in the **Value/Text Expression box**.

379) Enter **your initials** for Value. Click in the **Evaluated Value** box.

380) Click the **fifth row** under the Property Name.

381) Select **DrawnDate** from the Name drop-down menu.

382) Enter **today's date** for Value. Click in the **Value/Text Expression box**.

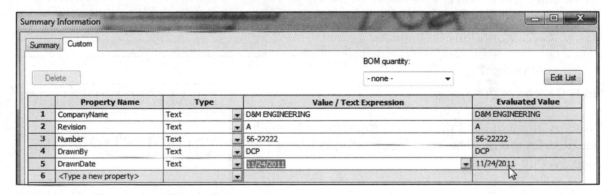

383) Click **OK** from the Summary Information dialog box.

The DrawnBy Property and DrawnDate Property are updated in the Title block. The current DWG NO. is the SolidWorks File Name. Assign the Number Property to the DWG NO. in the Title block.

Link the Properties to the Drawing Notes.
384) Click inside **Sheet1**.

385) Right-click **Edit Sheet Format**.

	NAME	DATE
DRAWN	DCP	11/24/2011
CHECKED		
ENG APPR.		
MFG APPR.		
Q.A.		
COMMENTS:		

Edit the DWG NO.

386) Double-click the **GUIDE** text in the DWG NO. box. The Formatting and Note PropertyManager is displayed.

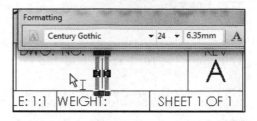

387) Delete **GUIDE**. Note: The Formatting dialog is still active.

388) Click the **Link to Property** icon in the Properties dialog box.

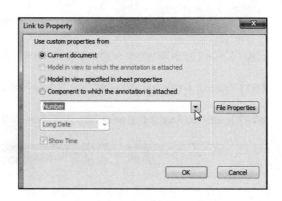

389) Select **Number** from the Link to Property drop-down menu.

390) Click **OK** from the Link to Property box. The number is displayed in the DWG. NO. box.

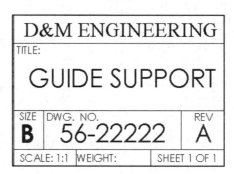

391) Click **OK** ✓ from the Note PropertyManager.

392) Right-click **Edit Sheet**.

Fit the drawing to the Graphics window.

393) Press the **f** key. The drawing remains in the Edit Sheet Format mode.

Save the Drawing.

394) Click **Save** 💾.

Note: Remain in the Edit Sheet Format mode through the next few steps. The Title block information is displayed in blue.

Define Custom Properties for Material and Surface Finish in the GUIDE part. Return to the GUIDE drawing to complete the Title block. The design team decided that the GUIDE part would be fabricated from 304 Stainless Steel. There are numerous types of Stainless steels for various applications. Select the correct material for the application. This is critical!

AISI 304 was set in the part. Open the GUIDE part and create the Material Custom Property. The AISI 304 value propagates to the drawing. Surface Finish is another important property in machining. The GUIDE utilizes an all around 0.80 μm high grade machined finish. The numerical value refers to the roughness height of the machined material. Surface finish adds cost and time to the part. Work with the manufacturer to specify the correct part finish for your application.

Create the Material Custom Property.
395) Open the GUIDE part.

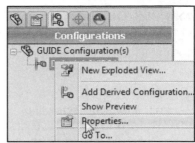

396) Click the **Configuration PropertyManager** icon.

397) Right-click **Default[GUIDE]**.

398) Click **Properties**. The Configuration Properties PropertyManager is displayed.

399) Click the **Custom Properties** button.

400) Click the **Configuration Specific** tab.

Specify the Material.
401) Click inside the **Property Name** box.

402) Select **Material** from the drop-down menu.

403) Click inside the **Value/Text Expression** box.

404) Select **Material** from the drop-down menu.

Determine the correct units before you create Custom Properties linked to Mass, Density, Volume, Center of Mass, Moment of Inertia, etc. Recall Weight = Mass *Gravity. Modify the Title block text to include "g" for gravity, SI units in the Weight box.

Specify the Finish.
405) Click inside the **second row** under the Property Name.

406) Select **Finish** from the drop-down menu.

407) Enter **0.80** for Value/Text Expression as illustrated.

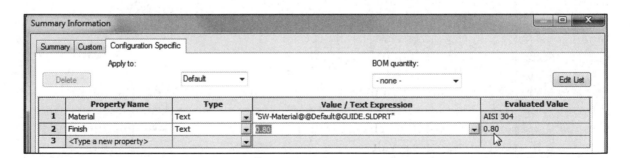

408) Click **OK** from the Summary Information box.

409) Click **OK** ✓ from the Configuration Properties PropertyManager.

Return to the GUIDE drawing.
410) **Return** to the GUIDE drawing. The Material Custom Property displays AISI 304 in the MATERIAL block and 0.80 in the FINISH block.

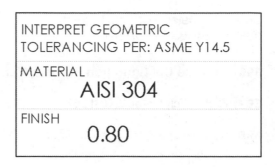

INTERPRET GEOMETRIC
TOLERANCING PER: ASME Y14.5

MATERIAL
AISI 304

FINISH
0.80

411) Right-click **Edit Sheet Format**.

Insert a micrometer symbol.
412) Double-click **0.80**. The Formatting toolbar is displayed.

413) Click **SWGrekc** from the Formatting toolbar for Font.

414) Press the **m** key to display the Greek letter μ. μ is the Greek letter for micro.

Formatting
SWGrekc   9
FINISH
0.80

415) Click **Century Gothic** for Font.

416) Enter **m** for meter. The text reads μm for micrometer.

417) Click a **position** outside the text box.

Formatting
Century Gothic   9
FINISH
0.80 μm

Return to the Edit Sheet mode.
418) Right-click in the **Graphics window**.

419) Click **Edit Sheet**. The Title block is displayed in black.

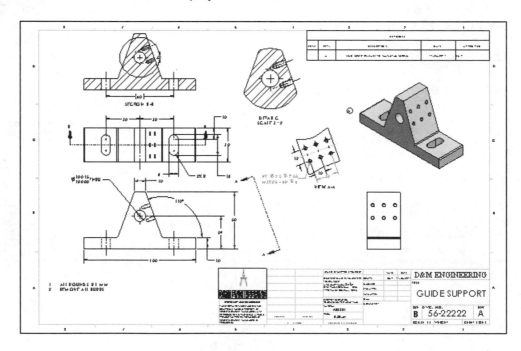

Fit the drawing to the Graphics window.
420) Press the **f** key.

421) Drag the **views** and **annotations** if required for spacing.

Save the GUIDE drawing.
422) Click **Save** 🖫.

💡 Establish Custom Properties early in the design process. The MATERIAL and FINISH Custom Properties are established in the part and propagate to the Title block in the drawing. MATERIAL Custom Property is also listed in the default Bill of Material template and the SolidWorks SimulationXpress Analysis tool.

💡 Create engineering procedures to define the location and values of Custom Properties. The REVISION Custom Property was created in the drawing in the REVISION Table. The REVISION Custom Property can also be established in the part and propagate to the drawing. Does the part or the drawing control the revision? Your company's engineering procedures determine the answer.

Additional annotations are required for drawings. Utilize the Surface Finish ∜ Surface Finish tool to apply symbols to individual faces/edges in the part or in the drawing. When an assembly contains mating parts, document their relationship. Add part numbers in the Used On section. Additional annotations are left as an exercise.

Exploded View

Add an Exploded view and Bill of Materials to the drawing. Add the GUIDE-ROD assembly Exploded view. The Bill of Materials reflects the components of the GUIDE-ROD assembly. Create a drawing with a Bill of Materials.

Perform the following steps:

- Create a new drawing from the B-ANSI-MM Drawing Template.

- Display the Exploded view of the assembly.

- Insert the Exploded view of the assembly into the drawing.

- Label each component with Balloon text.

- Create a Bill of Materials.

Activity: Exploded View

Close all parts and drawings.
423) Click **Windows**, **Close All** from the Menu bar.

Open the GUIDE-ROD assembly.
424) Open the **GUIDE-ROD** assembly from the PROJECTS folder. The GUIDE-ROD assembly is displayed in the Graphics window with the updates to the GUIDE.

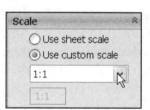

Create a new drawing.
425) Click **Make Drawing from Part/Assembly** from the Menu bar.

426) Double-click **B-ANSI-MM** from the MY-TEMPLATES tab. The Model View PropertyManager is displayed.

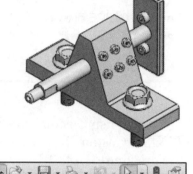

🔆 The View Palette drop-down menu lists the open documents in a session.

Insert the GUIDE-ROD assembly using the View Palette tool.
427) Click and drag the **Isometric view** from the View Palette onto Sheet1. The Drawing View1 PropertyManager is displayed.

Display a Shaded With Edges, 1:1 scale view.
428) Click **Shaded With Edges** in the Display Style box.

429) Click the **Use custom scale** box.

430) Select **1:1** from the drop-down menu.

431) Deactivate all **origins**.

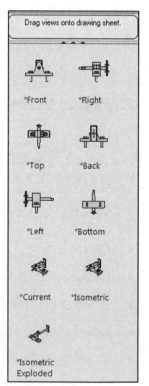

🔆 The View Palette contains images of Standard views, Annotation views, Section views, and Flat patterns (sheet metal parts) of the selected model.

🔆 Click the Browse button in the View Palette to locate additional models.

🔆 A key goal of this book is to expose the new user to various tools, techniques and procedures. It may not always use the most direct tool or process.

Display the Exploded view.

432) Click **inside** the Isometric view boundary. Right-click **Properties**.

433) Check **Show in exploded state**.

434) Click **OK** from the Drawing View Properties dialog box.

435) Click **OK** ✅ from the Drawing View1 PropertyManager.

Note: The Explode view was created in Project 2. The Show in exploded state option is visible if an Exploded view exists in the assembly.

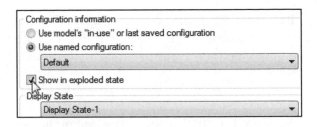

Fit the drawing to the Graphics window.

436) Press the **f** key. Move the **Isometric view** boundary into the drawing.

437) Click **OK** ✅ from the Drawing View1 PropertyManager.

The drawing filename is the same as the assembly filename.

Save the GUIDE-ROD drawing.

438) Click **Save As** from the Menu bar.

439) Select **ENGDESIGN-W-SOLIDWORKS\PROJECTS** for folder. GUIDE-ROD is displayed for the File name.

Save the drawing.

440) Click **Save**.

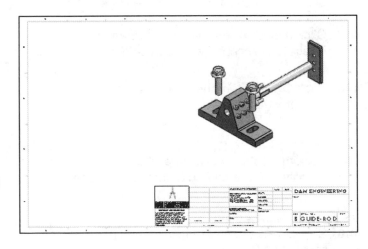

Balloons

Label each component with a unique item number. The item number is placed inside a circle. The circle is called Balloon text. List each item in a Bill of Materials table. Utilize the Auto Balloon tool to apply Balloon text to all components. Utilize the Bill of Materials tool to apply a BOM to the drawing.

The Circle Split Line option contains the Item Number and Quantity. Item number is determined by the order listed in the assembly FeatureManager. Quantity lists the number of instances in the assembly.

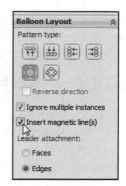

New for 2012 is the Insert magnetic lines(s) option. Magnetic lines are a convenient way to align balloons along a line at any angle. You attach balloons to magnetic lines, choose to space the balloons equally or not, and move the lines freely, at any angle, in the drawing.

Activity: Balloons

Insert the Automatic Balloons.

441) Click inside the **Isometric view** boundary.

442) Click **Auto Balloon** AutoBal... from the Annotation toolbar. The Auto Balloon PropertyManager is displayed. Accept the default settings.

443) Click and **align the magnetic lines** as illustrated. Accept the default settings.

444) Click **OK** ✔ from the Auto Balloon PropertyManager.

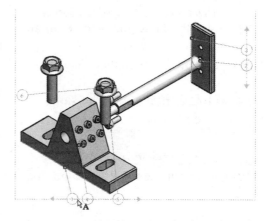

Reposition the Balloon text.

445) Click the **Balloon arrowhead attachment** to reposition the arrow on a component edge.

Display Item Number and Quantity.

446) Shift-Select the six **Balloon texts**. The Balloon PropertyManager is displayed.

447) Select **Circular Split Line** for Style.

448) Click **OK** ✔ from the Balloon PropertyManager.

449) Click **Save** 💾.

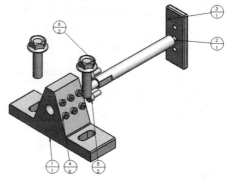

The symbol "?" is displayed when the Balloon attachment is not coincident with an edge or face. ASME Y14.2 defines the attachment display as an arrowhead for Edge and a dot for Face.

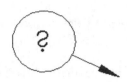

Arrowhead display is located under Options, Document Properties, Annotations or under the Leaders tab in the Dimension PropertyManager. Note: Edge/ vertex attachment is selected by default.

Each Balloon references a component with a SolidWorks File Name. The SolidWorks File Name is linked to the Part Name in the Bill of Materials by default. Customize both the Balloon text and the Bill of Materials according to your company's requirements.

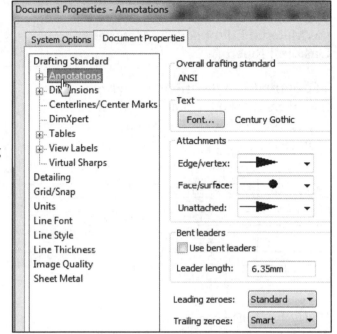

Bill of Materials

The Bill of Materials reflects the components of the GUIDE-ROD assembly. The Bill of Materials is a table in the assembly drawing that contains Item Number, Quantity, Part Number and Description by default. Insert additional columns to customize the Bill of Materials.

Part Number and Description are Custom Properties entered in the individual part documents. Insert additional Custom Properties to complete the Bill of Materials.

Activity: Bill of Materials

Create a Bill of Materials.
450) Click inside the **Isometric view** in the GUIDE-ROD drawing.

451) Click the **Bill of Materials** tool from the Annotation tab. The Bill of Materials PropertyManager is displayed.

452) Click the **Open Table Template for Bill of Materials** icon.

453) Double-click **bom-material.sldbomtbt** from the Select BOM Table dialog box.

454) Click **OK** ✔ from the Bill of Materials PropertyManager.

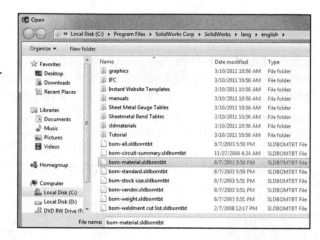

455) Click a **position** in the upper left corner of Sheet1.

The Bill of Materials requires some editing. The current part file name determines the PART NUMBER values. Material for the GUIDE was defined in the Material Editor and the Material Custom Property.

The current part description determines the DESCRIPTION values. Redefine the PART NUMBER for the Bill of Materials.

ITEM NO.	PART NUMBER	DESCRIPTION	MATERIAL	QTY.
1	GUIDE	GUIDE SUPPORT	AISI 304	1
2	ROD	ROD 10MM DIA x 100MM		1
3	PLATE	PLATE 56MM x 22MM		1
4	3MMCAPSCREW	CAP SCREW, 3MM		6
5	4MMCAPSCREW	CAP SCREW, 4MM		2
6	M8-1.25 x 30			2

Modify the GUIDE part number.
456) Right-click on the **GUIDE** in the Isometric view.

457) Click **Open Part**. The Guide FeatureManager is displayed.

458) Click the GUIDE **ConfigurationManager** tab.

459) Right-click **Default [GUIDE]** in the ConfigurationManager.

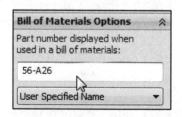

460) Click **Properties**. The Configuration Properties PropertyManager is displayed.

461) Select **User Specified Name** in the spin box from the Bill of Materials Options.

462) Enter **56-A26** for the Part Number in the Bill of Materials Options.

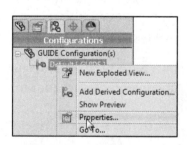

463) Click **OK** ✔ from the Configuration Properties PropertyManager. Default [56-A26] is displayed in the ConfigurationManager.

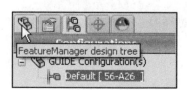

464) Return to the FeatureManager.

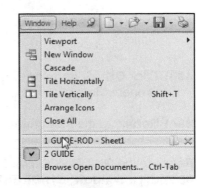

Return to the GUIDE-ROD drawing.
465) Click **Window**, **GUIDE-ROD - Sheet1** from the Menu bar.

Modify the ROD part number.
466) Right-click on the **ROD** in the Isometric view.

467) Click **Open Part**. The ROD FeatureManager is displayed.

468) Click the ROD **ConfigurationManager** tab.

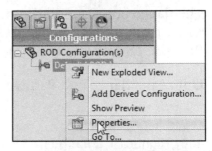

469) Right-click **Default [ROD]** from the ConfigurationManager.

470) Click **Properties**.

471) Select **User Specified Name** in the spin box from the Bill of Materials Options.

472) Enter **56-A27** for the Part Number in the Bill of Materials Options.

473) Click **OK** from the Configuration Properties PropertyManager.

474) Return to the FeatureManager.

Return to the GUIDE-ROD drawing.
475) Click **Window**, **GUIDE-ROD - Sheet1** from the Menu bar.

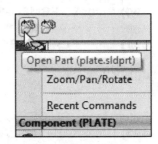

Modify the PLATE part number.
476) Right-click on the **PLATE** in the Isometric view.

477) Click **Open Part**.

478) Click the PLATE **ConfigurationManager** tab.

479) Right-click **Default [PLATE]** from the ConfigurationManager.

480) Click **Properties**.

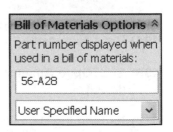

481) Select **User Specified Name** in the spin box from the Bill of Materials Options.

482) Enter **56-A28** for the Part Number in the Bill of Materials Options.

483) Click **OK** ✔ from the Configuration Properties PropertyManager.

484) Return to the FeatureManager.

Return to the GUIDE-ROD drawing.
485) Click **Window**, **GUIDE-ROD - Sheet1** from the Menu bar.

The flange bolt is a SolidWorks library part. Copy the part with a new name to the ENGDESIGN-W-SOLIDWORKS\ PROJECTS folder with the Save As command.

Utilize the flange bolt Part Number for File name.

Modify the flange bolt.
486) Right-click a **flange bolt** in the Isometric view.

487) Click **Open Part**.

Save the flange bolt.
488) Click **Save As** from the Menu bar.

489) Select the **ENGDESIGN-W-SOLIDWORKS\PROJECTS** folder.

490) Enter **99-FBM8-1-25** for the File name.

491) Enter **FLANGE BOLT M8x1.25x30** for Description.

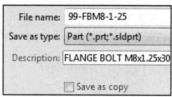

492) Click **Save**.

Return to the GUIDE-ROD drawing.
493) Press **Ctrl+Tab** to return to the GUIDE-ROD drawing.

Open the GUIDE-ROD assembly.
494) Right-click inside the **Isometric view** boundary on Sheet1.

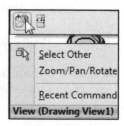

495) Click **Open Assembly**. The GUIDE-ROD assembly FeatureManager is displayed.

496) Expand the Hardware folder from the FeatureManager. The flange bolt displays the new name.

Return to the GUIDE-ROD drawing.
497) Press **Ctrl+Tab** to return to the GUIDE-ROD drawing.

Update the Bill of Materials.
498) Rebuild the model.

Fit the drawing to the Graphics window.
499) Press the **f** key.

If the Bill of Materials does not update when changes to a part have been made, then open the assembly. Return to the drawing. Issue a Rebuild.

As an exercise, update the Bill of Materials and fill in all of the columns.

ITEM NO.	PART NUMBER	DESCRIPTION	MATERIAL	QTY.
1	56-A26	GUIDE SUPPORT	AISI 304	1
2	55-66	ROD	AISI 304	1
3	56-A28	PLATE 56MM x 22MM	AISI 304	1
4	3MM CAPSCREW	CAP SCREW, 3MM	AISI 304	6
5	4MM CAPSCREW	CAP SCREW, 4MM	AISI 304	2
6	M8-1.25 x 30	FLANGE BOLT M8x1.25x30	AISI 304	2

Save the GUIDE-ROD drawing.
500) Click **Save**.

501) Click **Yes** to update. You are finished with this Project.

List assembly items such as: adhesives, oil, and labels in the Bill of Materials. Select Insert, Row. Enter the information in each cell.

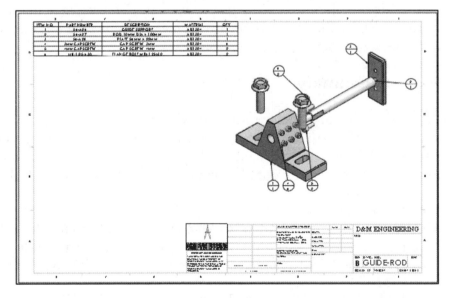

Review of the Parametric Notes, Revision Table and Bill of Materials

You created a Parametric note in the drawing by inserting a part dimension into the Note text box. The Revision Table was inserted into the drawing to maintain drawing history.

The Bill of Materials listed the Item Number, Part Number, Description, Material, and Quantity of components in the assembly.

You developed Custom Properties in the part and utilized the Properties in the drawing and Bill of Materials.

Drawings and Custom Properties are an integral part of the design process. Part, assemblies and drawings all work together.

From your initial design concepts, you created parts and drawings that fulfill the design requirements of your customer.

Refer to Help, Online Tutorial, Lesson3-Drawings exercise and Advanced Drawings exercise for additional information.

New for 2012 is the Insert magnetic lines(s) option. Magnetic lines are a convenient way to align balloons along a line at any angle. You attach balloons to magnetic lines, choose to space the balloons equally or not, and move the lines freely, at any angle, in the drawing.

Project Summary

In this Project, you developed two drawings: GUIDE drawing and the GUIDE-ROD assembly drawing. The drawings contained three standard views, (principle views) and an Isometric view. The drawings utilized a custom Sheet Format containing a Company logo, Title block and Custom Properties. You incorporated the GUIDE part dimensions into the drawing.

The Drawing toolbar contained the Model View tool and the Projected View tool to develop standard views. Additional views were required and utilized the Auxiliary, Detail and Section view tools. Dimension were inserted from the part and added to the drawing.

You used two major design modes in the drawings: Edit Sheet Format and Edit Sheet.

The detailed GUIDE drawing included annotations and Custom Properties. The GUIDE-ROD assembly drawing incorporated a Bill of Materials and additional Custom Properties.

Project Terminology

Alternate Position View: A drawing view in which one or more views are superimposed in phantom lines on the original view. Alternate position views are often used to show range of motion of an assembly.

Anchor Point: The end of a leader that attaches to the note, block, or other annotation.

Annotations: Text note or a symbol that adds specific design intent to a part, assembly, or drawing. Annotations in the drawing include note, hole callout, surface finish symbol, datum feature symbol, datum target, geometric tolerance symbol, weld symbol, balloon, stacked balloon, center mark, area hatch, and block.

Area Hatch: A crosshatch pattern or fill applied to a selected face or to a closed sketch in a drawing.

Associatively: The relationship between parts, assemblies, drawings and other SolidWorks documents that share a common file structure.

Auxiliary View: Displays a plane parallel to an angled plane with true dimensions. A primary Auxiliary View is hinged to one of the six principle views.

Balloon: Annotations on a drawing used to identify items in a Bill of Materials.

Baseline Dimensions: Sets of dimensions measured from the same edge or vertex in a drawing.

Bill of Materials: A table in an assembly drawing that contains the properties: Item Number, Part Number, Description and Quantity. Additional properties are added to a Bill of Materials.

Bottom-up Design: An assembly modeling technique where you create parts and then insert them into an assembly.

Centerlines: Are composed of alternating long and short dash lines. The lines identify the center of a circle, axes or cylindrical geometry.

Center Marks: Represents two perpendicular intersecting centerlines.

Crop View: Displays only a bounded area from a view. Sketch a Spline for the boundary.

Custom Properties: Variables shared between documents. Custom Properties are defined in the part and utilized in the drawing through Linked Notes. Configuration Specific Properties are defined in a part or assembly.

Design Table: An Excel spreadsheet that is used to create multiple configurations in a part or assembly document.

Detached Drawing: A drawing format that allows opening and working in a drawing without loading the corresponding models into memory. The models are loaded on an as-needed basis.

Detailed View: Detailed Views enlarge an area of an existing view. Specify location, shape and scale.

Dimension line: A linear dimension line references the dimension text to extension lines indicating the entity being measured. An angular dimension line references the dimension text directly to the measured object.

Drawing: A 2D representation of a 3D part or assembly. The extension for a SolidWorks drawing file name is .SLDDRW.

Drawing Sheets: The "paper sheets" used to hold the views, dimensions and annotations and create the drawing.

Drawing Template: The foundation of a SolidWorks drawing. Defines sheet size, drawing standards, company information, manufacturing and or assembly requirements, units, layers and other properties

Edit Sheet Format Mode: Provides the ability to change the Title block, incorporate a company logo, and insert Custom Properties. Remember: A part cannot be insert into a drawing when the Edit Sheet Format mode is selected. Edit Sheet Format displays all lines in blue.

Edit Sheet Mode: Provides the ability to insert and modify views and dimensions.

Exploded View: Shows an assembly with its components separated from one another, usually to show how to assemble the mechanism.

General Notes: Text utilized on a drawing. In engineering drawings notes are usually all upper case letters, left justification and the same size as dimension text.

Hole Callout: The Hole Callout function creates additional notes required to dimension the holes.

Isometric View: A graphical projection of a model in a drawing that displays the model with its x-, y-, and z- axes spaced 120° apart and with the vertical z-axis.

Layers: Contain dimensions, annotations and geometry. Layers are assigned display properties and styles.

Leader lines: Reference the size of the profile. A gap must exist between the profile lines and the leader lines.

Line types: A term describing the display of a line in a drawing including its type and thickness.

Multiple Drawing Sheets: The drawing can have multiple sheets, if required. To create an additional sheet, use Add Sheet. The size and format of the new sheet is copied from the original but can be edited and changed.

Notes: Text annotations inserted into a document. Notes are displayed with leaders or as a stand-alone text string. If an edge, face or vertex is selected prior to adding the note, a leader is created to that location. Link Notes to Custom Properties.

Projected View: A drawing view projected orthogonally from an existing view.

Revision Table: A table that documents the history of changes to a drawing. The Revision letter in the Table is linked to the Rev letter in the Title block.

Section line: A line or centerline sketched in a drawing view to create a section view.

Section View: Section Views display the interior features. Define a cutting plane with a sketched line in a view perpendicular to the Section View.

Sheet Format: The Sheet Format is incorporated into the Drawing Template. The Sheet Format contains the border, Title block information, revision block information, company name and or logo information, Custom Properties and SolidWorks Properties.

Standard 3 Views: The three orthographic views (front, right, and top) that are often the basis of a drawing.

Title block: Contains vital part or assembly information. Each company can have a unique version of a Title block.

The default location for storing drawing sheet formats is:

*   **Windows XP -** C:Documents and Settings\All Users\ApplicationData\SolidWorks \ version\lang\language\sheetformat

*   **Windows 7 -** C:ProgramData\SolidWorks\version\lang\language\sheetformat

*   **Windows Vista -** C:ProgramData\SolidWorks\version\lang\language\sheetformat

To change the default location, click Tools, Options, System Options, File Locations. In Show folder for, select Sheet Formats.

Questions

1. Describe a Bill of Materials and its contents in a drawing.

2. Name the two major design modes used to develop a drawing in SolidWorks.

3. Identify seven components that are commonly found in a Title block.

4. Describe a procedure to insert an Isometric view into a drawing.

5. In SolidWorks, Drawing file names end with a _____ suffix.

6. In SolidWorks, Part file names end with a _____ suffix.

7. Can a part and drawing have the same name?

8. True or False. In SolidWorks, if a part is modified, the drawing is automatically updated.

9. True or False. In SolidWorks, when a dimension in the drawing is modified, the part is automatically updated.

10. Name three guidelines to create General Notes in a drawing.

11. True or False. Most engineering drawings use the following font: Time New Roman – All small letters.

12. What are Leader lines? Provide an example.

13. Name the three ways that Holes and other circular geometry can be dimensioned.

14. Describe Center Marks. Provide an example.

15. How do you calculate the maximum and minimum variation?

16. Describe the differences between a Drawing Template and a Sheet Format.

17. Describe the key differences between a Detail view and a Section view.

18. Describe a Revision table and its contents.

19. Identify the name of the following drawing tool icons.

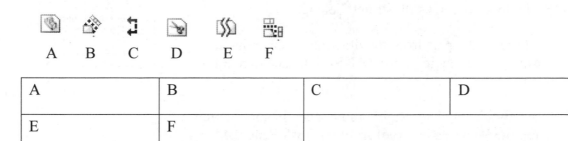

A   B   C   D   E   F

A	B	C	D
E	F		

Exercises

Exercise 3.1: L - BRACKET Drawing

Create the A (ANSI) Landscape - IPS - Third Angle L-BRACKET drawing as illustrated below.

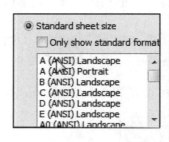

- First create the part from the drawing - then create the drawing.

- Insert the Front, Top, Right, and Shaded Isometric view using the View Palette tool from the Task Pane. Hide the Top view.

- Insert dimensions into the Sheet. Think about the require extension line gaps needed between the Feature lines. Think about the proper view for your dimensions. Use the default A (ANSI) Landscape Sheet Format/Size.

- Insert Custom Properties: Material, Description DrawnBy, DrawnDate, CompanyName, etc. Note: Material is 1060 Alloy.

- Insert Company and Third Angle projection icons. The icons are available in the homework folder.

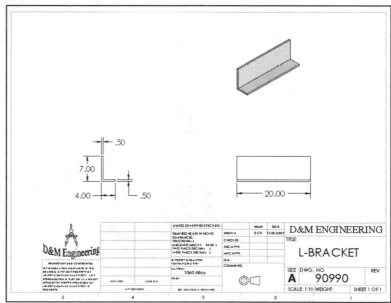

Exercise 3.2: T - SECTION Drawing

Create the A (ANSI) Landscape - IPS - Third Angle
T-SECTION drawing as illustrated below.

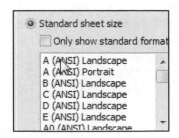

- First create the part from the drawing - then create the
  drawing. Use the default A (ANSI) Landscape Sheet
  Format/Size.

- Insert the Front, Top, Right, and Shaded Isometric view
  using the View Palette tool from the Task Pane. Hide the
  Top view.

- Insert dimensions into the Sheet. Think about the proper view for your dimensions.

- Think about the needed Extension line gaps between the Feature lines.

- Insert Custom Properties: Material, Description, DrawnBy, DrawnDate,
  CompanyName, etc. Note: Material is 1060 Alloy.

- Insert Company and Third Angle projection icons. The icons are available in the
  homework folder.

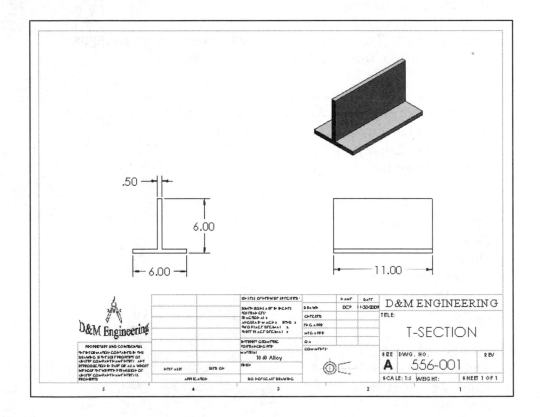

Exercise 3.3: FLAT BAR - 3HOLE Drawing

Create the A (ANSI) Landscape - IPS - Third Angle 3HOLES drawing as illustrated below. Do not display Tangent Edges. Do not dimension to Hidden Lines.

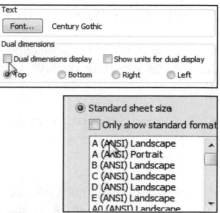

- First create the part from the drawing - then create the drawing. Use the default A (ANSI) Landscape Sheet Format/Size.

- Insert a Front, Top and a Shaded Isometric view as illustrated. Insert dimensions. Address all needed gaps. Insert the proper Display modes.

- Add a Smart (Linked) Parametric note for MATERIAL THICKNESS in the drawing as illustrated. Hide the dimension in the Top view. Insert needed Centerlines.

- Modify the Hole dimension text to include 3X EQ. SP. and 2X as illustrated.

- Insert Custom Properties: Material, Description, DrawnBy, DrawnDate, CompanyName, etc. Note: Material is 1060 Alloy

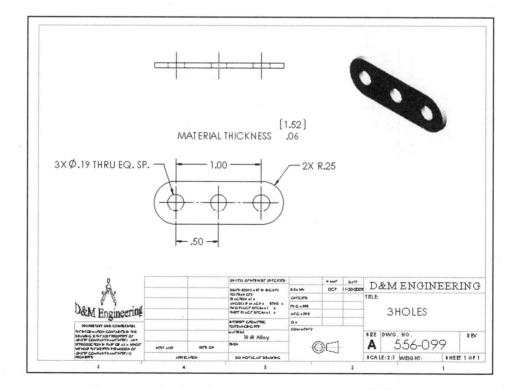

Exercise 3.4: CYLINDER Drawing

Create the A (ANSI) Landscape - IPS - Third Angle CYLINDER drawing as illustrated below. Do not display Tangent Edges. Do not dimension to Hidden lines.

- First create the part from the drawing - then create the drawing. Use the default A (ANSI) Landscape Sheet Format/Size.

- Inert the Front and Right view as illustrated. Insert dimensions. Think about the proper view for your dimensions. Address all needed extension line gaps. Insert the proper Display modes.

- Insert Company and Third Angle projection icons. The icons are available in the homework folder.

- Insert all needed Centerlines, Center Marks and annotations.

- Insert Custom Properties: Material, Description, DrawnBy, DrawnDate, CompanyName, etc. Note: Material is AISI 1020.

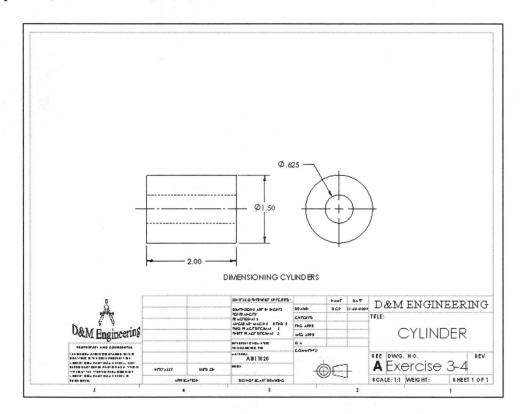

Exercise 3.5: PRESSURE PLATE Drawing

Create the A (ANSI) Landscape - IPS - Third Angle
PRESSURE PLATE drawing as illustrated below. Do not
display Tangent edges. Do not dimension to Hidden lines.

- First create the part from the drawing - then create the
  drawing. Use the default A (ANSI) Landscape Sheet
  Format/Size.

- Inert the Front and Right view as illustrated. Insert dimensions. Address all needed
  extension line gaps. Think about the proper view for your dimensions. Insert the
  proper Display modes.

- Insert Company and Third Angle projection icons. The icons are available in the
  homework folder.

- Insert all needed Centerlines, Center Marks and annotations.

- Insert Custom Properties: Material, Description, DrawnBy, DrawnDate,
  CompanyName, etc. Note: Material is 1060 Alloy.

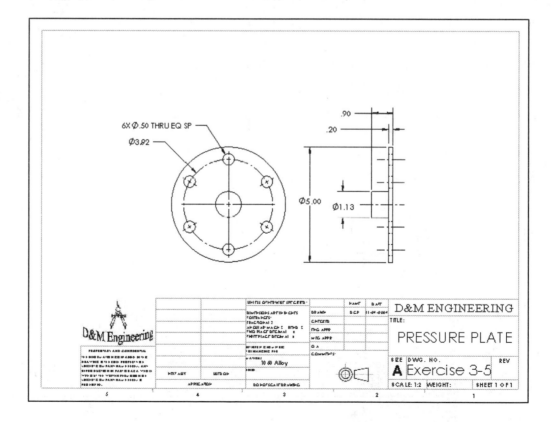

Exercise 3.6: PLATE-1 Drawing

Create the A (ANSI) Landscape - MMGS - Third Angle PLATE-1 drawing as illustrated below. Do not display Tangent Edges. Do not dimension to Hidden lines.

- First create the part from the drawing - then create the drawing. Use the default A (ANSI) Landscape Sheet Format/Size.

- Inert the Front and Right view as illustrated. Insert dimensions. Address all needed extension line gaps. Think about the **proper view** for your dimensions. Insert the proper Display modes.

- Insert Company and Third Angle projection icons. The icons are available in the homework folder.

- Insert all needed Centerlines, Center Marks and annotations.

- Insert Custom Properties: Material, Description, DrawnBy, DrawnDate, CompanyName, etc. Note: Material is 1060 Alloy.

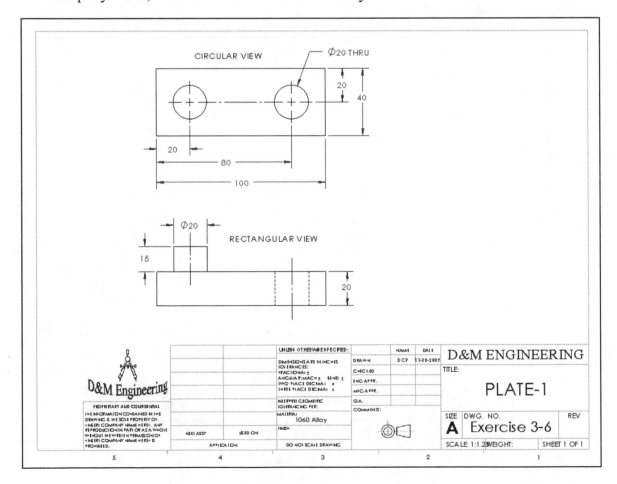

Exercise 3.7: SHAFT-1 Drawing

Create the A (ANSI) Landscape - IPS - Third Angle PLATE-1 drawing as illustrated below. Do not display Tangent Edges. Phantom lines are fine.

- First create the part from the drawing - then create the drawing. Use the default A (ANSI) Landscape Sheet Format/Size.

- Inert the Front, Right (Break), Isometric and Auxiliary Broken Crop view as illustrated. Insert dimensions. Address all needed extension line gaps. Think about the proper view for your dimensions. Insert the proper Display modes.

- Insert Company and Third Angle projection icons. The icons are available in the homework folder.

- Insert all needed Centerlines, Center Marks and annotations.

- Insert Custom Properties: Material, Description, DrawnBy, DrawnDate, CompanyName, etc. Note: Material is 1060 Alloy.

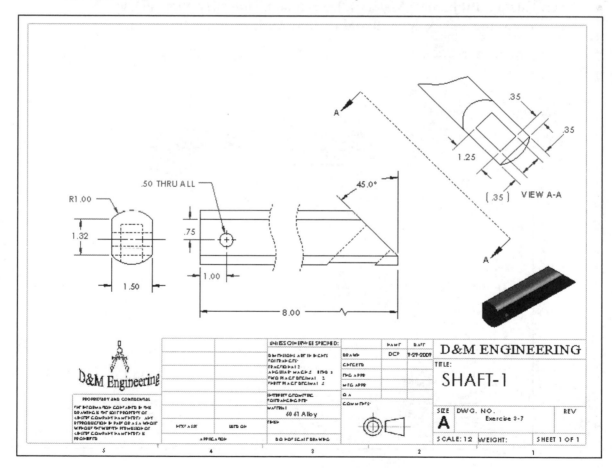

Exercise 3.8: TUBE-2 Drawing

Create the A (ANSI) Landscape - IPS - Third Angle TUBE-2 drawing as illustrated below. Do not display Tangent edges. Phantom lines are fine.

- First create the part from the drawing - then create the drawing. Use the default A (ANSI) Landscape Sheet Format/Size.

- Inert the Front, Top, Right, and Isometric views as illustrated. Insert dimensions. Address all needed extension line gaps. Think about the proper view for your dimensions! Hide the Right view. It does not help the drawing.

- Insert Company and Third Angle projection icons. The icons are available in the homework folder.

- Insert all needed Centerlines, Center Marks and annotatios.

- Insert the proper Display modes.

- Insert Custom Properties: Material, Description, DrawnBy, DrawnDate, CompanyName, etc. Note: Material is 1060 Alloy.

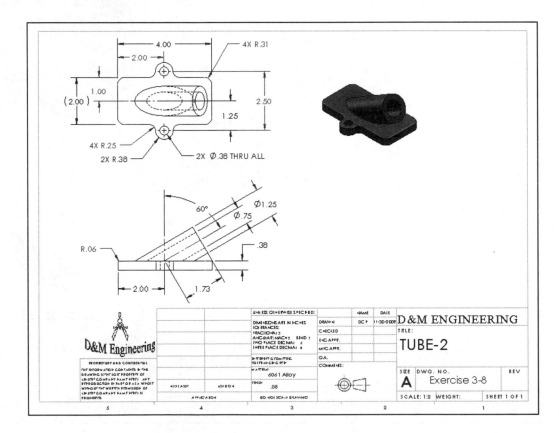

Exercise 3.9: FLAT-PLATE Drawing

Create the A (ANSI) Landscape - IPS - Third Angle FLAT-PLATE drawing as illustrated below. Do not display Tangent Edges. Phantom lines are fine.

- First create the part from the drawing - then create the drawing. Use the default A (ANSI) Landscape Sheet Format/Size.

- Inert the Front, Top, Right, and Isometric views as illustrated. Insert dimensions. Think about the proper view for your dimensions. Insert the proper Display modes.

- Insert Company and Third Angle projection icons. The icons are available in the homework folder.

- Insert all needed Centerlines, Center Marks and annotations.

- Insert Custom Properties: Material, Description, DrawnBy, DrawnDate, CompanyName, etc. Note: Material is 1060 Alloy.

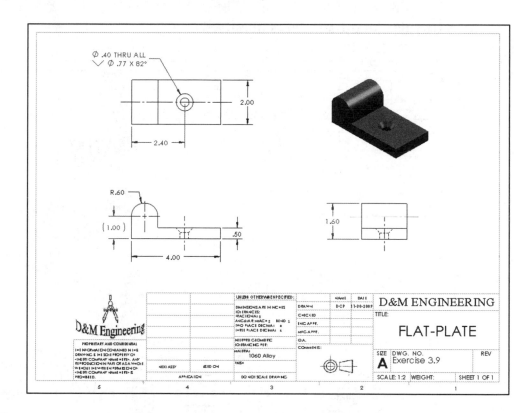

Exercise 3.10: FRONT-SUPPORT Assembly Drawing

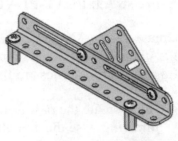

Create the A (ANSI) Landscape - Third Angle FRONT-SUPPORT Assembly drawing with a Bill of Materials and balloons.

- Open the FRONT-SUPPORT assembly. The FRONT-SUPPORT assembly is located in the Chapter3 - Homework folder in the book DVD.

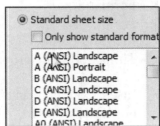

- Use the default A (ANSI) Landscape Sheet Format/Size. Insert an Isometric Shaded With Edges view.

- Insert a Bill of Materials. Display the top level sub-assemblies and parts only in the Bill of Materials.

- Select bom-material.sldbomtbt from the Select BOM Table dialog box.

- Resize the text in the Title box and DWG. NO. box. Insert Custom Properties: Description, DrawnBy, DrawnDate, CompanyName, etc.

- Insert Circular Auto-Balloons as illustrated.

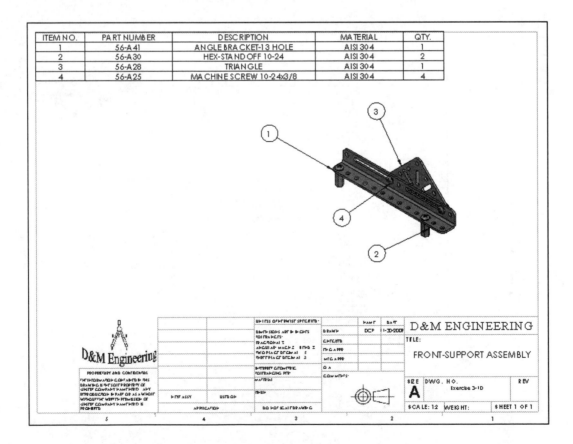

ITEM NO.	PART NUMBER	DESCRIPTION	MATERIAL	QTY.
1	56-A41	ANGLE BRACKET-13 HOLE	AISI 304	1
2	56-A30	HEX-STANDOFF 10-24	AISI 304	2
3	56-A28	TRIANGLE	AISI 304	1
4	56-A25	MACHINE SCREW 10-24x3/8	AISI 304	4

Exercise 3.10A: Modify the Bill of Materials

Modify the Bill of Materials of the FRONT-SUPPORT Assembly Drawing which you created in exercise 3.10. Input the illustrated Material and Part numbers to the items thought the individual parts (Custom Properties).

ITEM NO.	PART NUMBER	DESCRIPTION	MATERIAL	QTY.
1	56-A41	ANGLE BRACKET-13 HOLE	AISI 304	1
2	56-A30	HEX-STANDOFF 10-24	AISI 304	2
3	56-A28	TRIANGLE	AISI 304	1
4	56-A25	MACHINE SCREW 10-24x3/8	AISI 304	4

• Utilize the MATERIAL EDITOR to assign AISI 304 to all of the items.

• Add the Custom Property MATERIAL to the items.

• Utilize User Defined for Part Number when used in Bill of Materials.

Exercise 3.11: VALVE PLATE Drawing.
Create the A-ANSI Third Angle VALVE PLATE Drawing document according to the ASME 14.5M standard.

• Open the VALVE PLATE part. The VALVE PLATE part is located in the Chapter3 - Homework folder in the book DVD,

• Create the VALVE PLATE drawing.

• Insert three views: Front, Top and Right. Insert the proper Display modes.

• Utilize the Tolerance/Precision option to modify the decimal place value.

• Utilize Surface Finish, Geometric Tolerance and Datum Feature ⊢Ⓐ tools in the Annotation toolbar.

• Create the following drawing.

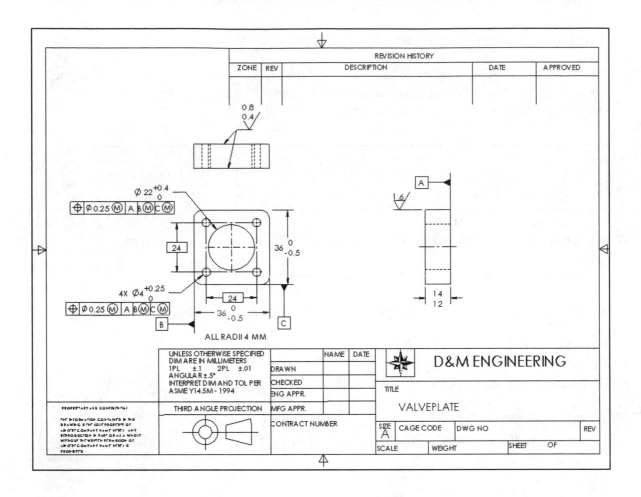

		REVISION HISTORY		
ZONE	REV	DESCRIPTION	DATE	APPROVED

0.8
0.4

Ø 22 +0.4 / 0

⊕ | Ø0.25 Ⓜ | A | B Ⓜ | C Ⓜ

24

36 0 / -0.5

4X Ø4 +0.25 / 0

⊕ | Ø0.25 Ⓜ | A | B Ⓜ | C Ⓜ

B

24

36 0 / -0.5

C

ALL RADII 4 MM

A

√6

14
12

UNLESS OTHERWISE SPECIFIED DIM ARE IN MILLIMETERS 1PL ±.1 2PL ±.01 ANGULAR ±.5° INTERPRET DIM AND TOL PER ASME Y14.5M - 1994		NAME	DATE	D&M ENGINEERING		
	DRAWN					
	CHECKED			TITLE		
	ENG APPR.			VALVEPLATE		
THIRD ANGLE PROJECTION	MFG APPR.					
	CONTRACT NUMBER			SIZE A	CAGE CODE	DWG NO	REV
				SCALE	WEIGHT	SHEET OF

Exercise 3.11: GUIDE eDrawing

Create the GUIDE eDrawing. A SolidWorks eDrawing is a compressed document that does not require the corresponding part or assembly. A SolidWorks eDrawing is animated to display multiple views and dimensions. Review the eDrawing SolidWorks Help Topics for additional functionality. The eDrawings Professional version contains additional options to mark up a drawing.

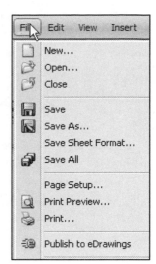

- Open the GUIDE drawing. Note: The Finished GUIDE drawing is located in the Chapter 3 Homework folder.

- Click File, Publish to eDrawing from the Main menu. Click the Play ᵖˡᵃʸ button to animate the drawing views. Click Stop. View the Menu features.

- Save the GUIDE eDrawing.

Refer to Help, Online Tutorial, eDrawings exercise for additional information.

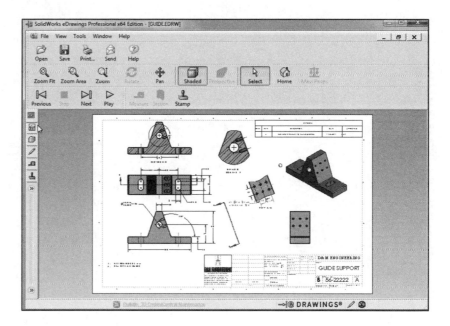

Notes:

Project 4

Extrude and Revolve Features

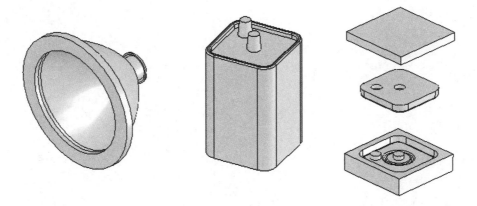

Below are the desired outcomes and usage competencies based on the completion of Project 4.

Project Desired Outcomes:	Usage Competencies:
• Obtain your customer's requirements for the FLASHLIGHT assembly.	• Ability to incorporate Design Intent into sketches, features, parts and assemblies.

• Two Part Templates:
  ○ PART-IN-ANSI
  ○ PART-MM-ISO
	• Aptitude to apply Document Properties in a custom Part Template.

• Four key parts:
  ○ BATTERY
  ○ BATTERYPLATE
  ○ LENS
  ○ BULB
	• Specific knowledge of the following features: Extruded Boss/Base, Instant3D, Extruded Cut, Revolved Boss/Base, Revolved Cut, Dome, Shell, Circular Pattern and Fillet.

• Core and Cavity Tooling for the BATTERYPLATE.	• Understanding of the Mold tools: Scale, Parting Lines, Parting Surfaces, Shut-off Surfaces, Tooling Split and Draft.

Notes:

Project 4 - Extrude and Revolve Features

Project Objective

Design a FLASHLIGHT assembly according to the customer's requirements. The FLASHLIGHT assembly will be cost effective, serviceable and flexible for future manufacturing revisions.

Design intent is the process in which the model is developed to accept future changes. Build design intent into the FLASHLIGHT sketches, features, parts and assemblies. Create two custom Part Templates. The Part Template is the foundation for the FLASHLIGHT parts.

Create the following parts:

- BATTERY
- BATTERYPLATE
- LENS
- BULB

The other parts for the FLASHLIGHT assembly are addressed in Project 5. Create the Core and Cavity mold tooling required for the BATTERYPLATE.

On the completion of this project, you will be able to:

- Apply design intent to sketches, features, parts, and assemblies.
- Select the best profile for a sketch.
- Select the proper Sketch planc.
- Create a template: English and Metric units.
- Set Document Properties.
- Customize the SolidWorks CommandManager toolbar.
- Insert/Edit dimensions.
- Insert/Edit relations.
- Use the following SolidWorks features:
  - Instant3D
  - Extruded Boss/Base
  - Extruded Cut

- o Revolved Boss/Base
- o Revolved Boss Thin
- o Revolved Cut Thin
- o Dome
- o Shell
- o Circular Pattern
- o Fillet

- Use the following Mold tools:

- o Draft
- o Scale
- o Parting Lines
- o Shut-off Surfaces
- o Parting Surfaces
- o Tooling Split

Project Overview

Start the design of the FLASHLIGHT assembly according to the customer's requirements. The FLASHLIGHT assembly will be cost effective, serviceable and flexible for future manufacturing revisions.

A template is the foundation for a SolidWorks document. A template contains document settings for units, dimensioning standards and other properties. Create two part templates for the FLASHLIGHT Project:

- PART-IN-ANSI
- PART-MM-ISO

Create four parts for the FLASHLIGHT assembly in this Project:

- BATTERY
- BATTERYPLATE
- LENS
- BULB

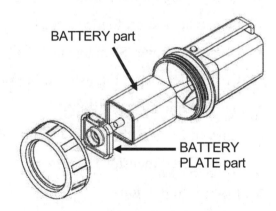

FLASHLIGHT Assembly

Parts models consist of 3D features. Features are the building blocks of a part.

A 2D Sketch Plane is required to create an Extruded feature. Utilize the sketch geometry and sketch tools to create the following features:

- Extruded Boss/Base

- Extruded Cut

Utilize existing faces and edges to create the following features:

- Fillet

- Chamfer

This project introduces you to the Revolved feature. Create two parts for the FLASHLIGHT assembly in this section:

- LENS

- BULB

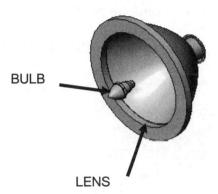

A Revolved feature requires a 2D sketch profile and a centerline. Utilize sketch geometry and sketch tools to create the following features:

- Revolved Boss/Base

- Revolved Boss-Thin

- Revolved Cut

Utilize existing faces to create the following features:

- Shell

- Dome

- Hole Wizard

Utilize the Extruded Cut feature to create a Circular Pattern.

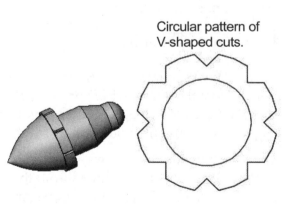

Circular pattern of V-shaped cuts.

Utilize the Mold tools to create the Cavity tooling plates for the BATTERYPLATE part.

Design Intent

The SolidWorks definition of design intent is the process in which the model is developed to accept future changes.

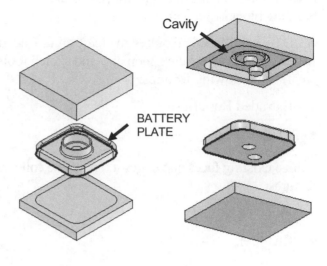

Cavity

BATTERY PLATE

Isometric view   Rotated
Mold Tools

Models behave differently when design changes occur. Design for change. Utilize geometry for symmetry, reuse common features and reuse common parts.

Build change into the following areas:

1. Sketch

2. Feature

3. Part

4. Assembly

5. Drawing

 See Project 8 (Intelligent modeling techniques) for additional information.

1. Design Intent in the Sketch

In SolidWorks, relations between sketch entities and model geometry, in either 2D or 3D sketches, are an important means of building in design intent. In this chapter - we will only address 2D sketches.

Apply design intent in a sketch as the profile is created. A profile is determined from the Sketch Entities. Example: Rectangle, Circle, Arc, Point, Slot etc.

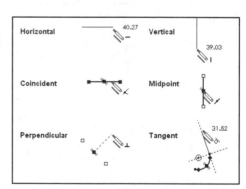

Develop design intent as you sketch with Geometric relations. Sketch relations are geometric constraints between sketch entities or between a sketch entity and a plane, axis, edge, or vertex. Relations can be added automatically or manually.

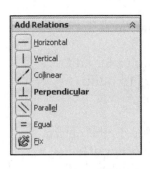
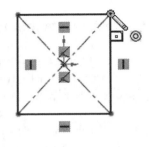

A rectangle contains Horizontal, Vertical, and Perpendicular automatic Geometric relations. Apply design intent using added Geometric relations. Example: Horizontal, Vertical, Collinear, Perpendicular, Parallel, etc.

Example A: Apply design intent to create a square profile. Sketch a rectangle. Apply the Center Rectangle tool. Note: No construction reference centerline or Midpoint relation is required with the Center Rectangle tool. Insert dimensions to define the square.

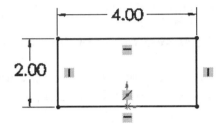

Example B: Develop a rectangular profile. Apply the Corner Rectangle tool. The bottom horizontal midpoint of the rectangular profile is located at the Origin. Add a Midpoint relation between the horizontal edge of the rectangle and the Origin. Insert two dimensions to define the width and height of the rectangle as illustrated.

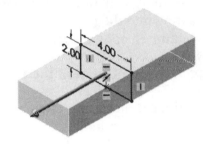

2. Design Intent in the Feature

Build design intent into a feature by addressing symmetry, feature selection, and the order of feature creation.

Example A: The Boss-Extrude1 feature (Base feature) remains symmetric about the Front Plane. Utilize the Mid Plane End Condition option in Direction 1. Modify the depth, and the feature remains symmetric about the Front Plane.

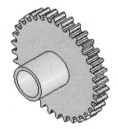

Example B: Do you create each tooth separate using the Extruded Cut feature? No. Create a single tooth and then apply the Circular Pattern feature. Create 34 teeth for a Circular Pattern feature. Modify the number of teeth from 32 to 24.

3. Design Intent in the Part

Utilize symmetry, feature order and reusing common features to build design intent into the part.

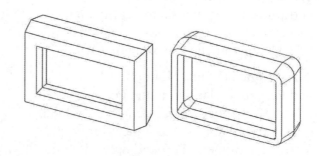

Example A: Feature order. Is the entire part symmetric? Feature order affects the part. Apply the Shell feature before the Fillet feature and the inside corners remain perpendicular.

4. Design Intent in the Assembly

Utilizing symmetry, reusing common parts and using the Mate relation between parts builds the design intent into an assembly.

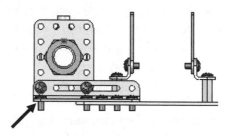

Example A: Reuse geometry in an assembly. The assembly contains a linear pattern of holes. Insert one screw into the first hole. Utilize the Component Pattern feature to copy the machine screw to the other holes.

5. Design Intent in the Drawing

Utilize dimensions, tolerance and notes in parts and assemblies to build the design intent into the Drawing.

Example A: Tolerance and material in the drawing.

Insert an outside diameter tolerance +.000/-.002 into the TUBE part. The tolerance propagates to the drawing.

Define the Custom Property MATERIAL in the part. The MATERIAL Custom Property propagates to the drawing.

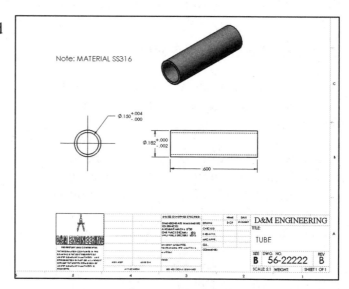

Project Situation

You work for a company that specializes in providing promotional tradeshow products. The company is expecting a sales order for 100,000 flashlights with a potential for 500,000 units next year. Prototype drawings of the flashlight are required in three weeks.

You are the design engineer responsible for the project. You contact the customer to discuss design options and product specifications. The customer informs you that the flashlights will be used in an international marketing promotional campaign. Key customer requirements:

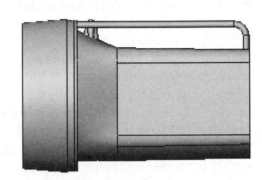

- Inexpensive reliable flashlight.

- Available advertising space of 10 square inches, 64.5 square centimeters.

- Lightweight semi indestructible body.

- Self standing with a handle.

Your company's standard product line does not address the above key customer requirements. The customer made it clear that there is no room for negotiation on the key product requirements.

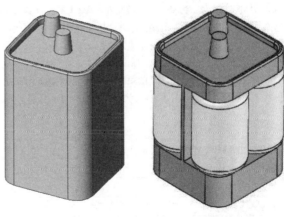

You contact the salesperson and obtain additional information on the customer and product. This is a very valuable customer with a long history of last minute product changes. The job has high visibility with great future potential.

In a design review meeting, you present a conceptual sketch. Your colleagues review the sketch. The team's consensus is to proceed with the conceptual design.

The first key design decision is the battery. The battery type directly affects the flashlight body size, bulb intensity, case structure integrity, weight, manufacturing complexity and cost.

Review two potential battery options:

- A single 6-volt lantern battery

- Four 1.5-volt D cell batteries

The two options affect the product design and specification. Think about it.

A single 6-volt lantern battery is approximately 25% higher in cost and 35% more in weight. The 6-volt lantern battery does provide higher current capabilities and longer battery life.

A special battery holder is required to incorporate the four 1.5 volt D cell configuration. This would directly add to the cost and design time of the FLASHLIGHT assembly.

Time is critical. For the prototype, you decide to use a standard 6-volt lantern battery. This eliminates the requirement to design and procure a special battery holder. However, you envision the four D cell battery model for the next product revision.

Design the FLASHLIGHT assembly to accommodate both battery design options. Battery dimensional information is required for the design. Where do you go? Potential sources: product catalogs, company web sites, professional standards organizations, design handbooks and colleagues.

The team decides to purchase the following parts: 6-volt BATTERY, LENS ASSEMBLY, SWITCH and an O-RING. Model the following purchased parts: BATTERY, LENS assembly, SWITCH and the O-RING. The LENS assembly consists of the LENS and the BULB.

Your company will design, model and manufacture the following parts: BATTERYPLATE, LENSCAP and HOUSING.

Purchased Parts:	Designed Parts:
BATTERY	BATTERYPLATE
LENS assembly	MOLD TOOLING
*SWITCH	*LENSCAP
*O-RING	*HOUSING

*Parts addressed in Project 5.

The BATTERYPLATE, LENSCAP and HOUSING are plastic parts. Review the injection molded manufacturing process and the SolidWorks Mold tools. Modify the part features to eject the part from the mold. Create the MOLD TOOLING for the BATTERYPLATE.

Part Template

Units are the measurement of physical quantities. Millimeter dimensioning and decimal inch dimensioning are the two most common unit types specified for engineering parts and drawings. The FLASHLIGHT project is designed in inch units and manufactured in millimeter units. Inch units are the primary unit and Millimeter units are the secondary unit.

Create two Part templates:

- PART-IN-ANSI

- PART-MM-ISO

Save the Part templates in the MY-TEMPLATES folder. System Options, File Locations option controls the file folder location of SolidWorks documents. Utilize the File Locations option to reference your Part templates in the MY-TEMPLATES folder. Add the MY-TEMPLATES folder path name to the Document Templates File Locations list.

Activity: Create Two Part Templates

Create a PART-IN-ANSI Template.

1) Click **New** from the Menu bar.

2) Double-click **Part** from the default Templates tab from the Menu bar.

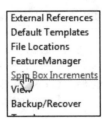

Set the Dimensioning Standard to ANSI.

3) Click **Options** .

4) Click the **System Options** tab.

5) Click **Spin Box Increments**. View the default settings.

6) Click **inside** the English units box.

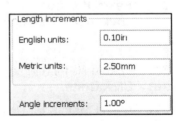

7) Enter **.10**in.

8) Click **inside** the Metric units box.

9) Enter **2.50**mm.

Set Document Properties.

10) Click the **Document Properties** tab.

11) Select **ANSI** from the Overall drafting standard drop-down menu.

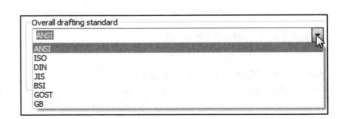

Set the part units for inch.

12) Click **Units**.

13) Select **IPS** for Unit system.

14) Select **.123** for Basic unit length decimal place.

15) Select **millimeters** for Dual dimension length unit.

16) Select **.12** for Basic unit decimal place.

17) Select **None** for Basic unit angle decimal place.

18) Click **OK** from the Document Properties - Units dialog box.

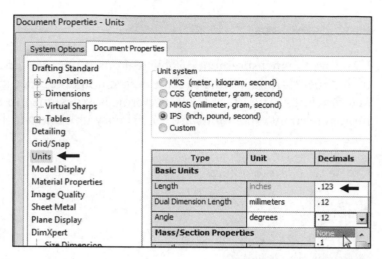

Save and name the part template.

19) Click **Save As** from the drop-down Menu bar.

20) Select **Part Templates (*.prtdot)** from the Save As type box.

21) Select **ENGDESIGN-W-SOLIDWORKS\MY-TEMPLATES** for the Save in folder.

22) Enter **PART-IN-ANSI** for File name.

23) Click **Save**.

Utilize the PART-IN-ANSI template to create the PART-MM-ISO template.

24) Click **Options** , **Document Properties** tab.

25) Select **ISO** from the Overall drafting standard drop-down menu.

Set the part units for millimeter.

26) Click **Units**.

27) Select **MMGS** for Unit system.

28) Select **.12** for Basic unit length decimal place.

29) Select **None** for Basic unit angle decimal place.

30) Click **OK** from the Document Properties - Units dialog box.

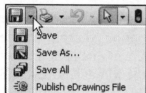

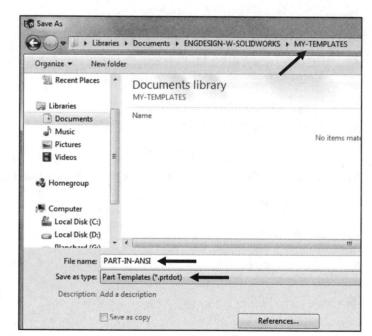

Save and name the part template.

31) Click **Save As** from the drop-down Menu bar.

32) Select **Part Templates (*.prtdot)** from the Save As type box.

33) Select **ENGDESIGN-W-SOLIDWORKS\MY-TEMPLATES** for the Save in folder.

34) Enter **PART-MM-ISO** for File name.

35) Click **Save**.

Set the System Options for File Locations to display in the New dialog box *if needed.*

36) Click **Options** ⊞ from the Menu bar.

37) Click **File Locations** from the System Options tab.

38) Select **Document Templates** from Show folders for.

39) Click the **Add** button.

40) Select the **MY-TEMPLATES** folder.

41) Click **OK** from the Browse for Folder dialog box.

42) Click **OK** from the System Options dialog box.

Close all documents.

43) Click **Windows**, **Close All** from the Menu bar.

Each folder listed in the System Options, File Locations, Document Templates, Show Folders For option produces a corresponding tab in the New SolidWorks Document dialog box. The order in the Document Templates box corresponds to the tab order in the New dialog box.

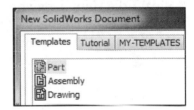

The MY-TEMPLATES tab is visible when the folder contains SolidWorks Template documents. Create the PART-MM-ANSI template as an exercise.

To remove Tangent edges on a model, click **Display/Selections** from the Options menu, check the **Removed** box.

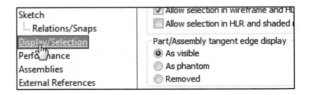

The PART-IN-ANSI template contains Document Properties settings for the parts contained in the FLASHLIGHT assembly. Substitute the PART-MM-ISO or PART-MM-ANSI template to create the identical parts in millimeters.

The primary units in this Project are IPS, (inch, pound, seconds).

The optional secondary units are MMGS (millimeters, grams, second) and are indicated in brackets [ ].

Illustrations are provided in both inches and millimeters. Utilize inches, millimeters or both.

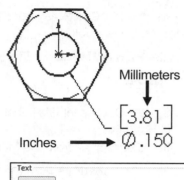

To set dual dimensions, select Options, Document Properties, Dimensions. Check the Dual dimensions display box as illustrated.

To set dual dimensions for an active document, check the Dual Dimension box in the Dimension PropertyManager.

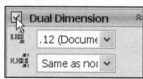

Enter Toolbars, Features in SolidWorks Help Search category to review the function of each Features toolbar.

Additional information on System Options, Document Properties, File Locations and Templates is found in SolidWorks Help. Keywords: Options (detailing, units), templates, Files (locations), menus and toolbars (features, sketch).

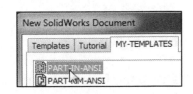

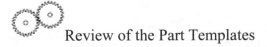 Review of the Part Templates

You created two Part templates: PART-MM-ANSI and PART-IN-ISO. Note: Other templates were created in the previous project. The Document Properties Dimensioning Standard, units and decimal places are stored in the Part Templates.

The File Locations System Option, Document Templates option controls the reference to the MY-TEMPLATES folder.

Note: In some network locations and school environments, the File Locations option must be set to MY-TEMPLATES for each session of SolidWorks.

You can exit SolidWorks at any time during this project. Save your document. Select File, Exit from the Menu bar.

BATTERY Part

The BATTERY is a simplified representation of a purchased OEM part. Represent the battery terminals as cylindrical extrusions. The battery dimensions are obtained from the ANSI standard 908D.

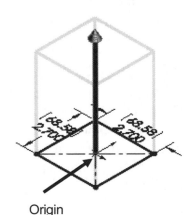

A 6-Volt lantern battery weighs approximately 1.38 pounds, (0.62kg). Locate the center of gravity closest to the center of the battery.

Create the BATTERY part. Use features to create parts. Features are building blocks that add or remove material.

Utilize the Instant3D tool to create the Extruded Boss/Base feature vs. using the Boss-Extrude PropertyManager. The Extrude Boss/Base features add material. The Base feature (Boss-Extrude1) is the first feature of the part. Note: The default End Condition for Instant3D is Blind.

Apply symmetry. Use the Center Rectangle Sketch tool on the Top Plane. The 2D Sketch profile is centered at the Origin.

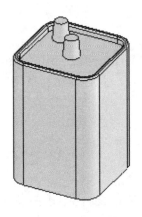

Origin

Extend the profile perpendicular (⊥) to the Top Plane.

Utilize the Fillet feature to round the four vertical edges.

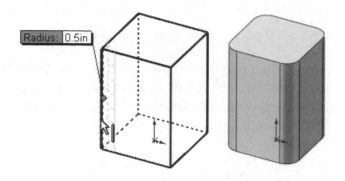

The Extruded Cut feature removes material from the top face. Utilize the top face for the Sketch plane. Utilize the Offset Entity Sketch tool to create the profile.

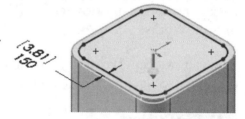

Utilize the Fillet feature to round the top narrow face.

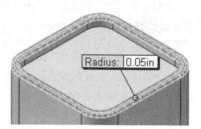

The Extruded Boss/Base feature adds material. Conserve design time. Represent each of the terminals as a cylindrical Extruded Boss feature.

Think design intent. When do you use the various End Conditions and Geometric sketch relations? What are you trying to do with the design? How does the component fit into an assembly? Design for change and flexibility.

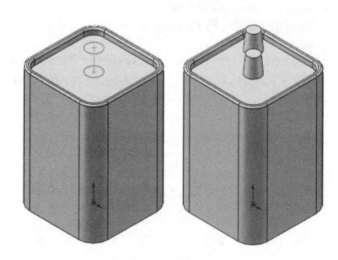

BATTERY Part-Extruded Boss/Base Feature

The Extruded Boss/Base feature requires:

- Sketch plane (Top)
- Sketch profile (Rectangle)
  - Geometric relations and dimensions
- End Condition Depth (Blind) in Direction 1

Create a new part named, BATTERY. Insert an Extruded Boss/Base feature. Extruded features require a Sketch plane. The Sketch plane determines the orientation of the Extruded Base feature. The Sketch plane locates the Sketch profile on any plane or face.

The Top Plane is the Sketch plane. The Sketch profile is a rectangle. The rectangle consists of two horizontal lines and two vertical lines.

Geometric relations and dimensions constrain the sketch in 3D space. The Blind End Condition in Direction 1 requires a depth value to extrude the 2D Sketch profile and to complete the 3D feature.

Alternate between the Features tab and the Sketch tab in the CommandManager to display the available Feature and Sketch tools for the Part document.

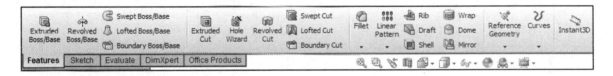

Activity: BATTERY Part-Create the Extruded Base Feature

Create a New part.

44) Click **New** ⬜ from the Menu bar.

45) Click the **MY-TEMPLATES** tab.

46) Double-click **PART-IN-ANSI**, [**PART-MM-ISO**].

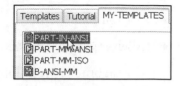

Save and name the empty part.

47) Click **Save** 💾.

48) Select **PROJECTS** for Save in folder.

49) Enter **BATTERY** for File name.

50) Enter **BATTERY**, **6-VOLT** for Description.

51) Click **Save**. The Battery FeatureManager is displayed.

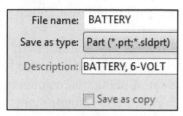

Select the Sketch plane.

52) Right-click **Top Plane** from the FeatureManager. This is your Sketch plane.

Sketch the 2D Sketch profile centered at the Origin.

53) Click **Sketch** ✏ from the Context toolbar. The Sketch toolbar is displayed.

54) Click the **Center Rectangle** ▭ Sketch tool. The Center Rectangle ▭ icon is displayed.

55) Click the **Origin**. This is your first point.

56) Drag and click the **second point** in the upper right quadrant as illustrated. The Origin is located in the center of the sketch profile. The Center Rectangle Sketch tool automatically applies equal relations to the two horizontal and two vertical lines. A midpoint relation is automatically applied to the Origin.

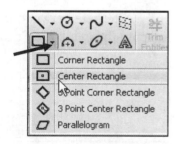

🔆 A goal of this book is to expose the new user to various tools, techniques and procedures. The text may not always use the most direct tool or process.

🔆 Click **View**, **Sketch Relations** from the Main menu to view sketch relations in the Graphics area.

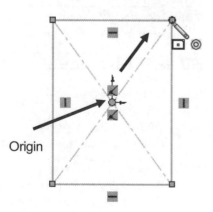

Origin

Dimension the sketch.

57) Click the **Smart Dimension** ✏ Sketch tool.

58) Click the **top horizontal line**.

59) Click a **position** above the horizontal line.

60) Enter **2.700**in, [**68.58**] for width.

61) Click the **Green Check mark** ✔ in the Modify dialog box.

62) Enter **2.700**in, [**68.58**] for height as illustrated.

63) Click the **Green Check mark** ✔ in the Modify dialog box. The black Sketch status is fully defined

64) Click **OK** ✔ from the Dimension PropertyManager.

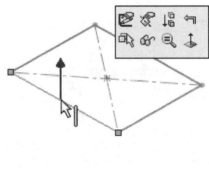

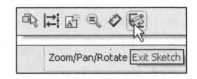

Exit the Sketch.
65) Click **Exit Sketch**.

Insert an Extruded Boss/Base feature. Apply the Instant3D tool. The Instant3D tool provides the ability to drag geometry and dimension manipulator points to resize or to create features directly in the Graphics window.

Use the on-screen ruler.
66) Click **Isometric view** from the Heads-up View toolbar.

67) Click the **front horizontal line** as illustrated. A green arrow is displayed.

68) Click and drag the **green/red arrow** upward.

69) Click the on-screen ruler at **4.1**in, [104.14] as illustrated. This is the depth in direction 1. The extrude direction is upwards. Boss-Extrude1 is displayed in the FeatureManager.

Check the Boss-Extrude1 feature depth dimension.
70) Right-click **Boss-Extrude1** from the FeatureManager.

71) Click **Edit Feature** 🖼 from the Context toolbar. 4.100in is displayed for depth. Blind is the default End Condition. Note: If you did not select the correct depth, input the depth in the Boss-Extrude1 PropertyManager.

72) Click **OK** ✔ from the Boss-Extrude1 PropertyManager.

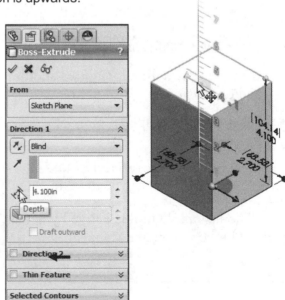

Modify the **Spin Box Increments** in System Options to display different increments in the on-screen ruler.

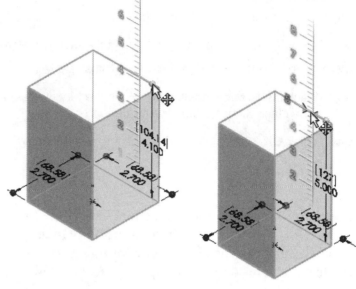

Fit the part to the Graphics window.
73) Press the **f** key.

Rename the Boss-Extrude1 feature.
74) Rename **Boss-Extrude1** to **Base Extrude**.

Save the BATTERY.
75) Click **Save** 💾.

Modify the BATTERY.
76) Click **Base Extrude** from the FeatureManager. Note: Instant3D is activated by default.

77) Drag the **manipulator point** upward and click the on-screen ruler to create a **5.000in, [127]** depth as illustrated. Blind is the default End Condition.

Return to the 4.100 depth.

78) Click the **Undo** 🔄 button from the Menu bar. The depth of the model is 4.100in, [104.14]. Blind is the default End Condition. Practice may be needed to select the correct on-screen ruler dimension.

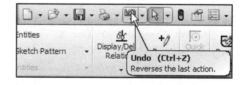

The color of the sketch indicates the sketch status.

- Light Blue - Currently selected.

- Blue - Under defined, requires additional geometric relations and dimensions.

- Black - Fully defined.

- Red - Over defined, requires geometric relations or dimensions to be deleted or redefined to solve the sketch.

The Instant3D tool is active by default in the Features toolbar located in the CommandManager.

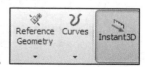

BATTERY Part-Fillet Feature Edge

Fillet features remove sharp edges. Utilize Hidden Lines Visible  from the Heads-up View toolbar to display hidden edges.

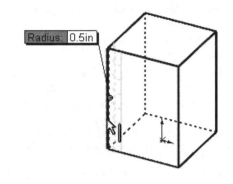

An edge Fillet feature requires:

• A selected edge

• Fillet radius

Select a vertical edge. Select the Fillet feature from the Features toolbar. Enter the Fillet radius. Add the other vertical edges to the Items To Fillet option.

The order of selection for the Fillet feature is not predetermined. Select edges to produce the correct result.

The Fillet feature uses the Fillet PropertyManager. The Fillet PropertyManager provides the ability to select either the *Manual* or *FilletXpert* tab.

Each tab has a separate menu and PropertyManager. The Fillet PropertyManager and FilletXpert PropertyManager displays the appropriate selections based on the type of fillet you create.

The FilletXpert automatically manages, organizes and reorders your fillets in the FeatureManager design tree. The FilletXpert PropertyManager provides the ability to add, change or corner fillets in your model. The PropertyManager remembers its last used state. View the SolidWorks tutorials for additional information on fillets.

The FilletXpert can ONLY create and edit Constant radius fillets.

Activity: BATTERY Part-Fillet Feature Edge

Display the hidden edges.

79) Click **Hidden Lines Visible** ⬜ from the Heads-up View toolbar.

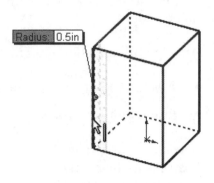

Insert a Fillet feature.

80) Click the **left front vertical edge** as illustrated. Note the mouse pointer edge ⬛ icon.

81) Click the **Fillet** ⬤ feature tool. The Fillet PropertyManager is displayed.

82) Click the **Manual** tab. Edge<1> is displayed in the Items To Fillet box. Constant radius is the default Fillet Type.

83) Click the remaining **3 vertical edges**. The selected entities are displayed in the Items To Fillet box

84) Enter **.500**in, [**12.7**] for Radius. Accept the default settings.

85) Click **OK** ✔ from the Fillet PropertyManager. Fillet1 is displayed in the FeatureManager.

86) Click **Isometric view** ⬛ from the Heads-up View toolbar.

87) Click **Shaded With Edges** ⬛ from the Heads-up View toolbar.

Rename the feature.

88) Rename **Fillet1** to **Side Fillets** in the FeatureManager.

Save the BATTERY.

89) Click **Save** 💾.

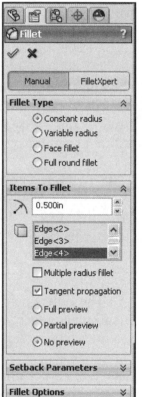

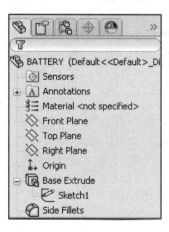

BATTERY Part-Extruded Cut Feature

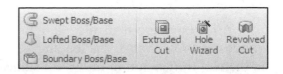

An Extruded Cut feature removes material. An Extruded Cut feature requires:

- Sketch plane (Top face)
- Sketch profile (Offset Entities)
- End Condition depth (Blind) in Direction 1

The Offset Entity Sketch tool uses existing geometry, extracts an edge or face and locates the geometry on the current Sketch plane.

Offset the existing Top face for the 2D sketch. Utilize the default Blind End Condition in Direction 1.

Activity: BATTERY Part-Extruded Cut Feature

Select the Sketch plane.
90) Right-click the **Top face** of the BATTERY in the Graphics window. Base Extruded is highlighted in the FeatureManager.

Create a sketch.
91) Click **Sketch** ✏ from the Context toolbar. The Sketch toolbar is displayed.

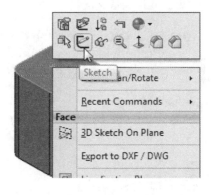

Display the face.
92) Click **Top view** ⬒ from the Heads-up View toolbar.

Offset the existing geometry from the boundary of the Sketch plane.
93) Click the **Offset Entities** 🗗 Sketch tool. The Offset Entities PropertyManager is displayed.

94) Enter **.150**in, [3.81] for the Offset Distance.

95) Check the **Reverse** box. The new Offset yellow profile displays inside the original profile.

96) Click **OK** ✔ from the Offset Entities PropertyManager.

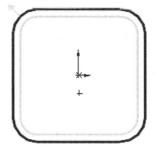

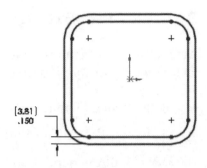

A leading zero is displayed in the spin box. For inch dimensions less than 1, the leading zero is not displayed in the part dimension in the ANSI standard.

Display the profile.

97) Click **Isometric view** from the Heads-up View toolbar.

98) Click **Hidden Lines Removed** from the Heads-up View toolbar.

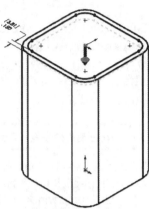

Insert an Extruded Cut feature. As an exercise, use the Instant3D tool to create the Extruded Cut feature. In this section, the Extruded-Cut PropertyManager is used. Note: With the Instant3D tool, you may lose the design intent of the model.

99) Click the **Extruded Cut** feature tool. The Cut-Extrude PropertyManager is displayed.

100) Enter **.200**in, **[5.08]** for Depth in Direction 1. Accept the default settings.

101) Click **OK** from the Cut-Extrude PropertyManager. Cut-Extrude1 is displayed in the FeatureManager.

Rename the feature.
102) Rename **Cut-Extrude1** to **Top Cut** in the FeatureManager.

Save the BATTERY
103) Click **Save**.

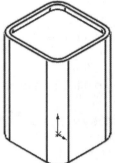

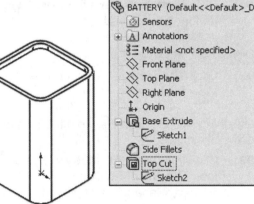

The Cut-Extrude PropertyManager contains numerous options. The Reverse Direction option determines the direction of the Extrude. The Extruded Cut feature is valid only when the direction arrow points into material to be removed.

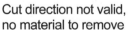

Cut direction not valid, no material to remove

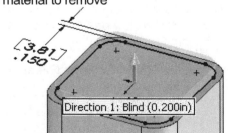

The Flip side to cut option determines if the cut is to the inside or outside of the Sketch profile. The Flip side to cut arrow points outward. The Extruded Cut feature occurs on the outside of the BATTERY.

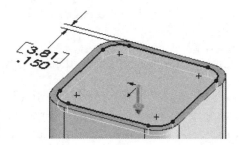

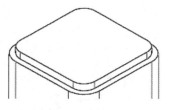

Extruded Cut with Flip side to cut option checked

BATTERY Part-Fillet Feature

The Fillet 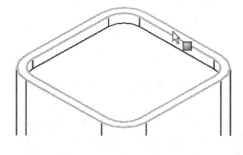 feature tool rounds sharp edges with a constant radius by selecting a face. A Fillet requires a:

- A selected face

- Fillet radius

Activity: BATTERY Part-Fillet Feature Face

Insert a Fillet feature on the top face.
104) Click the **top thin face** as illustrated. Note: The face

       icon feedback symbol.

105) Click the **Fillet** feature tool. The Fillet PropertyManager is displayed. Face<1> is displayed in the Items To Fillet box.

106) Click the **Manual** tab. Create a Constant radius for Fillet Type.

107) Enter .050in, [**1.27**] for Radius.

108) Click **OK** from the Fillet PropertyManager. Fillet2 is displayed in the FeatureManager.

Rename the feature.
109) Rename **Fillet2** to **Top Face Fillet**.

110) Press the **f** key.

Save the BATTERY.
111) Click **Save**.

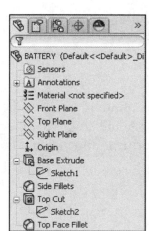

View the mouse pointer for feedback to select Edges or Faces for the fillet.

 Do not select a fillet radius which is larger then the surrounding geometry.

Example: The top edge face width is .150in, [3.81]. The fillet is created on both sides of the face. A common error is to enter a Fillet too large for the existing geometry. A minimum face width of .200in, [5.08] is required for a fillet radius of .100in, [2.54].

The following error occurs when the fillet radius is too large for the existing geometry:

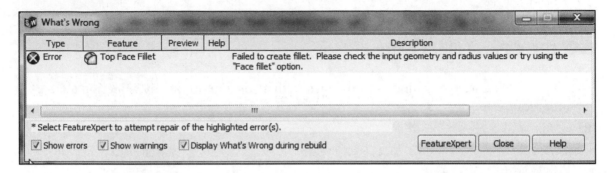

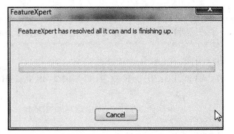

Avoid the fillet rebuild error. Use the FeatureXpert to address a constant radius fillet build error or manually enter a smaller fillet radius size.

BATTERY Part-Extruded Boss/Base Feature

The Extruded Boss feature requires a truncated cone shape to represent the geometry of the BATTERY terminals. The Draft Angle option creates the tapered shape.

Sketch the first circle on the Top face. Utilize the Ctrl key to copy the first circle.

The dimension between the center points is critical. Dimension the distance between the two center points with an aligned dimension. The dimension text toggles between linear and aligned. An aligned dimension is created when the dimension is positioned between the two circles.

An angular dimension is required between the Right Plane and the centerline. Acute angles are less than 90°. Acute angles are the preferred dimension standard. The overall BATTERY height is a critical dimension. The BATTERY height is 4.500in, [114.3].

Calculate the depth of the extrusion: For inches: 4.500in - (4.100in Base-Extrude height - .200in Offset cut depth) = .600in. The depth of the extrusion is .600in.

For millimeters: 114.3mm - (104.14mm Base-Extrude height - 5.08mm Offset cut depth) = 15.24mm. The depth of the extrusion is 15.24mm.

Activity: BATTERY Part-Extruded Boss Feature

Select the Sketch plane.

112) Right-click the **Top face** of the Top Cut feature in the Graphics window. This is your Sketch plane.

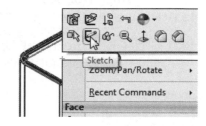

Create the sketch.

113) Click **Sketch** ✏ from the Context toolbar. The Sketch toolbar is displayed.

114) Click **Top view** ⬚ from the Heads-up View toolbar.

Sketch the Close profile.

115) Click the **Circle** ⊙ Sketch tool. The Circle PropertyManager is displayed.

116) Click the **center point** of the circle coincident to the

Origin ↳.

117) Drag and click the **mouse pointer** to the right of the Origin as illustrated.

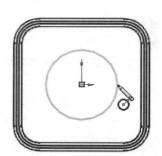

Add a dimension.

118) Click the **Smart Dimension** ✐ Sketch tool.

119) Click the **circumference** of the circle.

120) Click a **position** diagonally to the right.

121) Enter **.500**in, **[12.7]**.

122) Click the **Green Check mark** ✓ in the Modify dialog box. The black sketch is fully defined.

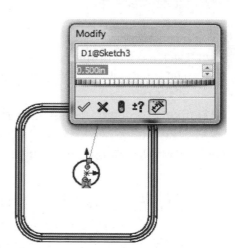

Copy the sketched circle.

123) Right-click **Select** to de-select the Smart Dimension Sketch tool.

124) Hold the **Ctrl** key down.

125) Click and drag the **circumference** of the circle to the upper left quadrant as illustrated.

126) Release the **mouse button**.

127) Release the **Ctrl** key. The second circle is selected and is displayed in blue.

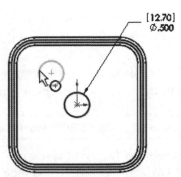

Add an Equal relation.
128) Hold the **Ctrl** key down.

129) Click the **circumference of the first circle**. The Properties PropertyManager is displayed. Both circles are selected and are displayed in green.

130) Release the **Ctrl** key.

131) Right-click **Make Equal** = from the Context toolbar.

132) Click **OK** ✓ from the Properties PropertyManager. The second circle remains selected.

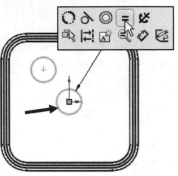

Show the Right Plane for the dimension reference.
133) Click **Right Plane** from the FeatureManager.

134) Click **Show**. The Right Plane is displayed in the Graphics window.

Add an aligned dimension.
135) Click the **Smart Dimension** ✏ Sketch tool.

136) Click the **two center points** of the two circles.

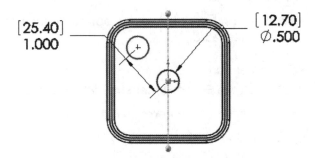

137) Click a **position** off the profile in the upper left corner.

138) Enter **1.000**in, [**25.4**] for the aligned dimension.

139) Click the **Green Check mark** ✓ in the Modify dialog box.

Insert a centerline.
140) Click the **Centerline** ┊ Sketch tool. The Insert Line PropertyManager is displayed.

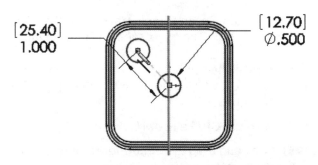

141) Sketch a centerline between the **two circle center points** as illustrated.

142) Right-click **Select** to end the line.

Double-click to end the centerline.

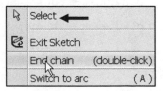

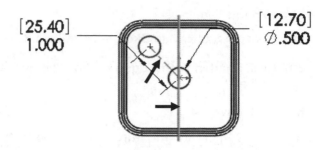

💡 Press the Enter key to accept the value in the Modify dialog box. The Enter key replaces the Green Check mark.

Add an angular dimension.

143) Click the **Smart Dimension** ✏ Sketch tool. Click the **centerline** between the two circles.

144) Click the **Right Plane** (vertical line) in the Graphics window. Note: You can also click Right Plane in the FeatureManager.

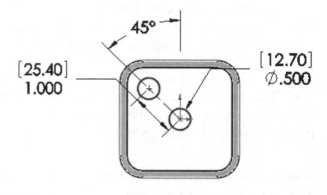

145) Click a **position** between the centerline and the Right Plane, off the profile.

146) Enter **45**.

147) Click **OK** ✔ from the Dimension PropertyManager.

Fit the model to the Graphics window.
148) Press the **f** key.

Hide the Right Plane.
149) Right-click **Right Plane** in the FeatureManager.

150) Click **Hide** from the Context toolbar. Click **Save** 🖫.

💡 Create an angular dimension between three points or two lines. Sketch a centerline/construction line when an additional point or line is required.

Insert an Extruded Boss feature.

151) Click **Isometric view** ⬛ from the Heads-up View toolbar.

152) Click the **Extruded Boss/Base** 🗒 feature tool. The Boss-Extrude PropertyManager is displayed. Blind is the default End Condition Type.

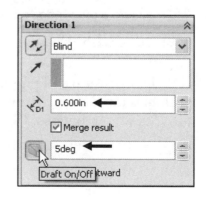

153) Enter **.600**in, **[15.24]** for Depth in Direction 1.

154) Click the **Draft ON/OFF** button.

155) Enter **5**deg in the Draft Angle box.

156) Click **OK** ✔ from the Boss-Extrude PropertyManager. The Boss-Extrude2 feature is displayed in the FeatureManager.

Rename the feature and sketch.

157) Rename **Boss-Extrude2** to **Terminals**.

158) **Expand** Terminals.

159) Rename **Sketch3** to **Sketch-TERMINALS**.

160) Click **Shaded With Edges** 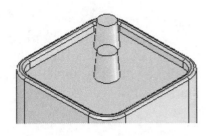 from the Heads-up View toolbar.

161) Click **Save** .

Each time you create a feature of the same feature type, the feature name is incremented by one. Example: Boss-Extrude1 is the first Extrude feature. Boss-Extrude2 is the second Extrude feature. If you delete a feature, rename a feature or exit a SolidWorks session, the feature numbers will vary from those illustrated in the text.

Rename your features with descriptive names. Standardize on feature names that are utilized in mating parts. Example: Mounting Holes.

Measure the overall BATTERY height.

162) Click **Front view** from the Heads-up View toolbar.

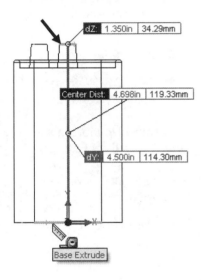

163) Click the **Measure** Measure tool from the Evaluate tab in the CommandManager. The Measure - BATTERY dialog box is displayed.

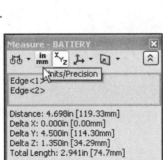

164) Click the **top edge** of the battery terminal as illustrated.

165) Click the **bottom edge** of the battery. The overall height, Delta Y is 4.500, [114.3]. Apply the Measure tool to ensure a proper design.

166) **Close** the Measure - BATTERY dialog box.

The Measure tool provides the ability to display custom settings. Click **Units/Precision** from the Measure dialog box. View your options. Click **OK**.

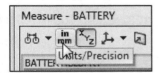

The Selection Filter 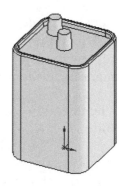 option toggles the Selection Filter toolbar. When Selection Filters are activated, the mouse pointer displays the Filter icon ⓚᵥ . The Clear All Filters ⓦᵥ tool removes the current Selection Filters. The Help Ⓟ icon displays the SolidWorks Online Users Guide.

Display the Trimetric view.
167) Click **Trimetric view** 🔲 from the Heads-up View toolbar.

Save the BATTERY.

168) Click **Save** 💾 .

⚲ Additional information on Extruded Boss/Base Extruded Cut and Fillets is located in SolidWorks Help Topics. Keywords: Extruded (Boss/Base, Cut), Fillet (Constant radius fillet), Geometric relations (sketch, equal, midpoint), Sketch (rectangle, circle), Offset Entities and Dimensions (angular).

Refer to the Help, SolidWorks Tutorials, Fillet exercise for additional information.

⚙⚙ Review of the BATTERY Part

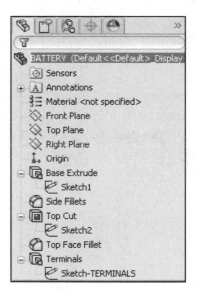

The BATTERY utilized a 2D Sketch profile located on the Top Plane. The 2D Sketch profile utilized the Center Rectangle Sketch tool. The Center Rectangle Sketch tool applied equal geometric relations to the two horizontal and two vertical lines. A midpoint relation was added to the Origin.

The Extruded Boss/Base feature was created using the Instant3D tool. Blind was the default End Condition. The Fillet feature rounded sharp edges. All four edges were selected to combine common geometry into the same Fillet feature. The Fillet feature also rounded the top face. The Sketch Offset Entity created the profile for the Extruded Cut feature.

The Terminals were created with an Extruded Boss feature. You sketched a circular profile and utilized the Ctrl key to copy the sketched geometry.

A centerline was required to locate the two holes with an angular dimension. The Draft Angle option tapered the Extruded Boss feature. All feature names were renamed.

Injection Molded Process

Lee Plastics of Sterling, MA is a precision injection molding company. Through the World Wide Web (www.leeplastics.com), review the injection molded manufacturing process.

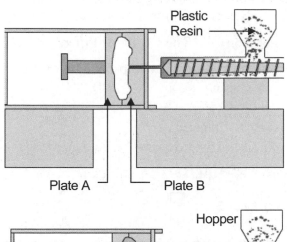

The injection molding process is as follows:

An operator pours the plastic resin in the form of small dry pellets, into a hopper. The hopper feeds a large augur screw. The screw pushes the pellets forward into a heated chamber. The resin melts and accumulates into the front of the screw.

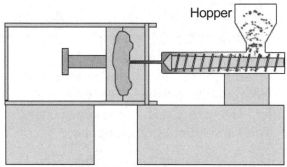

At high pressure, the screw pushes the molten plastic through a nozzle, to the gate and into a closed mold, (Plates A & B). Plates A and B are the machined plates that you will design in this project.

The plastic fills the part cavities through a narrow channel called a gate.

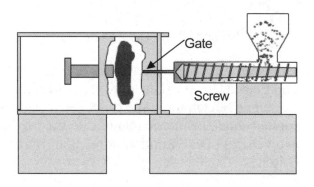

The plastic cools and forms a solid in the mold cavity. The mold opens, (along the parting line) and an ejection pin pushes the plastic part out of the mold into a slide.

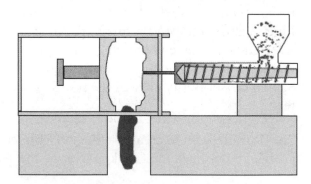

Injection Molded Process
(Courtesy of Lee Plastics, Inc.)

BATTERYPLATE Part

The BATTERYPLATE is a critical plastic part. The BATTERYPLATE:

- Aligns the LENS assembly

- Creates an electrical connection between the BATTERY and LENS

Design the BATTERYPLATE. Utilize features from the BATTERY to develop the BATTERYPLATE. The BATTERYPLATE is manufactured as an injection molded plastic part. Build Draft into the Extruded Boss/Base features.

Edit the BATTERY features. Create two holes from the original sketched circles. Apply the Instant3D tool to create an Extruded Cut feature.

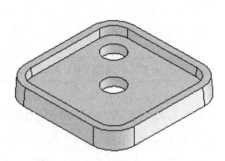

Modify the dimensions of the Base feature. Add a 3° draft angle.

☀ A sand pail contains a draft angle. The draft angle assists the sand to leave the pail when the pail is flipped upside down.

Insert an Extruded Boss/Base feature. Offset the center circular sketch.

The Extruded Boss/Base feature contains the LENS. Create an inside draft angle. The draft angle assists the LENS into the Holder.

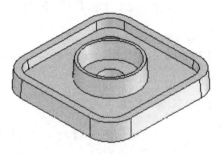

Insert a Face Fillet and a Multi-radius Edge Fillet to remove sharp edges. Plastic parts require smooth edges. Group Fillet features together into a folder.

Perform a Draft Analysis on the part and create the Core and Cavity mold tooling.

☀ Group fillets together into a folder to locate them quickly. Features listed in the FeatureManager must be continuous in order to be placed as a group into a folder.

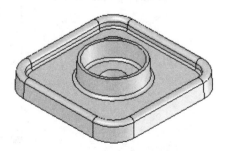

Save As, Delete, Edit Feature and Modify

Create the BATTERYPLATE part from the BATTERY part. Utilize the Save As tool from the Menu bar to copy the BATTERY part to the BATTERYPLATE part.

Reuse existing geometry. Create two holes. Delete the Terminals feature and reuse the circle sketch. Select the sketch in the FeatureManager. Create an Extruded Cut feature from the Sketch–TERMINALS using the Instant3D tool. Blind is the default End Condition. Edit the Bass-Extrude feature. Modify the overall depth. Rebuild the model.

 Sketch dimensions are displayed in black.

 Feature dimensions are displayed in blue.

Activity: BATTERYPLATE Part-Save As, Delete, Modify and Edit Feature

Create a new part.

169) Click **Save As** from the drop-down Menu bar.

170) Select **PROJECTS** for Save In folder.

171) Enter **BATTERYPLATE** for File name.

172) Enter **BATTERY PLATE, FOR 6-VOLT** for Description.

173) Click **Save**. The BATTERYPLATE FeatureManager is displayed. The BATTERY part is closed.

Delete the Terminals feature.

174) Right-click **Terminals** from the FeatureManager.

175) Click **Delete**.

176) Click **Yes** from the Confirm Delete dialog box. Do not delete the two-circle sketch, Sketch-TERMINALS.

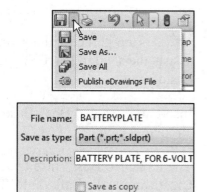

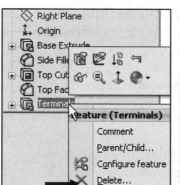

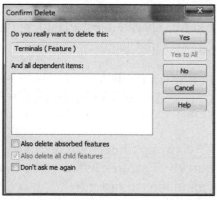

Create an Extruded Cut feature from the Sketch–
TERMINALS using Instant3D.

177) Click **Sketch-TERMINALS** from the
FeatureManager.

178) Click the **circumference** of the center
circle as illustrated. A green arrow is
display.

179) Hold the **Alt** key down. Drag the **green
arrow** downward below the model to
create a hole in Direction 1.

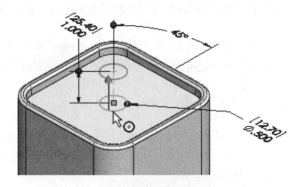

180) Release the mouse button on the **vertex** as
illustrated. This ensures a Through All End
Condition with model dimension changes.

181) Release the **Alt** key. Boss-Extrude1 is displayed in the
FeatureManager.

182) Rename the **Boss-Extrude1** feature to **Holes** in the
FeatureManager.

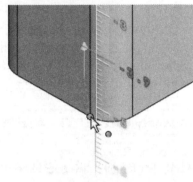

Edit the Base Extrude feature.

183) Right-click **Base Extrude** from the FeatureManager.

184) Click **Edit Feature** 🔧 from the Context toolbar. The
Base Extrude PropertyManager is displayed.

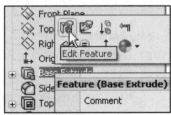

Modify the overall depth.

185) Enter **.400**in, **[10.16]** for Depth in Direction 1.

186) Click the **Draft ON/OFF** button.

187) Enter **3.00**deg in the Angle box.

188) Click **OK** ✔ from the Base Extrude PropertyManager.

Fit the model to the Graphics window.

189) Press the **f** key.

Save the BATTERYPLATE.

190) Click **Save** 💾.

💡 Modify the **Spin Box Increments** in System Options
to display different increments for the Instant3D on-screen ruler.

___

💡 To delete both the feature and the sketch at the same time,
select the Also delete absorbed features check box from the
Confirm Delete dialog box.

BATTERYPLATE Part-Extruded Boss Feature

The Holder is created with a circular Extruded Boss/Base feature. Utilize the Offset Entities ⮡ Sketch tool to create the second circle. Apply a draft angle of 3° in the Extruded Boss feature.

When applying the draft angle to the two concentric circles, the outside face tapers inwards and the inside face tapers outwards.

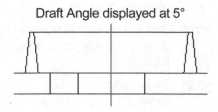

Draft Angle displayed at 5°

💡 Plastic parts require a draft angle. Rule of thumb; 1° to 5° is the draft angle. The draft angle is created in the direction of pull from the mold. This is defined by geometry, material selection, mold production and cosmetics. Always verify the draft with the mold designer and manufacturer.

Activity BATTERYPLATE Part-Extruded Boss Feature

Select the Sketch plane.

191) Right-click the **top face** of Top Cut. This is your Sketch plane.

Create the sketch.

192) Click **Sketch** ⮥ from the Context toolbar.

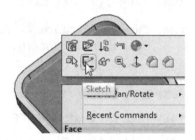

193) Click the **top circular edge** of the center hole. Note: Use the keyboard arrow keys or the middle mouse button to rotate the sketch if needed.

194) Click the **Offset Entities** ⮡ Sketch tool. The Offset Entities PropertyManager is displayed.

195) Enter **.300**in, [**7.62**] for Offset Distance. Accept the default settings.

196) Click **OK** ✔ from the Offset Entities PropertyManager.

197) Drag the **dimension** off the model.

Create the second offset circle.

198) Click the **offset circle** in the Graphics window.

199) Click the **Offset Entities** ⮡ Sketch tool. The Offset Entities PropertyManager is displayed.

200) Enter **.100**in, [**2.54**] for Offset Distance.

[7.62]
.300

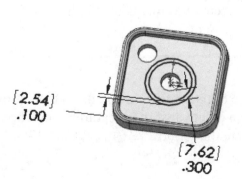

[2.54]
.100

[7.62]
.300

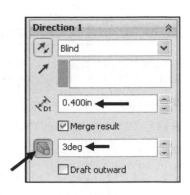

201) Click **OK** ✓ from the Offset Entities PropertyManager. Drag the dimension off the model. Two offset concentric circles define the sketch.

Insert an Extruded Boss/Base feature.

202) Click the **Extruded Boss/Base** 🗔 feature tool. The Boss-Extrude PropertyManager is displayed.

203) Enter **.400**in, **[10.16]** for Depth in Direction 1.

204) Click the **Draft ON/OFF** button.

205) Enter **3**deg in the Angle box.

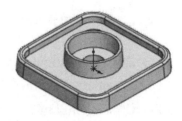

206) Click **OK** ✓ from the Boss-Extrude PropertyManager. The Boss-Extrude2 feature is displayed in the FeatureManager.

Rename the feature.
207) Rename the **Boss-Extrude2** feature to **Holder** in the FeatureManager.

Save the model.
208) Click **Save** 💾.

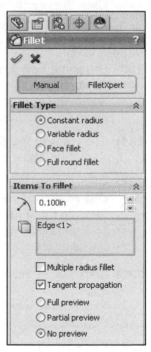

BATTERYPLATE Part-Fillet Features: Full Round and Multiple Radius Options

Use the Fillet feature 🖲 tool to smooth rough edges in a model. Plastic parts require fillet features on sharp edges. Create two Fillets. Utilize different techniques. The current Top Face Fillet produced a flat face. Delete the Top Face Fillet. The first Fillet feature is a Full round fillet. Insert a Full round fillet feature on the top face for a smooth rounded transition.

The second Fillet feature is a Multiple radius fillet. Select a different radius value for each edge in the set. Select the inside and outside edge of the Holder. Select all inside tangent edges of the Top Cut. A Multiple radius fillet is utilized next as an exercise. There are machining instances were radius must be reduced or enlarged to accommodate tooling. Note: There are other ways to create Fillets.

🔆 Group Fillet features into a Fillet folder. Placing Fillet features into a folder reduces the time spent for your mold designer or toolmaker to look for each Fillet feature in the FeatureManager.

Activity: BATTERYPLATE Part-Fillet Features: Full Round, Multiple Radius Options

---

Delete the Top Edge Fillet.

209) Right-click **Top Face Fillet** from the FeatureManager.

210) Click **Delete**.

211) Click **Yes** to confirm delete.

212) Drag the **Rollback** bar below Top Cut in the FeatureManager.

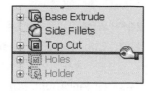

Create a Full round fillet feature.

213) Click **Hidden Lines Visible** ⬚ from the Heads-up View toolbar.

214) Click the **Fillet** 🗋 feature tool. The Fillet PropertyManager is displayed.

215) Click the **Manual** tab.

216) Click the **Full round fillet** box for Fillet Type.

217) Click the **inside Top Cut face** for Side Face Set 1 as illustrated.

218) Click **inside** the Center Face Set box.

219) Click the **top face** for Center Face Set as illustrated.

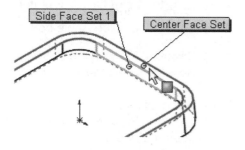

Rotate the part.

220) Press the **Left Arrow** key until you can select the outside Base Extrude face.

221) Click **inside** the Side Face Set 2 box.

222) Click the **outside Base Extrude face** for Side Face Set 2 as illustrated. Accept the default settings.

223) Click **OK** ✓ from the Fillet PropertyManager. Fillet1 is displayed in the FeatureManager.

224) Rename **Fillet1** to **TopFillet**.

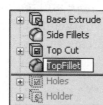

Save the BATTERYPLATE.

225) Click **Isometric view** 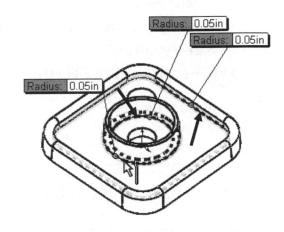 from the Heads-up View toolbar.

226) Click **Hidden Lines Removed** from the Heads-up View toolbar.

227) Drag the **Rollback bar** to the bottom of the FeatureManager.

228) Click **Save** .

Create a Multiple radius fillet feature.

229) Click the **bottom outside circular edge** of the Holder as illustrated.

230) Click the **Fillet** feature tool. The Fillet PropertyManager is displayed.

231) Click the **Constant radius** box.

232) Enter .050in, [**1.27**] for Radius.

233) Click the **bottom inside circular edge** of the Top Cut as illustrated.

234) Click the **inside edge** of the Top Cut.

235) Check the **Tangent propagation** box.

236) Check the **Multiple radius fillet** box.

Modify the Fillet values.

237) Click the **Radius** box for the Holder outside edge.

238) Enter **0.060**in, [**1.52**].

239) Click the **Radius** box for the Top Cut inside edge.

240) Enter **0.040**in, [**1.02**].

241) Click **OK** from the Fillet PropertyManager. Fillet2 is displayed in the FeatureManager.

242) Rename **Fillet2** to **HolderFillet**.

243) Click **Shaded With Edges** from the Heads-up View toolbar. View the results in the Graphics window.

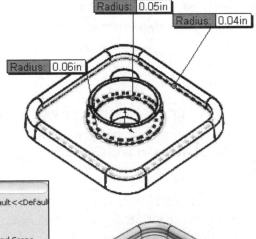

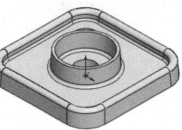

Group the Fillet features into a new folder.

244) Click **TopFillet** from the FeatureManager.

245) Drag the **TopFillet** feature directly above the HolderFillet feature in the FeatureManager.

246) Click **HolderFillet** in the FeatureManager.

247) Hold the **Ctrl** key down.

248) Click **TopFillet** in the FeatureManager.

249) Right-click **Add to New Folder**.

250) Release the **Ctrl** key.

251) Rename **Folder1** to **FilletFolder**.

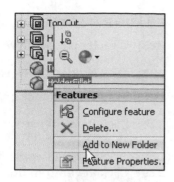

Save the BATTERYPLATE.

252) Click **Save** 💾.

Multi-body Parts and the Extruded Boss/Base Feature

A Multi-body part has separate solid bodies within the same part document.

A WRENCH consists of two cylindrical bodies. Each extrusion is a separate body. The oval profile is sketched on the right plane and extruded with the Up to Body End Condition option.

The BATTERY and BATTERYPLATE parts consisted of a solid body with one sketched profile. Each part is a single body part.

🔍 Additional information on Save, Extruded Boss/Base, Extruded Cut, Fillets, Copy Sketched Geometry, and Multi-body are located in SolidWorks Help Topics. Keywords: Save (Save as copy), Extruded (Boss/Base, Cut), Fillet (face blends, variable radius), Chamfer, Geometric relations (sketch), Copy (sketch entities), Multi-body (Extruded, Modeling techniques).

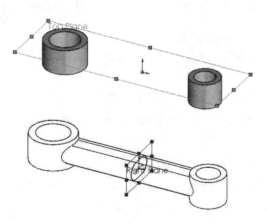

Multi-body part
Wrench

Refer to **Help, SolidWorks Tutorials, Special Types of Models, Multibody Parts** for additional information.

 Review of the BATTERYPLATE Part

The Save As option was utilized to copy the BATTERY part to the BATTERYPLATE part. You created a hole in the BATTERYPLATE using Instant3D and modified features using the PropertyManager.

The BATTERYPLATE is a plastic part. The Draft Angle option was added in the Extruded Base (Boss-Extrude1) feature.

The Holder Extruded Boss utilized a circular sketch and the Draft Angle option. The Sketch Offset tool created the circular ring profile. Multi radius Edge Fillets and Face Fillets removed sharp edges.

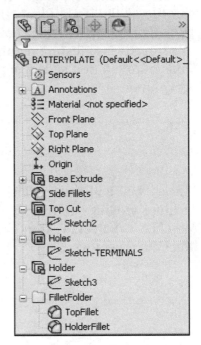

Similar Fillet features were grouped together into a folder. Features were renamed in the FeatureManager.

The BATTERY and BATTERYPLATE utilized an Extruded Boss/Base feature.

It is considered best practice to fully define all sketches in the model. However; there are times when this is not practical, generally when using the spline tool to create a complex freeform shape.

LENS Part

Create the LENS. The LENS is a purchased part. The LENS utilizes a Revolved Base feature.

Sketch a centerline and a closed profile on the Right Plane. Insert a Revolved Base feature. The Revolved Base feature requires an axis of revolution and an angle of revolution.

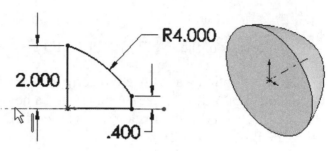

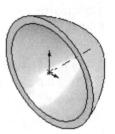

Insert the Shell feature. The Shell feature provides uniform wall thickness. Select the front face as the face to be removed.

Utilize the Convert Entities sketch tool to extract the back circular edge for the sketched profile. Insert an Extruded Boss feature from the back of the LENS.

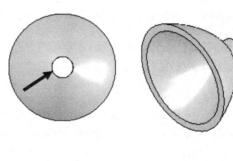

Sketch a single profile. Insert a Revolved Thin feature to connect the LENS to the BATTERYPLATE. The Revolved Thin feature requires thickness.

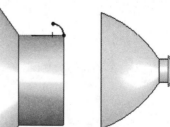

Insert a Counterbore Hole feature using the Hole Wizard tool. The BULB is located inside the Counterbore Hole.

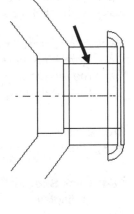

Insert the front Lens Cover with an Extruded Boss/Base feature. The Extruded Boss/Base feature is sketched on the Front Plane. Add a transparent Lens Shield with the Extruded Boss feature.

LENS Part-Revolved Base Feature

Create the LENS with a Revolved Base feature. The solid Revolved Base feature requires:

- Sketch plane (Right)

- Sketch profile

- Centerline

- Angle of Revolution (360°)

The profile lines reference the Top and Front Planes. Create the curve of the LENS with a 3-point arc.

Activity: LENS Part-Create a Revolved Base Feature

Create the New part.

253) Click **New** from the Menu bar.

254) Click the **MY-TEMPLATES** tab.

255) Double-click **PART-IN-ANSI**, **[PART-MM-ISO]** from the Template dialog box.

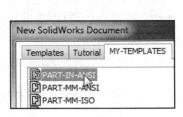

Save the part.

256) Click **Save** .

257) Select **PROJECTS** for Save in folder.

258) Enter **LENS** for File name.

259) Enter **LENS WITH SHIELD** for Description.

260) Click **Save**. The LENS FeatureManager is displayed.

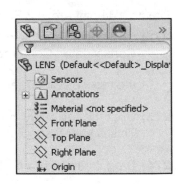

View the Front Plane.
261) Right click **Front Plane** from the FeatureManager.

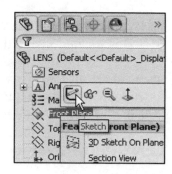

262) Click **Show** from the Context toolbar. Note: If you upgraded from 2009 versus a new installation, your planes and Origins maybe displayed by default. Hide unwanted planes in the FeatureManager if needed.

Create a sketch.
263) Right-click **Right Plane** from the FeatureManager.

264) Click **Sketch** ✎ from the Context toolbar. The Sketch toolbar is displayed.

265) Click the **Centerline** ⌇ Sketch tool. The Insert Line PropertyManager is displayed.

266) Sketch a **horizontal centerline** collinear to the Top Plane, through the Origin ↳ as illustrated.

Sketch the profile. Create three lines.
267) Click the **Line** ＼ Sketch tool. The Insert Line PropertyManager is displayed.

268) Sketch a **vertical line** collinear to the Front Plane coincident with the Origin.

269) Sketch a **horizontal line** coincident with the Top Plane.

270) Sketch a **vertical line** approximately 1/3 the length of the first line.

271) Right-click **End Chain**.

Create a 3 Point Arc. A 3 Point Arc requires three points.
272) Click the **3 Point Arc** ⌢ Sketch tool from the Consolidated Centerpoint Arc toolbar. Note: the mouse pointer feedback ⌢ icon.

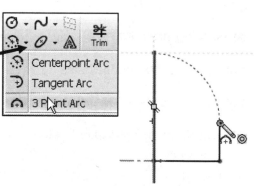

273) Click the **top point** on the left vertical line. This is your first point.

274) Drag the **mouse pointer** to the right.

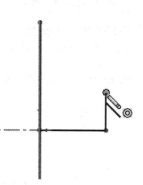

275) Click the **top point** on the right vertical line. This is your second point.

276) Drag the **mouse pointer** upward.

277) Click a **position** on the arc.

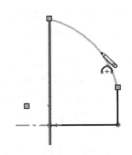

Add an Equal relation.
278) Right-click **Select** to deselect the sketch tool.

279) Click the **left vertical** line.

280) Hold the **Ctrl** key down.

281) Click the **horizontal** line. The Properties PropertyManager is displayed. The selected entities are displayed in the Selected Entities box.

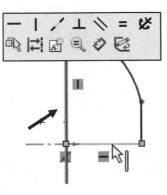

282) Release the **Ctrl** key.

283) Right-click **Make Equal** from the Context toolbar.

284) Click **OK** ✅ from the Properties PropertyManager.

Add dimensions.
285) Click the **Smart Dimension** 🖉 Sketch tool.

286) Click the **left vertical** line.

287) Click a **position** to the left of the profile.

288) Enter **2.000**in, [**50.8**].

289) Click the **right vertical** line.

290) Click a **position** to the right of the profile.

291) Enter **.400**in, [**10.16**]. Click the **arc**.

292) Click a **position** to the right of the profile.

293) Enter **4.000**in, [**101.6**]. The black sketch is fully defined.

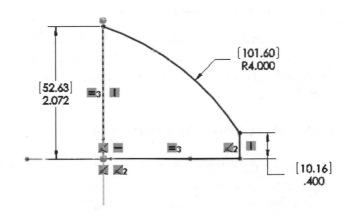

🔆 Utilize **Tools, Sketch Tools, Check Sketch for Feature** option to determine if a sketch is valid for a specific feature and to understand what is wrong with a sketch.

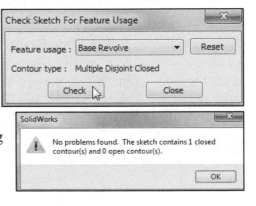

Insert the Revolved Base feature.

294) Click the **Revolved Boss/Base** ⊕ feature tool. The Revolve PropertyManager is displayed.

295) If needed, click the **horizontal centerline** for the axis of revolution. Note: The direction arrow points clockwise. Click **OK** ✓ from the Revolve PropertyManager. Revolve1 is displayed in the FeatureManager.

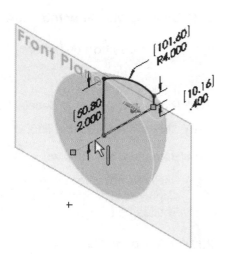

Rename the feature.
296) Rename **Revolve1** to **BaseRevolve**.

Save the model.
297) Click **Save** 💾.

Display the axis of revolution.
298) Click **View**; check **Temporary Axes** from the Menu bar.

Revolve features contain an axis of revolution. The axis of revolution utilizes a sketched centerline, edge or an existing feature/sketch or a Temporary Axis. The solid Revolved feature contains a closed profile. The Revolved thin feature contains an open or closed profile.

LENS Part-Shell Feature

The Revolved Base feature is a solid. Utilize the Shell feature to create a constant wall thickness around the front face. The Shell feature removes face material from a solid. The Shell feature requires a face and thickness. Use the Shell feature to create thin-walled parts.

Activity: LENS Part-Shell Feature

Insert the Shell feature.
299) Click the **front face** of the BaseRevolve feature.

300) Click the **Shell** 🔲 feature tool. The Shell1 PropertyManager is displayed. Face<1> is displayed in the Faces to Remove box.

301) Enter **.250**in, [**6.35**] for Thickness.

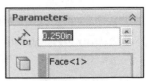

Display the Shell feature.

302) Click **OK** ✓ from the Shell1 PropertyManager. Shell1 is displayed in the FeatureManager.

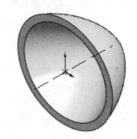

303) Right-click **Front Plane** from the FeatureManager.

304) Click **Hide** from the Context toolbar.

305) Rename **Shell1** to **LensShell**.

306) Click **Save** 🖫.

💡 To insert rounded corners inside a shelled part, apply the Fillet feature before the Shell feature. Select the Multi-thickness option to apply different thicknesses.

Extruded Boss/Base Feature and Convert Entities Sketch tool

Create the LensNeck. The LensNeck houses the BULB base and is connected to the BATTERYPLATE. Use the Extruded Boss/Base feature. The back face of the Revolved Base feature is the Sketch plane.

Utilize the Convert Entities Sketch tool to extract the back circular face to the Sketch plane. The new curve develops an On Edge relation. Modify the back face, and the extracted curve updates to reflect the change. No sketch dimensions are required.

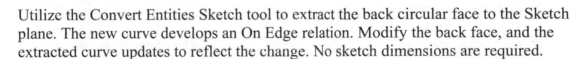

Activity: Extruded Boss Feature and Convert Entities Sketch tool

Rotate the LENS.

307) Rotate the LENS with the middle mouse button to display the back face as illustrated. Note: The Rotate ↻ icon is displayed.

Sketch the profile.

308) Right-click the **back face** for the Sketch plane. BaseRevolve is highlighted in the FeatureManager. This is your Sketch plane.

309) Click **Sketch** ✏ from the Context toolbar. The Sketch toolbar is displayed.

310) Click the **Convert Entities** ⬚ Sketch tool. The Convert Entities PropertyManager is displayed. Click **OK** ✔ from the Convert Entities PropertyManager.

Insert an Extruded Boss/Base feature.

311) Click the **Extruded Boss/Base** 🔂 feature tool. The Boss-Extrude PropertyManager is displayed.

312) Enter **.400**in, **[10.16]** for Depth in Direction 1. Accept the default settings.

313) Click **OK** ✔ from the Boss-Extrude PropertyManager. Boss-Extrude1 is displayed in the FeatureManager.

314) Click **Isometric view** 🔳 from the Heads-up View toolbar.

315) Rename **Boss-Extrude1** to **LensNeck**.

316) Click **Save** 💾.

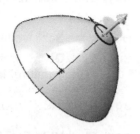

LENS Part-Hole Wizard

The LENS requires a Counterbore hole. Apply the Hole Wizard 🔲 feature. The Hole Wizard feature assists in creating complex and simple holes. The Hole Wizard Hole type categories are: *Counterbore, Countersink, Hole, Tapped, PipeTap* and *Legacy* (Holes created before SolidWorks 2000).

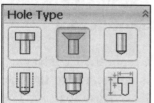

Specify the user parameters for the custom Counterbore hole. The parameters are: *Description, Standard, Screw Type, Hole Size, Fit, Counterbore diameter, Counterbore depth* and *End Condition*. Select the face or plane to locate the hole profile.

Insert a Coincident relation to position the hole center point. Dimensions for the Counterbore hole are provided in both inches and millimeters.

Activity: LENS Part-Hole Wizard Counterbore Hole Feature

Create the Counterbore hole.

317) Click **Front view** 🔲 from the Heads-up View toolbar.

318) Click the **Hole Wizard** 🔲 feature tool. The Hole Specification PropertyManager is displayed. Type is the default tab.

319) Click the **Counterbore** icon as illustrated.

Note: For a metric hole, skip the next few steps.

For inch Counterbore hole:
320) Select **Ansi Inch** for Standard. Select **Hex Bolt** for Type.

321) Select ½ for Size. Check the **Show custom sizing** box.

322) Click inside the **Counterbore Diameter** value box.

323) Enter **.600**in. Click inside the **Counterbore Depth** value box.

324) Enter **.200**in. Select **Through All** for End Condition.

325) Click the **Position** tab.

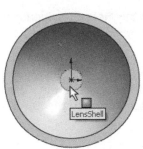

326) Click the small **inside back face** of the LensShell feature as illustrated. Do not select the Origin. LensShell is highlighted in the FeatureManager. The Point tool icon is displayed.

327) Click the **origin**. A Coincident relation is displayed.

Deselect the Point tool.
328) Right-click **Select** in the Graphics window.

Note: For an inch hole, skip the next few steps and start on the next page.

For millimeter Counterbore hole:
329) Select **Ansi Metric** for Standard.

330) Enter **Hex Bolt** for Type. Select **M5** for Size.

331) Click **Through All** for End Condition. Check the **Show custom sizing** box.

332) Click inside the **Counterbore Diameter** value box.

333) Enter **15.24**. Click inside the **Counterborebore Depth** value box.

334) Enter **5**. Click the **Position** tab. Click the small **inside back face** of the LensShell feature as illustrated. Do not select the Origin.

LensShell is highlighted in the FeatureManager. The Point tool icon is displayed.

335) Click the **origin**. A Coincident relation is displayed.

Deselect the Point tool.
336) Right-click **Select** in the Graphics window.

337) Click **OK** ✔ from the Hole Properties PropertyManager.

338) Click the **Type** tab.

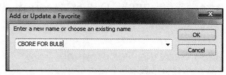

Add the new hole type to your Favorites list.
339) Expand the Favorites box.

340) Click the **Add or Update Favorite** ⭐➕ icon.

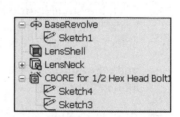

341) Enter **CBORE FOR BULB**.

342) Click **OK** from the Add or Update a Favorite dialog box.

343) Click **OK** ✔ from the Hole Specification PropertyManager.

Expand the Hole feature.
344) Expand the CBORE feature in the FeatureManager. Note: Sketch3 and Sketch4 created the CBORE feature.

Display the Section view.
345) Click **Right Plane** from the FeatureManager.

346) Click **Section view** 🔲 from the Heads-up View toolbar.

347) Click **OK** ✔ from the Section View PropertyManager.

348) Click **Isometric view** 🔲 from the Heads-up View toolbar.

Display the Full view.
349) Click **Section view** 🔲 from the Heads-up View toolbar.

350) Rename **CBORE for ½ Hex Head Bolt1** to **BulbHole**.

351) Click **Save** 💾.

LENS Part-Revolved Boss Thin Feature

Create a Revolved Boss Thin feature. Rotate an open sketched profile around an axis. The sketch profile must be open and cannot cross the axis. A Revolved Boss Thin feature requires:

- Sketch plane (Right Plane)

- Sketch profile (Center point arc)

- Axis of Revolution (Temporary axis)

- Angle of Rotation (360)

- Thickness .100in [2.54]

A Revolved feature produces silhouette edges in 2D views. A silhouette edge represents the extent of a cylindrical or curved face.

Select the Temporary Axis for Axis of Revolution. Select the Revolved Boss feature. Enter .100in [2.54] for Thickness in the Revolve PropertyManager. Enter 360° for Angle of Revolution.

Activity: LENS Part-Revolved Boss Thin Feature

Create a sketch.
352) Right-click **Right Plane** from the FeatureManager.

353) Click **Sketch** ✏ from the Context toolbar.

354) Click **Right view** ⬜ from the Heads-up View toolbar.

355) Zoom in on the LensNeck.

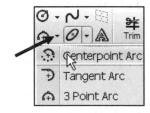

356) Click the **Centerpoint Arc** ⌒ Sketch tool.

357) Click the **top horizontal edge** of the LensNeck. Do not select the midpoint of the silhouette edge.

358) Click the **top right corner** of the LensNeck.

359) Drag the **mouse pointer** counterclockwise to the left.

360) Click a **position** directly above the first point as illustrated.

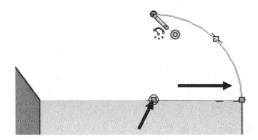

Add a dimension.
361) Click the **Smart Dimension** ✏ Sketch tool.

362) Click the **arc**. Click a **position** to the right of the profile.

363) Enter **.100**in, **[2.54]**. The Sketch is fully defined.

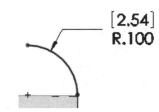

Insert a Revolved Thin feature.

364) Click the **Revolved Boss/Base** ⊕ feature tool. The Revolve PropertyManager is displayed.

365) Select **Mid-Plane** for Revolve Type in the Thin Feature box.

366) Enter **.050**in, [**1.27**] for Direction1 Thickness.

367) Click the **Temporary Axis** for Axis of Revolution.

368) Click **OK** ✓ from the Revolve PropertyManager.

369) Rename **Revolve-Thin1** to **LensConnector**.

Fit the model to the Graphics window.
370) Press the **f** key.

371) Click **Save** 🖫.

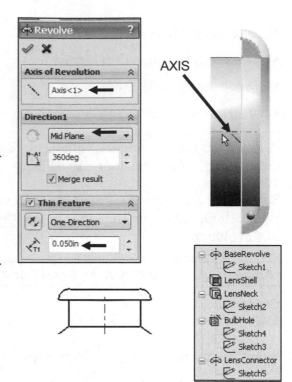

💡 A Revolved sketch that remains open results in a Revolve

Thin feature ———⌐ . A Revolved sketch that is automatically closed, results in a line drawn from the start point to the end point of the sketch. The sketch is closed and results in a non-

Revolve Thin feature ——⌐ .

LENS Part-Extruded Boss/Base Feature and Offset Entities

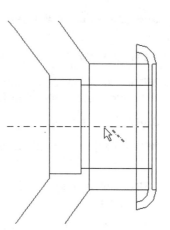

Use the Extruded Boss/Base feature tool to create the front LensCover. Utilize the Offset Entities Sketch tool to offset the outside circular edge of the Revolved feature. The Sketch plane for the Extruded Boss feature is the front circular face.

The Offset Entities Sketch tool requires an Offset Distance and direction. Utilize the Bi-direction option to create a circular sketch in both directions. The extrude direction is away from the Front Plane.

Activity: LENS Part-Extruded Boss Feature and Offset Entities

Create the sketch.

372) Click **Isometric view** 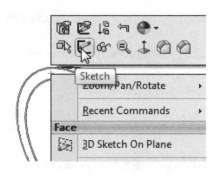 from the Heads-up View toolbar.

373) Click **Hidden Lines Removed** from the Heads-up View toolbar.

374) Right-click the **front circular face** for the Sketch plane.

375) Click **Sketch** from the Context toolbar.

376) Click **Front view** from the Heads-up View toolbar.

Offset the selected edge.

377) Click the **outside circular edge** of the LENS in the Graphics window.

378) Click the **Offset Entities** Sketch tool. The Offset Entities PropertyManager is displayed.

379) Check the **Bi-directional** box.

380) Enter **.250**in, **[6.35]** for Offset Distance. Accept the default settings.

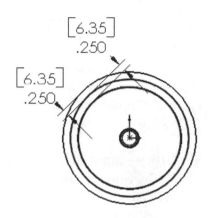

381) Click **OK** from the Offset Entities PropertyManager.

382) Click **Isometric view** from the Heads-up View toolbar.

383) Click **Shaded With Edges** from the Heads-up View toolbar.

Insert an Extruded Boss/Base feature.

384) Click the **Extruded Boss/Base** feature tool. The Boss-Extrude PropertyManager is displayed.

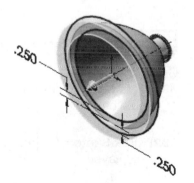

385) Enter **.250**in, **[6.35]** for Depth in Direction 1. Accept the default settings.

386) Click **OK** from the Boss-Extrude PropertyManager. Boss-Extrude2 is displayed in the PropertyManager.

Verify the position of the extruded feature.

387) Click the **Top view** 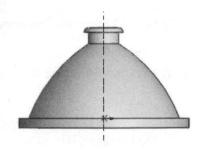. View the extruded feature.

388) Rename **Boss-Extrude2** to **LensCover**.

389) Click **Isometric view** .

390) Click **Save** .

LENS Part-Extruded Boss Feature and Transparency

Apply the Extruded Boss/Base feature to create the LensShield. Utilize the Convert Entities Sketch tool to extract the inside circular edge of the LensCover and place it on the Front plane.

Apply the Transparent Optical property to the LensShield to control the ability for light to pass through the surface. Transparency is an Optical Property found in the Color PropertyManager. Control the following properties:

- **Diffuse amount, Specular amount, Specular spread, Reflection amount, Transparent amount and Luminous intensity**.

Activity: LENS Part-Extruded Boss Feature and Transparency

Create the sketch.

391) Right-click **Front Plane** from the FeatureManager. This is your Sketch plane.

392) Click **Sketch** from the Context toolbar. The Sketch toolbar is displayed.

393) Click **Isometric view** from the Heads-up View toolbar.

394) Click the **front inner circular edge** of the LensCover (Boss-Extrude2) as illustrated.

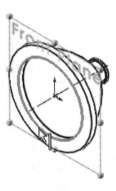

395) Click the **Convert Entities**  Sketch tool. The circle is projected onto the Front Plane.

396) Click **OK** ✔ from the Convert Entities PropertyManager.

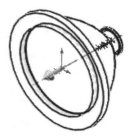

Insert an Extruded Boss feature.

397) Click the **Extruded Boss/Base** 📇 feature tool. The Boss-Extrude FeatureManager is displayed.

398) Enter **.100**in, [**2.54**] for Depth in Direction 1.

399) Click **OK** ✔ from the Boss-Extrude PropertyManager. Boss-Extrude3 is displayed in the FeatureManager.

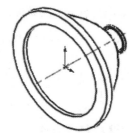

400) Rename **Boss-Extrude3** to **LensShield**.

401) Click **Save** 💾.

Change Transparency of the LensShield.
402) Right-click **LensShield** in the FeatureManager.

403) Click **Change Transparency**. View the results.

Save the model.
404) Click **Save** 💾.

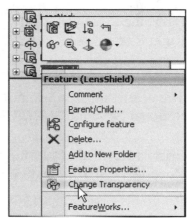

🔍 Additional information on Revolved Boss/Base, Shell, Hole Wizard and Appearance is located in SolidWorks Help Topics. Keywords: Revolved (features), Shells, Hole Wizard (Counterbore) and Appearances.

Refer to Help, SolidWorks Tutorials, Revolve and Swept for additional information.

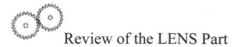 Review of the LENS Part

The LENS feature utilized a Revolved Base feature. A Revolved feature required an axis, profile and an angle of revolution. The Shell feature created a uniform wall thickness.

You utilized the Convert Entities Sketch tool to create the Extruded Boss feature for the LensNeck. The Counterbore hole was created using the Hole Wizard feature.

The Revolved Thin feature utilized a single 3 Point Arc. Geometric relations were added to the silhouette edge to define the arc. The LensCover and LensShield utilized existing geometry to Offset and Convert the geometry to the sketch. The Color and Optics PropertyManager determined the LensShield transparency.

BULB Part

The BULB fits inside the LENS. Use the Revolved feature as the Base feature for the BULB.

Insert the Revolved Base 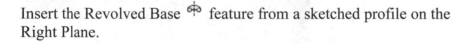 feature from a sketched profile on the Right Plane.

Insert a Revolved Boss feature using a Spline sketched profile. The profile utilizes a complex curve called a Spline (Non-Uniform Rational B-Spline or NURB). Draw Splines with control points.

Insert a Revolved Cut Thin feature at the base of the BULB. A Revolved Cut Thin feature removes material by rotating an open sketch profile about an axis.

Insert a Dome feature at the base of the BULB. A Dome feature creates spherical or elliptical shaped geometry. Use the Dome feature to create the Connector feature of the BULB. The Dome feature requires a face and a height value.

Insert a Circular Pattern feature from an Extruded Cut feature.

BULB Part-Revolved Base Feature

Create the new part, BULB. The BULB utilizes a solid Revolved Base feature.

The solid Revolved Base feature requires a:

- Sketch plane (Right Plane)

- Sketch profile (Lines)

- Axis of Revolution (Centerline)

- Angle of Rotation (360°)

Utilize the centerline to create a diameter dimension for the profile. The flange of the BULB is located inside the Counterbore hole of the LENS. Align the bottom of the flange with the Front Plane. The Front Plane mates against the Counterbore face.

Activity: BULB Part-Revolved Base Feature

Create a New part.
405) Click **New** ⬚ from the Menu bar.

406) Click the **MY-TEMPLATES** tab.

407) Double-click **PART-IN-ANSI**, [**PART-MM-ISO**].

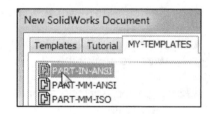

Save the part.
408) Click **Save** 🖫.

409) Select **PROJECTS** for Save in folder.

410) Enter **BULB** for File name.

411) Enter **BULB FOR LENS** for Description.

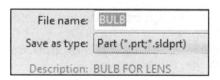

412) Click **Save**. The BULB FeatureManager is displayed.

Select the Sketch plane.
413) Right-click **Right Plane** from the FeatureManager. This is your Sketch plane.

Create the sketch.
414) Click **Sketch** ✎ from the Context toolbar. The Sketch toolbar is displayed.

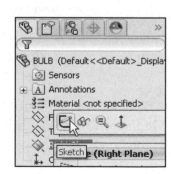

Sketch a centerline.

415) Click the **Centerline** ⋮ Sketch tool. The Insert Line PropertyManager is displayed.

416) Sketch a **horizontal centerline** through the Origin ⌞.

Create six profile lines.

417) Click the **Line** ⟍ Sketch tool. The Insert Line PropertyManager is displayed.

418) Sketch a **vertical line** to the left of the Front Plane.

419) Sketch a **horizontal line** with the endpoint coincident to the Front Plane.

420) Sketch a short **vertical line** towards the centerline, collinear with the Front Plane.

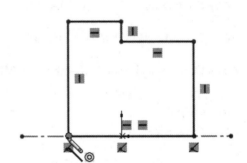

421) Sketch a **horizontal line** to the right.

422) Sketch a **vertical line** with the endpoint collinear with the centerline.

423) Sketch a **horizontal line** to the first point to close the profile.

Add dimensions.

424) Click the **Smart Dimension** ⟋ Sketch tool.

425) Click the **centerline**. (Note: Select the centerline).

426) Click the **top right horizontal line** as illustrated.

427) Click a **position** below the centerline and to the right.

428) Enter **.400**in, [**10.016**].

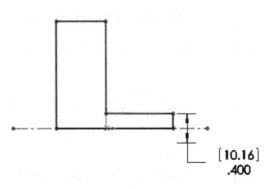

[10.16]
.400

☀ Click **View, Sketch Relations** from the Menu bar to display the relations of the model in the Graphics window.

429) Click the **centerline**. (Note: Select the centerline).

430) Click the **top left horizontal line**.

431) Click a **position** below the centerline and to the left.

432) Enter **.590**in, [**14.99**].

433) Click the **top left horizontal line**.

434) Click a **position** above the profile.

435) Enter **.100**in, [**2.54**].

436) Click the **top right horizontal line**.

437) Click a **position** above the profile.

438) Enter **.500**in, [**12.7**].

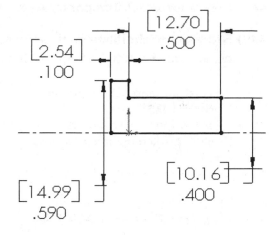

Fit the model to the Graphics window.
439) Press the **f** key.

Insert a Revolved Base feature.
440) Click the **Revolved Boss/Base** ⌖ feature tool. The Revolve PropertyManager is displayed. Accept the default settings.

441) Click **OK** ✔ from the Revolve PropertyManager. Revolve1 is displayed in the FeatureManager.

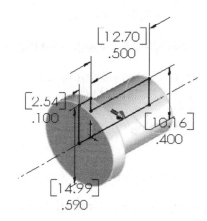

442) Click **Isometric view** ⬔ from the Heads-up View toolbar.

443) Click **Save** 💾.

BULB Part-Revolved Boss Feature and Spline Sketch Tool

The BULB requires a second solid Revolved feature. The profile utilizes a complex curve called a Spline (Non-Uniform Rational B-Spline or NURB). Draw Splines with control points. Adjust the shape of the curve by dragging the control points.

💡 For additional flexibility, deactivate the Snaps option in Document Properties for this model

Activity: BULB Part-Revolved Boss Feature and Spline Sketch Tool

Create the sketch.

444) Click **View**; check **Temporary Axes** from the Menu bar.

445) Right-click **Right Plane** from the FeatureManager for the Sketch plane. Click **Sketch** ✎ from the Context toolbar.

446) Click **Right view** ⊞ . The Temporary Axis is displayed as a horizontal line.

447) Press the **z** key approximately five times to view the left vertical edge.

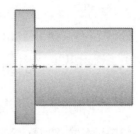

Sketch the profile.

448) Click the **Spline** ∿ Sketch tool.

449) Click the **left vertical edge** of the Base feature for the Start point.

450) Drag the **mouse pointer** to the left.

451) Click a **position** above the Temporary Axis for the Control point.

452) Double-click the **Temporary Axis** to create the End point and to end the Spline.

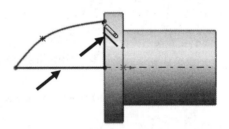

453) Click the **Line** ╲ Sketch tool.

454) Sketch a **horizontal line** from the Spline endpoint to the left edge of the Revolved feature.

455) Sketch a **vertical line** to the Spline start point, collinear with the left edge of the Revolved feature. Note: Dimensions are not required to create a feature.

Insert a Revolved Boss feature.

456) De-select the Line Sketch tool. Right-click **Select**.

457) Click the **Temporary Axis** from the Graphics window as illustrated.

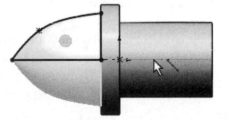

458) Click the **Revolved Boss/Base** ⚲ feature tool. The Revolve PropertyManager is displayed. Accept the default settings.

459) Click **OK** ✔ from the Revolve PropertyManager. Revolve2 is displayed in the FeatureManager.

460) Click **Isometric view** 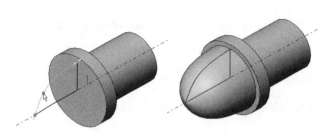 from the Heads-up View toolbar.

461) Click **Save** .

The points of the Spline dictate the shape of the Spline. Edit the control points in the sketch to produce different shapes for the Revolved Boss feature.

BULB Part-Revolved Cut Thin Feature

A Revolved Cut Thin feature removes material by rotating an open sketch profile around an axis. Sketch an open profile on the Right Plane. Add a Coincident relation to the silhouette and vertical edge. Insert dimensions.

Sketch a Centerline to create a diameter dimension for a revolved profile. The Temporary axis does not produce a diameter dimension.

Note: If lines snap to grid intersections, uncheck Tools, Sketch Settings, Enable Snapping for the next activity.

Activity: BULB Part-Revolved Cut Thin Feature

Create the sketch.
462) Right-click **Right Plane** from the FeatureManager.

463) Click **Sketch** from the Context toolbar.

464) Click **Right view** from the Heads-up View toolbar.

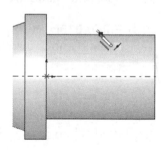

465) Click the **Line** Sketch tool.

466) Click the **midpoint** of the top silhouette edge.

467) Sketch a **line** downward and to the right as illustrated.

468) Sketch a **horizontal line** to the right vertical edge.

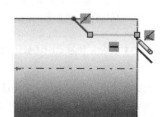

469) De-select the Line Sketch tool. Right-click **Select**.

If needed, add a Coincident relation.
470) Click the **endpoint** of the line.

471) Hold the **Ctrl** key down.

472) Click the right **vertical edge**.

473) Release the **Ctrl** key.

474) Click **Coincident** ⟋.

475) Click **OK** ✓ from the Properties PropertyManager.

Sketch a centerline.
476) Click **View**; uncheck **Temporary Axes** from the Menu bar.

477) Click the **Centerline** ┊ Sketch tool.

478) Sketch a **horizontal centerline** through the Origin.

Add dimensions.
479) Click the **Smart Dimension** ✑ Sketch tool.

480) Click the **horizontal centerline**.

481) Click the **short horizontal line**.

482) Click a **position** below the profile to create a diameter dimension.

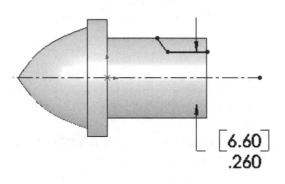

483) Enter **.260**in, **[6.6]**.

484) Click the **short horizontal line**.

485) Click a **position** above the profile to create a horizontal dimension.

486) Enter **.070**in, **[1.78]**. The Sketch is fully defined and is displayed in black.

☼ For Revolved features, the ∅ symbol is not displayed in the part. The ∅ symbol is displayed when inserted into the drawing.

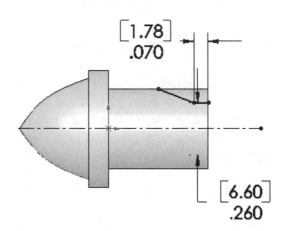

Insert the Revolved Cut Thin feature.
487) De-select the Smart Dimension Sketch tool. Right-click **Select**.

488) Click the **centerline** in the Graphics window.

489) Click the **Revolved Cut** ⋒ feature tool. The Cut-Revolve PropertyManager is displayed.

490) Click **No** to the Warning Message, "Would you like the sketch to be automatically closed?" The Cut-Revolve PropertyManager is displayed.

491) Check the **Thin Feature** box.

492) Enter **.150**in, **[3.81]** for Thickness.

493) Click the **Reverse Direction** box.

494) Click **OK** ✅ from the Cut-Revolve PropertyManager. Cut-Revolve-Thin1

is displayed in the FeatureManager.

495) Click **Save** 💾.

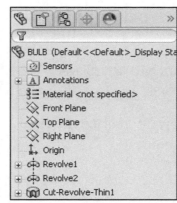

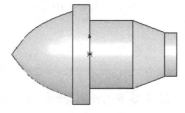

BULB Part-Dome Feature

The Dome 🔘 feature creates spherical or elliptical shaped geometry. Use the Dome feature to create the Connector feature of the BULB. The Dome feature requires a face and a height/distance value.

Activity: BULB Part-Dome Feature

Insert the Dome feature.

496) Click the **back circular face** of Revolve1. Revolve1 is highlighted in the FeatureManager.

497) Click the **Dome** 🔘 feature tool. The Dome PropertyManager is displayed. Face1 is displayed in the Parameters box.

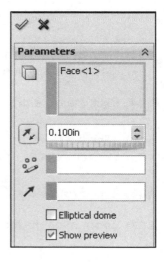

498) Enter .100in, [2.54] for Distance.

499) Click **OK** ✔ from the Dome PropertyManager. Dome1 is displayed in the FeatureManager.

500) Click **Isometric view** 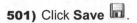 from the Heads-up View toolbar.

501) Click **Save** 💾.

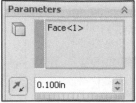

💡 Before creating sketches that use Geometric relations, check the Enable Snapping option in the Document Properties dialog box.

BULB Part-Circular Pattern Feature

A Pattern feature creates one or more instances of a feature or a group of features. The Circular Pattern feature places the instances around an axis of revolution.

Seed   Pattern

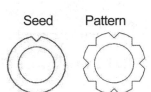

The Circular Pattern 🔧 feature requires a seed feature. The seed feature is the first feature in the pattern. The seed feature in this section is a V-shaped Extruded Cut feature.

Activity: BULB Part-Circular Pattern Feature

Create the Seed Cut feature.
502) Right-click the **front face** of the Base feature, Revolve1 in the Graphics window for the Sketch plane as illustrated. Revolve1 is highlighted in the FeatureManager.

503) Click **Sketch** ✏ from the Context toolbar. The Sketch toolbar is displayed.

504) Click the **outside circular edge** of the BULB.

505) Click the **Convert Entities** 🗗 Sketch tool. Click **OK** ✔ from the Convert Entities PropertyManager.

506) Click **Front view** 🗗 from the Heads-up View toolbar.

507) **Zoom in** on the top half of the BULB.

Sketch a centerline.

508) Click the **Centerline** ⫶ Sketch tool.

509) Sketch a **vertical centerline** coincident with the top and bottom circular circles and coincident with the Right plane.

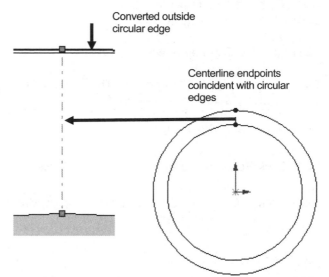

Converted outside circular edge

Centerline endpoints coincident with circular edges

Sketch a V-shaped line.

510) Click **Tools**, **Sketch Tools**, **Dynamic Mirror** from the Menu bar. The Mirror PropertyManager is displayed.

511) Click the **centerline** from the Graphics window.

512) Click the **Line** ⟍ Sketch tool.

513) Click the **midpoint** of the centerline.

514) Click the coincident **outside circle edge to the left** of the centerline.

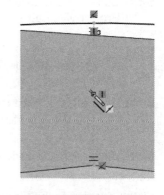

Deactivate the Dynamic Mirror tool.

515) Click **Tools**, **Sketch Tools**, **Dynamic Mirror** from the Menu bar.

Trim unwanted geometry.

516) Click the **Trim Entities** ⊁ Sketch tool. The Trim PropertyManager is displayed.

517) Click **Power trim** from the Options box.

518) Click a **position** in the Graphics window and drag the mouse pointer until it intersects the circle circumference.

519) Click **OK** ✓ from the Trim PropertyManager.

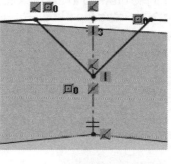

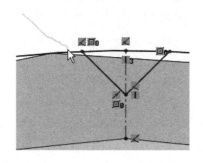

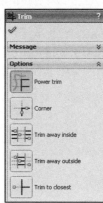

Add a Perpendicular relation.
520) Click the **left V shape** line.

521) Hold the **Ctrl** key down.

522) Click the **right V shape** line. The Properties PropertyManager is displayed.

523) Release the **Ctrl** key.

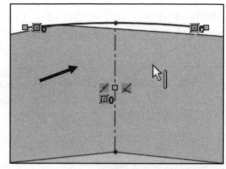

524) Click **Perpendicular** ⊥ from the Add Relations box.

525) Click **OK** ✔ from the Properties PropertyManager. The sketch is fully defined.

Create an Extruded Cut feature.
526) Click the **Extruded Cut** 🔲 feature tool. The Cut-Extrude PropertyManager is displayed.

527) Click **Through All** for End Condition in Direction 1. Accept the default settings.

528) Click **OK** ✔ from the Cut-Extrude PropertyManager. The Cut-Extrude1 feature is displayed in the FeatureManager.

529) Click **Isometric view** 🧊 from the Heads-up View toolbar.

Fit the drawing to the Graphics window.
530) Press the **f** key.

531) Click **Save** 💾.

🔅 Reuse Geometry in the feature. The Cut-Extrude1 feature utilized the centerline, Mirror Entity, and geometric relations to create a sketch with no dimensions.

The Cut-Extrude1 feature is the seed feature for the pattern. Create four copies of the seed feature. A copy of a feature is called an Instance. Modify the four copies Instances to eight.

Insert the Circular Pattern feature.
532) Click the **Cut-Extrude1** feature from the FeatureManager.

533) Click the **Circular Pattern**  feature tool. The Circular Pattern PropertyManager is displayed. Cut-Extrude1 is displayed in the Features to Pattern box.

534) Click **View**; check **Temporary Axes** from the Menu bar.

535) Click the **Temporary Axis** from the Graphics window. Axis<1> is displayed in the Pattern Axis box.

536) Enter **4** in the Number of Instances box. Check the **Equal spacing** box.

537) Check the **Geometry pattern** box. Accept the default settings. Click **OK** 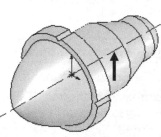 from the Circular Pattern PropertyManager. CirPattern1 is displayed in the FeatureManager.

Edit the Circular Pattern feature.
538) Right-click **CirPattern1** from the FeatureManager. Click **Edit Feature** from the Context toolbar. The CirPattern1 PropertyManager is displayed.

539) Enter **8** in the Number of Instances box. Click **OK**  from the CirPattern1 PropertyManager.

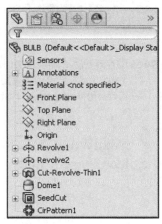

Rename the feature
540) Rename **Cut-Extrude1** to **SeedCut**.

Hide the reference geometry.
541) Click **View**; uncheck **Temporary Axes** from the Menu bar.

Save the model.
542) Click **Save** 🖫.

Rename the seed feature of a pattern to locate it quickly for future assembly.

Customizing Toolbars and Short Cut Keys

The default toolbars contain numerous icons that represent basic functions. Additional features and functions are available that are not displayed on the default toolbars.

You have utilized the z key for Zoom In/Out, the f key for Zoom to Fit and Ctrl-C/Ctrl-V to Copy/Paste. Short Cut keys save time.

Assign a key to execute a SolidWorks function. Create a Short Cut key for the Temporary Axis.

Activity: Customizing Toolbars and Short Cut Keys

Customize the toolbar.

543) Click **Tools**, **Customize** from the Menu bar. The Customize dialog box is displayed.

Place the Freeform icon on the Features toolbar.

544) Click the **Commands** tab.

545) Click **Features** from the category text box.

546) Drag the **Freeform** feature 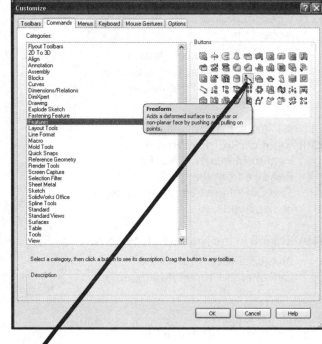 icon into the Features toolbar. The Freeform feature option is displayed.

 SolidWorks 2012 does not support the older Shape feature.

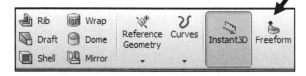

Customize the keyboard for the
Temporary Axes.

547) Click the **Keyboard** tab
from the Customize dialog
box.

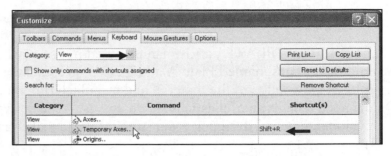

548) Select **View** for Categories.

549) Select **Temporary Axes** for
Commands.

550) Enter **R** for Shortcut(s) key. Click **OK**.

551) Press the **R** key to toggle the display of the Temporary Axes.

Test the proposed Short cut key, before you customize
your keyboard. Refer to the default Keyboard Short cut table in
the Appendix.

Additional information on Revolved Boss/Base, Spline,
Circular Pattern, Dome, Line / Arc Sketching is located in
SolidWorks Help Topics. Keywords: Revolved (features),
Spline, Pattern (Circular) and Dome.

Refer to Help, SolidWorks Tutorials, Pattern Features for
additional information.

 Review of the BULB Part

The Revolved Base feature utilized a sketched profile on the Right
plane and a centerline. The Revolved Boss feature utilized a
Spline sketched profile. A Spline is a complex curve.

You created the Revolved Cut Thin feature at the base of the
BULB to remove material. A centerline was inserted to add a
diameter dimension. The Dome feature was inserted on the back
face of the BULB. The Circular Pattern feature was created from
an Extruded Cut feature. The Extruded Cut feature utilized
existing geometry and required no dimensions.

Toolbars and keyboards were customized to save time. Always
verify that a Short cut key is not predefined in SolidWorks.

Design Checklist and Goals before Plastic Manufacturing

The BATTERYPLATE part is manufactured from a plastic resin. A plastic part requires mold base plates, named Plate A and Plate B, to create the mold. A plastic part also requires knowledge of the material and manufacturing process. Critical decisions need to be made. What type of plastic resin do you use? The answer is derived from the development team: designer, resin supplier, mold maker and mold base, (Plate A & Plate B) supplier.

Ticona (www.ticona.com) supplies their customers with a designer's checklist to assist the designer in selecting the correct material.

	TICONA-Celanese Typical Design Checklist
1.	What is the function of the part?
2.	What is the expected lifetime of the part?
3.	What agency approvals are required? (UL, FDA, USDA, NSF, USP, SAE, MIL spec)
4.	What electrical characteristics are required and at what temperatures?
5.	What temperature will the part see and for how long?
6.	What chemicals will the part be exposed to?
7.	Is moisture resistance necessary?
8.	How will the part be assembled? Can parts be combined into one plastic part?
9.	Is the assembly going to be permanent or one time only?
10.	Will adhesives be used? Some resins require special adhesives.
11.	Will fasteners be used? Will threads be molded in?
12.	Does the part have a snap fit? Glass filled materials will require more force to close the snap fit, but will deflect less.
13.	Will the part be subjected to impact? If so, radius the corners.
14.	Is surface appearance important? If so, beware of weld lines, parting lines, ejector location and gate location.
15.	What color is required for the part? Is a specific match required or will the part be color-coded? Some glass or mineral filled materials do not color as well as unfilled materials.
16.	Will the part be painted? Is primer required? Will the part go through a high temperature paint oven?
17.	Is weathering or UV exposure a factor?
18.	What are the required tolerances? Can they be relaxed to make molding more economical?
19.	What is the expected weight of the part? Will it be too light (or too heavy)?
20.	Is wear resistance required?
21.	Does the part need to be sterilized? With what methods (chemical, steam, radiation)?
22.	Will the part be insert molded or have a metal piece press fit in the plastic part? Both methods result in continuous stress in the part.
23.	Is there a living hinge designed in the part? Be careful with living hinges designed for crystalline materials such as acetal.
24.	What loading and resulting stress will the part see? And, at what temperature and environment.
25.	Will the part be loaded continuously or intermittently? Will permanent deformation or creep be an issue?
26.	What deflections are acceptable?
27.	Is the part moldable? Are there undercuts? Are there sections that are too thick or thin?
28.	Will the part be machined?
29.	What is the worst possible situation the part will be in? Worst-case environment.

In "Designing With Plastic - The Fundamentals", Ticona lists three design goals for creating injection molded parts:

Goal 1: Maximize Functionality

Mold bases are costly. Design functionality into the part. A single plastic chassis replaces several sheet metal components.

Reduce assembly time and part weight whenever it is appropriate.

Sheet Metal
Assembly

Injection Molded
Thermoplastic

Goal 2: Optimize Material Selection

For material selection, consider the following elements: part functionality, operating environment, cost/price constraint and any special cosmetic requirements.

Several materials should be selected with a developed list of advantages and disadvantages for review.

For maximum product performance at a competitive cost in this project, use Celanese® (Nylon 6/6) (P/A 6/6), where:

- Celanese® is the registered trademark name.

- Nylon 6/6 is the common plastic name.

- P/A 6/6 (Polyhexamethyleneadipamide) is the chemical name.

Goal 3: Minimize Material Use

Optimize part wall thickness for cost and performance.

The minimum volume of plastic that satisfies the structural, functional, appearance and molding requirements of the application is usually the best choice.

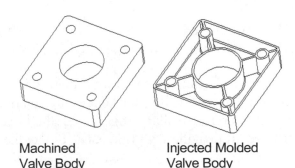

Machined
Valve Body

Injected Molded
Valve Body

Mold Base

Designing a custom mold base is expensive and time consuming. An example of a mold base supplier is Progressive Components (www.procomps.com).

The mold base positions the mold cavities. The mold base plates are machined to create the mold cavities. The mold base is designed to withstand high pressure, heating and cooling sequences during the fabrication process.

The mold base assembly is composed of the following two plates and a variety of support plates, ejector plates, pins and bushings:

- PLATE A

- PLATE B

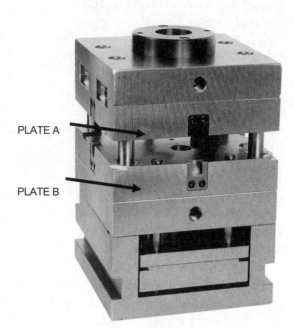

PLATE A

PLATE B

Courtesy of Progressive Components
Wauconda, IL USA

Applying SolidWorks Features for Mold Tooling Design

SolidWorks features such as Draft, Fillet and Shell assist in the design of plastic parts. Utilize the Draft tool to analyze the part before creating the Mold Tooling, PLATE A and PLATE B.

Most molded parts require Draft. Plastic molded injection parts require a draft angle to eject the part from the mold. To properly eject the part, design parts with a draft angle in the direction of the mold movement. A draft angle of 1° - 3° is the design rule for most injection molded parts. There are exceptions based on material type and draw depth. For illustration purposes, a draft angle of 1° is used in the BATTERYPLATE.

The draft angle is an option in both the Extruded Boss/Base and Extruded Cut features. The Draft feature adds a specified Draft Angle to faces of an existing feature.

Use the Fillet feature to remove sharp edges from plastic parts. For thin walled parts, ensure that the inside sharp edges are removed. Sketch Arcs and 2D Fillets in the 2D profile to remove sharp edges. The Shell feature, Extruded Cut feature or Extruded Thin feature all provide wall thickness. Select the correct wall thickness for a successful part.

If the wall thickness is too thin, the part will structurally fail. If the wall thickness is too thick, the part will be overweight and too costly to manufacture. This will increase cycle time and decrease profits.

There are numerous types of plastic molds. SolidWorks contains a variety of Mold tools to assist the designer, mold maker and mold manufacturer. For a simple Cavity Mold follow the following steps:

- *Draft analysis*. A draft analysis examines the faces of the model for sufficient draft, to ensure that the part ejects properly from the tooling.

- *Undercut detection*. Identifies trapped areas that prevent the part from ejecting.

- *Parting Lines*. This tool has two functions:

   o   Verifies that you have draft on your model, based on the angle you specify.

   o   Creates a parting line from which you create a parting surface. The Parting Lines tool includes the option to select an edge and have the system Propagate to all edges.

- *Shut-off Surfaces*. Creates surface patches to close up through holes in the molded part.

- *Parting Surfaces*. Extrude from the parting line to separate mold cavity from core. You can also use a parting surface to create an interlock surface.

- *Ruled Surface*. Adds draft to surfaces on imported models. You can also use the Ruled Surface tool to create an interlock surface.

- *Tooling Split*. Creates the core and cavity bodies, based on the steps above.

Activity: SolidWorks Features for Mold Tooling Design

Create the New part.
552) Click **New** from the Menu bar.

553) Click the **MY-TEMPLATES** tab.

554) Double-click **PART-IN-ANSI**, [**PART-MM-ISO**].

Save the part.
555) Click **Save**. Select **PROJECTS** for Save in folder.

556) Enter **MOLD-TOOLING** for File name.

557) Enter **MOLD FEATURES FOR TOOLING** for Description. Click **Save**. The MOLD-TOOLING FeatureManager is displayed.

Insert a part.

558) Click **Insert, Part** from the Menu bar.

559) Double-click **BATTERYPLATE** from the PROJECTS folder. The Insert Part PropertyManager is displayed.

560) Click **OK** ✔ from the Insert Part PropertyManager. The BATTERYPLATE part is the first feature in the MOLD-TOOLING part.

💡 Utilize the Insert Part option to maintain the original part and a simplified FeatureManager. The Insert Part maintains an External reference to the original part ->. Modify the original part and the Insert Part updates on a Rebuild. Apply the Insert Part option to SolidWorks parts and imported parts.

Display the Mold toolbar.

561) Right-click in the **gray area** of the Feature toolbar and click **Mold Tools**. The Molds toolbar is displayed.

Perform a Draft Analysis.

562) Click the **Draft** 🔲 tool from the Mold toolbar. The Draft PropertyManager is displayed. Note: This is the same Draft tool as displayed in the Features toolbar.

563) Click the **DraftXpert** tab. The DraftXpert PropertyManager is displayed.

564) Click the **Add** tab.

Create the Direction of Pull.

565) **Rotate** the part and click the **bottom face**. Face<1> is displayed in the Neutral Plane box.

566) Click the **Reverse Direction** arrow. The direction of pull arrow points upwards. Display an **Isometric view**.

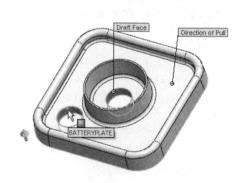

Select a face to draft.

567) Zoom in and click the **inside hole face** of the center hole. Zoom in and click the **inside hole face** of the small hole. The selected faces are displayed.

568) Enter **1**deg for Draft angle. Check the **Auto Paint** box.

569) Click the **Apply** button.

570) Rotate the **part** to display the faces that require a draft. The faces are displayed in different colors. View the results in the Graphics window.

571) Click **OK** ✔ from the DraftXpert PropertyManager. Draft1 is displayed.

By default, the faces that require draft are displayed in red. The Top Cut, Extruded Cut, and the Holes Extruded Cut requires a draft. The bottom flat surface will become the Parting surface.

Modify the BATTERYPLATE part In-Context of the MOLD-TOOLING part. Both parts are open during the modification session.

Modify the BATTERYPLATE In-Context.
572) Right-click **BATTERYPLATE** in the FeatureManager.

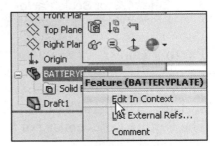

573) Click **Edit In Context**. The BATTERYPLATE FeatureManager is displayed.

Insert Draft into the Top Cut.
574) Right-click **Top Cut** in the BATTERYPLATE FeatureManager.

575) Click **Edit Feature**. The Top Cut PropertyManager is displayed.

576) If needed, click the **Draft on/off** button. Enter **1deg** for Draft Angle.

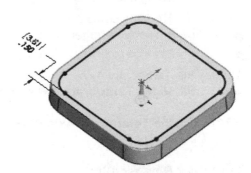

577) Click **OK** ✔ from the Top Cut PropertyManager.

Insert a draft into the holes.
578) Right-click **Holes** in the BATTERYPLATE FeatureManager.

579) Click **Edit Feature**. The Holes PropertyManager is displayed.

580) If needed, click the **Draft on/off** button.

581) Enter **1deg** for Draft Angle.

582) Click **OK** ✔ from the Holes PropertyManager.

Return to the MOLD-TOOLING part.
583) Press **Ctrl-Tab**. The MOLD TOOLING FeatureManager is displayed.

584) **Rebuild** the model.

Scale:

The Scale feature increases or decreases the part's volume by a specified factor. The hot thermoplastic utilized in this part, cools during the molding process, hardens and shrinks. Increase the part with the Scale feature to compensate for shrinkage.

Insert the Scale feature.

585) Click **Scale** from the Mold toolbar. The Scale PropertyManager is displayed.

586) Enter **1.2** for Scale Factor. Click **OK** from the Scale PropertyManager. Scale1 is displayed in the FeatureManager.

Parting Lines:

The Parting Lines are the boundary between the core and cavity surfaces. The Parting Lines create the edges of the molded part. The Parting Lines consist of the bottom eight edges.

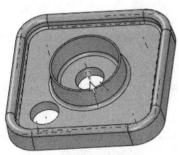

Insert the Parting Lines feature.

587) Click **Parting Lines** from the Mold toolbar. The Parting Line PropertyManager is displayed.

Display the Temporary Axis.
588) Click **View**, **Temporary Axes** from the Menu bar.

589) Click the **Temporary Axis** displayed through the center hole for the Direction of Pull.

590) Click **Reverse Direction** if required. The direction arrow points upward.

591) Enter **1**deg for Draft Angle.

592) Click the **Draft Analysis** button. The Parting Lines box contains the eight bottom edges of the BATTERYPLATE.

593) Click **OK** from the Parting Line PropertyManager. Parting Line1 is displayed in the FeatureManager.

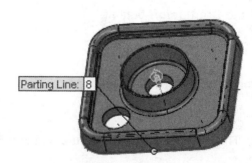

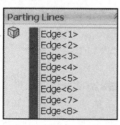

Hide the Temporary Axes.
594) Click **View**; uncheck **Temporary Axes** from the Menu bar.

Shut-off Surfaces:

Holes and windows that penetrate through an entire part require separate mold tooling. A shut-off area in a mold is where two pieces of the mold tooling contact each other. These areas require a Shut-off Surface. There are two Thru Holes in the BATTERYPLATE.

Insert the Shut-off Surfaces feature.

595) Click **Shut-off Surfaces** from the Mold toolbar. The Shut-Off Surface PropertyManager is displayed. Accept the defaults settings.

596) Click **OK** from the Shut-Off Surface PropertyManager. A surface closes off the opening of each hole. Shut-Off Surface1 is displayed in the FeatureManager.

597) **Rotate** the model and view the closed surface.

Parting Surface:

A Parting Surface is utilized to separate the mold tooling components. A Parting Surface extends radially, a specified distance from the Parting Lines. Usually, the surface is normal to the direction of pull for simple molds. The boundary of the Parting Surface must be larger than the profile for the Tooling Split.

Insert the Parting Suface feature.
598) Click **Parting Surfaces** from the Mold toolbar. The Parting Surface PropertyManager is displayed.

599) Click **Top view** from the Heads-up View toolbar.

600) Enter **1.000**in, **[25.4]** for Distance. Accept the default settings.

601) Click **OK** from the Parting Surface PropertyManager. A Parting Surface (Radiate Surface) is created from the Parting Lines and the Distance value.

Fit the model to the Graphics window.
602) Press the **f** key.

603) Click **Save** 🖫 .

Tooling Split:

The Tooling Split separates the solid bodies from the parting surfaces. The solid bodies become the mold tooling and the molded part. The Tooling Split for the BATTERYPLATE creates two mold base plates named, Tooling Split1 and Tooling Split2.

Insert the Tooling Split feature.
604) Click **Parting Surface1** in the FeatureManager.

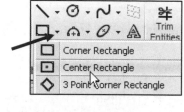

605) Click **Tooling Split** 🔺 from the Mold toolbar.

606) Click the **front face** of Parting Surface1.

607) Click the **Center Rectangle** ▢ Sketch tool.

608) Click the **Origin**.

609) Sketch a **rectangle** as illustrated about the Origin.

De-select the Center Rectangle Sketch tool.
610) Right-click **Select**.

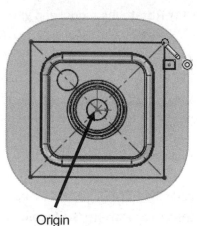

Add an Equal relation.
611) Click the **top horizontal line** of the rectangle.

612) Hold the **Ctrl key** down.

613) Click the **left vertical line, right vertical line,** and **bottom horizontal line** of the rectangle.

Origin

614) Release the **Ctrl key**.

615) Click **Equal** ＝ .

616) Click **OK** ✓ from the Properties PropertyManager.

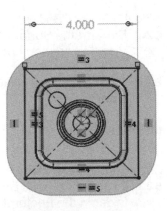

Add a dimension.

617) Click the **Smart Dimension** ✦ Sketch tool.

618) Click the **top horizontal** line.

619) Enter **4.000**in, **[101.60]**.

620) Click **Exit Sketch**. The Tooling Split PropertyManager is displayed.

Enter values for the Block Size.

621) Click **Front view** 🔲 from the Heads-up View toolbar.

622) Enter **1.000**in, **[25.4]** for Depth in Direction 1.

623) Enter **.500**in, **[12.7]** for Depth in Direction 2.

624) Click **Isometric view** ⬦ from the Heads-up View toolbar.

625) Click **OK** ✓ from the Tooling Split PropertyManager. Tooling Split1 is displayed in the FeatureManager.

The Mold tools produce Surface Bodies and Solid Bodies. The Solid Bodies are utilized in an assembly for the actual mold base. The Surface Bodies were utilized as interim features to create the mold tooling parts. Hide the Surface Bodies to display the solid mold tooling parts.

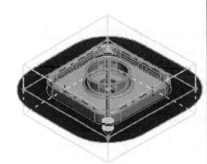

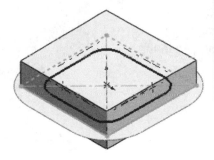

Hide the Surface Bodies.
626) Expand the Surface Bodies folder in the FeatureManager. View the created bodies.

627) Right-click the **Surface Bodies** folder.

628) Click **Hide**.

Display the Solid Bodies.
629) Expand the Solid Bodies folder in the FeatureManager.

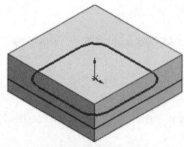

View each entry.
630) Click each **entry** in the Solid Bodies folder.

Separate Tooling Split1.
631) Click **Tooling Split1** from the FeatureManager.

632) Click **Insert**, **Features**, **Move/Copy**  Move/Copy... from the Menu bar. The Move/Copy Body PropertyManager is displayed.

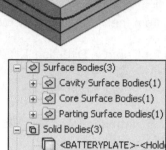

633) Click and drag the **vertical axis of the Triad** downward approximately 3inches (75mm) to display the BATTERYPLATE part.

634) Click **OK** ✔ from the PropertyManager. Body-Move/Copy1 is displayed in the FeatureManager.

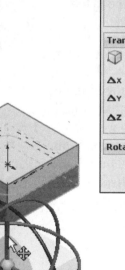

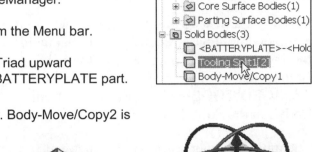

Separate Tooling Split1[2].

635) Click **Tooling Split1[2]** from the FeatureManager.

636) Click **Insert**, **Features**, **Move/Copy** from the Menu bar.

637) Click and drag the **vertical axis** of the Triad upward approximately 4in, [100] to display the BATTERYPLATE part.

638) Click **OK** ✓ from the PropertyManager. Body-Move/Copy2 is displayed in the FeatureManager.

Rotate the view.

639) Click the **Up Arrow key** to display the model as illustrated.

Save the MOLD-TOOLING part.

640) Click **Save** 🖫.

641) Click **Yes** to save the referenced part (BATTERYPLATE).

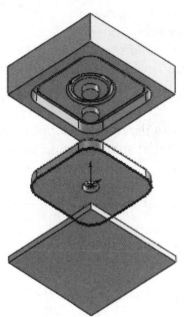

Manufacturing Design Issues

For the experienced mold designer, the complete mold base requires additional mold tooling, lines and fittings. Utilize the Solid Bodies, Insert into New Part option to create the MOLD-TOOLING-CAVITY part from Body2 in the Solid Bodies folder.

The mold designer works with a material supplier and mold manufacturer. For example, Bohler-Uddeholm provides mold steels in rod, bar and block form. Cutting operations include Milling, Drilling, Sawing, Turning and Grinding. A wrong material selection is a costly mistake.

As a designer, you work with the material supplier to determine the best material for the project. For example, a material supplier application engineer asks the following questions:

- What is the plastic material utilized in this application?

- Will the material be corrosive or abrasive?

- Will the mold be textured?

- How important is the surface finish?

- What are the quantities of the part to be produced?

- Are part tolerances held within close limits?

- Have you inserted fillets on all sharp corners?

- Did you utilize adequate wall thickness and overall dimensions?

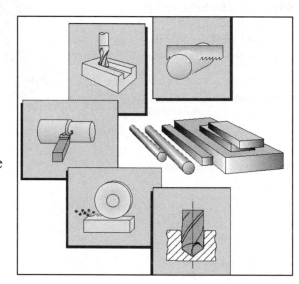

Cutting Operations for Tool Steels
Courtesy of Bohler-Uddeholm
www.bucorp.com

You send an eDrawing of your MOLD-TOOLING part for advice to the application engineer. Take the time early in the design process to determine the tooling material.

Project Summary

You are designing a FLASHLIGHT assembly that is cost effective, serviceable and flexible for future design revisions. The FLASHLIGHT assembly consists of various parts. The BATTERY, BATTERYPLATE, LENS and BULB parts were modeled in this project.

This Project concentrated on the Extruded Boss/Base feature and the Revolved Boss/Base feature. The Extruded Boss/Base feature required a Sketch plane, Sketch profile and End Condition (Depth). The BATTERY and BATTERYPLATE parts incorporated an Extruded Boss/Base (Boss-Extrude1) feature. A Revolved feature requires a Sketch plane, Sketch profile, Axis of Revolution and an Angle of Rotation.

You also utilized the Extruded Boss/Base, Extruded Cut, Fillet, Chamfer, Revolved Cut, Revolved Thin-Cut, Shell, Hole Wizard, Dome and Circular Pattern features.

You addressed the following Sketch tools in this project: Sketch, Smart Dimension, Line, Rectangle, Circle, Tangent Arc and Centerline. You addressed additional Sketch tools that utilized existing geometry: Add Relations, Display/Delete Relations, Mirror Entities, Convert Entities and Offset Entities. Geometric relations were utilized to build symmetry into the sketches.

The BATTERYPLATE utilized various Mold tools to develop the Core and Cavity tooling plates required in the manufacturing process. Plastic parts require Draft. The Draft Analysis tool determined areas of positive and negative draft. Features were modified to accommodate the manufacturing process.

Practice these concepts with the project exercises. The other parts for the FLASHLIGHT assembly are addressed in Project 5.

Use the Defeature tool, to remove details from a part or assembly and save the results to a new file in which the details are replaced by dumb solids (that is, solids without feature definition or history). You can then share the new file without revealing all the design details of the model.

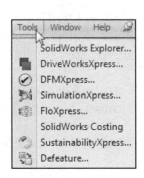

Project Terminology

Assembly: An assembly is a document in which components, features and other assemblies (sub-assemblies) are put together. The filename extension for a SolidWorks assembly file name is *.SLDASM. The FLASHLIGHT is an assembly. The BATTERY is a component in the FLASHLIGHT assembly.

Base Feature: The first feature listed in the FeatureManager.

Base Sketch: The first sketch listed in the FeatureManager.

Box-select: A selection method used in parts, assemblies, and drawings to select entities by dragging a selection box with the mouse pointer. Dragging from left to right will select all items that fall within the area of the box.

Centerpoint Arc: A Sketch tool that requires a centerpoint, start point and end point to create a centerpoint arc.

Chamfer: A feature that bevels sharp edges or faces by a specified distance and angle or by two specified distances.

Closed Profile: Also called a closed contour, it is a sketch or sketch entity with no exposed endpoints: for example, a circle or polygon.

Circular Pattern: A feature that creates a pattern of features or faces in a circular array about an axis.

Coincident: A geometric condition where two or more entities share the same point in space.

Convert Entities: A sketch tool that projects one or more curves onto the current sketch plane. Select an edge, loop, face, curve, or external sketch contour, set of edges, or set of sketch curves.

Cursor Feedback: Feedback is provided by a symbol attached to the cursor arrow indicating your selection.

Dimension: A value indicating the size of feature geometry.

Dimensioning Standard: A set of drawing and detailing options developed by national and international organizations. A few key dimensioning standard options are: ANSI, ISO, DIN, JIS, BSI, GOST and GB.

Design Intent: How a model reacts to changes and updates.

Display Pane: A portion of the FeatureManager that can be expanded to view the various display settings in a part, assembly or drawing document.

Document Templates: A part, assembly, or drawing document that includes user-defined parameters or standard default parameters that is used in the creation of a new SolidWorks document.

Dome: A feature used to add a spherical or elliptical dome to a selected face.

Draft angle: A draft angle is the degree of taper applied to a face. Draft angles are usually applied to molds or castings.

Edit Feature: A tool utilized to modify existing feature parameters. Right-click the feature in the FeatureManager. Click Edit Feature.

Edit Sketch: A tool utilized to modify existing sketch geometry. Right-click the feature in the FeatureManager. Click Edit Sketch.

End Condition: An option located in the PropertyManager. End Condition determines how a feature extends and the design intent of the model.

Extruded Boss/Base: A feature that adds material utilizing a 2D sketch profile and a depth perpendicular to the sketch plane. The Base feature is the first feature in the part.

Extruded Cut: A feature that removes material utilizing a 2D sketch profile and a depth perpendicular to the sketch plane.

Features: Features are geometry building blocks. Features add or remove material. Features are created from sketched profiles or from edges and faces of existing geometry.

Fillet: A feature that rounds sharp edges or faces by a specified radius.

Fully Defined: A sketch where all lines and curves in the sketch, and their positions, are described by dimensions or relations, or both, and cannot be moved. Fully defined sketch entities are displayed in black.

Geometric Relations: A relation is a geometric constraint between sketch entities or between a sketch entity and a plane, axis, edge, or vertex. Relations force a behavior on a sketch element to capture the design intent.

Graphics Window: The largest working window within SolidWorks where parts, assemblies and drawings are displayed.

Heads-up View Toolbar: A transparent toolbar located in the upper portion of the Graphics window by default.

Hidden Geometry: Geometry that is not displayed. Utilize Hide/Show to control display of components in an assembly.

Hidden Lines: A line type used in 2D drafting to represent edges that can't be seen normally in the view - which are hidden by other geometry.

Hole Wizard: The Hole Wizard feature is used to create specialized holes in a solid. The HoleWizard creates simple, tapped, counterbore and countersunk holes using a step-by-step procedure.

Mold Tools: The options available to develop the Parting Lines, Parting Surfaces and Shut-off Surfaces to create the core and cavity plates and additional mold tooling.

Menus: Menus provide access to the commands that the SolidWorks software offers.

Mirror Entities: A sketch tool that mirrors sketch geometry to the opposite side of a sketched centerline.

Mouse Buttons: The left and right mouse buttons have distinct meanings in SolidWorks. The left mouse button is utilized to select geometry. The right-mouse button is utilized to invoke commands.

Offset Entities: A sketch tool utilized to create sketch curves offset by a specified distance. Select sketch entities, edges, loops, faces, curves, set of edges or a set of curves.

Part: A part is a single 3D object that consists of various features. The filename extension for a SolidWorks part is .SLDPRT.

Plane: Planes are flat and infinite. Planes are represented on the screen with visible edges.

Relation: A relation is a geometric constraint between sketch entities or between a sketch entity and a plane, axis, edge or vertex. Utilize Add Relations to manually connect related geometry.

Revolved Base/Boss: A feature used to add material by revolutions. A Revolved feature requires a centerline, a sketch on a sketch plane and an angle of revolution. The sketch is revolved around the centerline.

Revolved Cut: A feature used to remove material by revolutions. A Revolved Cut requires a centerline, a sketch on a sketch plane and angle of revolution. The sketch is revolved around the centerline.

Shell: A feature used to remove faces of a part by a specified wall thickness.

Silhouette Edge: The imaginary edge of a cylinder or cylindrical face.

Sketch: The name to describe a 2D profile is called a sketch. 2D sketches are created on flat faces and planes within the model. Typical geometry types are lines, arcs, rectangles, circles, polygons and ellipses.

Spline: A complex sketch curve.

States of a Sketch: There are four key states that are utilized in this Project:

- Fully Defined: Has complete information, (Black).
- Over Defined: Has duplicate dimensions, (Red).
- Under Defined: There is inadequate definition of the sketch, (Blue).
- Selected: The current selected entity, (Green).

Template: A template is the foundation of a SolidWorks document. A Part Template contains the Document Properties such as: Dimensioning Standard, Units, Grid/Snap, Precision, Line Style and Note Font.

Thin option: The Thin option for the Revolved Boss and Revolved Cut utilizes an open sketch to add or remove material, respectively.

Toolbars: The toolbars provide shortcuts enabling you to access the most frequently used commands.

Trim Entities: A sketch tool used to delete selected sketched geometry.

Units: Used in the measurement of physical quantities. Decimal inch dimensioning and Millimeter dimensioning are the two types of common units specified for engineering parts and drawings.

Questions

1. Provide a short explanation of the following features:

Fillet	Extruded Cut	Draft	Scale
Extruded Boss	Revolved Base	Shell	Move/Copy
Revolved Cut Thin	Circular Pattern	Hole Wizard	Parting Surface

2. True of False. Design intent exists only in the sketch.

3. Describe a Symmetric relation.

4. Describe an Angular dimension.

5. Describe a draft angle? When do you apply a draft angle in a model?

6. When do you apply the Mirror feature? Utilize the SolidWorks Online Users Guide to determine the difference between the Sketch Mirror and the Dynamic Mirror Sketch tool.

7. Describe the Spline Sketch tool.

8. Describe the procedure for adding a feature icon to the default Feature toolbar.

9. Describe the procedure to create a Short Cut key to Show/Hide Planes.

10. For a simple mold, describe the procedure to create the core and cavity mold tooling.

11. Identify the following Sketch tool icons.

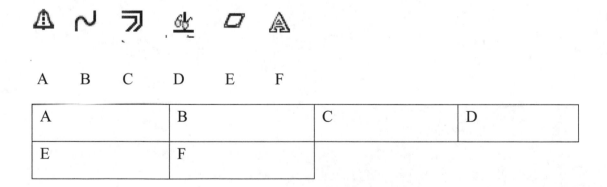

A   B   C   D   E   F

A	B	C	D
E	F		

12. Identify the following Mold tool icons.

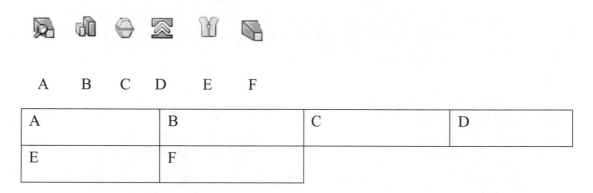

A	B	C	D
E	F		

Exercise 4.1: FLAT PLATE

- Create the ANSI FLAT PLATE Part on the Top Plane. The FLAT PLATE is machined from 0.090in, [2.3mm] 6061 Alloy stock. Primanry Unit system: IPS.

- The 8.688in, [220.68mm] x 5.688in, [144.48mm] FLAT PLATE contains a Linear Pattern of ∅.190in, [4.83mm] Thru Holes.

- The holes are equally spaced, .500in, [12.70mm] apart.

- Utilize the Geometric Pattern option for the Linear Pattern feature.

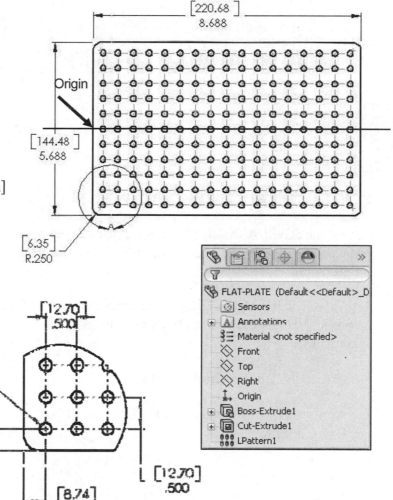

Exercise 4.2: IM15-MOUNT Part

Create the ANSI IM15-MOUNT Part as illustrated on the Right plane. The IM15-MOUNT Part is machined from 0.060in, [1.5mm] 304 Stainless Steel flat stock. Primary Unit system: IPS.

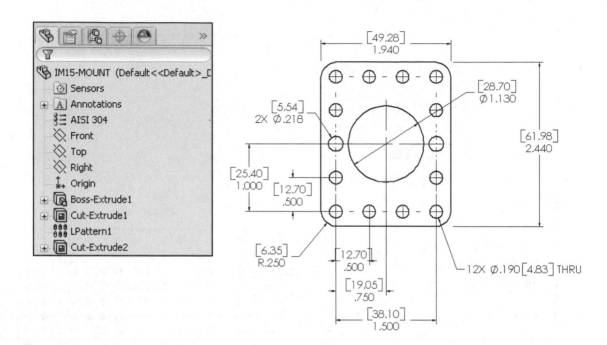

Exercise 4.3: ANGLE BRACKET Part

Create the ANSI ANGLE BRACKET Part. The Boss-Extrude1 feature is sketched with an L-Shaped profile on the Right Plane.

The ANGLE BRACKET Part is machined from 0.060in, [1.5mm] 304 Stainless Steel flat stock. Primary Unit system: IPS.

- Create the Boss-Extrude1 feature.

- Apply the Mid Plane option for Direction 1 End Condition.

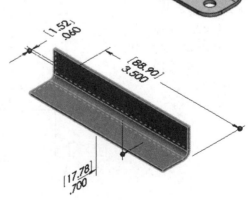

- Create the second Extrude feature on the Top view for the hole.

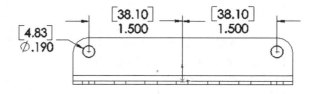

- Use the Linear Pattern feature to create six additional holes along the ANGLE BRACKET.

- Add Fillets to the four edges, .250in.

- Create the third Extrude feature.

- Apply the Linear Pattern feature.

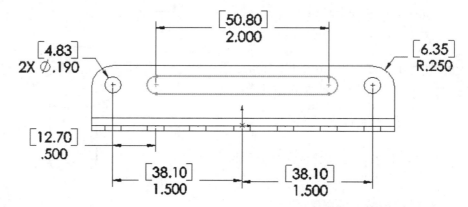

- Create the fourth Extrude feature for the center slot.

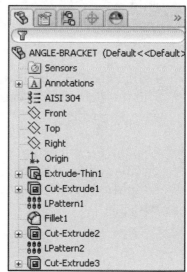

Exercise 4.4: SCREW

Create the ANSI 10-24 x 3/8 SCREW as illustrated. Note: A Simplified version. Primary Unit system: IPS.

- Sketch a centerline on the Front Plane.

- Sketch a closed profile.

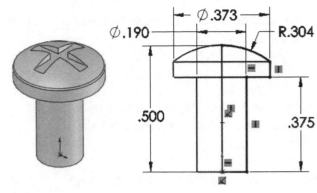

- Utilize a Revolved Feature. For metric size, utilize an M4x10 machine screw.

- Edit the Revolved Base Sketch. Apply the Tangent Arc tool with Trim Entities. Enter an Arc dimension of .304in, [7.72].

- Utilize an Extruded Cut feature with the Mid Plane option to create the Top Cut.
  Depth = .050in.

- The Top Cut is sketched on the Front Plane.

- Utilize the Convert Entities Sketch tool to extract the left edge of the profile.

- Utilize the Circular Pattern feature and the Temporary Axis to create four Top Cuts.

- Use the Fillet Feature on the top circular edge, .01in to finish the simplified version of the SCREW part.

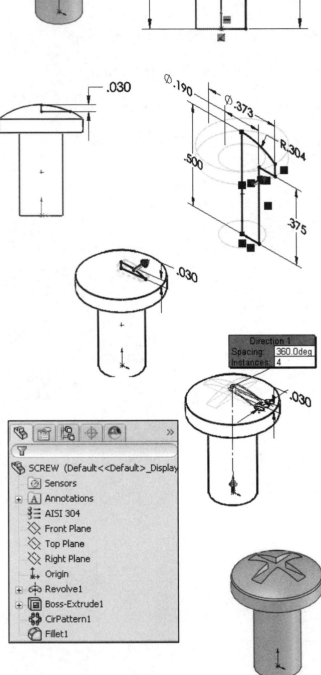

Exercise 4.5

Create the illustrated ANSI part. All edges of the model are not located on perpendicular planes. Think about the steps required to build the model. Insert two features: Extruded Boss/Base and Extruded Cut.

Note the location of the Origin. Select the Right Plane as the Sketch plane. Apply construction geometry. Insert the required geometric relations and dimensions for Sketch1. Calculate the overall mass of the part.

Given:
A = 3.00, B = 1.00
Material: 6061 Alloy
Density = .097 lb/in^3
Units: IPS
Decimal places = 2

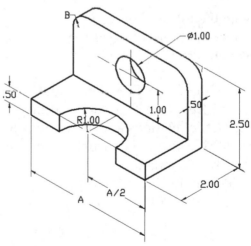

Origin

There are numerous ways to build the models in this section.

Exercise 4.6

Create the illustrated ANSI part. Think about the required steps to build this part. Insert four features: Boss-Extrude1, Cut-Extrude1, Cut-Extrude2 and Fillet. Note the location of the Origin.

Select the Right Plane as the Sketch plane. The part Origin is located in the lower left corner of the sketch. Calculate the overall mass of the part.

Given:
A = 4.00
B = R.50
Material: 6061 Alloy
Density = .0975 lb/in^3
Units: IPS
Decimal places = 2

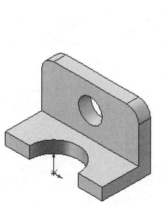

Exercise 4.7

Create the illustrated ANSI part. Think about the required steps to build this part. Insert three features, Boss-Extrude1, Boss-Extrude2, and Mirror. Three holes are displayed with a 1.00in diameter.

Note the location of the Origin.

Create Sketch1. Select the Top Plane as the Sketch plane. Apply the Tangent Arc and Line Sketch tool. Insert the required Geometric relations and dimensions. Calculate the overall mass of the part.

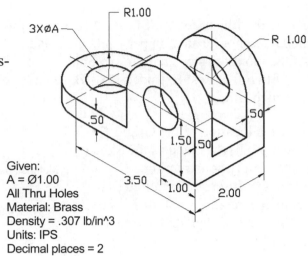

Given:
A = Ø1.00
All Thru Holes
Material: Brass
Density = .307 lb/in^3
Units: IPS
Decimal places = 2

Exercise 4.8

Create the illustrated ANSI part. Think about the required steps to build this part. Insert a Revolved Base feature and an Extruded Cut feature to build this part.

Note the location of the Origin. Select the Front Plane as the Sketch plane.

Given:
A = Ø12
Material: Cast Alloy Steel
Density = .0073 g/mm^3
Units: MMGS

Apply the Centerline Sketch tool for the Revolve1 feature. Insert the required geometric relations and dimensions for Sketch1. Sketch1 is the profile for the Revolve1 feature.
Calculate the overall mass of the part.

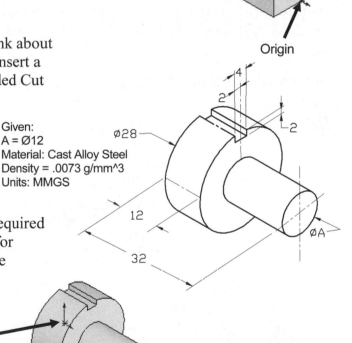

Exercise 4.9

Create the illustrated ANSI part. Think about the required steps to build this part.

Insert two features: Boss-Extrude1 and Revolve1.

Note the location of the Origin.

Select the Top Plane as the Sketch plane.

Apply construction geometry.

Apply the Tangent Arc and Line Sketch tool.

Insert the required geometric relations and dimensions. Calculate the overall mass of the part.

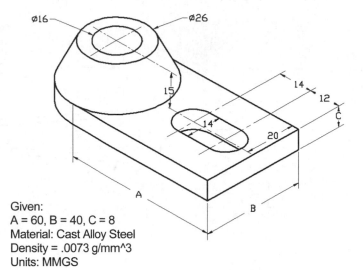

Given:
A = 60, B = 40, C = 8
Material: Cast Alloy Steel
Density = .0073 g/mm^3
Units: MMGS

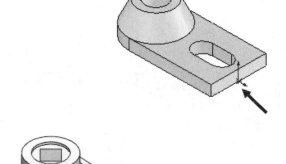

Exercise 4.10

Create the illustrated ANSI part. Think about the required steps to build this part. Apply the Extruded Boss/Base, Extruded Cut and Shell features. View the provided drawing to obtain a few dimensions (next page). As an engineer, you may not have all of the needed information. You are the designer. Fill in the gaps.

Units: MMGS.

Decimal Places: 2

Part Origin is arbitrary in this example.

Material: Plain Carbon Steel.

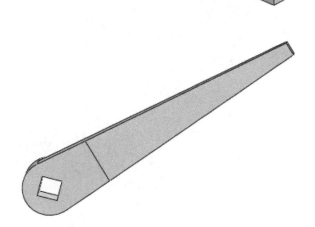

Calculate the overall mass of the part in grams.

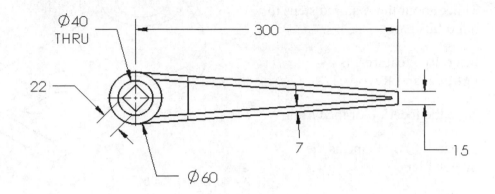

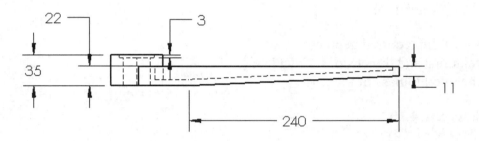

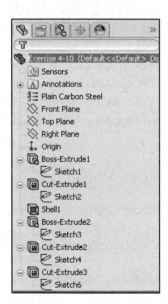

Exercise 4.11: Ice Cream Cone

Create a traditional Ice Cream Cone as illustrated. This is a common item that you see all of the time. Think about where you would start. Think about the design features that create this model. Why does the cone use ribs? Ribs are used for structural integrity.

Use a standard Cake Ice Cream Cone and measure all dimensions (approximately). Do your best.

View the sample FeatureManager. Your FeatureManager can (should) be different. This is just ONE way to create this part. You are the designer. Be creative. Use the Wrap feature and/or the Text Sketch entities tool.

Below are sample models from my ES-1310 Freshman Engineering class.

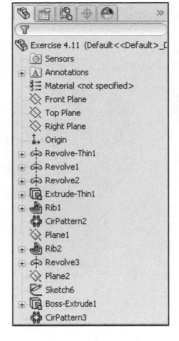

Below are sample models from my ES-1310 Freshman Engineering (Cont:).

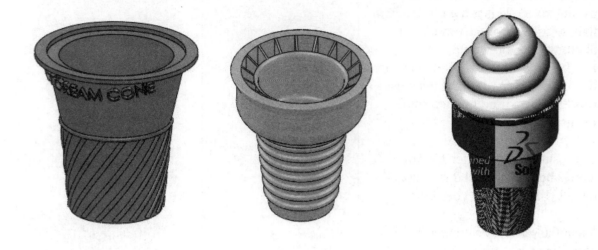

Exercise 4.12: Gem® Paper clip

Create a simple paper clip. You see this common item every day. Create an ANSI - IPS model. Apply material to the model. Think about where you would start. What is your Base Sketch?

What are the dimensions? Measure or estimate all needed dimensions from a small gem paper clip. You are the designer.

Note: The paper clip uses a circle as the profile and (lines and arcs) as the path.

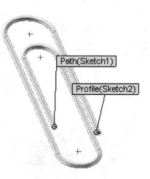

Exercise 4.13: Anvil Spring

Create a Variable Pitch Helix.
Create an Anvil Spring
with 6 coils, two active as
illustrated.

First create a Variable Pitch
Helix feature. Sketch a circle,
Coincident to the Origin on the
Top plane.

Apply Smart Dimension –
Diameter = Ø.235.

Create a Variable Pitch Helix/Spiral.

Enter the following information as illustrated in the Region
Parameters table. The spring has 6 coils. Coils 1,2,5 & 6
are the closed ends of the spring. The spring has a
diameter of .020in. The pitch needs to slightly lager than the
wire. Enter .021in for Pitch. Enter .080in for the free state of
the two active coils.

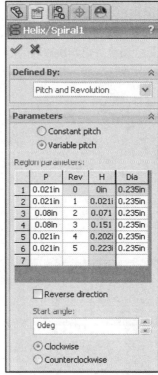

	P	Rev	H	Dia
1	0.021in	0	0in	0.235in
2	0.021in	1	0.021i	0.235in
3	0.08in	2	0.071	0.235in
4	0.08in	3	0.151	0.235in
5	0.021in	4	0.202i	0.235in
6	0.021in	5	0.223i	0.235in
7				

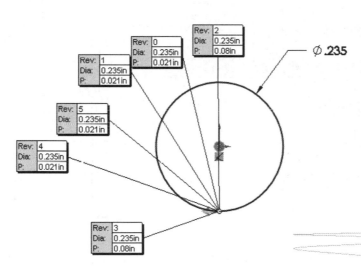

Second, create the profile for the spring.

Third, create the Sweep feature. You are the
designer; estimate any needed dimensions to create
the model.

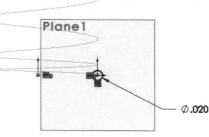

Exercise 4.14: Water Bottle

Create the container as illustrated. You see this common item every day. Create an ANSI - IPS model.

Apply material to the model. Think about where you would start.

What is your Base Sketch? What is your Base Feature?

View the sample FeatureManager. Your FeatureManager can (should) be different. This is just ONE way to create this part. You are the designer. Be creative. Estimate any needed dimensions.

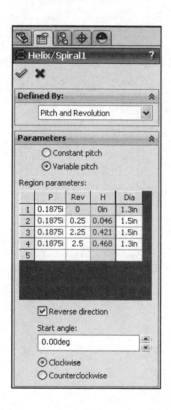

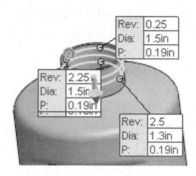

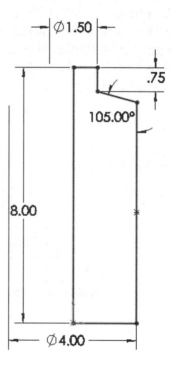

Exercise 4.15: Explicit Equation Driven Curve tool

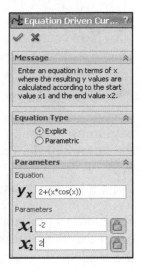

Create an Explicit Equation Driven Curve on the Front plane. Revolve the curve. Calculate the volume of the solid.

Create a New part. Use the default ANSI, IPS Part template.

Create a 2D Sketch on the Front Plane.

Active the Equation Driven Curve Sketch ♈ tool from the Consolidated drop-down menu.

Enter the Equation y_x as illustrated.

Enter the parameters x_1, x_2 that define the lower and upper bounds of the equation as illustrated. View the curve in the Graphics window.

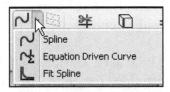

Size the curve in the Graphics window. The Sketch is under defined.

Insert three lines to close the profile as illustrated. Fully define your sketch. Enter your own dimensions.

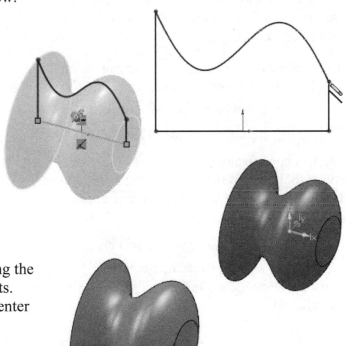

Create the Revolved feature. View the results in the Graphics window. Revolve1 is displayed. Utilize the Section tool, parallel with the Right plane to view how each cross section is a circle.

Apply Brass for material.

Calculate the volume of the part using the Mass Properties tool. View the results. Also note the surface area and the Center of mass.

You can create parametric (in addition to explicit) equation-driven curves in both 2D and 3D sketches.

Use regular mathematical notation and order of operations to write an equation. x_1 and x_2 are for the beginning and end of the curve. Use the transform options at the bottom of the PropertyManager to move the entire curve in x-, y- or rotation. To specify $x = f(y)$ instead of $y = f(x)$, use a 90 degree transform.

Print and submit a hard copy of an Isometric view and the expanded FeatureManager along with the Mass Properties of the model.

Exercise 4.16: Curve Through XYZ Points tool to create a CAM.

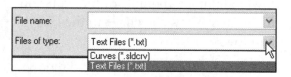

The Curve Through XYZ Points $\mathcal{U}$ feature provides the ability to either type in (using the Curve File dialog box) or click Browse and import a text file with x-, y-, z-, coordinates for points on a curve.

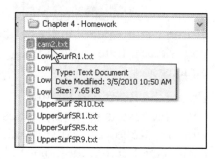

A text file can be generated by any program which creates columns of numbers. The Curve $\mathcal{U}$ feature reacts like a default spline that is fully defined.

Create a curve using the Curve Through XYZ Points tool. Import the x-, y-, z- data. Verify that the first and last points in the curve file are the same for a closed profile.

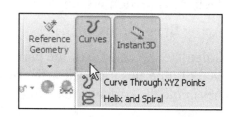

Create a new part. Click the Curve Through XYZ Points $\mathcal{U}$ tool from the Features CommandManager. The Curve File dialog box is displayed.

Import the curve data. Click Browse from the Curve File dialog box.

Browse to the downloaded folder location Chapter 4 - Homework folder location on the DVD in the book. Note: Copy all information from the DVD to your working folder. Set Files of type to Text Files.

Double-click cam2.text. View the data in the Curve File dialog box. View the sketch in the Graphics window. Review the data points in the dialog box.

Click OK from the Curve File dialog box. Curve1 is displayed in the FeatureManager. You created a curve using the Curve Through XYZ Points tool with imported x-, y-, z- data from a CAM program. This Curve can now be used to create a sketch (closed profile), in this case a CAM. Let's view the final model.

Close the model. Open Curve Through XYZ points from the Chapter 4 - Homework folder to view the final Cam model.

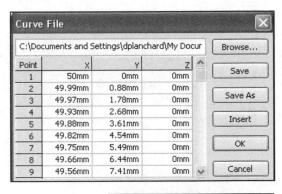

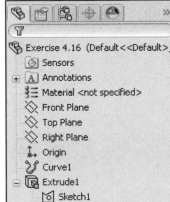

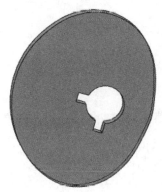

Notes:

Project 5

Swept, Lofted and Additional Features

Below are the desired outcomes and usage competencies based on the completion of Project 5.

Project Desired Outcomes:	Usage Competencies:

• Four FLASHLIGHT parts:
  o O-RING
  o SWITCH
  o LENSCAP
  o HOUSING
	• Ability to apply the following 3D features: Extruded Boss/Base, Extruded Cut, Swept Boss/Base, Shell, Lofted Boss/Base, Draft, Rib, Linear Pattern, Circular Pattern, Mirror Dome and Revolved Cut.

• Three O-RING configurations: o Small, Medium and Large	• Create and edit a Design Table for various configurations.

• Four assemblies:
  o LENSANDBULB assembly
  o CAPANDLENS assembly
  o BATTERYANDPLATE assembly
  o FLASHLIGHT assembly
	• Skill to combine multiple features to create components.

	• Knowledge of the Bottom-up assembly modeling method using Standard mates.

Notes:

Project 5 - Swept, Lofted and Additional Features

Project Objective

Create four parts: O-RING, SWITCH, LENSCAP and HOUSING of the FLASHLIGHT assembly. Create four assemblies: LENSANDBULB, CAPANDLENS, BATTERYANDPLATE and the final FLASHLIGHT assembly.

On the completion of this project, you will be able to:

- Select the best Sketch profile.

- Choose the proper Sketch plane.

- Develop three O-RING configurations using a Design Table.

- Create an Assembly Template with Document Properties.

- Apply Bottom-up assembly modeling techniques.

- Calculate assembly interference.

- Export SolidWorks files.

- Utilize the following SolidWorks features:

  o Swept Boss/Base

  o Lofted Boss/Base

  o Extruded Boss/Base

  o Revolved Boss/Base

  o Extruded Cut

  o Revolved Cut

  o Draft

  o Dome

  o Rib

  o Circular Pattern and Linear Pattern

  o Mirror

Project Overview

Project 5 introduces the Swept Boss/Base and Lofted Boss/Base features. The O-RING utilizes a Swept Base feature. The SWITCH utilizes the Lofted Base feature. The LENSCAP and HOUSING utilize the Swept Boss and Lofted Boss feature.

A Swept feature requires a minimum of two sketches: *Path* and *Profile*.

Sketch the path and profile on different planes. The profile follows the path to create the following Swept features:

- Swept Base

- Swept Boss

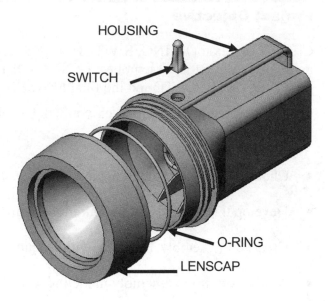

FLASHLIGHT assembly

The Lofted feature requires a minimum of two profiles sketched on different planes. The profiles are blended together to create the following Lofted features:

- Lofted Base

- Lofted Boss

Utilize existing features to create the Rib, Linear Pattern and Mirror features.

Utilize existing faces to create the Draft and Dome features.

The LENSCAP and HOUSING combines the Extruded Boss/Base, Extruded Cut, Revolved Cut Thin, Shell and Circular Pattern with the Swept and Lofted features.

Features in the SWITCH, LENSCAP and HOUSING have been simplified for educational purposes.

Use the Defeature tool, to remove details from a part or assembly and save the results to a new file in which the details are replaced by dumb solids (that is, solids without feature definition or history). You can then share the new file without revealing all the design details of the model.

Create four assemblies in this project:

1. LENSANDBULB assembly

2. CAPANDLENS assembly

3. BATTERYANDPLATE assembly

4. FLASHLIGHT assembly

Create an inch and metric Assembly template.

- ASM-IN-ANSI

- ASM-MM-ISO

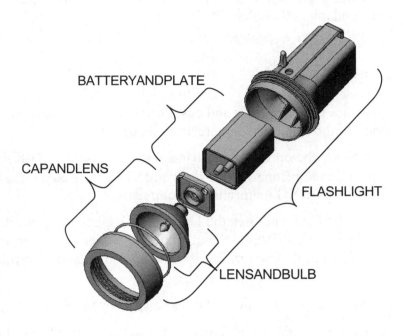

Develop an understanding of assembly modeling techniques.

Combine the LENSANDBULB assembly, CAPANDLENS assembly, BATTERYANDPLATE assembly, HOUSING part and SWITCH part to create the FLASHLIGHT assembly.

Create the following Standard Mate types: *Coincident*, *Concentric* and *Distance*.

Utilize the following tools: *Insert Component* , *Hide/Show* , *Suppress* , *Mate* , *Move Component* , *Rotate Component* , *Exploded View* and *Interference Detection* .

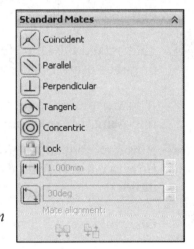

Project Situation

Communication is a major component to a successful development program. Provide frequent status reports to the customer and to your team members.

Communicate with suppliers. Ask questions. Check on details. What is the delivery time for the BATTERY, LENS and SWITCH parts?

Talk to colleagues to obtain useful manufacturing suggestions and ideas. Your team decided that plastic injection molding is the most cost effective method to produce large quantities of the desired part.

Investigate surface finishes that support the customers advertising requirement. You have two fundamental choices:

- Adhesive label

- Silk-screen

There are time, quantity and cost design constraints. For the prototype, use an adhesive label. In the final product, create a silk screen.

Investigate the options on O-Ring material. Common O-Ring materials for this application are Buna-N (Nitrile®) and Viton®. Be cognizant of compatibility issues between O-RING materials and lubricants.

The LENSCAP encloses the LENS. The HOUSING protects the BATTERY. How do you design the LENSCAP and HOUSING to ease the transition from development to manufacturing? Answer: Review the fundamental design rules behind the plastic injection manufacturing process:

- Maintain a constant wall thickness. Inconsistent wall thickness creates stress.

- Create a radius on all corners. No sharp edges. Sharp edges create vacuum issues when removing the mold. Filleting inside corners on plastic parts will strengthen the part and improve plastic flow characteristics around corners in the mold.

- Understand the Draft feature. The Draft feature is commonly referred to as: plus draft or minus draft. Plus draft adds material. Minus draft removes material.

- Allow a minimum draft angle of 1°. Draft sides and internal ribs. Draft angles assist in removing the part from the mold.

- Ribs are often added to improve strength. Ribs must also be added to improve plastic flow to entirely fill the cavity and to avoid a short shot; not enough material to fill the mold.

- Generally use a 2:3 ratio for rib thickness compared to overall part thickness.

- Shrinkage of the cooling plastic requires that the cavity is larger than the finished part. Shrinkage around the core will cause the part to bind onto the core. The Draft feature helps eject the part from the mold. Shrinkage on the cavity side will cause the part to pull from the mold. Less draft is needed on the cavity faces than on the core faces.

- Sink marks occur where there is an area with thicker material than the rest of the part wall thickness. A problem may occur where a tall rib with draft meets a face of the part.

- Mold cavities should be vented to allow trapped gases to escape through the channels.

Obtain additional information on material and manufacturing from material suppliers or mold manufacturers. Example: GE Plastics of Pittsfield, MA provides design guidelines for selecting raw materials and creating plastic parts and components.

O-RING Part-Swept Base Feature

The O-RING part is positioned between the LENSCAP and the LENS.

Create the O-RING with the Swept Base 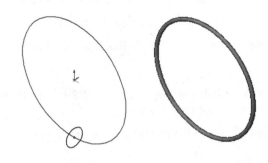 feature.

The Swept Base feature uses:

- A circular path sketched on the Front Plane.

- A small cross section profile sketched on the Right Plane.

Path & Profile   Swept feature

The Pierce relation positions the center of the cross section profile on the sketched path.

Utilize the PART-IN-ANSI Template created in Project 1 for inch units. Utilize the PART-MM-ISO Template for millimeter units. Millimeter dimensions are provided in brackets [x].

Activity: O-RING Part-Swept Base Feature

Create a New part.

1) Click **New** ⬜ from the Menu bar.

2) Click the **MY-TEMPLATES** tab.

3) Double-click **PART-IN-ANSI**, [**PART-MM-ISO**].

4) Click **Save**.

5) Select **PROJECTS** for the Save in folder.

6) Enter **O-RING** for File name.

7) Enter **O-RING FOR LENS** for Description.

8) Click **Save**. The O-RING FeatureManager is displayed.

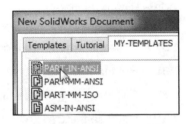

Create the Swept path.

9) Right-click **Front Plane** from the FeatureManager for the Sketch plane. This is your Sketch plane.

10) Click **Sketch** ✏ from the Context toolbar. The Sketch toolbar is displayed.

11) Click the **Circle** ⊘ Sketch tool. The Circle PropertyManager is displayed.

12)   Sketch a circle centered at the **Origin** 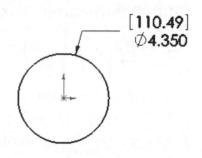.

Add a dimension.

13)   Click the **Smart Dimension** ✏ Sketch tool.

14)   Click the **circumference** of the circle.

15)   Click a **position** off the profile.

16)   Enter **4.350**in, [**110.49**] as illustrated.

[110.49]
⌀4.350

Close the sketch.

17)   **Rebuild** the model. Sketch1 is displayed in the FeatureManager.

18)   Rename **Sketch1** to **Sketch-path**.

Create the Swept profile.

19)   Click **Isometric view** 🔲 from the Heads-up View toolbar.

20)   Right-click **Right Plane** from the FeatureManager. This is your Sketch plane.

21)   Click **Sketch** ✏ from the Context toolbar. The Sketch toolbar is displayed.

22)   Click the **Circle** ⊙ Sketch tool.

23)   Create a **small circle** left of the Sketch-path on the Right Plane as illustrated.

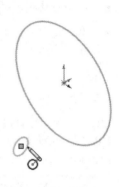

Add a Pierce relation.

24)   Right-click **Select** to deselect the Circle Sketch tool.

25)   Click the **small circle centerpoint**.

26)   Hold the **Ctrl** key down.

27)   Click the **large circle circumference**.

28)   Release the **Ctrl** key. The selected entities are displayed in the Selected Entities box.

29)   Right-click **Make Pierce** ⬚ from the Context toolbar. The centerpoint of the small circle pierces the Sketch-path, (large circle).

30)   Click **OK** ✅ from the Properties PropertyManager.

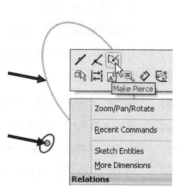

Add a dimension.

31)   Click the **Smart Dimension** ✏ Sketch tool.

32)   Click the **circumference** of the small circle.

33)   Click a **position** to the left of the profile.

34)   Enter **.125**in, [**3.18**].

35)   Click **Isometric view** 🔲 from the Heads-up View toolbar.

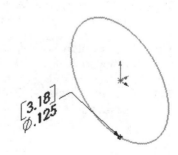

[3.18]
⌀.125

Swept, Lofted and Additional Features

Exit the sketch.

36) Right-click **Exit Sketch**. Sketch2 is displayed in the FeatureManager.

37) Rename **Sketch2** to **Sketch-profile**.

The FeatureManager displays two sketches: Sketch-path and Sketch-profile. Create the path before the profile. The profile requires a Pierce relation to the path.

Improve visibility. Small profiles are difficult to dimension on large paths. Perform the following steps to create a detailed small profile:

- Create a large cross section profile that contains the required dimensions and relationships. The black profile is fully defined.

- Pierce the profile to the path. Add dimensions to reflect the true size.

- Rename the profile and path to quickly locate sketches in the FeatureManager.

Insert the Swept feature.

38) Click the **Swept Boss/Base** feature tool. The Sweep PropertyManager is displayed.

39) **Expand** O-RING from the fly-out FeatureManager.

40) Click **Sketch-profile** from the fly-out FeatureManager. Sketch-profile is displayed in the Profile box.

41) Click inside the **Path** box.

42) Click **Sketch-path** from the fly-out FeatureManager. Sketch-path is displayed in the Path box.

43) Click **OK** from the Sweep PropertyManager. Sweep1 is displayed in the FeatureManager.

44) Rename **Sweep1** to **Base-Sweep**.

45) Click **Isometric view** from the Heads-up View toolbar.

46) Click **Save**.

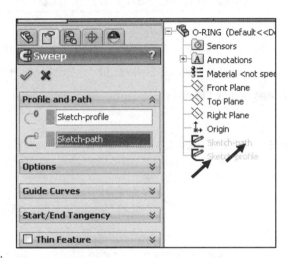

O-RING Part-Design Table

A Design Table is a spreadsheet used to create multiple configurations in a part or assembly. The Design Table controls the dimensions and parameters in the part. Utilize the Design Table to modify the overall path diameter and profile diameter of the O-RING.

PAGE 5 - 9

Create three configurations of the O-RING:

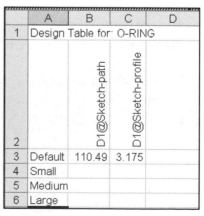

- *Small*

- *Medium*

- *Large*

Note: The O-RING contains two dimension names in the Design Tables. Parts contain hundreds of dimensions and values. Rename dimension names for clarity.

The part was initially designed in inches. You are required to manufacture the configurations in millimeters. Modify the part units to millimeters.

Activity: O-RING Part-Design Table

Modify the Primary Units.

47) Click **Options**, **Document Properties** tab from the Menu bar. Select **Units**.

48) Select **MMGS**. Select **.12** for Basic unit length decimal place.

49) Click **OK** from the Documents Properties - Units box.

Insert a Design Table.

50) Click **Insert**, **Tables**, **Design Table** from the Menu bar. The Auto-create option is selected. Accept the default settings.

51) Click **OK** from the Design Table PropertyManager.

52) Click **D1@Sketch-path**. Hold the **Ctrl** key down.

53) Click **D1@Sketch-profile**. Release the **Ctrl** key.

54) Click **OK** from the Dimensions dialog box.

Note: The dimension variable name will be different if sketches or features were deleted.

The input dimension names and default values are automatically entered into the Design Table. The value Default is entered in Cell A3.

The values for the O-RING are entered in Cells B3 through C6. The sketch-path diameter is controlled in Column B. The sketch-profile diameter is controlled in Column C.

Enter the three configuration names.

55) Click **Cell A4**. Enter **Small**.

56) Click **Cell A5**. Enter **Medium**.

57) Click **Cell A6**. Enter **Large**.

	A	B	C	D
1	Design Table for: O-RING			
2		D1@Sketch-path	D1@Sketch-profile	
3	Default	110.49	3.175	
4	Small			
5	Medium			
6	Large			

Enter the dimension values for the Small configuration.
58) Click **Cell B4**. Enter **100**. Click **Cell C4**. Enter **3**.

Enter the dimension values for the Medium configuration.
59) Click **Cell B5**. Enter **150**. Click **Cell C5**. Enter **4**.

Enter the dimension values for the Large configuration.
60) Click **Cell B6**. Enter **200**. Click **Cell C6**. Enter **10**.

Build the three configurations.
61) Click a **position** inside the Graphics window.

62) Click **OK** to generate the three configurations. The Design Table icon is displayed in the FeatureManager.

63) **Rebuild** the model.

Display the configurations.
64) Click the **ConfigurationManager** 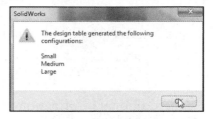 tab. View the Design Table icon.

View the three configurations.
65) Double-click **Small**.

66) Double-click **Medium**.

67) Double-click **Large**.

68) Double-click **Default**.

Return to the FeatureManager.
69) Click the **FeatureManager** tab.

70) Click **Save** .

	A	B	C
1	Design Table for: O-RING		
2		D1@Sketch-path	D1@Sketch-profile
3	Default	110.49	3.175
4	Small	100	3
5	Medium	150	4
6	Large	200	10

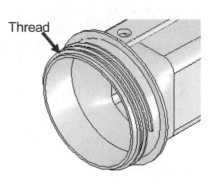

The O-RING is classified as a simple Swept feature. A complex Swept feature utilizes 3D curves and Guide Curves. Investigate additional Swept features later in the project.

An example of a complex Swept is a Thread. A Thread requires a Helical/Spiral curve for the path and a circular profile.

Another example of a complex Swept is the violin body. The violin body requires Guide Curves to control the Swept geometry. Without Guide Curves the profile follows the straight path to produce a rectangular shape.

Thread

With Guide Curves, the profile follows the path and the Guide Curve geometry to produce the violin body.

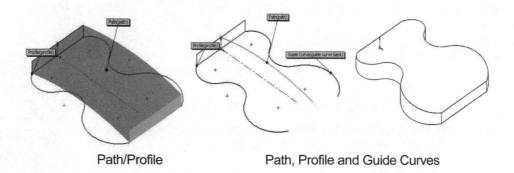

Path/Profile	Path, Profile and Guide Curves

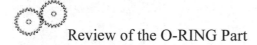 Additional information on Swept and Pierce are found in SolidWorks Help Topics. Keywords: Swept (overview, simple sweep) and Pierce (relations). Refer to Help, SolidWorks Tutorials, Revolves and Sweeps for additional information.

Review of the O-RING Part

The O-RING utilized the Swept feature. The Swept feature required a Sketched path and a Sketched profile. The path was a large circle sketched on the Front Plane. The profile was a small circle sketched on the Right Plane. The Pierce relation was utilized to attach the Profile to the Path.

The Swept feature required a minimum of two sketches. You created a simple Swept feature. Swept features can be simple or complex. Small, Medium and Large configurations of the O-RING were developed with a Design Table. The Design Table icon displayed in the ConfigurationManager is an EXCEL based spread sheet utilized to create variations of dimensions and features.

Recognize the properties and understand the order of the sketches to create successful simple Swept features. The following are the steps to create successful Sweeps:

- Create the path as a separate sketch. The path is open or closed. The path is a set curves contained in one sketch. The path can also be one curve or a set of model edges.

- Create each profile as a separate sketch. The profile is closed for a Swept Boss/Base.

- Fully define each sketch. The sketch is displayed in black.

- Sketch the profile last before inserting the Swept feature.

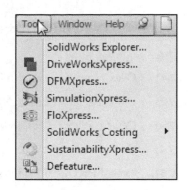

- Position the start point of the path on the plane of the profile.

- Path, profile and solid geometry cannot intersect themselves.

☼ The Sketch toolbar contains two areas: Sketch Entities and Sketch Tools. There are numerous ways to access these areas: Select the Sketch ✏ icon from the Sketch toolbar, right-click and select the Sketch icon from the Context toolbar or click Tools, Sketch Entities or Sketch Tools from the Menu bar.

SWITCH Part-Lofted Base Feature

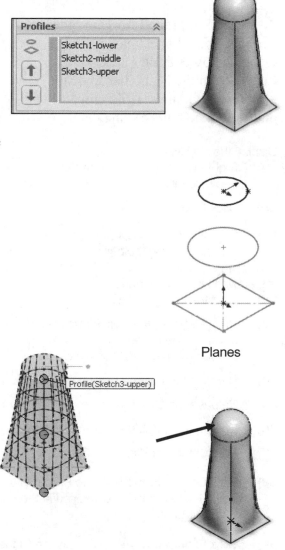

The SWITCH is a purchased part. The SWITCH is a complex assembly. Create the outside casing of the SWITCH as a simplified part. Create the SWITCH with the Lofted Base ⬧ feature.

The orientation of the SWITCH is based on the position in the assembly. The SWITCH is comprised of three cross section profiles. Sketch each profile on a different plane.

The first plane is the Top Plane. Create two reference planes parallel to the Top Plane.

Sketch one profile on each plane. The design intent of the sketch is to reduce the number of dimensions.

Utilize symmetry, construction geometry and Geometric relations to control three sketches with one dimension.

Insert a Lofted feature. Select the profiles to create the Loft feature.

Insert the Dome feature to the top face of the Loft and modify the Loft Base feature.

☼ Specify an elliptical dome for cylindrical or conical models or a continuous dome for polygonal models.

Planes

Activity: Switch Part-Loft Base Feature

Create a New part.

71) Click **New** ⬜ from the Menu bar.

72) Click the **MY-TEMPLATES** tab.

73) Double-click **PART-IN-ANSI**, [**PART-MM-ISO**].

74) Click **Save**.

75) Select **PROJECTS** for the Save in folder.

76) Enter **SWITCH** for File name.

77) Enter **BUTTON STYLE** for Description.

78) Click **Save**. The SWITCH FeatureManager is displayed.

💡 If you upgraded from an older version; your Planes may be displayed by default. Click View, Planes from the Menu bar to display all reference planes. Hide unwanted planes in the FeatureManager.

Display the Top Plane.

79) Right-click **Top Plane** from the FeatureManager.

80) Click **Show**.

81) Click **Isometric view** ⬦ from the Heads-up View toolbar.

Insert two Reference planes.

82) Hold the **Ctrl** key down.

83) Click and drag the **Top Plane** upward. The Plane PropertyManager is displayed.

84) Release the **mouse button**.

85) Release the **Ctrl** key.

86) Enter **.500**in, [**12.7**] for Offset Distance.

87) Enter **2** for # of Planes to Create.

88) Click **OK** ✔ from the Plane PropertyManager. Plane1 and Plane2 is displayed in the FeatureManager.

89) Click **Front view** ⬚ to display the Plane1 and Plane2 offset from the Top plane.

Hold the Ctrl key down and drag the Top Plane upward.
Pick an edge, not the handles.

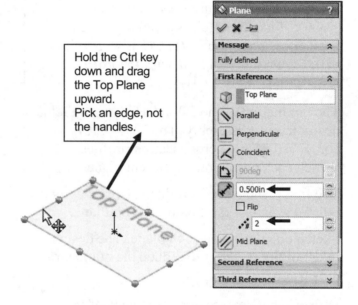

90) Click **Isometric view** from the Heads-up View toolbar.

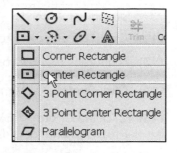

Insert Sketch1. Sketch1 is a square on the Top Plane centered about the Origin.

91) Right-click **Top Plane** from the FeatureManager.

92) Click **Sketch** from the Context toolbar.

93) Click the **Center Rectangle** tool from the Consolidated Sketch toolbar.

94) Click the **Origin**.

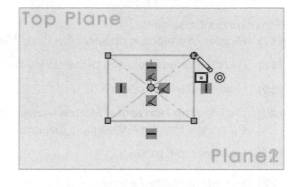

95) Click a **point** in the upper right hand of the window as illustrated. The Center Rectangle Sketch tool automatically inserts a midpoint relation about the Origin and an equal relation between the two horizontal and two vertical lines.

96) Click **OK** from the Rectangle PropertyManager.

Add an Equal relation between the four edges.

97) Click the **left vertical line**.

98) Hold the **Ctrl** key down.

99) Click the **top horizontal line**.

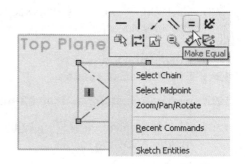

100) Click the **right vertical line**.

101) Click the **bottom horizontal line**.

102) Release the **Ctrl** key.

103) Right-click **Make Equal** = from the Context toolbar.

Add a dimension.

104) Click the **Smart Dimension** Sketch tool.

105) Click the **top horizontal line**.

106) Click a **position** above the profile.

107) Enter **.500**in, [**12.7**].

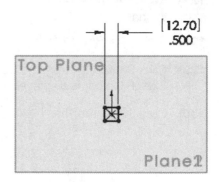

Close, fit and rename the sketch.

108) **Rebuild** the model.

109) **Fit** the sketch to the Graphics area.

110) Rename **Sketch1** to **Sketch1-lower**.

Save the model.

111) Click **Save**.

Insert Sketch2 on Plane1.

112) Click **Top view** ⬚ from the Heads-up View toolbar.

113) Right-click **Plane1** from the FeatureManager

114) Click **Sketch** ✏ from the Context toolbar.

115) Click the **Circle** ⊘ Sketch tool. The Circle PropertyManager is displayed.

116) Sketch a **Circle** centered at the Origin as illustrated.

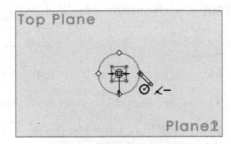

Add a Tangent relation.

117) Right-click **Select** to deselect the Circle Sketch tool.

118) Click the **circumference** of the circle.

119) Hold the **Ctrl** key down.

120) Click the **top horizontal Sketch1-lower line**. The Properties PropertyManager is displayed.

121) Release the **Ctrl** key.

122) Right-click **Make Tangent** ◠ from the Pop-up Context toolbar.

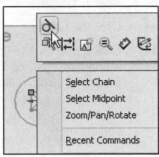

123) Click **OK** ✔ from the Properties PropertyManager.

Close and rename the sketch.

124) Right-click **Exit Sketch**.

125) Rename **Sketch2** to **Sketch2-middle**.

126) Click **Isometric view** ⬚ from the Heads-up View toolbar. View the results.

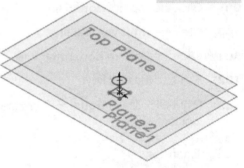

Insert Sketch3 on Plane2.

127) Click **Top view** ⬚ from the Heads-up View toolbar.

128) Right-click **Plane2** from the FeatureManager.

129) Click **Sketch** ✏ from the Context toolbar. The Sketch toolbar is displayed.

130) Click the **Centerline** ⋮ Sketch tool.

131) Sketch a **centerline** coincident with the Origin and the upper right corner point as illustrated.

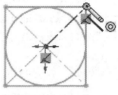

132) Click the **Point** ✳ Sketch tool.

133) Click the **midpoint** of the right diagonal centerline. The Point PropertyManager is displayed.

134) Click **Circle** ⊘ from the Sketch toolbar.

135) Sketch a **Circle** centered at the Origin to the midpoint of the diagonal centerline.

Close and rename the sketch.

136) Right-click **Exit Sketch**.

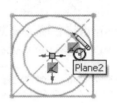

137) Rename **Sketch3** to **Sketch3**-upper.

Insert the Lofted Base feature.

138) Click the **Lofted Boss/Base** ⟁ feature tool. The Loft PropertyManager is displayed.

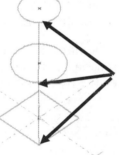

139) Click **Isometric view** ⬙ from the Heads-up View toolbar.

140) Right-click in the **Profiles box**.

Click the front of Sketch1-lower, Sketch2-middle and Sketch3-upper.

141) Click **Clear Selections**.

142) Click the **front corner** of Sketch1-lower as illustrated.

143) Click **Sketch2-middle** as illustrated.

144) Click **Sketch3-upper** as illustrated. The selected sketch entities are displayed in the Profiles box.

145) Click **OK** ✔ from the Loft PropertyManager. Loft1 is displayed in the FeatureManager.

Rename the feature.

146) Rename **Loft1** to **Base Loft**.

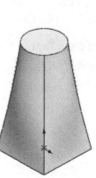

Hide the planes.

147) Click **View**; uncheck **Planes** from the Menu bar.

Save the part.

148) Click **Save**.

The system displays a preview curve and loft as you select the profiles. Use the Up button and Down button in the Loft PropertyManager to rearrange the order of the profiles.

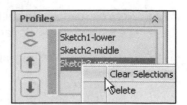

☀ Redefine incorrect selections efficiently. Right-click in the Graphics window, click Clear Selections to remove selected profiles. Select the correct profiles.

SWITCH Part-Dome Feature

Insert the Dome feature on the top face of the Lofted Base feature. The Dome feature forms the top surface of the SWITCH. Note: You can specify an elliptical dome for cylindrical or conical models and a continuous dome for polygonal models. A continuous dome's shape slopes upwards, evenly on all sides.

Activity: SWITCH Part-Dome Feature

Insert the Dome feature.

149) Click the **top face** of the Base Loft feature in the Graphics window.

150) Click the **Dome** ⬒ feature tool. The Dome PropertyManager is displayed. Face<1> is displayed in the Faces to Dome box.

Enter distance.

151) Enter .20in, [5.08] for Distance. View the dome feature in the Graphics window.

152) Click **OK** ✓ from the Dome PropertyManager.

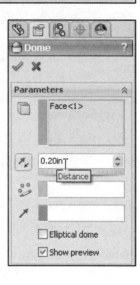

Experiment with the Dome feature to display different results. Click Insert, Features, Freeform ⬒ to view the Freeform PropertyManager. As an exercise, replace the Dome feature with a Freeform feature.

In the next section, modify the offset distance between the Top Plane and Plane1.

Modify the Loft Base feature.

153) **Expand** the Base Loft feature.

154) Right-click on **Annotations** in the FeatureManager.

155) Click **Show Feature Dimensions**.

156) Click on the Plane1 offset dimension, **.500**in, [**12.700**].

157) Enter **.125**in, [**3.180**]. Click **Rebuild**.

158) Click **OK** ✅ from the Dimension PropertyManager.

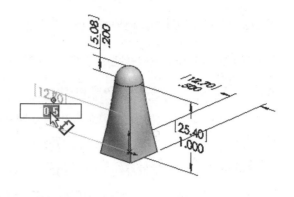

Hide Feature dimensions.

159) Right-click on **Annotations** in the FeatureManager.

160) Un-check **Show Feature Dimensions**.

Display Feature Statistics.

161) Click **Statistics** 🔧 from the Evaluate tab in the CommandManager. View the results.

View the SWITCH feature statistics.

162) Click **Close** from the Feature Statistics dialog box.

163) Click **Save**.

Feature Statistics

Print... | Copy | Refresh | Close

SWITCH
Features 7, Solids 1, Surfaces 0
Total rebuild time in seconds: 0.41

Feature Order	Time %	Time(s)
Sketch1-lower	42.36	0.17
Dome1	34.48	0.14
Base Loft	15.52	0.06
Sketch3-upper	7.64	0.03
Plane1	0.00	0.00
Plane2	0.00	0.00
Sketch2-middle	0.00	0.00

The Dome feature created the top for the SWITCH. The Feature Statistics report displays the rebuild time for the Dome feature and the other SWITCH features. As feature geometry becomes more complex, the rebuild time increases.

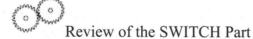 Review of the SWITCH Part

The SWITCH utilized the Lofted Base and Dome feature. The Lofted Base feature required three planes: Sketch1-lower, Sketch2-middle and Sketch3-upper. A profile was sketched on each plane. The three profiles were combined to create the Lofted Base feature.

The Dome feature created the final feature for the SWITCH.

The SWITCH utilized a simple Lofted Base feature. Lofts become more complex with additional Guide Curves. Complex Lofts can contain hundreds of profiles.

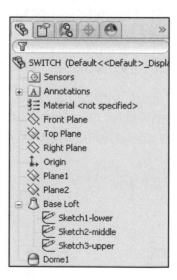

SWITCH (Default<<Default>_Displ
- Sensors
- Annotations
- Material <not specified>
- Front Plane
- Top Plane
- Right Plane
- Origin
- Plane1
- Plane2
- Base Loft
  - Sketch1-lower
  - Sketch2-middle
  - Sketch3-upper
- Dome1

Four Major Categories of Solid Features

The LENSCAP and HOUSING combine the four major categories of solid features:

- Extrude: Requires one profile

- Revolve: Requires one profile and axis of revolution

- Swept: Requires one profile and one path sketched on different planes

- Lofted: Requires two or more profiles sketched on different planes

Identify the simple features of the LENSCAP and HOUSING. Extrude and Revolve are simple features. Only a single sketch profile is required. Swept and Loft are more complex features. Two or more sketches are required.

Example: The O-RING was created as a Swept.

Could the O-RING utilize an Extruded feature?

Answer: No. Extruding a circular profile produces a cylinder.

Can the O-RING utilize a Revolved feature? Answer: Yes. Revolving a circular profile about a centerline creates the O-RING.

Revolved feature   Sweep feature

A Swept feature is required if the O-RING contained a non-circular path. Example: A Revolved feature does not work with an elliptical path or a more complex curve as in a paper clip. Combine the four major features and additional features to create the LENSCAP and HOUSING.

LENSCAP Part

The LENSCAP is a plastic part used to position the LENS to the HOUSING. The LENSCAP utilizes an Extruded Boss/Base, Extruded Cut, Extruded Thin Cut, Shell, Revolved Cut and Swept features.

The design intent for the LENSCAP requires that the Draft Angle be incorporated into the Extruded Boss/Base and Revolved Cut feature. Create the Revolved Cut feature by referencing the Extrude Base feature geometry. If the Draft angle changes, the Revolved Cut also changes.

Insert an Extruded Boss/Base feature with a circular profile on the Front Plane. Use a Draft option in the Boss-Extrude PropertyManager. Enter 5deg for Draft angle.

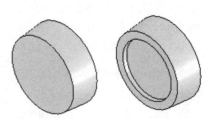

Insert an Extruded Cut  feature. The
Extruded Cut feature should be equal to the
diameter of the LENS Revolved Base feature.

Insert a Shell  feature. Use the Shell feature
for a constant wall thickness.

Insert a Revolved Cut  feature on the back
face. Sketch a single line on the Silhouette edge
of the Extruded Base. Utilize the Thin Feature option in the Cut-
Revolve PropertyManager.

Utilize a Swept 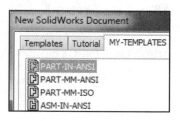 feature for the thread. Insert a new reference
plane for the start of the thread. Insert a Helical Curve for the path.
Sketch a trapezoid for the profile.

LENSCAP Part-Extruded Boss/Base, Extruded Cut and Shell Features

Create the LENSCAP. Review the Extruded Boss/Base, Extruded Cut and Shell features
introduced in Project 4. The first feature is a Boss-Extrude feature. Select the Front Plane
for the Sketch plane. Sketch a circle centered at the Origin for the profile. Utilize a Draft
angle of 5deg.

Create an Extruded Cut feature on the front face of the Base feature. The diameter of the
Extruded Cut equals the diameter of the Revolved Base feature of the LENS. The Shell
feature removes the front and back face from the solid LENSCAP.

Activity: LENSCAP Part-Extruded Base, Extruded Cut and Shell Features

Create a New part.

164) Click **New** from the Menu bar.

165) Click the **MY-TEMPLATES** tab.

166) Double-click **PART-IN-ANSI, [PART-MM-ISO]**.

167) Click **Save**.

168) Select **PROJECTS** for the Save in folder.

169) Enter **LENSCAP** for File name.

170) Enter **LENSCAP for 6V FLASHLIGHT** for Description.

171) Click **Save**. The LENSCAP FeatureManager is displayed.

Create the sketch for the Extruded Base feature.

172) Right-click **Front Plane** from the FeatureManager. Click **Sketch** ⬉ from the Context toolbar.

173) Click the **Circle** ⊙ Sketch tool. The Circle PropertyManager is displayed.

174) Sketch a **circle** centered at the Origin ⊥.

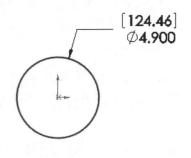

Add a dimension.

175) Click the **Smart Dimension** ✦ Sketch tool.

176) Click the **circumference** of the circle.

177) Click a **position** off the profile. Enter **4.900**in, **[124.46]**.

Insert an Extruded Boss/Base feature.

178) Click the **Extruded Boss/Base** ▨. feature tool. The Boss-Extrude PropertyManager is displayed. Blind is the default End Condition in Direction 1.

179) Click the **Reverse Direction** box.

180) Enter **1.725**in, **[43.82]** for Depth in Direction 1.

181) Click the **Draft On/Off** button.

182) Enter **5**deg for Angle. Click the **Draft outward** box.

183) Click **OK** ✓ from the Boss-Extrude PropertyManager. Boss-Extrude1 is displayed in the FeatureManager.

184) Rename **Boss-Extrude1** to **Base Extrude**.

185) Click **Save**.

Create the sketch for the Extruded Cut feature.

186) Right-click the **front face** for the Sketch plane.

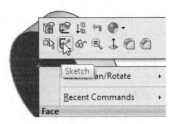

187) Click **Sketch** ⬉ from the Context toolbar. The Sketch toolbar is displayed.

188) Click the **Circle** ⊙ Sketch tool. The Circle PropertyManager is displayed. Sketch a **circle** centered at the Origin ⊥.

Add a dimension.

189) Click the **Smart Dimension** ✦ Sketch tool.

190) Click the **circumference** of the circle.

191) Click a **position** off the profile.

192) Enter **3.875**in, **[98.43]**.

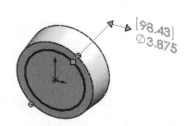

Insert an Extruded Cut feature.

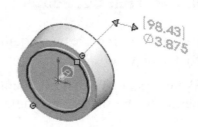

193) Click the **Extruded Cut** feature tool. The Cut-Extrude PropertyManager is displayed. Blind is the default End Condition.

194) Enter **.275**in, [6.99] for Depth in Direction 1.

195) Click the **Draft On/Off** button.

196) Enter **5**deg for Angle. Accept the default settings.

197) Click **OK** from the Cut-Extrude PropertyManager. Cut-Extrude1 is displayed in the FeatureManager.

198) Rename **Cut-Extrude1** to **Front-Cut**.

Insert the Shell feature.

199) Click the **Shell** feature tool. The Shell1 PropertyManager is displayed.

200) Click the **front face** of the Front-Cut as illustrated.

201) Press the **left arrow** approximately 8 times to view the back face.

202) Click the **back face** of the Base Extrude.

203) Enter **.150**in, [3.81] for Thickness.

204) Click **OK** from the Shell1 PropertyManager. Shell1 is displayed in the FeatureManager.

205) Click **Isometric view** from the Heads-up View toolbar.

Display the inside of the Shell.

206) Click **Right view** from the Heads-up View toolbar.

207) Click **Hidden Lines Visible** from the Heads-up View toolbar.

208) Click **Save**.

Use the inside gap created by the Shell feature to seat the O-RING in the assembly.

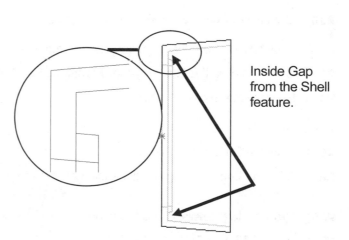

Inside Gap from the Shell feature.

LENSCAP Part-Revolved Thin Cut Feature

The Revolved Thin Cut 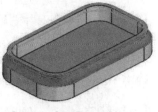 feature removes material by rotating a sketched profile around a centerline.

The Right Plane is the Sketch plane. The design intent requires that the Revolved Cut maintains the same Draft angle as the Extruded Base feature.

Utilize the Convert Entities Sketch tool to create the profile. Small thin cuts are utilized in plastic parts. Utilize the Revolved Thin Cut feature for cylindrical geometry in the next activity.

Utilize a Swept Cut for non-cylindrical geometry. The semi-circular Swept Cut profile is explored in the project exercises.

Sweep Cut Example

Activity: LENSCAP Part-Revolved Thin Cut Feature

Create the sketch.

209) Right-click **Right Plane** from the FeatureManager. This is your Sketch plane.

210) Click **Sketch** ✏ from the Context toolbar. The Sketch toolbar is displayed.

Sketch a centerline.

211) Click the **Centerline** ┆ Sketch tool. The Insert Line PropertyManager is displayed.

212) Sketch a **horizontal centerline** through the Origin as illustrated.

Create the profile.

213) Right-click **Select** to deselect the Centerline Sketch tool.

214) Click the **top silhouette outside edge** of Base Extrude as illustrated.

215) Click the **Convert Entities** 🗗 Sketch tool.

216) Click **OK** ✓ from the Convert Entities PropertyManager.

217) Click and drag the **left endpoint 2/3** towards the right endpoint.

218) Release the **mouse button**.

Add a dimension.

219) Click the **Smart Dimension** ✐ Sketch tool.

220) Click the **line**. The aligned dimension arrows are parallel to the profile line.

221) Drag the **text upward** and to the right.

222) Enter **.250**in, [**6.35**].

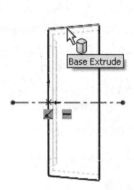

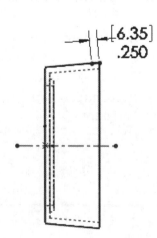

[6.35]
.250

Insert a Revolved Cut feature.

223) Click **Revolved Cut** 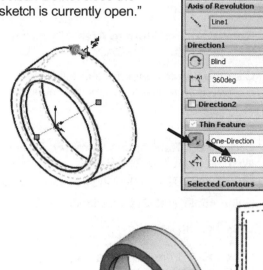 from the Features toolbar. Do not close the Sketch. The warning message states; "The sketch is currently open."

224) Click **No**. The Cut-Revolve PropertyManager is displayed.

225) Click the **Reverse Direction** box in the Thin Feature box. The arrow points counterclockwise.

226) Enter **.050**in, **[1.27]** for Direction 1 Thickness.

227) Click **OK** ✔ from the Cut-Revolve PropertyManager. Cut-Revolve-Thin1 is

displayed in the FeatureManager.

Display the Revolved Thin Cut feature.
228) **Rotate** the part to view the back face.

229) Click **Isometric view** ⬦.

230) Click **Shaded With Edges** ⬦.

231) Rename **Cut-Revolve-Thin1** to **BackCut**.

232) Click **Save**.

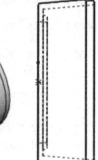

LENSCAP Part-Thread, Swept Feature and Helix/Spiral Curve

Utilize the Swept feature 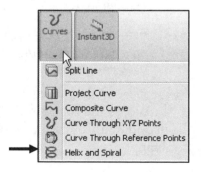 to create the required threads. The thread requires a spiral path. This path is called the ThreadPath. The thread requires a Sketched profile. This cross section profile is called the ThreadSection.

The plastic thread on the LENSCAP requires a smooth lead in. The thread is not flush with the back face. Use an Offset plane to start the thread. There are numerous steps required to create a thread:

- Create a new plane for the start of the thread.

- Create the Thread path. Utilize Convert Entities and Insert, Curve, Helix/Spiral.

- Create a large thread cross section profile for improved visibility.

- Insert the Swept feature.

- Reduce the size of the thread cross section.

💡 Obtain the Helix and Spiral tool from the Features tab, Curves Consolidated toolbar.

Activity: LENSCAP Part-Thread, Swept Feature, and Helix/Spiral Curve

Create the offset plane.

233) **Rotate** 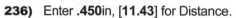 and **Zoom to Area** on the back face of the LENSCAP.

234) Click the **narrow back face** of the Base Extrude feature. Note the mouse feedback icon.

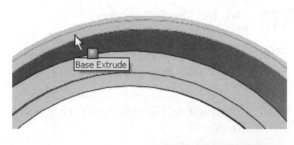

235) Click **Insert, Reference Geometry, Plane** from the Menu bar. The Plane PropertyManager is displayed.

236) Enter .450in, [**11.43**] for Distance.

237) Click the **Flip** box.

238) Click **OK** from the Plane PropertyManager. Plane1 is displayed in the FeatureManager.

239) Rename **Plane1** to **ThreadPlane**.

Display the Isometric view with Hidden Lines Removed.

240) Click **Isometric view**.

241) Click **Hidden Lines Removed**.

242) Click **Save**.

Utilize the Convert Entities Sketch tool to extract the back circular edge of the LENSCAP to the ThreadPlane.

Create the Thread path.

243) Right-click **ThreadPlane** from the FeatureManager.

244) Click **Sketch** from the Context toolbar.

245) Click the **back inside circular edge** of the Shell as illustrated.

246) Click the **Convert Entities** Sketch tool.

247) Click **OK** from the Convert Entities PropertyManager.

248) Click **Top view**. The circular edge is displayed on the ThreadPlane.

249) Click **Hidden Lines Visible** from the Heads-up View toolbar. View the results.

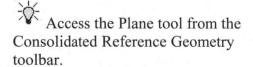

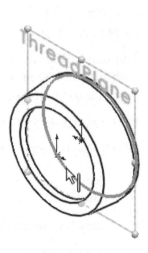

Access the Plane tool from the Consolidated Reference Geometry toolbar.

Insert the Helix/Spiral curve path.

250) Click **Insert, Curve, Helix/Spiral** from the Menu bar. The Helix/Spiral PropertyManager is displayed.

251) Enter **.250**in, [**6.35**] for Pitch.

252) Check the **Reverse direction** box.

253) Enter **2.5** for Revolutions.

254) Enter **0**deg for Starting angle. The Helix start point and end point are Coincident with the Top Plane.

255) Click the **Clockwise** box.

256) Click the **Taper Helix** box.

257) Enter **5**deg for Angle.

258) Uncheck the **Taper outward** box.

259) Click **OK** from the Helix/Spiral PropertyManager.

260) Rename **Helix/Spiral1** to **ThreadPath**.

261) Click **Save**.

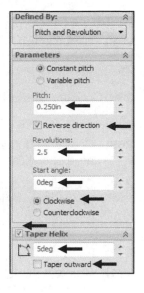

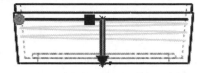

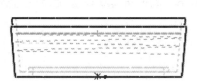

The Helix tapers with the inside wall of the LENSCAP. Position the Helix within the wall thickness to prevent errors in the Swept.

Sketch the profile on the Top Plane. Position the profile to the Top right of the LENSCAP in order to pierce to the ThreadPath in the correct location.

If required, hide the ThreadPlane.

262) Right-click **ThreadPlane** from the FeatureManager.

263) Click **Hide** from the Context toolbar.

264) Click **Hidden Lines Removed** from the Heads-up View toolbar.

Select the Plane for the Thread.

265) Right-click **Top Plane** from the FeatureManager.

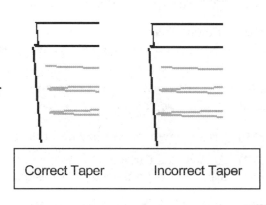

Correct Taper   Incorrect Taper

Sketch to the Top right →

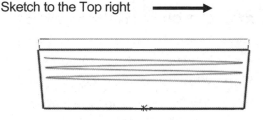

Sketch the profile.

266) Click **Sketch** from the Context toolbar.

267) Click **Top view** from the Heads-up View toolbar.

268) Click the **Centerline** Sketch tool.

269) Create a short **vertical centerline** off to the upper top area of the ThreadPath feature.

270) Create a second **centerline** horizontal from the Midpoint to the left of the vertical line.

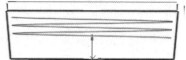

271) Create a third centerline coincident with the left horizontal endpoint. Drag the **centerline upward** until it is approximately the same size as the right vertical line.

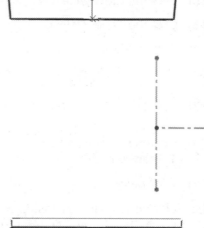

272) Create a fourth **centerline** coincident with the left horizontal endpoint. Drag the **centerline** downward until it is approximately the same size as the left vertical line as illustrated.

Add an Equal relation.

273) Right-click **Select** to de-select the Centerline Sketch tool.

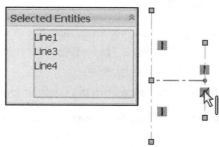

274) Click the **right vertical centerline**.

275) Hold the **Ctrl** key down.

276) Click the **two left vertical centerlines**.

277) Release the **Ctrl** key. The selected sketch entities are displayed in the Selected Entities box.

278) Click **Equal** =

279) Click **OK** from the Properties PropertyManager.

Utilize centerlines and construction geometry with geometric relations to maintain relationships with minimal dimensions.

Check **View, Sketch Relations** from the Menu bar to show/hide sketch relation symbols.

Add a dimension.

280) Click the **Smart Dimension** ✎ Sketch tool.

281) Click the two **left vertical endpoints**. Click a **position** to the left.

282) Enter **.500**in, **[12.7]**.

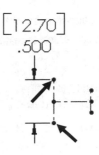

Sketch the profile. The profile is a trapezoid.

283) Click the **Line** ＼ Sketch tool.

284) Click the **endpoints** of the vertical centerlines to create the trapezoid as illustrated.

285) Right-click **Select** to de-select the Line Sketch tool.

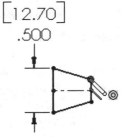

Add an Equal relation.

286) Click the **left vertical line**.

287) Hold the **Ctrl** key down.

288) Click the **top** and **bottom lines** of the trapezoid.

289) Release the **Ctrl** key.

290) Click **Equal** ＝ .

291) Click **OK** ✔ from the Properties PropertyManager.

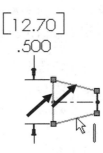

Click and drag the sketch to a position above the top right corner of the LENSCAP.

Add a Pierce relation.

292) Click the **left midpoint** of the trapezoid.

293) Hold the **Ctrl** key down.

294) Click the **starting left back edge** of the ThreadPath.

295) Release the **Ctrl** key.

296) Click **Pierce** 🖰 from the Add Relations box. The sketch is fully defined.

297) Click **OK** ✔ from the Properties PropertyManager.

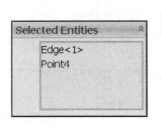

Selected Entities
Edge<1>
Point4

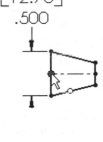

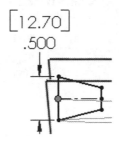

Select edge on the left side

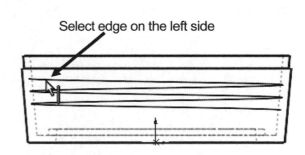

Display the sketch in an Isometric view.

298) Click **Isometric view** from the Heads-up View toolbar.

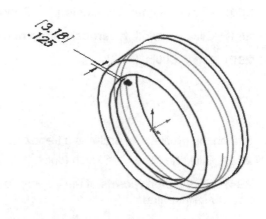

Modify the dimension.

299) Double-click the **.500** dimension text.

300) Enter **.125**in, [**3.18**].

Close the sketch.

301) **Rebuild** the model.

302) Rename **Sketch5** to **ThreadSection**.

303) Click **Save**.

Insert the Swept feature.

304) Click the **Swept Boss/Base** feature tool. The Sweep PropertyManager is displayed. If required, click **ThreadSection** for the Profile from the flyout FeatureManager.

305) Click inside the **Path** box.

306) Click **ThreadPath** from the flyout FeatureManager. ThreadPath is displayed in the Path box.

307) Click **OK** from the Sweep PropertyManager. Sweep1 is displayed in the FeatureManager.

308) Rename **Sweep1** to **Thread**.

309) Click **Shaded With Edges** from the Heads-up View toolbar.

310) Click **Save**.

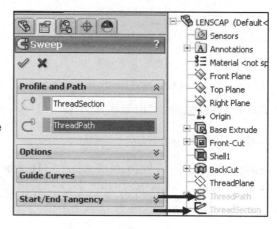

Swept geometry cannot intersect itself. If the ThreadSection geometry intersects itself, the cross section is too large. Reduce the cross section size and recreate the Swept feature.

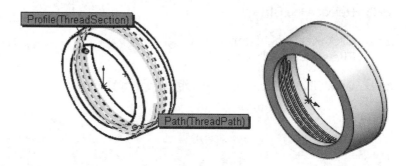

The Thread feature is composed of the following: ThreadSection and ThreadPath.

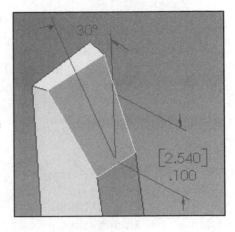

The ThreadPath contains the circular sketch and the helical curve.

Most threads require a beveled edge or smooth edge for the thread part start point. A 30° Chamfer feature can be utilized on the starting edge of the trapezoid face. This action is left as an exercise.

Create continuous Swept features in a single step. Pierce the cross section profile at the start of the swept path for a continuous Swept feature.

Un-suppress the Pattern feature to resolve both the Pattern feature and the seed feature at the same time.

The LENSCAP is complete. Review the LENSCAP before moving onto the last part of the FLASHLIGHT.

Additional information on Extruded Base/Boss, Extruded Cut, Swept, Helix/Spiral, Circular Pattern and Reference Planes are found in SolidWorks Help Topics.

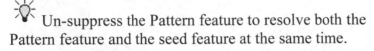

Review of the LENSCAP Part

The LENSCAP utilized the Extruded Base feature with the Draft Angle option. The Extruded Cut feature created an opening for the LENS. You utilized the Shell feature with constant wall thickness to remove the front and back faces.

The Revolved Thin Cut feature created the back cut with a single line. The line utilized the Convert Entities tool to maintain the same draft angle as the Extruded Base feature.

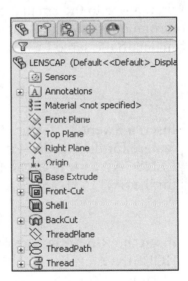

You utilized a Swept feature with a Helical Curve and Thread profile to create the thread.

HOUSING Part

The HOUSING is a plastic part utilized to contain the BATTERY and to support the LENS. The HOUSING utilizes an Extruded Boss/Base, Lofted Boss, Extruded Cut, Draft, Swept, Rib, Mirror and Linear Pattern features.

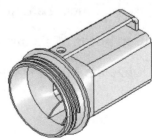

Insert an Extruded Boss/Base (Boss-Extrude1 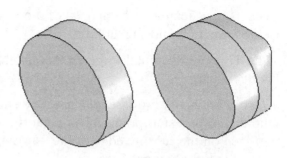 feature centered at the Origin.

Insert a Lofted Boss feature. The first profile is the converted circular edge of the Extruded Base. The second profile is a sketch on the BatteryLoftPlane.

Insert the second Extruded Boss/Base (Boss-Extrude2) feature. The sketch is a converted edge from the Loft Boss. The depth is determined from the height of the BATTERY.

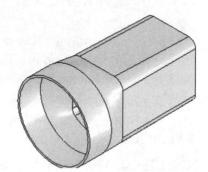

Insert a Shell feature to create a thin walled part.

Insert the third Extruded Boss (Boss-Extrude3) feature. Create a solid circular ring on the back circular face of the Boss-Extrude1 feature. Insert the Draft feature to add a draft angle to the circular face of the HOUSING. The design intent for the Boss-Extrude1 feature requires you to maintain the same LENSCAP draft angle.

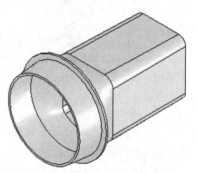

Insert a Swept feature for the Thread. Insert a Swept feature for the Handle. Reuse the Thread profile from the LENSCAP part. Insert an Extruded Cut to create the hole for the SWITCH.

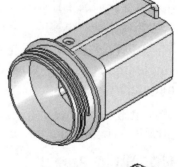

Insert the Rib feature on the back face of the HOUSING.

Insert a Linear Pattern feature to create a row of Ribs.

Insert a Rib feature along the bottom of the HOUSING.

Utilize the Mirror feature to create the second Rib.

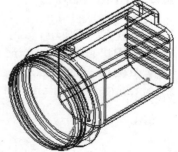

💡 Reuse geometry between parts. The LENSCAP thread is the same as the HOUSING thread. Copy the ThreadSection from the LENSCAP to the HOUSING.

💡 Reuse geometry between features. The Linear Pattern and Mirror Pattern utilized existing features.

Activity: HOUSING Part-Extruded Base Feature

Create the New part.

311) Click **New** 🗋 from the Menu bar.

312) Click the **MY-TEMPLATES** tab.

313) Double-click **PART-IN-ANSI**, [**PART-MM-ISO**].

314) Click **Save**. Select **PROJECTS** for the Save in folder.

315) Enter **HOUSING** for File name. Enter **HOUSING FOR 6VOLT FLASHLIGHT** for Description.

316) Click **Save**. The HOUSING FeatureManager is displayed.

Create the sketch.

317) Right-click **Front Plane** from the FeatureManager. This is your Sketch plane.

318) Click **Sketch** ✏ from the Context toolbar.

319) Click the **Circle** ⊙ Sketch tool. The Circle PropertyManager is displayed.

320) Sketch a circle centered at the **Origin** ↳ as illustrated.

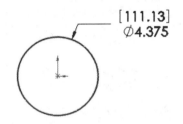

[111.13]
⌀**4.375**

Add a dimension.

321) Click the **Smart Dimension** ⌖ Sketch tool.

322) Click the **circumference**.

323) Enter **4.375**in, [**111.13**].

Insert an Extruded Boss/Base feature.

324) Click the **Extruded Boss/Base** 🗔 feature tool. The Boss-Extrude PropertyManager is displayed.

325) Enter **1.300**, [**33.02**] for Depth in Direction 1. Accept the default settings.

326) Click **OK** ✓ from the Boss-Extrude1 PropertyManager.

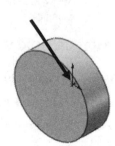

327) Click **Isometric view** 🗔. Note the location of the Origin.

328) Rename **Boss-Extrude1** to **Base Extrude**.

329) Click **Save**.

HOUSING Part-Lofted Boss Feature

The Lofted Boss feature is composed of two profiles. The first sketch is named Sketch-Circle. The second sketch is named Sketch-Square.

Create the first profile from the back face of the Extruded feature. Utilize the Convert Entities sketch tool to extract the circular geometry to the back face.

Create the second profile on an Offset Plane. The FLASHLIGHT components must remain aligned to a common centerline. Insert dimensions that reference the Origin and build symmetry into the sketch. Utilize the Mirror Entities Sketch tool.

Activity: HOUSING Part-Lofted Boss Feature

Create the first profile.
330) Right-click the **back face** of the Base Extrude feature. This is your Sketch plane.

331) Click **Sketch** ✒ from the Context toolbar. The Sketch toolbar is displayed.

332) Click the **Convert Entities** ◻ Sketch tool to extract the face to the Sketch plane.

333) Click **OK** ✔ from the Convert Entities PropertyManager.

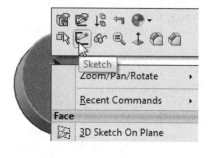

Close and rename the sketch.
334) Right-click **Exit Sketch**.

335) Rename **Sketch2** to **SketchCircle**.

Create an offset plane.
336) Click the **back face** of the Base Extrude feature.

337) Click **Plane** from the Consolidated Reference Geometry Features toolbar. The Plane PropertyManager is displayed.

338) Enter **1.300**in, [**33.02**] for Distance.

339) Click **Top view** ⬜ from the Heads-up View toolbar to verify the Plane position.

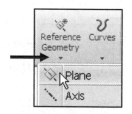

340) Click **OK** ✔ from the Plane PropertyManager. Plane1 is displayed in the FeatureManager.

341) Rename **Plane1** to **BatteryLoftPlane**.

342) **Rebuild** the model.

343) Click **Save**.

Create the second profile.

344) Right-click **BatteryLoftPlane** in the FeatureManager.

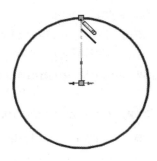

345) Click **Sketch** ✎ from the Context toolbar. The Sketch toolbar is displayed.

346) Click **Back view** ⬚ from the Heads-up View toolbar.

347) Click the **circumference** of the circle.

348) Click the **Convert Entities** ▣ Sketch tool.

349) Click **OK** ✓ from the Convert Entities PropertyManager.

350) Click the **Centerline** ⫶ Sketch tool.

351) Sketch a **vertical centerline** coincident to the Origin and to the top edge of the circle as illustrated.

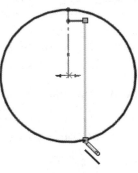

352) Click the **Line** ＼ Sketch tool.

353) Sketch a **horizontal line** to the right side of the centerline as illustrated.

354) Sketch a **vertical line** down to the circumference.

355) Click the **Sketch Fillet** ⌐ Sketch tool. The Sketch Fillet PropertyManager is displayed.

356) Click the **horizontal line** in the Graphics window.

357) Click the **vertical line** in the Graphics window.

358) Enter **.1**in [**2.54**] for Radius.

359) Click **OK** ✓ from the Sketch Fillet PropertyManager. View the Sketch Fillet in the Graphics window.

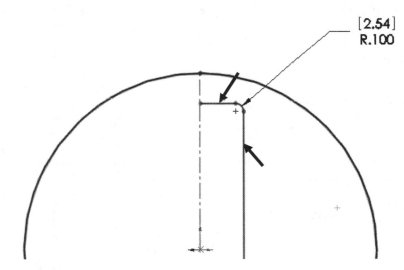

[2.54]
R.100

Mirror the profile.

360) Click the **Mirror Entities** ⚠ Sketch tool. The Mirror PropertyManager is displayed.

361) Click the **horizontal line**, **fillet** and **vertical line**. The selected entities are displayed in the Entities to mirror box.

362) Click inside the **Mirror about** box.

363) Click the **centerline** from the Graphics window.

364) Click **OK** ✔ from the Mirror PropertyManager.

Trim unwanted geometry.

365) Click the **Trim Entities** ⚑ Sketch tool. The Trim PropertyManager is displayed.

366) Click **PowerTrim** ⊞ from the Options box.

367) Click a **position** to the far right of the circle.

368) Drag the **mouse pointer** to intersect the circle.

369) Perform the same **actions** on the left side of the circle.

370) Click **OK** ✔ from the Trim PropertyManager.

Add dimensions.

371) Click the **Smart Dimension** ✎ Sketch tool.

Create the horizontal dimension.
372) Click the **left vertical** line.

373) Click the **right vertical** line.

374) Click a **position** above the profile.

375) Enter **3.100**in, [**78.74**]. View the results.

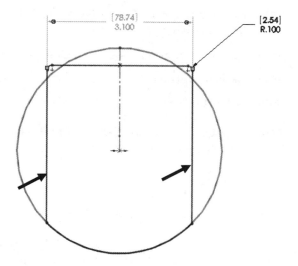

Create the vertical dimension.

376) Click the **Origin** ⌐.

377) Click the **top horizontal** line.

378) Click a **position** to the right of the profile.

379) Enter **1.600**in, [**40.64**].

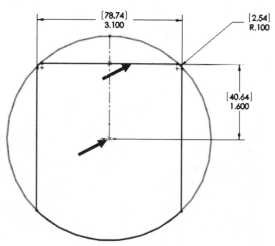

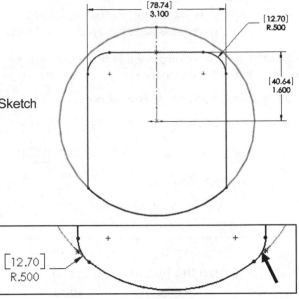

Modify the fillet dimension.

380) Double-click the **.100** fillet dimension.

381) Enter **.500**in, **[12.7]**.

Remove all sharp edges.

382) Click the **Sketch Fillet** Sketch tool. The Sketch Fillet PropertyManager is displayed.

383) Enter **.500**in, **[12.7]** for Radius.

384) Click the **lower left corner point**.

385) Click the **lower right corner point**.

386) Click **OK** from the Sketch Fillet PropertyManager.

Close and rename the sketch.

387) Right-click **Exit Sketch**.

388) Rename the **Sketch3** to **SketchSquare**.

389) Click **Save**.

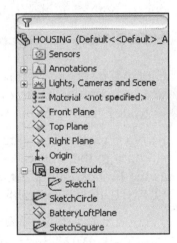

The Loft feature is composed of the SketchSquare and the SketchCircle. Select two individual profiles to create the Loft. The Isometric view provides clarity when selecting Loft profiles.

Display an Isometric view.

390) Click **Isometric view** from the Heads-up View toolbar.

Insert a Lofted Boss feature.

391) Click the **Lofted Boss/Base** feature tool. The Loft PropertyManager is displayed.

392) Click the **upper right side** of the SketchCircle as illustrated.

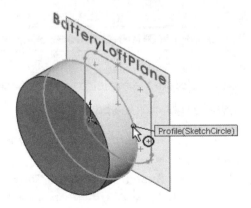

393) Click the **upper right side** of the SketchSquare as illustrated or from the Fly-out FeatureManager. The selected entities are displayed in the Profiles box.

394) Click **OK** ✔ from the Loft PropertyManager.

395) Rename **Loft1** to **Boss-Loft1**.

396) Click **Save**.

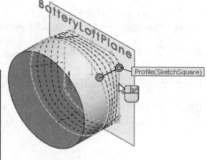

☀ Organize the FeatureManager to locate Loft profiles and planes. Insert the Loft reference planes directly before the Loft feature. Rename the planes, profiles and guide curves with clear descriptive names.

HOUSING Part-Second Extruded Boss/Base Feature

Create the second Extruded Boss/Base feature from the square face of the Loft. How do you estimate the depth of the second Extruded Boss/Base feature? Answer: The Boss-Extrude1 feature of the BATTERY is 4.100in, [104.14mm].

Ribs are required to support the BATTERY. Design for Rib construction. Ribs add strength to the HOUSING and support the BATTERY. Use a 4.400in, [111.76mm] depth as the first estimate. Adjust the estimated depth dimension later if required in the FLASHLIGHT assembly.

The Extruded Boss/Base feature is symmetric about the Right Plane. Utilize Convert Entities to extract the back face of the Loft Base feature. No sketch dimensions are required.

Activity: HOUSING Part-Second Extruded Boss/Base Feature

Select the Sketch plane.
397) **Rotate** the model to view the back.

398) Right-click the **back face** of Boss-Loft1. This is your Sketch plane.

Create the sketch.
399) Click **Sketch** ✏ from the Context toolbar. The Sketch toolbar is displayed.

400) Click the **Convert Entities** ⬚ Sketch tool.

401) Click **OK** ✔ from the Convert Entities PropertyManager.

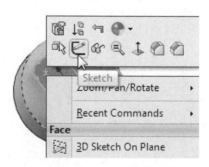

Insert the second Extruded Boss/Base feature.

402) Click the **Extruded Boss/Base** 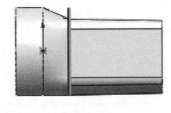 feature tool. The Boss-Extrude PropertyManager is displayed.

403) Enter **4.400**in, **[111.76]** for Depth in Direction 1.

404) Click the **Draft On/Off** box.

405) Enter **1**deg for Draft Angle.

406) Click **OK** ✔ from the Boss-Extrude PropertyManager.

407) Click **Right view** ⬚ from the Heads-up View toolbar.

408) Rename **Boss-Extrude2** to **Boss-Battery**.

409) Click **Save**.

HOUSING Part-Shell Feature

The Shell feature removes material. Use the Shell feature to remove the front face of the HOUSING. In the injection-molded process, the body wall thickness remains constant.

☼ A dialog box is displayed if the thickness value is greater than the Minimum radius of Curvature.

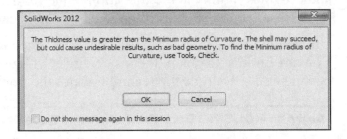

Activity: HOUSING Part-Shell Feature

Insert the Shell feature.

410) Click **Isometric view** ⬡ from the Heads-up View toolbar. If needed **Show** the BatteryLoftPlane from the FeatureManager as illustrated or click View, Planes from the Main menu.

411) Click the **Shell** ⬚ feature tool. The Shell1 PropertyManager is displayed.

412) Click the **front face** of the Base Extrude feature as illustrated.

413) Enter **.100**in, **[2.54]** for Thickness.

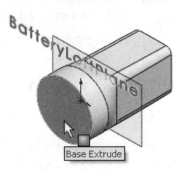

414) Click **OK** ✔ from the Shell1 PropertyManager. Shell1 is displayed in the FeatureManager.

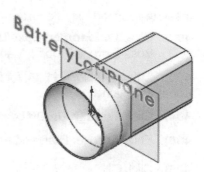

💡 The Shell feature position in the FeatureManager determines the geometry of additional features. Features created before the Shell contained the wall thickness specified in the Thickness option. Position features of different thickness such as the Rib feature and Thread Swept feature after the Shell. Features inserted after the Shell remain solid.

HOUSING Part-Third Extruded Boss/Base Feature

The third Extruded Boss/Base feature creates a solid circular ring on the back circular face of the Boss-Extrude1 feature. The solid ring is a cosmetic stop for the LENSCAP and provides rigidity at the transition of the HOUSING. Design for change. The Extruded Boss/Base feature updates if the Shell thickness changes.

Utilize the Front plane for the sketch. Select the inside circular edge of the Shell. Utilize Convert Entities to obtain the inside circle. Utilize the Circle Sketch tool to create the outside circle. Extrude the feature towards the front face.

Activity: HOUSING Part-Third Extruded Boss Feature

Select the Sketch plane.
415) Right-click **Front Plane** from the FeatureManager. This is your Sketch plane.

Create the sketch.
416) Click **Sketch** ✏ from the Context toolbar. The Sketch toolbar is displayed.

417) Zoom-in and click the **front inside circular edge** of Shell1 as illustrated.

418) Click the **Convert Entities** 🗗 Sketch tool.

419) Click **OK** ✔ from the Convert Entities PropertyManager.

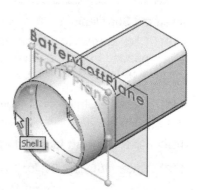

Create the outside circle.
420) Click **Front view** 🔲 from the Heads-up View toolbar.

421) Click the **Circle** ⊘ Sketch tool. The Circle PropertyManager is displayed.

422) Sketch a **circle** centered at the Origin.

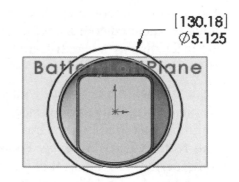

[130.18]
⌀5.125

Add a dimension.

423) Click the **Smart Dimension** ✧ Sketch tool.

424) Click the **circumference** of the circle.

425) Enter **5.125**in, [**130.18**].

Insert the third Extruded Boss/Base feature.

426) Click the **Extruded Boss/Base** 🗔 feature tool. The Boss-Extrude PropertyManager is displayed.

427) Enter **.100**in, [**2.54**] for Depth in Direction 1. The extrude arrow points to the front.

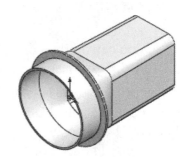

428) Click **OK** ✔ from the Boss-Extrude PropertyManager.

429) Click **Isometric view** 🔲 from the Heads-up View toolbar.

430) Rename **Boss-Extrude3** to **Boss-Stop**.

431) **Hide** the BatteryLoftPlane as illustrated.

432) Click **Save**.

HOUSING Part-Draft Feature

The Draft feature tapers selected model faces by a specified angle by utilizing a Neutral Plane or Parting Line. The Neutral Plane option utilizes a plane or face to determine the pull direction when creating a mold.

The Parting Line option drafts surfaces around a parting line of a mold. Utilize the Parting Line option for non-planar surfaces. Apply the Draft feature to solid and surface models.

A 5deg draft is required to ensure proper thread mating between the LENSCAP and the HOUSING. The LENSCAP Extruded Boss/Base (Boss-Extrude1) feature has a 5deg draft angle.

The outside face of the Extruded Boss/Base (Boss-Extrude1) feature HOUSING requires a 5° draft angle. The inside HOUSING wall does not require a draft angle. The Extruded Boss/Base (Boss-Extrude1) feature has a 5° draft angle. Use the Draft feature to create the draft angle. The front circular face is the Neutral Plane. The outside cylindrical surface is the face to draft.

You created the Extruded Boss/Base and Extruded Cut features with the Draft Angle option. The Draft feature differs from the Extruded feature, Draft Angle option. The Draft feature allows you to select multiple faces to taper.

In order for a model to eject from a mold, all faces must draft away from the parting line which divides the core from the cavity. Cavity side faces display a positive draft and core side faces display a negative draft. Design specifications include a minimum draft angle, usually less than 5deg.

For the model to eject successfully, all faces must display a draft angle greater than the minimum specified by the Draft Angle. The Draft feature, Draft Analysis Tools and DraftXpert utilize the draft angle to determine what faces require additional draft base on the direction of pull.

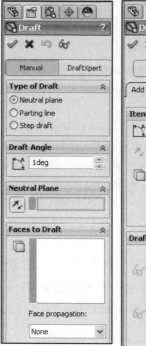

You can apply a draft angle as a part of an Extruded Boss/Base or Extruded Cut feature.

The Draft PropertyManager provides the ability to select either the *Manual* or *DraftXpert* tab. Each tab has a separate menu and option selections. The Draft PropertyManager displays the appropriate selections based on the type of draft you create.

The DraftXpert PropertyManager provides the ability to manage the creation and modification of all Neutral Plane drafts. Select the draft angle and the references to the draft. The DraftXpert manages the rest.

Activity: HOUSING Part-Draft Feature

Insert the Draft feature.

433) Click the **Draft** feature tool. The Draft PropertyManager is displayed.

434) Click the **Manual** tab.

435) Zoom-in and click the thin **front circular face** of Base Extrude. The front circular face is displayed in the Neutral Plane box. Note: The face feedback icon. Face<1> is displayed.

436) Click inside the **Faces to draft** box.

437) Click the **outside cylindrical face** as illustrated.

Note: The face feedback icon.

438) Enter **5**deg for Draft Angle.

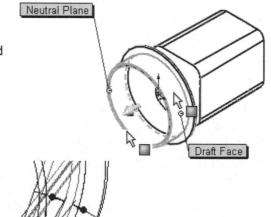

439) Click **OK** ✔ from the Draft PropertyManager. Draft1 is displayed in the FeatureManager.

Display the draft angle and the straight interior.

440) Click **Right view** ⬚ from the Heads-up View toolbar.

441) Click **Hidden Lines Visible** ⬚ from the Heads-up View toolbar.

442) Click **Save**.

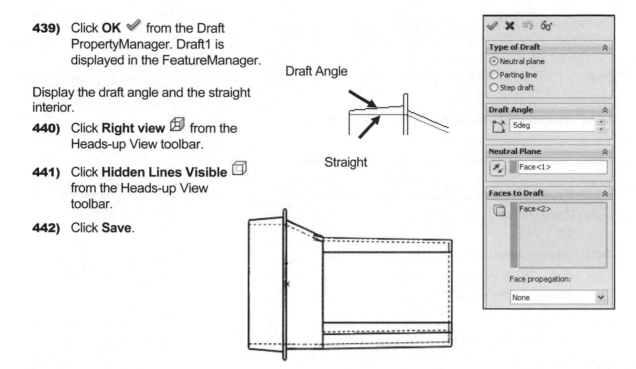

Draft Angle

Straight

💡 The order of feature creation is important. Apply threads after the Draft feature for plastic parts to maintain a constant thread thickness.

HOUSING Part-Thread with Swept Feature

The HOUSING requires a thread. Create the threads for the HOUSING on the outside of the Draft feature. Create the thread with the Swept feature. The thread requires two sketches: ThreadPath and ThreadSection. The LENSCAP and HOUSING Thread utilize the same technique. Create a ThreadPlane. Utilize Convert Entities to create a circular sketch referencing the HOUSING Extruded Boss\Base (Boss-Extrude1) feature. Insert a Helix/Spiral curve to create the path.

Reuse geometry between parts. The ThreadSection is copied from the LENSCAP and is inserted into the HOUSING Top Plane.

Activity: HOUSING Part-Thread with Swept Feature

Insert the ThreadPlane.

443) Click **Isometric view** 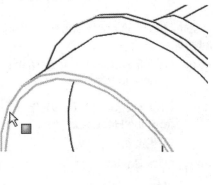 from the Heads-up View toolbar.

444) Click **Hidden Lines Removed** ⬜ from the Heads-up View toolbar.

445) Click the **thin front circular face**, Base Extrude.

446) Click **Plane** from the Consolidated Reference Geometry toolbar. The Plane PropertyManager is displayed.

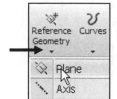

447) Check the **Flip** box.

448) Enter **.125**in, **[3.18]** for Distance. Accept the default settings.

449) Click **OK** ✔ from the Plane PropertyManager. Plane2 is displayed in the FeatureManager.

450) Click **Save**.

451) Rename **Plane2** to **ThreadPlane**.

Insert the ThreadPath.

452) Right-click **ThreadPlane** from the FeatureManager. This is your Sketch plane.

453) Click **Sketch** ✏ from the Context toolbar. The Sketch toolbar is displayed.

454) Click the **front outside circular edge** of the Base Extrude feature as illustrated.

455) Click the **Convert Entities** ⬜ Sketch tool. The circular edge is displayed on the ThreadPlane.

456) Click **OK** ✔ from the Convert Entities PropertyManager.

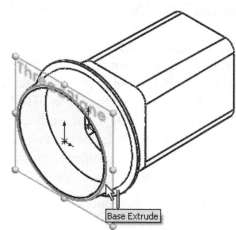

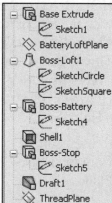

Insert the Helix/Spiral curve.

457) Click the **Helix and Spiral** tool from the Consolidated Curves toolbar as illustrated. The Helix/Spiral PropertyManager is displayed.

458) Enter .**250**in, [**6.35**] for Pitch.

459) Click the **Reverse Direction** box.

460) Enter **2.5** for Revolution.

461) Enter **180** in the Start angle box. The Helix start point and end point are Coincident with the Top Plane.

462) Click the **Taper Helix** box.

463) Enter **5**deg for Angle.

464) Check the **Taper outward** box.

465) Click **OK** ✔ from the Helix/Spiral PropertyManager. Hexlix/Spiral1 is displayed in the FeatureManager.

466) Rename **Helix/Spiral1** to **ThreadPath**.

467) Click **Isometric view** ⬦ from the Heads-up View toolbar.

468) Click **Save**.

Copy the LENSCAP ThreadSection.

469) **Open** the LENSCAP part. The LENSCAP FeatureManager is displayed.

470) **Expand** the Thread feature from the FeatureManager.

471) Click the **ThreadSection** sketch. ThreadSection is highlighted.

472) Click **Edit**, **Copy** from the Menu bar.

473) **Close** the LENSCAP.

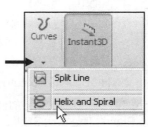

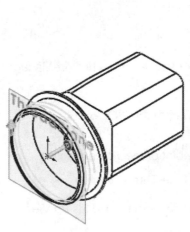

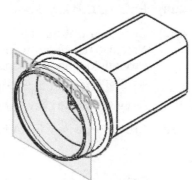

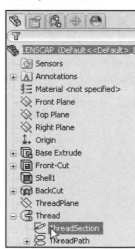

Open the HOUSING.

474) **Return** to the Housing.

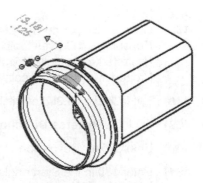

Paste the LENSCAP ThreadSection.

475) Click **Top Plane** from the HOUSING FeatureManager.

476) Click **Edit**, **Paste** from the Menu bar. The ThreadSection is displayed on the Top Plane. The new Sketch7 name is added to the bottom of the FeatureManager.

477) **Hide** ThreadPlane.

478) Rename **Sketch7** to **ThreadSection**.

479) Click **Save**.

Add a Pierce relation.

480) Right-click **ThreadSection** from the FeatureManager.

481) Click **Edit Sketch**.

482) Click **ThreadSection** from the HOUSING FeatureManager.

483) **Zoom in** on the Midpoint of the ThreadSection.

484) Click the **Midpoint** of the ThreadSection.

485) Click **Isometric view** from the Heads-up View toolbar.

486) Hold the **Ctrl** key down.

487) Click the **right back edge of the ThreadPath**. Note: Do not click the end point. The Properties PropertyManager is displayed. The selected entities are displayed in the Selected Entities box.

488) Release the **Ctrl** key.

489) Click **Pierce** from the Add Relations box.

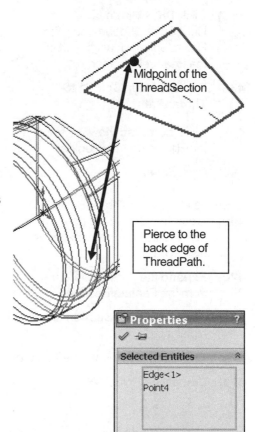

Midpoint of the ThreadSection

Pierce to the back edge of ThreadPath.

490) Click **OK** ✔ from the Properties PropertyManager.

Caution: Do not click the front edge of the Thread path. The Thread is then created out of the HOUSING.

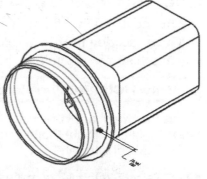

Close the sketch.
491) Right-click **Exit Sketch**.

Insert the Swept feature.

492) Click the **Swept Boss/Base** ⌐ feature tool. The Swept PropertyManager is displayed.

493) **Expand** HOUSING from the fly-out FeatureManager.

494) Click inside the **Profile** box.

495) Click **ThreadSection** from the fly-out FeatureManager.

496) Click **ThreadPath** from the fly-out FeatureManager.

497) Click **OK** ✔ from the Sweep PropertyManager. Sweep1 is displayed in the FeatureManager.

498) Rename **Sweep1** to **Thread**.

499) Click **Save**.

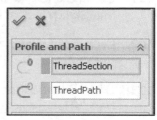

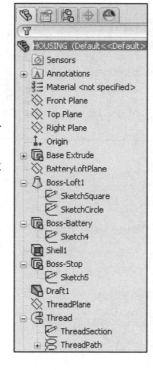

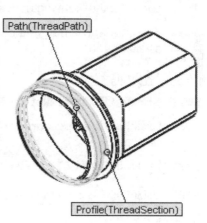

Creating a ThreadPlane provides flexibility to the design. The ThreadPlane allows for a smoother lead. Utilize the ThreadPlane offset dimension to adjust the start of the thread.

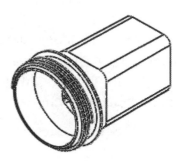

HOUSING Part-Handle with Swept Feature

Create the handle with the Swept feature. The Swept feature consists of a sketched path and cross section profile. Sketch the path on the Right Plane. The sketch uses edges from existing features. Sketch the profile on the back circular face of the Boss-Stop feature.

Activity: HOUSING Part-Handle with Swept Feature

Create the Swept path sketch.

500) Right-click **Right Plane** from the FeatureManager.

501) Select **Sketch**  from the Context toolbar. The Sketch toolbar is displayed.

502) Click **Right view** from the Heads-up View toolbar.

503) Click **Hidden Lines Removed** from the Heads-up View toolbar.

504) Click the **Line** \ Sketch tool.

505) Sketch a **vertical line** from the right top corner of the Housing upward.

506) Sketch a **horizontal line** below the top of the Boss Stop as illustrated.

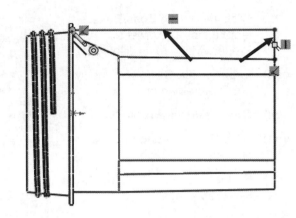

Insert a 2D Fillet.

507) Click the **Sketch Fillet** Sketch tool.

508) Click the **right top corner** of the sketch lines as illustrated.

509) Enter **.500**in, **[12.7]** for Radius.

510) Click **OK** from the Sketch Fillet PropertyManager. Click **OK**.

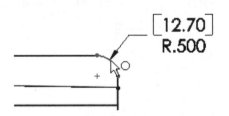

If needed, add a Coincident relation.

511) Click the **left end point** of the horizontal line. Note: the mouse feedback icon

512) Hold the **Ctrl** key down.

513) Click the **right vertical edge** of the Boss Stop.

514) Release the **Ctrl** key.

515) Click **Coincident** from the Add Relations box.

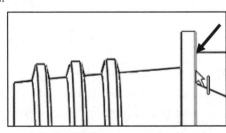

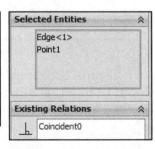

516) Click **OK** ✔ from the Properties PropertyManager.

Add an Intersection relation
517) Click the **bottom end point** of the vertical line.

518) Hold the **Ctrl** key down.

519) Click the **right vertical edge** of the Housing.

520) Click the **horizontal edge** of the Housing.

521) Release the **Ctrl** key.

522) Click **Intersection** ✕ from the Add Relations box.

523) Click **OK** ✔ from the Properties PropertyManager.

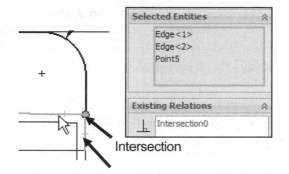

Intersection

Add a dimension.
524) Click the **Smart Dimension** ✏ Sketch tool.

525) Click the **Origin**.

526) Click the **horizontal line**.

527) Click a **position** to the right.

528) Enter **2.500**in, [63.5].

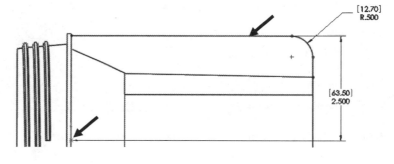

Close and rename the sketch.
529) Right-click **Exit Sketch**.

530) Rename **Sketch8** to **HandlePath**.

531) Click **Save**.

Create the Swept Profile.
532) Click **Back view** ▱ from the Heads-up View toolbar.

533) Right-click the **back circular face** of the Boss-Stop feature as illustrated.

534) Click **Sketch** ✎ from the Context toolbar. The Sketch toolbar is displayed.

💡 The book is design to expose the user to various methods in creating sketches and features. In the next section, create a slot **without using** the Slot Sketch tool.

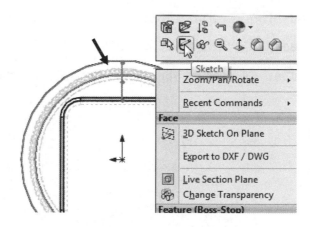

535) Click the **Centerline** ⋮ Sketch tool.

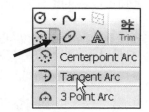

536) Sketch a **vertical centerline** collinear with the Right Plane, coincident to the Origin.

537) Sketch a **horizontal centerline**. The left end point of the centerline is coincident with the vertical centerline on the Boss-Stop feature. Do not select existing feature geometry.

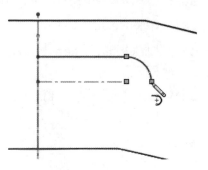

538) **Zoom in** on the top of the Boss-Stop.

539) Click the **Line** ＼ Sketch tool.

540) Sketch a **line** above the horizontal centerline as illustrated.

541) Click the **Tangent Arc** Sketch tool.

542) Sketch a **90° arc**.

543) Right-click **Select** to exit the Tangent Arc tool.

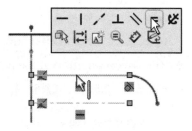

Add an Equal relation.
544) Click the **horizontal centerline**.

545) Hold the **Ctrl** key down.

546) Click the **horizontal line**.

547) Release the **Ctrl** key.

548) Right-click **Make Equal** = from the Context toolbar.

549) Click **OK** ✔ from the Properties PropertyManager.

Add a Horizontal relation.
550) Click the **right end point** of the tangent arc.

551) Hold the **Ctrl** key down.

552) Click the **arc center point**.

553) Click the **left end point** of the centerline.

554) Release the **Ctrl** key.

555) Right-click **Make Horizontal** ━ from the Context toolbar.

556) Click **OK** ✔ from the Properties PropertyManager.

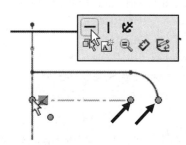

Mirror about the horizontal centerline.

557) Click the **Mirror Entities** ⚠ Sketch tool. The Mirror PropertyManager is displayed.

558) Click the **horizontal** line.

559) Click the **90° arc**. The selected entities are displayed in the Entities to mirror box.

560) Click inside the **Mirror about** box.

561) Click the **horizontal centerline**.

562) Click **OK** ✔ from the Mirror PropertyManager.

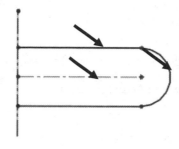

Mirror about the vertical centerline.

563) Click the **Mirror Entities** ⚠ Sketch tool. The Mirror PropertyManager is displayed.

564) Window-Select the **two horizontal lines**, the **horizontal centerline** and the **90° arc** for Entities to mirror. The selected entities are displayed in the Entities to mirror box.

565) Click inside the **Mirror about** box.

566) Click the **vertical centerline**.

567) Click **OK** ✔ from the Mirror PropertyManager.

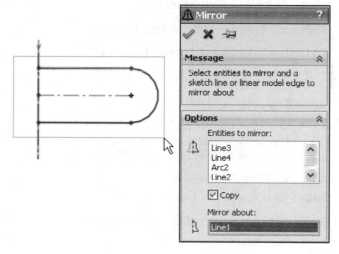

Add dimensions.

568) Click the **Smart Dimension** ✏ Sketch tool.

569) Enter **1.000**in, **[25.4]** between the arc center points.

570) Enter **.100**in, **[2.54]** for Radius.

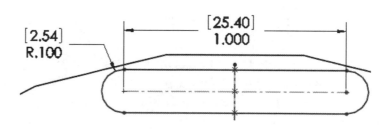

Add a Pierce relation.

571) Right-click **Select**.

572) Click **Isometric view** 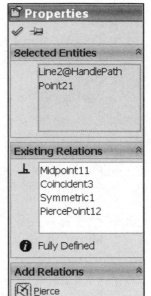 from the Heads-up View toolbar.

573) Click the **top midpoint** of the Sketch profile.

574) Hold the **Ctrl** key down.

575) Click the **line** from the Handle Path.

576) Release the **Ctrl** key.

577) Click **Pierce** from the Add Relations box. The sketch is fully defined.

578) Click **OK** from the Properties PropertyManager.

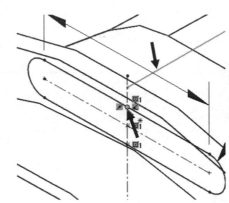

Close and rename the sketch.

579) Right-click **Exit Sketch**.

580) Rename **Sketch8** to **HandleProfile**.

581) If needed, **Hide** ThreadPlane and BatterlyLoftPlane.

Insert the Swept feature.

582) Click the **Swept Boss/Base** feature tool. The Sweep PropertyManager is displayed. HandleProfile is the Sweep profile.

583) Click inside the **Profile** box.

584) Click **HandleProfile** from the fly-out FeatureManager.

585) Click the **HandlePath** from the fly-out FeatureManager.

586) Click **OK** from the Sweep PropertyManager. Sweep2 is displayed in the FeatureManager.

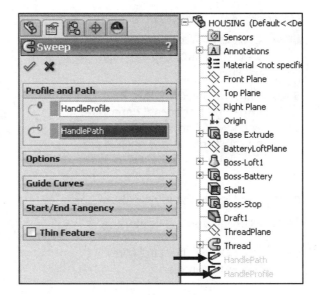

Fit the profile to the Graphics window.

587) Press the **f** key.

588) Click **Shaded With Edges** from the Heads-up View toolbar.

589) Rename **Sweep2** to **Handle**.

590) Click **Save**.

How does the Handle Swept feature interact with other parts in the FLASHLIGHT assembly? Answer: The Handle requires an Extruded Cut to insert the SWITCH.

HOUSING Part-Extruded Cut Feature with Up To Surface

Create an Extruded Cut in the Handle for the SWITCH. Utilize the top face of the Handle for the Sketch plane. Create a circular sketch centered on the Handle.

Utilize the Up To Surface End Condition in Direction 1. Select the inside surface of the HOUSING for the reference surface.

Activity: HOUSING Part-Extruded Cut Feature with Up To Surface

Select the Sketch plane.
591) Right-click the **top face** of the Handle. Handle is highlighted in the FeatureManager. This is your Sketch plane.

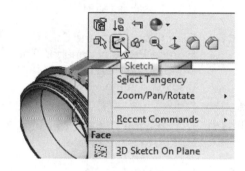

Create the sketch.
592) Click **Sketch** ⌁ from the Context toolbar. The Sketch toolbar is displayed.

593) Click **Top view** ⬚ from the Heads-up View toolbar.

594) Click **Circle** ⊘ from the Sketch toolbar. The Circle PropertyManager is displayed.

595) Sketch a **circle** on the Handle near the front as illustrated.

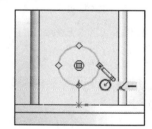

Deselect the circle sketch tool.
596) Right-click **Select**.

Add a Vertical relation.
597) Click the **Origin**.

598) Hold the **Ctrl** key down.

599) Click the **center point** of the circle. The Properties PropertyManager is displayed. The selected entities are displayed in the Selected Entities box.

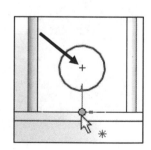

600) Release the **Ctrl** key.

601) Click **Vertical** | .

602) Click **OK** ✓ from the Properties PropertyManager.

Add dimensions.

603) Click the **Smart Dimension** ✎ Sketch tool.

604) Enter **.510**in, **[12.95]** for diameter.

605) Enter **.450**in, **[11.43]** for the distance from the Origin.

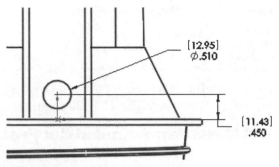

Insert an Extruded Cut feature.

606) **Rotate** the model to view the inside Shell1.

607) Click the **Extruded Cut** 🔳 feature tool. The Cut-Extrude PropertyManager is displayed.

608) Select the **Up To Surface** End Condition in Direction 1.

609) Click the **top inside face** of the Shell1 as illustrated.

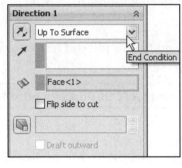

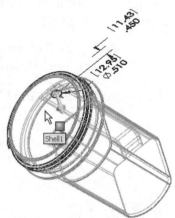

610) Click **OK** ✔ from the Cut-Extrude PropertyManager. The Cut-Extrude1 feature is displayed in the FeatureManager.

611) Rename the **Cut-Extrude1** feature to **SwitchHole**.

612) Click **Isometric view** 🔲 from the Heads-up View toolbar.

613) Click **Save**.

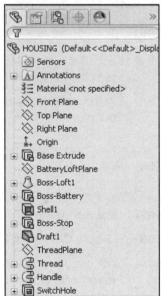

HOUSING Part-First Rib and Linear Pattern Feature

The Rib 🔲 feature adds material between contours of existing geometry. Use Ribs to add structural integrity to a part.

A Rib requires:

- A sketch
- Thickness
- Extrusion direction

The first Rib profile is sketched on the Top Plane. A 1° draft angle is required for manufacturing. Determine the Rib thickness by the manufacturing process and the material.

🔅 Rule of thumb states that the Rib thickness is ½ the part wall thickness. The Rib thickness dimension is .100 inches [2.54mm] for illustration purposes.

The HOUSING requires multiple Ribs to support the BATTERY. A Linear Pattern feature creates multiple instances of a feature along a straight line. Create the Linear Pattern feature in two directions along the same vertical edge of the HOUSING.

Activity: HOUSING Part-First Rib and Linear Pattern Feature

Display all hidden lines.

614) Click **Hidden Lines Visible** 🔲 from the Heads-up View toolbar.

Create the sketch.

615) Right-click **Top Plane** from the FeatureManager. This is your Sketch plane.

616) Click **Sketch** 🔲 from the Context toolbar.

617) Click **Top view** 🔲 from the Heads-up View toolbar.

618) Click the **Line** ＼ Sketch tool.

619) Sketch a **horizontal line** as illustrated. The endpoints are located on either side of the Handle.

Add a dimension.

620) Click the **Smart Dimension** ✏️ Sketch tool.

621) Click the **inner back edge**.

622) Click the **horizontal line**.

623) Click a **position** to the right off the profile.

624) Enter **.175**in, **[4.45]**.

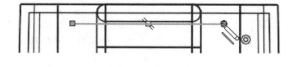

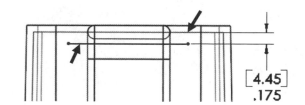

Insert the Rib feature.

625) Click the **Rib** feature tool. The Rib PropertyManager is displayed.

626) Click the **Both Sides** button.

627) Enter **.100**in, [**2.54**] for Rib Thickness.

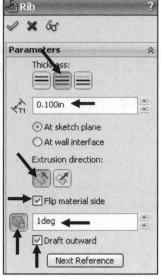

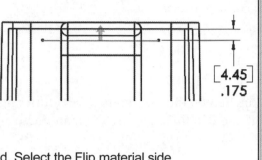

628) Click the **Parallel to Sketch** button. The Rib direction arrow points to the back. Flip the material side if required. Select the Flip material side check box if the direction arrow does not point towards the back.

629) Click the **Draft On/Off** box.

630) Enter **1**deg for Draft Angle.

631) Click **Front view** from the Heads-up View toolbar.

632) Click the **back inside face** of the HOUSING for the Body.

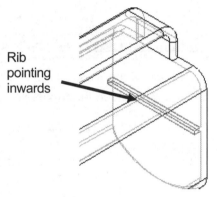

Rib pointing inwards

633) Click **OK** from the Rib PropertyManager. Rib1 is displayed in the FeatureManager.

634) Click **Isometric view** from the Heads-up View toolbar.

635) Click **Save**.

Existing geometry defines the Rib boundaries. The Rib does not penetrate through the wall.

Insert the Linear Pattern feature.

636) **Zoom to Area** on Rib1.

637) Click **Rib1** from the FeatureManager.

638) Click the **Linear Pattern** feature tool. Rib1 is displayed in the Features to Pattern box.

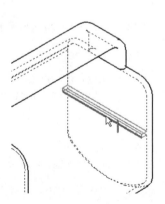

639) Click inside the **Direction 1 Pattern Direction** box.

640) Click the **hidden upper back vertical edge** of Shell1 in the Graphics window. The direction arrow points upward. Click the Reverse direction button if required.

641) Enter .**500**in, [**12.7**] for Spacing.

642) Enter **3** for Number of Instances.

643) Click inside the **Direction 2 Pattern Direction** box.

644) Click the hidden **lower back vertical edge** of Shell1 in the Graphics window. The direction arrow points downward. Click the Reverse direction button if required.

645) Enter .**500**in, [**12.7**] for Spacing.

646) Enter **3** for Number of Instances.

647) Click the **Pattern seed only** box.

648) Drag the Linear Pattern **Scroll bar** downward to display the Options box.

649) Check the **Geometry pattern** box. Accept the default values.

650) Click **OK** ✓ from the Linear Pattern PropertyManager. LPattern1 is displayed in the FeatureManager.

651) Click **Isometric view** 🔲 from the Heads-up View toolbar.

652) Click **Save**.

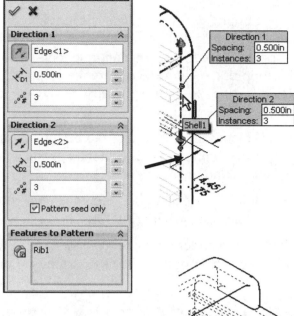

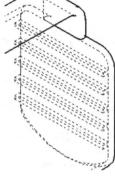

💡 Utilize the Geometry pattern option to efficiently create and rebuild patterns. Know when to check the Geometry pattern.

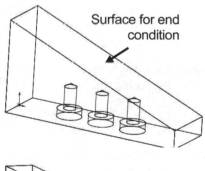

Check Geometry pattern. You require an exact copy of the seed feature. Each instance is an exact copy of the faces and edges of the original feature. End conditions are not calculated. This option saves rebuild time.

Uncheck Geometry pattern. You require the end condition to vary. Each instance will have a different end condition. Each instance is offset from the selected surface by the same amount.

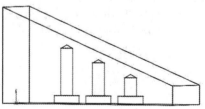

Suppress Patterns when not required. Patterns contain repetitive geometry that takes time to rebuild. Pattern features also clutter the part during the model creation process. Suppress patterns as you continue to create more complex features in the part. Unsuppress a feature to restore the display and load into memory for future calculations. Hide features to improve clarity. Show feature to display hidden features.

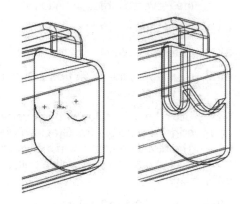

Rib sketches are not required to be fully defined. The Linear Rib option blends sketched geometry into existing contours of the model.

Example: Create an offset reference plane from the inside back face of the HOUSING.

Sketch two under defined arcs. Insert a Rib feature with the Linear option. The Rib extends to the Shell walls.

HOUSING Part-Second Rib Feature

The Second Rib feature supports and centers the BATTERY. The Rib is sketched on a reference plane created through a point on the Handle and parallel with the Right Plane. The Rib sketch references the Origin and existing geometry in the HOUSING. Utilize an Intersection and Coincident relation to define the sketch.

Activity: HOUSING Part-Second Rib Feature

Insert a Reference plane for the second Rib feature. Create a Parallel Plane at Point.

653) Click **Wireframe** from the Heads-up View toolbar.

654) **Zoom to Area** on the back right side of the Handle.

655) Click **Plane** from the Features toolbar. The Plane PropertyManager is displayed.

656) Click **Right Plane** from the fly-out FeatureManager.

657) Click the **vertex** (point) at the back right of the handle as illustrated.

658) Click **OK** from the Plane PropertyManager. Plane1 is displayed in the FeatureManager.

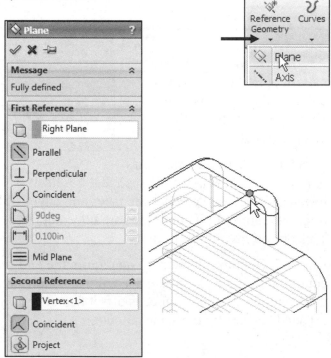

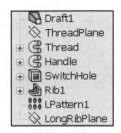

659) Rename **Plane1** to **LongRibPlane**.

Fit to the Graphics window.
660) Press the **f** key.

661) Click **Save**.

Create the second Rib.
662) Right-click **LongRibPlane** from the
FeatureManager.

Create the sketch.
663) Click **Sketch** ✎ from the Context toolbar.

664) Click **Right view** ⬚ from the Heads-up View
toolbar.

665) Click the **Line** ＼ Sketch tool.

666) Sketch a **horizontal line**. Do not select the
edges of the Shell1 feature.

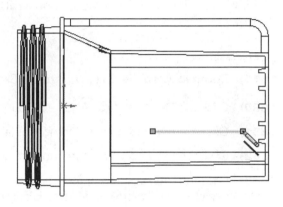

Deselect the Line Sketch tool.
667) Right-click **Select**.

Add a Coincident relation.
668) Click the **left end point** of
the horizontal sketch line.

669) Hold the **Ctrl** key down.

670) Click **BatteryLoftPlane**
from the FeatureManager.

671) Release the **Ctrl** key.

672) Click **Coincident** ⟨.

673) Click **OK** ✔ from the
Properties
PropertyManager.

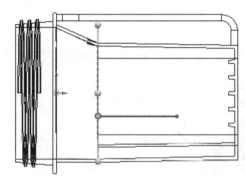

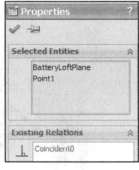

Add a dimension.
674) Click the **Smart
Dimension** ✐ Sketch tool.

675) Click the **horizontal line**.

676) Click the **Origin**.

677) Click a **position** for the vertical
linear dimension text.

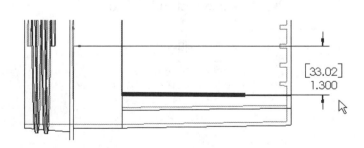

678) Enter **1.300**in, [**33.02**].

☀ When the sketch and reference geometry become complex, create dimensions by selecting Reference planes and the Origin in the FeatureManager.

☀ Dimension the Rib from the Origin, not from an edge or surface for design flexibility. The Origin remains constant. Modify edges and surfaces with the Fillet feature.

Sketch an arc.

679) **Zoom to Area** 🔍 on the horizontal Sketch line.

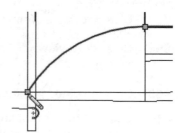

680) Click the **Tangent Arc** ⊃ Sketch tool.

681) Click the **left end** point of the horizontal line.

682) Click the **left end** point of the horizontal line.

683) Click the **intersection** of the Shell1 and Boss Stop features. The sketch is displayed in black and is fully defined. If needed, add an Intersection relation between the endpoint of the Tangent Arc, the left vertical Boss-Stop edge and the Shell1 Silhouette edge of the lower horizontal inside wall.

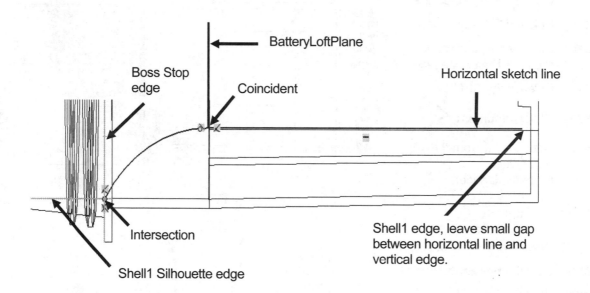

Insert the Rib feature.

684) Click the **Rib** ✋ feature tool. The Rib PropertyManager is displayed.

685) Click the **Both Sides** box.

686) Enter .075in, [**1.91**] for Rib Thickness.

687) Click the **Draft On/Off** box.

688) Enter **1**deg for Angle.

689) Click the **Draft outward** box.

690) Click the **Flip material side** box if required. The direction arrow points towards the bottom.

691) Rotate the model and click the **inside body** as illustrated.

692) Click **OK** ✅ from the Rib PropertyManager. Rib2 is displayed in the FeatureManager.

693) **Hide** LongRibPlane from the FeatureManager.

694) Click **Trimetric view** 📦 from the Heads-up View toolbar.

695) Click **Shaded With Edges** 📦 from the Heads-up View toolbar. View the created Rib feature.

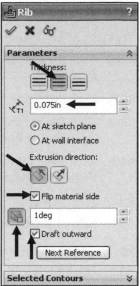

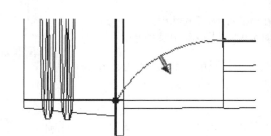

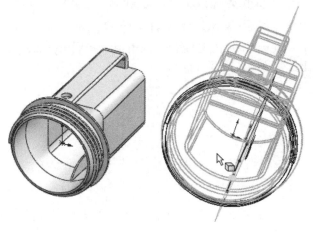

HOUSING Part-Mirror Feature

An additional Rib is required to support the BATTERY. Reuse features with the Mirror feature to create a Rib symmetric about the Right Plane.

The Mirror feature requires:

- Mirror Face or Plane reference

- Features or Faces to Mirror

Utilize the Mirror feature. Select the Right Plane for the Mirror Plane. Select the second Rib for the Features to Mirror.

Activity: HOUSING Part-Mirror Feature

Insert the Mirror feature.

696) Click the **Mirror** feature tool. The Mirror PropertyManager is displayed.

697) Click inside the **Mirror Face/Plane** box.

698) Click **Right Plane** from the fly-out FeatureManager.

699) Click **Rib2** for Features to Mirror from the fly-out FeatureManager.

700) Click **OK** ✅ from the Mirror PropertyManager. Mirror1 is displayed in the FeatureManager.

701) Click **Trimetric view** from the Heads-Up View toolbar.

702) Click **Save**.

Close all parts.
703) Click **Window**, **Close All** from the Menu bar.

The parts for the FLASHLIGHT are complete! Review the HOUSING before moving on to the FLASHLIGHT assembly.

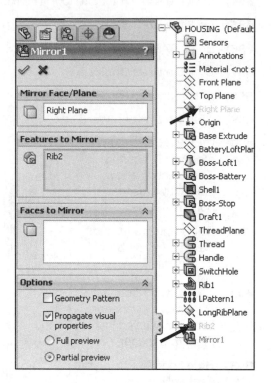

Additional information on Extrude Boss/Base, Extrude Cut, Swept, Loft, Helix/Spiral, Rib, Mirror and Reference Planes are found in SolidWorks Help Topics.

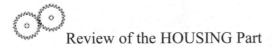 Review of the HOUSING Part

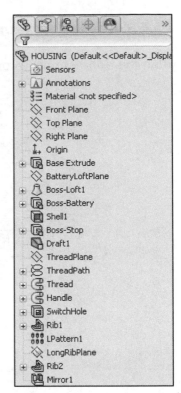

The HOUSING utilized the Extruded Boss/Base feature with the Draft Angle option. The Lofted Boss feature was created to blend the circular face of the LENS with the rectangular face of the BATTERY. The Shell feature removed material with a constant wall thickness. The Draft feature utilized the front face as the Neutral plane.

You created a Thread similar to the LENSCAP Thread. The Thread profile was copied from the LENSCAP and inserted into the Top Plane of the HOUSING. The Extruded Cut feature was utilized to create a hole for the Switch. The Rib features were utilized in a Linear Pattern and Mirror feature.

Each feature has additional options that are applied to create different geometry. The Offset From Surface option creates an Extruded Cut on the curved surface of the HOUSING and LENSCAP. The Reverse offset and Translate surface options produce a cut depth constant throughout the curved surface.

Utilize Tools, Sketch Entities, Text to create the text profile on an Offset Plane.

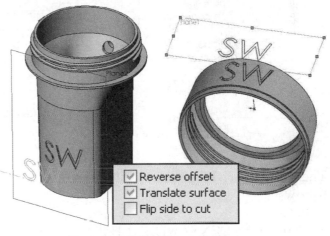

Example: Offset From Surface

FLASHLIGHT Assembly

Plan the Sub-assembly component layout diagram.

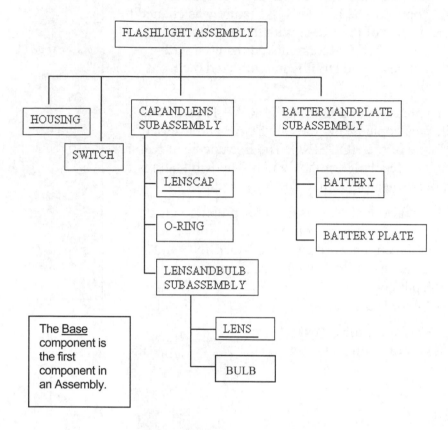

Assembly Layout Structure

The FLASHLIGHT assembly steps are as follows:

- Create the LENSANDBULB sub-assembly from the LENS and BULB components. The LENS is the Base component.

- Create the BATTERYANDPLATE sub-assembly from the BATTERY and BATTERYPLATE.

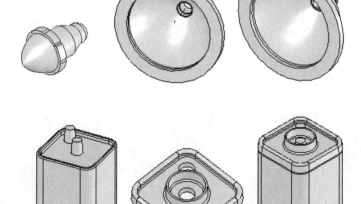

- Create the CAPANDLENS sub-assembly from the LENSCAP, O-RING, and LENSANDBULB sub-assembly. The LENSCAP is the Base component.

- Create the FLASHLIGHT assembly. The HOUSING is the Base component. Insert the SWITCH, CAPANDLENS and BATTERYANDPLATE.

- Modify the dimensions to complete the FLASHLIGHT assembly.

Assembly Template

An Assembly Document template is the foundation of the assembly. Create an Assembly template for the FLASHLIGHT assembly and its sub-assemblies. Create the ASM-IN-ANSI and ASM-MM-ISO Assembly templates in the Project.

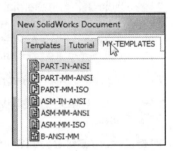

Save the "Click OK" step. Double-click on the Template icon to open the Template in one step instead of two. Double-click on a part or assembly name in the Open Dialog box to open the document.

LENSANDBULB Sub-assembly

Create the LENSANDBULB sub-assembly. Utilize three Coincident mates to assemble the BULB component to the LENS component. When geometry is complex, select the planes in the Mate Selection box.

Suppress the Lens Shield feature to view all inside surfaces during the mate process.

Activity: Create two Assembly Templates

Create an Assembly Template.

704) Click **New** ⬜ from the Menu bar.

705) Double-click the **Assembly** icon from the default Templates tab.

706) Click **Cancel** ✖ from the Begin Assembly PropertyManager. The Assem1 FeatureManager is displayed.

Set the Assembly Document Template options.

707) Click **Options** 📋 , **Document Properties** tab from the Menu bar.

Create an ANSI -IPS Assembly template.

708) Select **ANSI** for Overall drafting standard.

709) Click **Units**. Select **IPS** for Unit system.

710) Select **.123** for basic units length decimal places.

711) Select **None** for basic units angle decimal places.

712) Click **OK** from the dialog box.

Type	Unit	Decimals	Fractions	More
Basic Units				
Length	inches	.123		...
Dual Dimension Length	inches	.12		...
Angle	degrees	None		

Save the Assembly template.

713) Click **Save As** from the Menu bar.

714) Click **Assembly Templates (*asmdot)** from the Save As type box.

715) Select ENGDESIGN-W-SOLIDWORKS\MY-TEMPLATES for the Save in folder.

716) Enter **ASM-IN-ANSI** in the File name. Click **Save**.

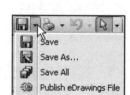

Create an ANSI-MMGS Assembly template.

717) Click **New** ⬜ from the Menu bar.

718) Double-click **Assembly** from the default Templates tab.

719) Click **Cancel** ✖ from the Begin Assembly PropertyManager.

Set the Assembly Document Template options.

720) Click **Options** 📋 , **Document Properties** tab from the Menu bar.

721) Select **ISO** for Overall drafting standard.

722) Click **Units**. Click **MMGS** for Unit system.

723) Select **.12** for basic units length decimal places. Select **None** for basic units angle decimal places.

724) Click **OK**.

Save the Assembly template.

725) Click **Save As** from the Menu bar menu.

726) Select **Assembly Templates (*asmdot)** from the Save as type drop-down menu.

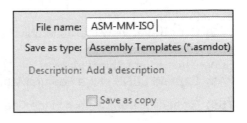

727) Select ENGDESIGN-W-SOLIDWORKS\MY-TEMPLATES for the Save in folder.

728) Enter **ASM-MM-ISO** for File name.

729) Click **Save**.

Activity: LENSANDBULB Sub-assembly

Close all documents.

730) Click **Windows**, **Close All** from the Menu bar.

Create the LENSANDBULB sub-assembly.

731) Click **New** ⬜ from the Menu bar.

732) Click the **MY-TEMPLATES** tab.

733) Double-click **ASM-IN-ANSI** [ASM-MM-ISO].

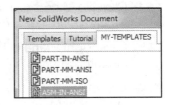

Insert the LENS.

734) Click the **Browse** button.

735) Select **Part** for file type in the PROJECTS folder.

736) Double-click **LENS**.

737) Click **OK** ✅ from the Begin Assembly PropertyManager. The LENS is fixed to the Origin.

738) Click **Save**.

739) Enter **LENSANDBULB** for File name in the PROJECTS folder.

740) Enter **LENS AND BULB ASSEMBLY** for Description.

741) Click **Save**.

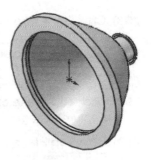

Insert the BULB.

742) Click **Insert Components** 📷 from the Assembly toolbar.

743) Click the **Browse** button.

744) Double-click **BULB** from the PROJECTS folder.

745) Click a **position** in front of the LENS as illustrated.

Fit the model to the Graphics window.

746) Press the **f** key.

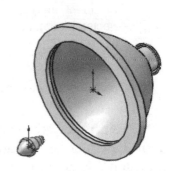

Move the BULB.

747) Click and drag the **BULB** in the Graphics window.

Save the LENSANDBULB.

748) Click **Save**. View the Assembly FeatureManager.

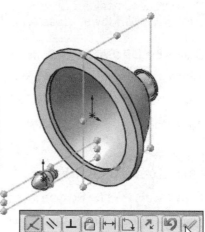

Suppress the LensShield feature.

749) **Expand** LENS in the FeatureManager.

750) Right-click **LensShield** in the FeatureManager.

751) Click Feature Properties.

752) Check the **Suppressed** box.

753) Click **OK** from the Feature Properties dialog box.

Insert a Coincident mate.

754) Click the **Mate** Assembly tool. The Mate PropertyManager is displayed.

755) Click **Right Plane** of the LENS from the fly-out FeatureManager.

756) **Expand** BULB in the fly-out FeatureManager.

757) Click **Right Plane** of the BULB. Coincident mate is selected by default.

758) Click **OK** from the Mate dialog box. Coincident1 is created.

Insert the second Coincident mate.

759) Click **Top Plane** of the LENS from the fly-out FeatureManager.

760) Click **Top Plane** of the BULB from the fly-out FeatureManager. Coincident mate is selected by default.

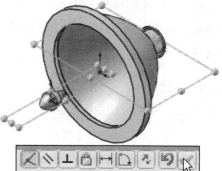

761) Click **OK** from the Mate dialog box. Coincident2 is created.

Select face geometry efficiently. Position the mouse pointer in the middle of the face. Do not position the mouse pointer near the edge of the face. Zoom in on geometry. Utilize the Face Selection Filter for narrow faces.

Activate the Face Selection Filter.

762) Click **View**, **Toolbars**, **Selection Filter**. The Selection Filter toolbar is displayed.

763) Click **Filter Faces**. The Filter icon is displayed.

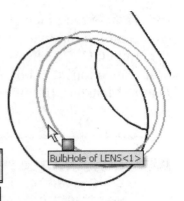

Only faces are selected until the Face Selection Filter is deactivated. Select Clear All Filters from the Selection Filter toolbar to deactivate all filters.

Insert a Coincident mate.

764) **Zoom to Area** and **Rotate** on the CBORE.

765) Click the **BulbHole face** of the LENS in the Graphics window as illustrated.

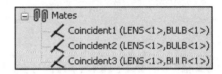

766) Click the **bottom back flat face, Rovolve1** of the BULB. The Coincident Mate is selected by default.

767) Click **OK** from the Mate dialog box. Coincident3 is created. Close the Mate PropertyManager.

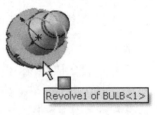

768) Click **OK** from the Mate PropertyManager. The LENSANDBULB is fully defined.

Clear the Face filter.

769) Click **Clear All Filters** from the Selection Filter toolbar.

Display the Mate types.
770) **Expand** the Mates folder in the FeatureManager. View the inserted mates. Note: The Mates under each sub-component in the FeatureManager are displayed.

771) Click **Right view** from the Heads-up View toolbar.

772) Click **Wireframe** from the Heads-up View toolbar.

Save the LENSANDBULB.
773) Click **Isometric view** from the Heads-up View toolbar.

774) Click **Shaded With Edges** from the Heads-up View toolbar.

775) Click **Save**. View the results in the Graphics window.

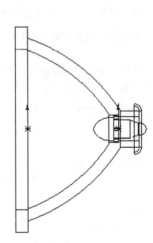

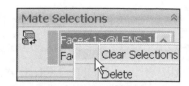

💡 If the wrong face or edge is selected, click the face or edge again to remove it from the Mate Selections text box. Right-click Clear Selections to remove all geometry from the Mate Selections text box. To delete a mate from the FeatureManager, right-click on the mate, click Delete.

BATTERYANDPLATE Sub-assembly

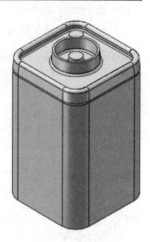

Create the BATTERYANDPLATE sub-assembly. Utilize two Coincident mates and one Concentric mate to assemble the BATTERYPLATE component to the BATTERY component.

💡 Use the Pack and Go option to save an assembly or drawing with references. The Pack and Go tool saves either to a folder or creates a zip file to e-mail. View SolidWorks help for additional information.

Activity: BATTERYANDPLATE Sub-assembly

Create the BATTERYANDPLATE sub-assembly.

776) Click **New** ⬜ from the Menu bar.

777) Click the **MY-TEMPLATES** tab.

778) Double-click **ASM-IN-ANSI**.

Insert the BATTERY part.
779) Click the **Browse** button.

780) Double-click **BATTERY** from the PROJECTS folder.

Place the BATTERY.
781) Click **OK** ✔ from the Begin Assembly PropertyManager. The BATTERY is fixed to the Origin.

Save the BATTERYANDPLATE sub-assembly.
782) Click **Save**.

783) Select the **PROJECTS** folder.

784) Enter **BATTERYANDPLATE** for File name.

785) Enter BATTERY AND PLATE FOR 6-VOLT FLASHLIGHT for Description.

786) Click **Save**. The BATTERYANDPLATE FeatureManager is displayed.

Insert the BATTERYPLATE part.

787) Click **Insert Components**  from the Assembly toolbar.

788) Click the **Browse** button.

789) Double-click **BATTERYPLATE** from the PROJECTS folder.

790) Click a **position** above the BATTERY as illustrated.

Insert a Coincident mate.

791) Click the **Mate** Assembly tool. The Mate PropertyManager is displayed.

792) Click the **outside bottom face** of the BATTERYPLATE.

793) Click the **top narrow flat face** of the BATTERY Base Extrude feature as illustrated. Coincident mate is selected by default.

794) Click **OK** from the Mate dialog box. Coincident1 is created.

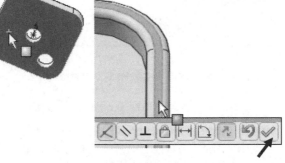

Insert a Coincident mate.

795) Click **Right Plane** of the BATTERY from the fly-out FeatureManager.

796) Click **Right Plane** of the BATTERYPLATE from the fly-out FeatureManager. Coincident mate is selected by default.

797) Click **OK** from the Mate dialog box. Coincident2 is created.

Insert a Concentric mate.

798) Click the center Terminal feature **cylindrical face** of the BATTERY as illustrated.

799) Click the Holder feature **cylindrical face** of the BATTERYPLATE. Concentric mate is selected by default.

800) Click **OK** from the Mate dialog box. Concentric1 is created.

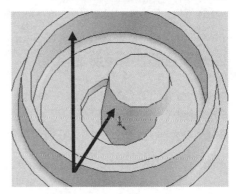

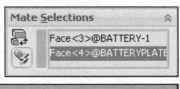

801) Click **OK** ✅ from the Mate PropertyManager. Note: If required, deactivate view Sketches from the Menu bar.

802) Expand the Mates folder. View the created mates.

Save the BATTERYANDPLATE.

803) Click **Isometric view** 🔲 from the Heads-up View toolbar.

804) Click **Save**.

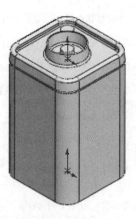

CAPANDLENS Sub-assembly

Create the CAPANDLENS sub-assembly. Utilize two Coincident mates and one Distance mate to assemble the O-RING to the LENSCAP. Utilize three Coincident mates to assemble the LENSANDBULB sub-assembly to the LENSCAP component.

Caution: Select the correct reference. Expand the LENSCAP and O-RING. Click the Right Plane within the LENSCAP. Click the Right plane within the O-RING.

Activity: CAPANDLENS Sub-assembly

Create the CAPANDLENS sub-assembly.

805) Click **New** 🔲 from the Menu bar.

806) Click the **MY-TEMPLATES** tab.

807) Double-click **ASM-IN-ANSI**.

Insert the LENSCAP sub-assembly.
808) Click the **Browse** button.

809) Double-click **LENSCAP** from the PROJECTS folder.

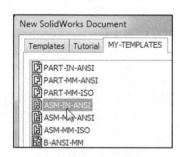

810) Click **OK** ✅ from the Begin Assembly PropertyManager. The LENSCAP is fixed to the Origin.

Save the CAPANDLENS assembly.
811) Click **Save**.

812) Select the **PROJECTS** folder.

813) Enter **CAPANDLENS** for File name.

814) Enter **LENSCAP AND LENS** for Description.

815) Click **Save**. The CAPANDLENS FeatureManager is displayed.

Insert the O-RING part.

816) Click **Insert Components**  from the Assembly toolbar.

817) Click the **Browse** button.

818) Double-click **O-RING** from the PROJECTS folder.

819) Click a **position** behind the LENSCAP as illustrated.

Insert the LENSANDBULB assembly.

820) Click **Insert Components** from the Assembly toolbar.

821) Click the **Browse** button.

822) Select **Assembly** for Files of type in the PROJECTS folder.

823) Double-click **LENSANDBULB**.

824) Click a **position** behind the O-RING as illustrated.

825) Click Isometric view .

Move and hide components.
826) Click and drag the **O-RING** and **LENSANDBULB** as illustrated in the Graphics window.

827) Right-click **LENSANDBULB** in the FeatureManager.

828) Click **Hide components** from the Context toolbar.

Insert three mates between the LENSCAP and O-RING.
829) Click the **Mate** Assembly tool. The Mate PropertyManager is displayed.

Insert a Coincident mate.
830) Click **Right Plane** of the LENSCAP in the fly-out FeatureManager.

831) Click **Right Plane** of the O-RING in the fly-out FeatureManager. Coincident mate is selected by default.

832) Click **OK** from the Mate dialog box. Coincident1 is created.

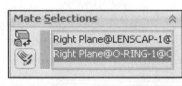

Insert a second Coincident mate.
833) Click **Top Plane** of the LENSCAP in the fly-out FeatureManager.

834) Click **Top Plane** of the O-RING in the fly-out FeatureManager. Coincident mate is selected by default.

835) Click **OK** from the Mate dialog box. Coincident2 is created.

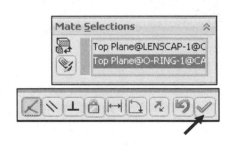

Insert a Distance mate.

836) Click the Shell1 **back inside face** of the LENSCAP as illustrated.

837) Click **Front Plane** of the O-RING in the fly-out FeatureManager.

838) Click **Distance**.

839) Enter **.125/2**in, [**3.175/2mm**].

840) Click **OK** from the Mate dialog box.

841) Click **OK** from the Mate PropertyManager.

842) Click **Isometric view** from the Heads-up View toolbar.

843) **Expand** the Mates folder. View the created mates. Note: View the created mates under each sub-component.

844) Click **Save**.

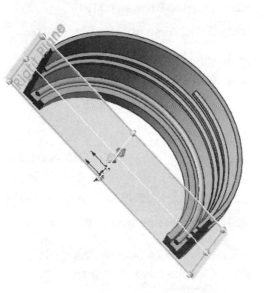

How is the Distance mate, .0625in, [1.588] calculated? Answer:

O-RING Radius (.1250in/2) = .0625in.

O-RING Radius [3.175mm/2] = [1.588mm].

Utilize a Section view to locate internal geometry for mating and verify position of components.

Build flexibility into the mate. A Distance mate offers additional flexibility over a Coincident mate. You can modify the value of a Distance mate.

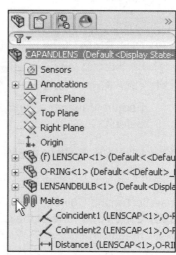

Show the LENSANDBULB.

845) Right-click **LENSANDBULB** in the FeatureManager.

846) Click **Show components** from the Contexts toolbar.

Fit the model to the Graphics window.

847) Press the **f** key.

848) Click **Mate** 🖉 from the Assembly toolbar. The Mate PropertyManager is displayed.

Insert a Coincident mate.

849) Click **Right Plane** of the LENSCAP in the fly-out FeatureManager.

850) Click **Right Plane** of the LENSANDBULB in the fly-out FeatureManager. Coincident mate ⦨ is selected by default.

851) Click **OK** ✅ from the Mate dialog box.

Insert a Coincident Mate.

852) Click **Top Plane** of the LENSCAP in the fly-out FeatureManager.

853) Click **Top Plane** of the LENSANDBULB in the fly-out FeatureManager. Coincident mate ⦨ is selected by default.

854) Click **OK** ✅ from the Mate dialog box.

Insert a Coincident Mate.

855) Click the flat inside **narrow back face** of the LENSCAP.

856) Click the **front flat face** of the LENSANDBULB. Coincident ⦨ is selected by default.

857) Click **OK** ✅ from the Mate dialog box.

858) Click **OK** ✅ from the Mate PropertyManager. View the created mates.

859) **Expand** the Mates folder. View the created mates.

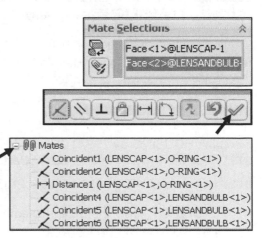

Confirm the location of the O-RING.

860) Click **Right Plane** of the CAPANDLENS from the FeatureManager.

861) Click **Section view** 🔲 from the Heads-up View toolbar.

862) Click **Isometric view** 🔷 from the Heads-up View toolbar.

863) **Expand** the Section 2 box.

864) Click **inside** the Reference Section Plane box.

865) Click **Top Plane** of the CAPANDLENS from the fly-out FeatureManager.

866) Click **OK** ✔ from the Section View PropertyManager.

Save the CAPANDLENS sub-assembly. Return to a full view.

867) Click **Section view** 🔲 from the Heads-up View toolbar.

868) Click **Save**.

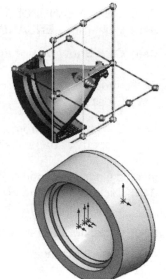

The LENSANDBULB, BATTERYANDPLATE and CAPANDLENS sub-assemblies are complete. The components in each assembly are fully defined. No minus (-) sign or red error flags exist in the FeatureManager. Insert the sub-assemblies into the final FLASHLIGHT assembly.

FLASHLIGHT Assembly

Create the FLASHLIGHT assembly. The HOUSING is the Base component. The FLASHLIGHT assembly mates the HOUSING to the SWITCH component. The FLASHLIGHT assembly mates the CAPANDLENS and BATTERYANDPLATE.

Activity: FLASHLIGHT Assembly

Create the FLASHLIGHT assembly.

869) Click **New** ☐ from the Menu bar.

870) Click the **MY-TEMPLATES** tab.

871) Double-click **ASM-IN-ANSI**.

Insert the HOUSING and SWITCH.

872) Click the **Browse** button. Select **Parts** for file type.

873) Double-click **HOUSING** from the PROJECTS folder.

874) Click **OK** ✔ from the Begin Assembly PropertyManager. The HOUSING is fixed to the Origin.

875) Click **Insert Components**  from the Assembly toolbar.

876) Click the **Browse** button.

877) Double-click **SWITCH** from the PROJECTS folder.

878) Click a **position** in front of the HOUSING as illustrated.

Save the FLASHLIGHT assembly.
879) Click **Save**. Select the **PROJECTS** folder.

880) Enter **FLASHLIGHT** for File name.

881) Enter **FLASHLIGHT ASSEMBLY** for Description.

882) Click **Save**. The FLASHLIGHT FeatureManager is displayed.

Insert a Coincident mate.
883) Click the **Mate** Assembly tool.

884) Click **Right Plane** of the HOUSING from the fly-out FeatureManager.

885) Click **Right Plane** of the SWITCH from the fly-out FeatureManager. Coincident mate is selected by default.

886) Click **OK** from the Mate dialog box.

Insert a Coincident mate.
887) Click **View**; check **Temporary Axes** from the Menu bar.

888) Click the **Temporary axis** inside the Switch Hole of the HOUSING.

889) Click **Front Plane** of the SWITCH from the fly-out FeatureManager. Coincident mate is selected by default.

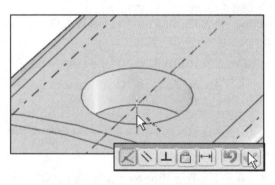

890) Click **OK** from the Mate dialog box.

Insert a Distance mate.
891) Click the **top face** of the Handle.

892) Click the **Vertex** on the Loft top face of the SWITCH.

893) Click **Distance**.

894) Enter **.100**in, **[2.54]**.

895) Check the **Flip Direction** box.

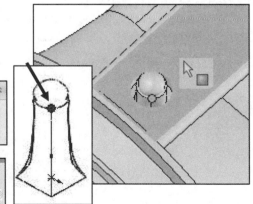

896) Click **OK** ✓ from the Mate dialog box.

897) Click **OK** ✓ from the Mate PropertyManager.

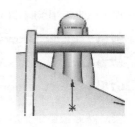

Insert the CAPANDLENS assembly.
898) Click **View**; uncheck **Temporary Axis** from the Menu bar.

899) Click **View**; un-check **Origins** from the Menu bar.

900) Click **Insert Components** 📦 from the Assembly toolbar.

901) Click the **Browse** button.

902) Select **Assembly** for Files of type.

903) Double-click **CAPANDLENS** from the PROJECTS folder.

Place the sub-assembly.
904) Click a **position** in front of the HOUSING as illustrated.

Insert Mates between the HOUSING component and the CAPANDLENS sub-assembly.
905) Click the **Mate** ✎ Assembly tool. The Mate PropertyManager is displayed.

Insert a Coincident mate.
906) Click **Right Plane** of the HOUSING from the fly-out FeatureManager.

907) Click **Right Plane** of the CAPANDLENS from the fly-out FeatureManager. Coincident mate ⟋ is selected by default.

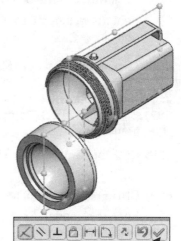

908) Click **OK** ✓ from the Mate dialog box.

Insert a Coincident mate.
909) Click **Top Plane** of the HOUSING from the fly-out FeatureManager.

910) Click **Top Plane** of the CAPANDLENS from the fly-out FeatureManager. Coincident mate ⟋ is selected by default.

911) Click **OK** ✓ from the Mate dialog box.

Insert a Coincident mate.

912) Click the **front face** of the Boss-Stop on the HOUSING.

Rotate the view.

913) Press the **Left arrow key** to view the back face.

914) Click the **back face** of the CAPANDLENS. Coincident mate ✕ is selected by default.

915) Click **OK** ✓ from the Mate dialog box.

916) Click **OK** ✓ from the Mate PropertyManager.

Save the FLASHLIGHT assembly.

917) Click Isometric view 🔲.

918) Click **Save**.

Insert the BATTERYANDPLATE sub-assembly.

919) Click **Insert Components** 📦 from the Assembly toolbar.

920) Click the **Browse** button.

921) Select **Assembly** for Files of type.

922) Double-click **BATTERYANDPLATE** from the PROJECTS folder.

923) Click a **position** to the left of the HOUSING as illustrated.

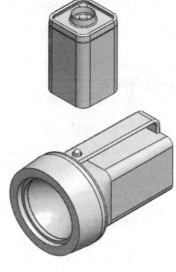

Rotate the part.

924) Click **BATTERYANDPLATE** in the FeatureManager.

925) Click **Rotate Component** 🔄. from the Assembly toolbar.

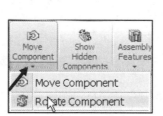

926) Rotate the **BATTERYANDPLATE** until it is approximately parallel with the HOUSING.

927) Click **OK** ✓ from the Rotate Component PropertyManager.

Insert a Coincident mate.

928) Click the **Mate** ✎ Assembly tool.

929) Click **Right Plane** of the HOUSING from the fly-out FeatureManager.

930) Click **Front Plane** of the BATTERYANDPLATE from the fly-out FeatureManager. Coincident is selected by default.

931) Click **OK** ✓ from the Mate dialog box.

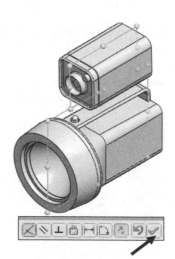

932) Move the **BATTERYANDYPLATE** in front of the HOUSING.

Insert a Coincident mate.
933) Click **Top Plane** of the HOUSING in the fly-out FeatureManager.

934) Click **Right Plane** of the BATTERYANDPLATE in the fly-out FeatureManager. Coincident mate ⟨ is selected by default.

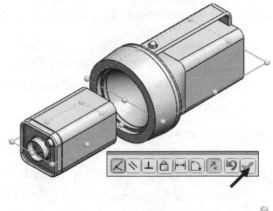

935) Click **OK** ✔ from the Mate dialog box.

936) Click **OK** ✔ from the Mate PropertyManager.

Display the Section view.
937) Click **Right Plane** in the FLASHLIGHT Assembly FeatureManager.

938) Click **Section view** from the Heads-up View toolbar.

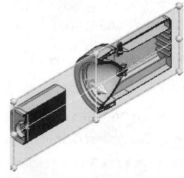

939) Click **OK** ✔ from the Section View PropertyManager.

Move the BATTERYANDPLATE in front of the HOUSING.
940) Click and drag the **BATTERYANDPLATE** in front of the HOUSING as illustrated.

Insert a Coincident mate.
941) Click the **Mate** Assembly tool. The Mate PropertyManager is displayed.

942) Click the **back center Rib1 face** of the HOUSING.

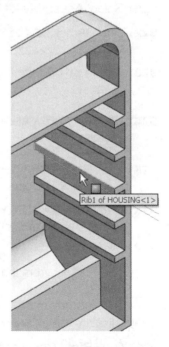

943) Click the **bottom face** of the BATTERYANDPLATE. Coincident ⟨ is selected by default.

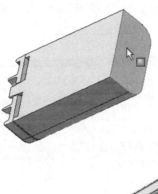

944) Click **OK** ✔ from the Mate dialog box.

945) Click **Isometric view** from the Heads-up View toolbar.

946) Click **OK** ✔ from the Mate PropertyManager.

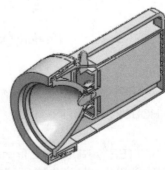

Display the Full view.

947) Click **Section view** 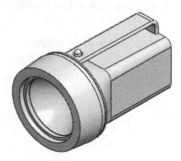 from the Heads-up View toolbar.

Save the FLASHLIGHT.
948) Click **Save**.

 Additional information on Assembly, Move Component, Rotate Component and Mates is available in SolidWorks Help Topics.

Review of the FLASHLIGHT assembly

The FLASHLIGHT assembly consisted of the HOUSING part, SWITCH part, CAPANDLENS sub-assembly and the BATTERYANDPLATE sub-assembly.

The CAPANDLENS sub-assembly contained the BULBANDLENS sub-assembly, the O-RING and the LENSCAP part. The BATTERANDPLATE sub-assembly contained the BATTERY and BATTERYPLATE part.

You inserted eight Coincident mates and a Distance mate. Through the Assembly Layout illustration you simplified the number of components into a series of smaller assemblies. You also enhanced your modeling techniques and skills.

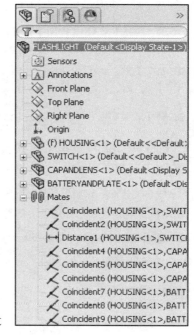

You still have a few more areas to address. One of the biggest design issues in assembly modeling is interference. Let's investigate the FLASHLIGHT assembly.

Clearance Verification checks the minimum distance between components and reports any value that fails to meet your input value of the minimum clearance. View SolidWorks Help for additional information.

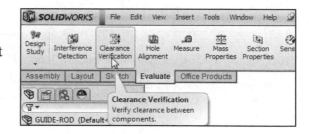

Addressing Interference Issues

There is an interference issue between the FLASHLIGHT components. Address the design issue. Adjust Rib2 on the HOUSING. Test with the Interference Check command. The FLASHLIGHT assembly is illustrated in inches.

Activity: Addressing Interference Issues

Check for interference.

949) Click the **Interference Detection** tool from the Evaluate tab in the CommandManager.

950) Delete **FLASHLIGHT.SLDASM** from the Selected Components box.

951) Click **BATTERYANDPLATE** from the fly-out FeatureManager.

952) Click **HOUSING** from the fly-out FeatureManager.

953) Click the **Calculate** button. The interference is displayed in red in the Graphics window.

954) Click each **Interference** in the Results box to view the interference in red with Rib2 of the HOUSING.

955) Click **OK** from the Interference Detection PropertyManager.

Modify the Rib2 dimension to address the interference issue.
956) **Expand** the HOUSING in the FeatureManager.

957) Double-click on the **Rib2** feature.

958) Double click **1.300**in, **[33.02]**.

959) Enter **1.350**in, **[34.29]**.

960) **Rebuild** the model.

961) Click **OK** .

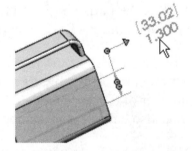

Recheck for Interference.

962) Click **Interference Detection** from the Evaluate tab. The Interference dialog box is displayed.

963) Delete **FLASHLIGHT.SLDASM** from the Selected Components box.

964) Click **BATTERYANDPLATE** from the FeatureManager.

965) Click **HOUSING** from the FeatureManager.

966) Click the **Calculate** button. No Interference is displayed in the Results box. The FLASHLIGHT design is complete.

967) Click **OK** ✓ from the Interference Detection PropertyManager.

Save the FLASHLIGHT.
968) Click **Save**.

Export Files and eDrawings

You receive a call from the sales department. They inform you that the customer increased the initial order by 200,000 units. However, the customer requires a prototype to verify the design in six days. What do you do? Answer: Contact a Rapid Prototype supplier. You export three SolidWorks files:

- HOUSING

- LENSCAP

- BATTERYPLATE

Use the Stereo Lithography (STL) format. Email the three files to a Rapid Prototype supplier. Example: Paperless Parts Inc. (www.paperlessparts.com). A Stereolithography (SLA) supplier provides physical models from 3D drawings. 2D drawings are not required. Export the HOUSING. SolidWorks eDrawings provides a facility for you to animate, view and create compressed documents to send to colleagues, customers and vendors. Publish an eDrawing of the FLASHLIGHT assembly.

Activity: Export Files and eDrawings

Open and Export the HOUSING.
969) Right-click **HOUSING** from the FeatureManager.

970) Click **Open Part** from the Context toolbar.

971) Click **Save As** from Menu bar.

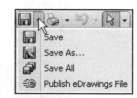

972) Select **STL (*.stl)** from the Save as type drop-down menu. The Save dialog box is displayed.

973) Click the **Options** button.

974) Drag the **dialog box** to the left to view the model in the Graphics window.

975) Click the **Binary** box from the Output as format box.

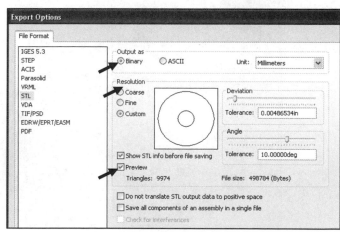

976) Click the **Coarse** box for Resolution.

Display the STL triangular faceted model.
977) Check the **Preview** box.

978) Drag the **dialog box** to the left to view the Graphics window.

Create the binary STL file.
979) Click **OK** from the Export Options dialog box.

980) Click **Save** from the Save dialog box. A status report is provided.

981) Click **Yes**.

Publish an eDrawing and email the document to a colleague.

Create the eDrawing and animation.
982) Click **File**, **Publish to eDrawings** from the Menu bar.

983) Click the **Play** button.

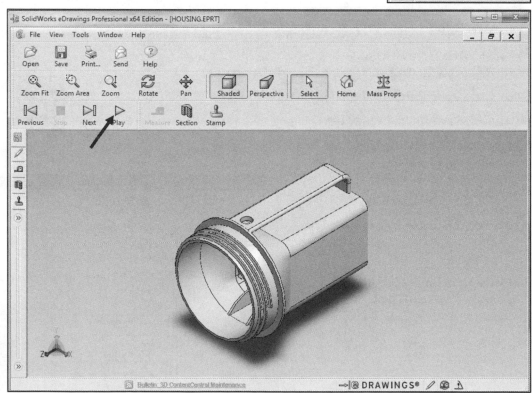

Stop the animation.
984) Click the **Stop** button.

Save the eDrawing.

985) Click **Save** 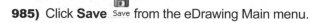 Save from the eDrawing Main menu.

986) Select the **PROJECTS** folder.

987) Enter **HOUSING** for File name.

988) Click **Save**.

989) Close ☒ the eDrawing dialog box.

990) Close all models in the session.

It is time to go home. The telephone rings. The customer is ready to place the order. Tomorrow you will receive the purchase order.

The customer also discusses a new purchase order that requires a major design change to the handle. You work with your industrial designer and discuss the two options. The first option utilizes Guide Curves on a Sweep feature.

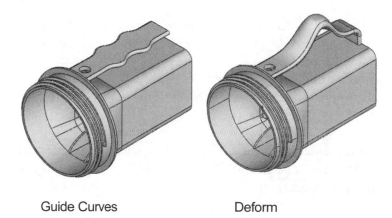

Guide Curves   Deform

The second option utilizes the Deform feature. These features are explored in a discussion on the Multi-media DVD contained in the text.

You contact your mold maker and send an eDrawing of the LENSCAP.

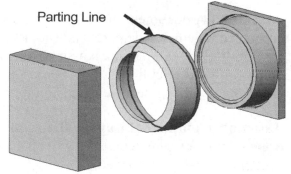

Parting Line

The mold maker recommends placing the parting line at the edge of the Revolved Cut surface and reversing the Draft Angle direction. The mold maker also recommends a snap fit versus a thread to reduce cost. The Core-Cavity mold tooling is explored in the project exercises.

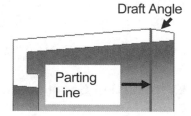

Draft Angle

Parting Line

Q Additional information on Interference Detection, eDrawings, STL files (stereolithography), Guide Curves, Deform and Mold Tools are available in SolidWorks Help Topics.

Project Summary

The FLASHLIGHT assembly contains over 100 features, reference planes, sketches and components. You organized the features in each part. You developed an assembly layout structure to organize your components.

The O-RING utilized a Swept Base feature. The SWITCH utilized a Loft Base feature. The simple Swept feature requires two sketches: a path and a profile. A complex Swept requires multiple sketches and guide curves. The Loft feature requires two or more sketches created on different planes.

The LENSCAP and HOUSING utilized a variety of features. You applied design intent to reuse geometry through Geometric relationships, symmetry and patterns.

The assembly required an Assembly Template. You utilized the ASM-IN-ANSI Template to create the LENSANDBULB, CAPANDLENS, BATTERYANDPLATE and FLASHLIGHT assemblies.

You created an STL file of the Housing and an eDrawing of the Housing to communicate with your vendor, mold maker and customer. Review the project exercises before moving on to Project 6.

Project Terminology

Assembly: A document in which parts, features, and other assemblies (sub-assemblies) are mated together. The parts and sub-assemblies exist in documents separate from the assembly. For example, in an assembly, a piston can be mated to other parts, such as a connecting rod or cylinder. This new assembly can then be used as a sub-assembly in an assembly of an engine. The extension for a SolidWorks assembly file name is .SLDASM.

Assembly Component Layout Diagram: A diagram used to plan the top-level assembly organization. Organize parts into smaller subassemblies. Create a flow chart or manual sketch to classify components.

Bottom-up Design: An assembly modeling technique where you create parts and then insert them into an assembly.

Component: A part or assembly inserted into a new assembly.

Construction Geometry: The characteristic of a sketch entity that the entity is used in creating other geometry but is not itself used in creating features.

Draft: A feature used to add a specified draft angle to a face. You utilized a Draft feature with the Neutral Plane option.

eDrawings: A compressed document used to animate and view SolidWorks documents.

Exploded View: Shows an assembly with its components separated from one another, usually to show how to assemble the mechanism.

Extruded Thin Cut: A feature used to remove material by extruding an open profile.

Heads-up View Toolbar: A transparent toolbar located in the upper portion of the Graphics window by default.

Helix/Spiral Curve: A Helix is a curve with pitch. The Helix is created about an axis. You utilized a Helix Curve to create a thread for the LENSCAP.

Hidden Geometry: Geometry that is not displayed. Utilize Hide/Show to control display of components in an assembly.

Hidden Lines: A line type used in 2D drafting to represent edges that can't be seen normally in the view - which are hidden by other geometry.

Hole Wizard: The Hole Wizard feature is used to create specialized holes in a solid. The Hole Wizard creates simple, tapped, counterbore and countersunk holes using a step-by-step procedure.

Interference Detection: The amount of interference between components in an assembly is calculated with the Interference Detection tool.

Loft: A feature used to blend two or more profiles on separate Planes. A Loft Boss adds material. A Loft Cut removes material. The HOUSING part utilized a Loft Boss feature to transition a circular profile of the LENSCAP to a square profile of the BATTERY.

Mates: A Mate is a geometric relationship between components in an assembly that constrain rotational and translational motion.

Mirror: The feature used to create a symmetric feature about a Mirror Plane. The Mirror feature created a second Rib, symmetric about the Right Plane.

Origin: The point where the three axes intersect at the 0,0,0 coordinate of the model or sketch. Always know the location of the origin when modeling.

Orthographic Projection: A single view that is meant to represent a 3D object as a flat 2D object.

Over Defined: A sketch is over defined when dimensions or relations are either in conflict or redundant.

Part: A single 3D object made up of features. A part can become a component in an assembly, and it can be represented in 2D in a drawing. Examples of parts are bolt, pin, plate, and so on. The extension for a SolidWorks part file name is .SLDPRT.

Pattern: A Pattern creates one or more instances of a feature or group of features. A Circular Pattern creates instances of a seed feature about an axis of revolution. A Linear Pattern creates instances of a seed feature in a rectangular array, along one or two edges. The Linear Pattern was utilized to create multiple instances of the HOUSING Rib1 feature.

Plastic injection manufacturing: Review the fundamental design rules behind the plastic injection manufacturing process:

- Maintain a constant wall thickness. Inconsistent wall thickness creates stress. Utilize the Shell feature to create constant wall thickness.

- Create a radius on all corners. No sharp edges. Sharp edges create vacuum issues when removing the mold. Utilize the Fillet feature to remove sharp edges.

- Allow a minimum draft angle of 1 degree. Draft sides and internal ribs. Draft angles assist in removing the part from the mold. Utilize the Draft feature or Rib Draft angle option.

Revolved Cut Thin: A feature used to remove material by rotating a sketched open profile around a centerline.

Rib: A feature used to add material between contours of existing geometry. Use Ribs to add structural integrity to a part.

Shape: A feature used to deform a surface created by expanding, constraining, and tightening a selected surface. A deformed surface is flexible, much like a membrane.

Suppressed: A feature or component not loaded into memory. A feature is suppressed or unsuppressed. A component is suppressed or resolved. Suppress features and components to improve model rebuild time.

Stereo Lithography (STL) format: STL format is the type of file format requested by Rapid Prototype manufacturers.

Swept: A Swept Boss/Base feature adds material. A Swept Cut removes material. A Swept feature requires a profile sketch and a path sketch. A Swept feature moves a profile along a path. The Swept Boss tool is located on the Features toolbar. The Swept Cut tool is located in the Insert menu.

Questions

1. Identify the function of the following features:
   - Swept
   - Revolved Thin Cut
   - Lofted
   - Rib
   - Draft
   - Linear Pattern
   - Wrap

2. Describe a Suppressed feature. Why would you suppress a feature?

3. The Rib features require a sketch, thickness and a _____ direction.

4. What is a Pierce relation?

5. Describe how to create a thread using the Swept feature. Provide an example.

6. Explain how to create a Linear Pattern feature. Provide an example.

7. Identify two advantages of utilizing Convert Entities in a sketch to obtain the profile.

8. How is symmetry built into a:
   - Sketch. Provide a few examples.
   - Feature. Provide a few examples.

9. Define a Guide Curve. Identify the features that utilize Guide Curves.

10. Identify the differences between a Draft feature and the Draft Angle option in the Extruded Boss/Base feature.

11. Describe the differences between a Circular Pattern feature and a Linear Pattern feature.

12. Identify the advantages of the Convert Entities tool.

13. True or False. A Loft feature can only be inserted as the first feature in a part. Explain your answer.

14. A Swept feature adds and removes material. Identify the location on the Main Pull down menu that contains the Swept Cut feature. Hint: SolidWorks Help Topics.

15. Describe the difference between the Distance Mate option and the Coincident Mate option. Provide an example.

Exercises

Exercise 5.1: Simple Wrap feature using the Sketch Entities text tool.

Create a simple Wrap feature using the Deboss option and the Sketch Text tool.

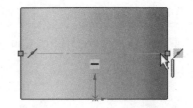

- Copy and open Wrap 5-1 from the Chapter 5 Homework folder on the DVD in the book. The Wrap 5-1 FeatureManager is displayed.

- Create a sketch on the Right plane with a horizontal construction line midpoint on Extrude1 to center the Wrap Sketch text.

Create text for the Wrap feature.
- Click the Tools, Sketch Entities, Text from the Main menu. The Sketch Text PropertyManager is displayed.

- Click the construction line from the Graphics window. Line1 is displayed in the Curves box. Enter Made in USA in the Text box.

- Click the Center Align button. Uncheck the Use document font box. Enter 150% in the Width Factor box. Enter 120% in the Spacing box.

- Click OK from the Sketch Text PropertyManager. Rebuild the model.

- Click Sketch2 from the FeatureManager. Click the Wrap Features tool. The Wrap PropertyManager is displayed. Sketch2 is displayed in the Source Sketch box.

- Select the Deboss option.

- Click the right cylindrical face of Extrude1. Face<1> is displayed in the Face for Wrap Sketch box. Enter 10mm for Depth.

- Click OK from the Wrap PropertyManager. View the results.

- Close the model.

This feature wraps a sketch onto a planar or non-planar face. You can create a planar face from cylindrical, conical, or extruded models. You can also select a planar profile to add multiple, closed spline sketches. The Wrap feature supports contour selection and sketch reuse.

Exercise 5.2: Wrap feature with Circular pattern.

Create a Wrap feature using the Emboss option and the Sketch Text tool.

- Copy and open Wrap 5-2 from the Chapter 5 Homework folder on the DVD in the book.

- Create a sketch on Plane1. This is your Sketch plane. Click Sketch.

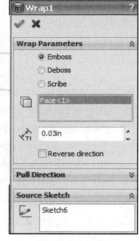

Create text for the Wrap feature around the top section of the cone.

- Click the Tools, Sketch Entities, Text from the Main menu. The Sketch Text PropertyManager is displayed.

- Click inside the Text box. Enter Eat it all in the Text box.

- Click the top section (face) of the cone (shell3).

- Click OK from the Sketch Text PropertyManager. Rebuild the model.

Create the Wrap feature.

- Click the Wrap Features tool. Click Sketch6 from the fly-out FeatureManager. The Wrap PropertyManager is displayed. Sketch6 is displayed in the Source Sketch box.

- Select the Emboss option. Click the top section of the cone (shell3). Face<1> is displayed in the Face for Wrap Sketch box.

- Enter .03in for thickness.

Create a Circular pattern of the Wrap feature.

- Click Circular Pattern from the Features tab.

- Check the Equal spacing box. Enter 5 instances. Click inside the Features to Pattern box. Click Wrap1 from the fly-out FeatureManager.

- Click inside the Pattern Access box. Click the Temporary Axis. Click OK from the Circular Pattern PropertyManager. View the results.

- Close the model.

Exercise 5.3: HOOK Part

Create the HOOK part. Create the HOOK with a Swept feature. View the illustrated FeatureManager. Note: Not all dimensions are provided. Your HOOK Part will vary.

A Swept feature adds material by moving a profile along a path. A simple Swept feature requires two sketches. The first sketch is called the path. The second sketch is called the profile. The profile and path are sketched on perpendicular planes. Remove all Tangent edges in the final model.

The Swept feature uses:

- A path sketched on the Right Plane.

- A profile sketched on the Top Plane.

- Utilize a Dome feature to create a spherical feature on a circular face. A Swept Cut feature removes material.

- Utilize a Swept Cut feature to create the Thread for the HOOK part.

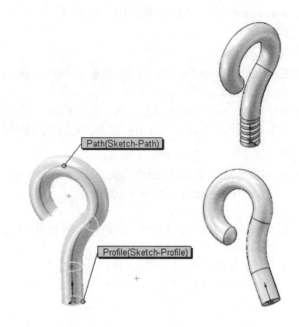

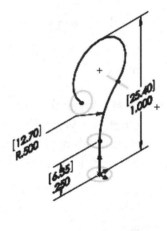

- Create the Swept Path.

- Create the Swept Profile.

- Insert a Swept feature.

- Use the Dome feature, .050in, [1.27].

- Create the Threads. Sketch a circle ∅.020in, [.51mm] on the Right Plane for the Thread profile.

- Use Helix Curve feature for the path. Pitch 0.050in, Revolution 4.0, Starting Angle 0.0deg. Chamfer the bottom face.

- Create the Thread Profile.

- Insert The Swept Cut featue.

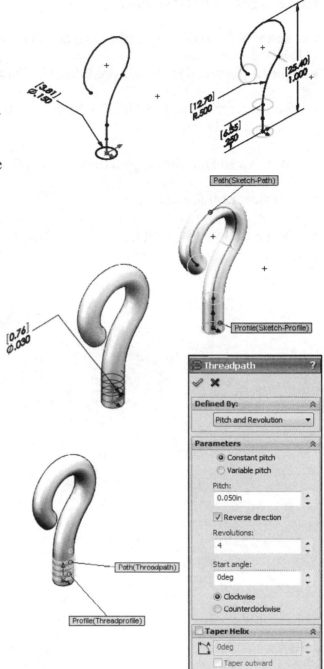

Exercise 5.4: WEIGHT Part

Create the WEIGHT part. Utilize the Loft Base Feature.

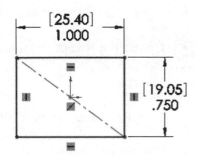

- The Top Plane and Plane1 are 0.5in, [12.7mm] apart.

- Sketch a rectangle 1.000in, [25.4mm] x .750in, [19.05] on the Top Plane.

- Sketch a square .500in, [12.7mm]on Plane1.

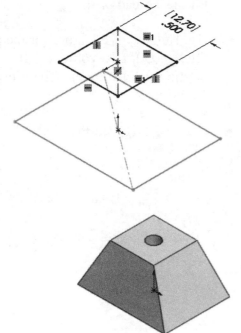

- Create a Loft feature.

- Add a centered ∅.150in, [3.81mm] Thru Hole.

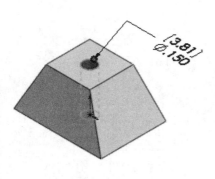

Exercise 5.5: Hole Wizard, Rib and Linear Pattern Features

Create the part from the illustrated A-ANSI Third Angle drawing: Front, Top, Right and Isometric views.

- Apply 6061 Alloy material.

- Calculate the volume of the part and locate the Center of mass.

- Think about the steps that you would take to build the model! **Note: ANSI standard states, "Dimensioning to hidden lines should be avoided wherever possible. However, sometimes it is necessary as below.**

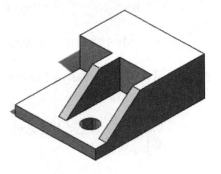

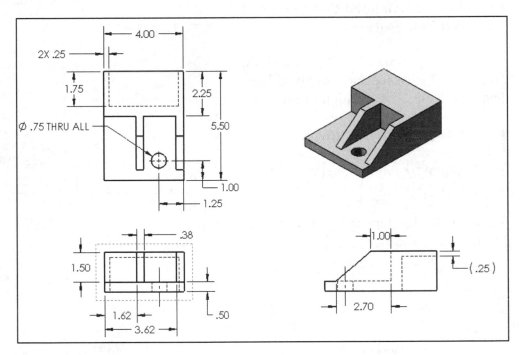

Exercise 5.6: Shell feature

Create the illustrated part with the Extruded Boss/Base, Fillet and Shell features. Note: The location of the Origin.

- Dimensions are not provided. Design your case to hold a bar of soap. Apply ABS material to the model. Think about the steps that you would take to build the model.

Exercise 5.7: Revolved Base, Hole Wizard, and Circular Pattern features

Create the illustrated ANSI part with the Revolved Base, Hole Wizard, and Circular Pattern features. Note: The location of the Origin.

- Dimensions are not provided.

- Apply PBT General Purpose Plastic material to the model.

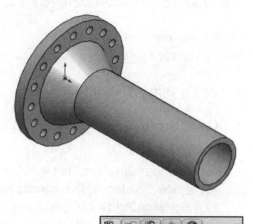

Think about the steps that you would take to build the model.

Exercise 5.8: Extruded Boss/Base and Revolved Boss features

Create the illustrated ANSI part with the Extruded Boss/Base and Revolved Boss features. Note: The location of the Origin.

- Dimensions are not provided.

- Apply 6061 Alloy material to the model.

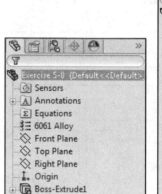

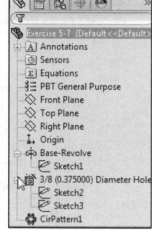

Think about the steps that you would take to build the model.

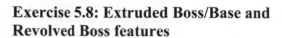

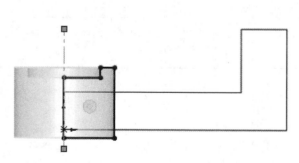

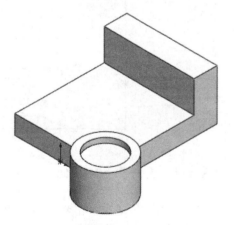

Exercise 5.9: - Exploded View Drawing

Create an Exploded View of the
FLASHLIGHT assembly. Create a
FLASHLIGHT assembly drawing.

- Insert a Bill of Materials into the
  drawing.

- Insert Balloons.

- Utilize Custom Properties to add Part
  No. and Material for each
  FLASHLIGHT component.

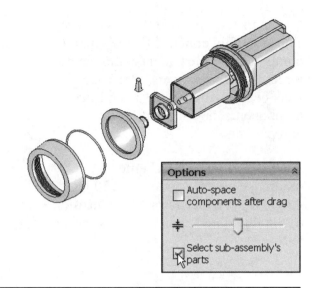

ITEM NO.	PART NUMBER	DESCRIPTION	MATERIAL	QTY.
1	A1-1-10	HOUSING FOR 6VOLT FLASHLIGHT		1
2	A1-1-C10	BUTTON STYLE		1
3	B1-200-1	LENS CAP for 6V FLASHLIGHT		1
4	C1-100-510	O-RING FOR LENS		1
5	A-20-5200-1	LENS WITH SHIELD		1
6	B-22-33	BULB FOR LENS		1
7	D-445-510	BATTERY, 6-VOLT		1
8	D-445-155	BATTERY PLATE, FOR 6-VOLT		1

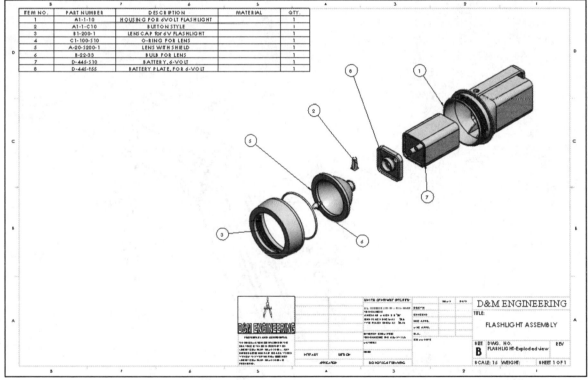

Exercise 5.10: LINKAGE Assembly

Create the ANSI LINKAGE assembly.
The assembly contains the following
components: One Clevis component,
three Axle components, two 5 Hole
Link components, two 3 Hole Link
components, and six Collar
components.

All holes Ø.190 THRU unless
otherwise noted. Angle A = 150deg.
Angle B = 120deg. Note: The location
of the illustrated coordinate system:
(+X, +Y, +Z).

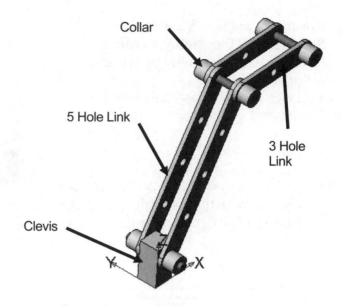

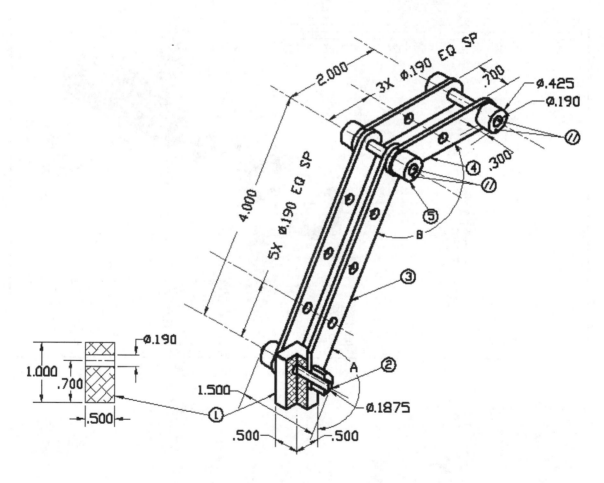

- Clevis, (Item 1): Material: 6061 Alloy. The two 5 Hole Link components are positioned with equal Angle mates, (150deg) to the Clevis component.

- Axle, (Item 2): Material: AISI 304. The first Axle component is mated Concentric and Coincident to the Clevis. The second and third Axle components are mated Concentric and Coincident to the 5 Hole Link and the 3 Hole Link components respectively.

- 5 Hole Link, (Item 3): Material: 6061 Alloy. Material thickness = .100in. Radius = .250in. Five holes located 1in. on center. The 5 Hole Link components are positioned with equal Angle mates, (120deg) to the 3 Hole Link components.

- 3 Hole Link, (Item 4): Material: 6061 Alloy. Material thickness = .100in. Radius = .250in. Three holes located 1in. on center. The 3 Hole Link components are positioned with equal Angle mates, (120deg) to the 5 Hole Link components.

- Collar, (Item 5): Material: 6061 Alloy. The Collar components are mated Concentric and Coincident to the Axle and the 5 Hole Link and 3 Hole Link components respectively.

Think about the steps that you would take to build the illustrated LINKAGE assembly. Identify and build the required parts. Identify the first fixed component. Position the Base component features in the part so they are in the correct orientation in the assembly. Insert the required Standard mates. Locate the Center of mass of the model with respect to the illustrated coordinate system.

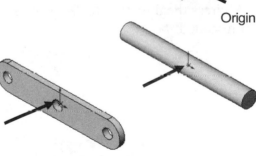

Origin

- Start with the Clevis part. Create the Clevis part.

- Create the Axle part.

- Create the 3 Hole Link part.

- Create the 5 Hole Link part. Apply the Save as Copy command to save the 5 Hole Link. Conserve design time. You can either: create a new part, apply the ConfigurationManager, or use the Save as Copy command to modify the existing 3 Hole Link to the 5 Hole Link part

- Create the Collar part.

- Create the LINKAGE assembly. The illustrated assembly contains the following: One Clevis component, three Axle components, two 5 Hole Link components, two 3 Hole Link components, and six Collar components. All holes Ø.190 THRU unless otherwise noted. Angle A = 150deg. Angle B = 120deg.

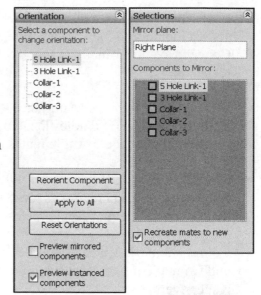

- Insert the required parts and standard mates. Note: create one side of the assembly, and then apply the Mirror Components tool.

- Create the coordinate system location for the assembly.

- Select the front right vertex of the Clevis component as illustrated.

- Click Insert, Reference Geometry, Coordinate System from the Menu bar.

- Click the right bottom edge of the Clevis component.

- Click the front bottom edge of the Clevis component as illustrated.

- Address the direction for X, Y Z as illustrated.

Exercise 5.11: PULLEY Assembly

Create the illustrated ANSI
PULLEY assembly. Locate the
Center of mass using the illustrated
coordinate system.

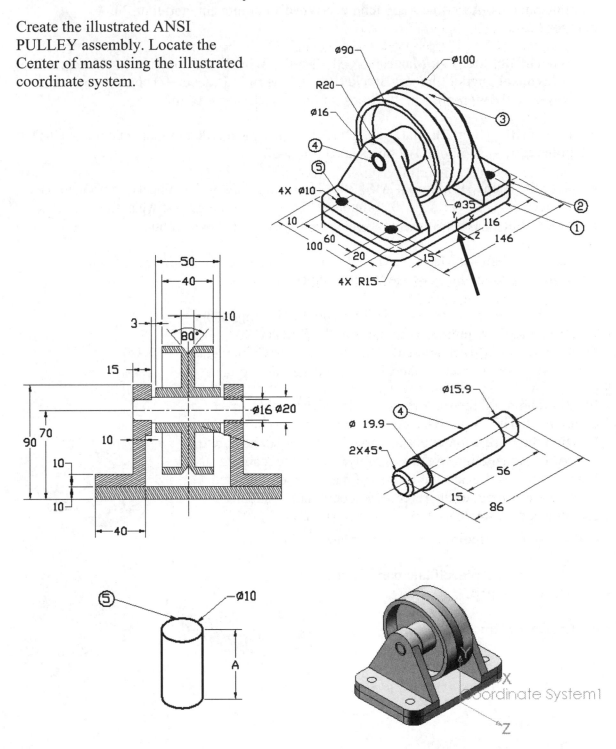

The assembly contains the following: One WheelPlate component, two Bracket100 components, one Axle40 component, one Wheel1 component, and four Pin-4 components.

- WheelPlate, (Item 1): Material: AISI 304. The WheelPlate contains 4-Ø10 holes. The holes are aligned to the left Bracket100 and the right Bracket100 components. All holes are through-all. The thickness of the WheelPlate = 10 mm.

- Bracket100, (Item 2): Material: AISI 304. The Bracket100 component contains 2-Ø10 holes and 1- Ø16 hole. All holes are through-all.

- Wheel1, (Item 3): Material AISI 304: The center hole of the Wheel1 component is Concentric with the Axle40 component. There is a 3mm gap between the inside faces of the Bracket100 components and the end faces of the Wheel hub.

- Axle40, (Item 4): Material AISI 304: The end faces of the Axle40 are Coincident with the outside faces of the Bracket100 components.

- Pin-4, (Item 5): Material AISI 304: The Pin-4 components are mated Concentric to the holes of the Bracket100 components, (no clearance). The end faces are Coincident to the WheelPlate bottom face and the Bracket100 top face.

Identify and build the required parts. Identify the first fixed component. This is the Base component of the assembly. Position the Base component features in the part so they are in the correct orientation in the assembly. Insert the required Standard mates. Locate the Center of mass of the illustrated model with respect to the referenced coordinate system. The referenced coordinate system is located at the bottom, right, midpoint of the Wheelplate.

- Start with the WheelPlate part. Create the WheelPlate part.

- Create the Bracket100 part.

- Create the Axle40 part.

- Create the Wheel1 part. Insert two features: Revolve1 and Cut-Revolve1. Think about the steps that you would take to build this part. Identify the location of the part Origin.

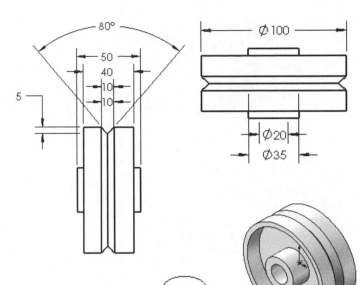

- Create the Pin-4 part.

- Create the PULLEY assembly. Remember, The assembly contains the following: one WheelPlate component, two Bracket100 components, one Axle40 component, one Wheel1 component, and four Pin-4 components.

- Insert the required parts and standard mates.

- Create the coordinate system location for the assembly as illustrated.

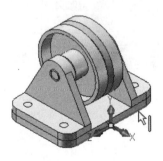

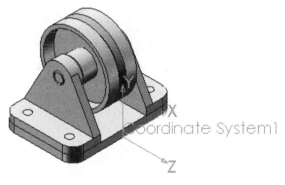

Notes:

Project 6

Top-Down Assembly Modeling and Sheet Metal Parts

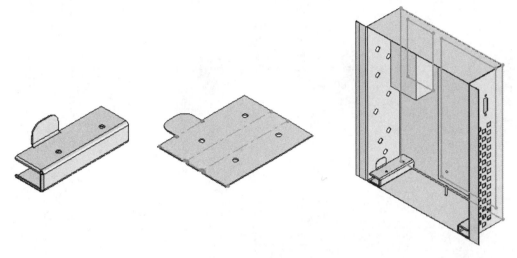

Below are the desired outcomes and usage competencies based on the completion of Project 6.

Project Desired Outcomes:	Usage Competencies:

• Three BOX configurations:
 	
   o Small
 	
   o Medium
 	
   o Large
	• Understand three key methods for the Top-down assembly modeling. Develop components In-Context with InPlace Mates.

	• Ability to import parts using the Top-down assembly method.
	• Knowledge to develop a Sheet Metal Design Table and equations.

• CABINET drawing with two configurations:
 	
   o Default - 3D formed state
 	
   o Flat - 2D flatten state
	• Ability to convert a solid part into a Sheet Metal part and insert Sheet Metal features.

	• Knowledge of Sheet Metal configurations in the drawing: 3D Formed and 2D Flat Pattern states.
• BRACKET part.	• Ability to insert Sheet Metal features: Base, Edge, Miter Flange, Hem and Flat Pattern.

Notes:

Project 6 - Top-Down Assembly Modeling and Sheet Metal Parts

Project Objective

Create three BOX configurations utilizing the Top-down assembly approach. The Top-down approach is a conceptual approach used to develop products from within the assembly. Major design requirements are translated into sub-assemblies or individual components and key relationships.

Model Sheet metal components in their 3D formed state. Manufacture the Sheet metal components in a flatten state. Control the formed state and flatten state through configurations.

Create the parameters and equations utilized to control the configurations in a Design Table.

On the completion of this project, you will be able to:

- Set Document Properties and System Options.

- Create a new Top-down assembly and sketch.

- Choose the proper Sketch plane.

- Insert Sheet Metal bends to transform a solid part into a sheet metal part.

- Use the Linear Pattern feature to create holes.

- Insert IGES format PEM® self-clinching fasteners.

- Use the Component Pattern feature to maintain a relationship between the holes and the fasteners.

- Create Global Values.

- Create and edit Equations.

- Insert a Sheet Metal Library feature and Forming tool feature.

- Address and understand general sheet metal manufacturing considerations.

- Insert Geometric relations.

- Insert sheet metal assembly features.

- Replace components in an assembly.

- Insert a Design Table.

- Create a sheet metal part by starting with a solid part and inserting a Shell feature, Rip feature, and Sheet Metal Bends feature.

- Create a sheet metal part by starting with a Base Flange feature.

- Use the following SolidWorks features: Extruded Boss/Base, Extruded Cut, Shell, Hole Wizard, Linear Pattern and Component Pattern, Rip, Insert Bends, Flatten, Hem, Edge Flange, Jog, Flat Pattern, Break Corner/Corner Trim, Miter Flange, Base Flange/Tab, Unfold/Fold, Mirror Components and Replace Component.

Project Situation

You now work for a different company. Life is full with opportunities. You are part of a global project design team that is required to create a family of electrical boxes for general industrial use. You are the project manager.

You receive a customer request from the Sales department for three different size electrical boxes.

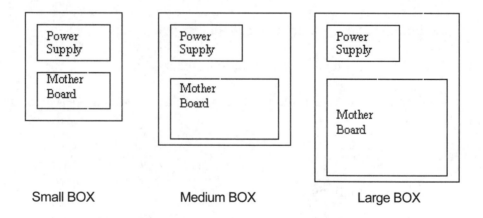

Delivery time to the customer is a concern. You work in a concurrent engineering environment. Your company is expecting a sales order for 5,000 units in each requested BOX configuration.

The BOX contains the following key components:

- Power supply

- Motherboard

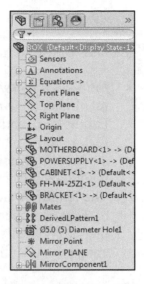

The size of the power supply is the same for all three boxes. The size of the power supply is the constant. The size of the BOX is a variable. The available space for the motherboard is dependent on the size of the BOX. The depth of the three boxes is 100mm.

You contact the customer to discuss and obtain design options and product specifications. Key customer requirements:

Three different BOX sizes:

- 300mm x 400mm x 100mm   Small

- 400mm x 500mm x 100mm   Medium

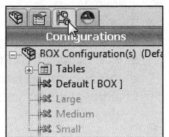

- 550mm x 600mm x100mm   Large

- Adequate spacing between the power supply, motherboard and internal walls

- Field serviceable

You are responsible to produce a sketched layout from the provided critical dimensions. You are also required to design the three boxes. The BOX is used in an outside environment.

Top-Down Assembly Modeling

Top-down assembly is a conceptual method used to develop products from within the assembly. In Top-down assembly modeling, one or more features of the part are defined by something in the assembly. Example: A Layout sketch or the geometry of another part. The design intent of the model, examples: the size of the features, location of the components in the assembly, etc. take place from the top level of the assembly and translates downward from the assembly to the component.

A few advantages of the Top-down assembly modeling are that design details of all components are not required and much less rework is required when a design change is needed. The model requires individual relationships between components. The parts know how to update themselves based on the way you created them.

Designers usually use Top-down assembly modeling to lay their assemblies out and to capture key design aspects of custom parts specific in the assemblies. There are three key methods to use for the Top-down assembly modeling approach. They are:

- *Individual features method.* The Individual features method provides the ability to reference the various components and sub-components in an existing assembly. Example: Creating a structural brace in a box by using the Extruded Boss/Base feature tool. You might use the Up to Surface option for the End Condition and select the bottom of the box, which is a different part. The Individual features method maintains the correct support brace length, even if you modify the box in the future. The length of the structural brace is defined in the assembly. The length is not defined by a static dimension in the part. The Individual features method is useful for parts that are typically static but have various features which interface with other assembly components in the model.

- *Entire assembly method.* The Entire assembly method provides the ability to create an assembly from a layout sketch. The layout sketch defines the component locations, key dimensions, etc. A major advantage of designing an assembly using a layout sketch is that if you modify the layout sketch, the assembly and its related parts are automatically updated. The entire assembly method is useful when you create changes quickly, and in a single location.

You can create an assembly and its components from a layout of sketch blocks.

- *Complete parts method.* The Complete parts method provides the ability to build your model by creating new components In-Context of the assembly. The component you build is actually mated to another existing component in the assembly. The geometry for the component you build is based upon the existing component.

This method is useful for parts like brackets and fixtures, which are mostly or completely dependent on other parts to define their shape and size.

Whenever you create a part or feature using the Top-Down approach, external references are created to the geometry you referenced.

The Top-down assembly approach is also referred to as "In-Context design".

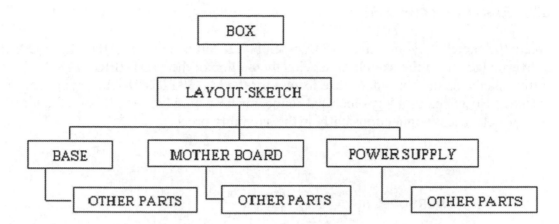

Conceptual Top-down design approach

Use a combination of the three assembly methods to create the BOX assembly. Consider the following in a preliminary design product specification:

- What are the major design components?

- The motherboard and power supply are the major design components.

- What are the key design constraints?

The customer specified three different BOX sizes. How does each part relate to the other? From experience and discussions with the electrical engineering department, a 25mm physical gap is required between the power supply and the motherboard. A 20mm physical gap is required between the internal components and the side of the BOX.

How will the customer use the product? The customer does not disclose the specific usage of the BOX. The customer is in a very competitive market.

What is the most cost-effective material for the product? Aluminum is cost-effective, strong, relatively easy to fabricate, corrosion resistant, and non-magnetic.

Incorporate the design specifications into the BOX assembly. Use a sketch, solid parts and sheet metal parts. Obtain additional parts from your vendors.

BOX Assembly Overview

Create the sketch "layout" in the BOX assembly. Insert a new part, MOTHERBOARD. Convert edges from the sketch to develop the outline of the MOTHERBOARD. Use the outline sketch as the Extruded Base feature for the MOTHERBOARD component. An Extruded Boss locates a key electrical connector for a wire harness. Design the location of major electrical connections early in the assembly process.

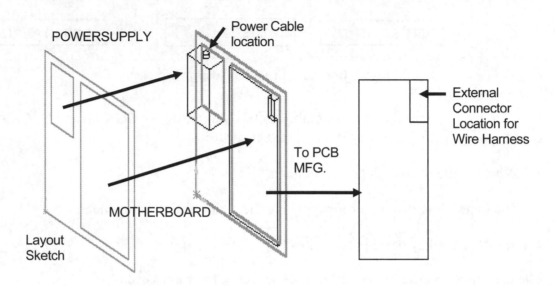

Insert a new part, POWERSUPPLY. Convert edges from the sketch to develop the outline of the POWERSUPPLY. Use the outline as the Extruded Boss/Base (Boss-Extude1) feature for the POWERSUPPLY component.

An Extruded Boss/Base feature represents the location of the power cable.

Add additional features to complete the POWERSUPPLY.

POWER SUPPLY:
Send the part to a colleague to
add additional features.

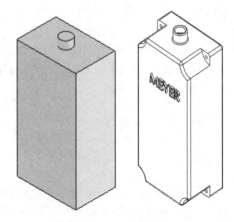

Insert a new part, CABINET. Convert edges from the sketch.

Use the outline sketch as the Extruded Base feature for the CABINET component.

Shell the Extruded feature. Add a Rip feature to the four CABINET corners.

Insert sheet metal bends to transform a solid part into a sheet metal part.

Sheet metal components utilize features to create flanges, cuts and forms.

Save Model sheet metal components in their 3D formed state.

Manufacture sheet metal components in their 2D flatten state.

Work with your sheet metal manufacturer. Discuss cost effective options.

Create a sketched pattern to add square cuts instead of formed louvers.

Create a Linear Pattern feature of holes for the CABINET.

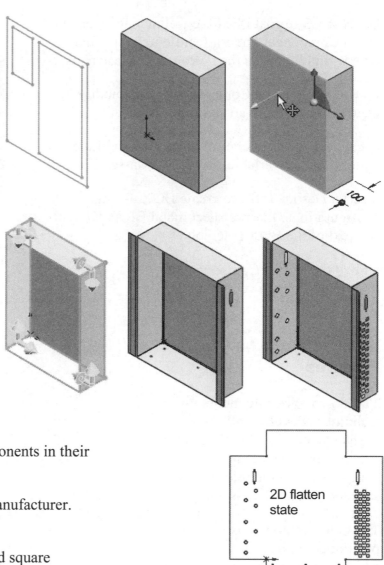

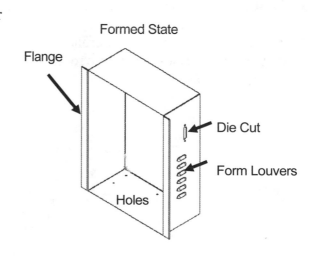

2D flatten state

Formed State

Flange

Die Cut

Form Louvers

Holes

Insert IGES format PEM® self-clinching fasteners. Use a Component Pattern feature to maintain a relationship between the holes and the fasteners.

Utilize the Replace Component tool to modify the Ø5.0mm fastener to an Ø 4.0mm fastener.

Create equations to control the sketch and Linear Pattern feature variables in the BOX assembly.

Insert a Design Table to create BOX assembly configurations. Insert a sheet metal BRACKET In-Context of the BOX assembly.

Utilize the Mirror Components feature to create a right and left hand version of the BRACKET.

PEM
Fasteners

Linear step
& repeat
sketched
pattern

InPlace Mates and In-Context Features

Top-down assembly modeling techniques develop InPlace mates. An InPlace mate is a Coincident mate between the Front Plane of the new component and the selected plane or selected face of the assembly.

The new component is fully positioned with an InPlace mate. No additional mates are required. The component is now dependent on the assembly.

An InPlace mate creates an External reference. The component references geometry in the assembly.

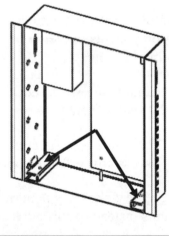

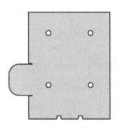

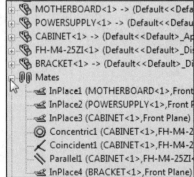

If the referenced document changes, the dependent document changes. The InPlace mates are listed under the Mates folder in the FeatureManager. Example:

⊟ 🔗 Mates
    🔗 InPlace1 (MOTHERBOARD<1>,Front Plane) .

In a Top-down assembly, you create In-Context features. An In-Context feature is a feature that references geometry from another component. The In-Context feature has an External reference to that component. If you change geometry on the referenced component, the associated In-Context feature also changes.

An In-Context feature is indicated in the FeatureManager with the ">" symbol. The External references exist with the feature, Sketch plane or sketch geometry. The update path to the referenced component is contained in the assembly. When the update path is not valid or the referenced component is not found the "?" symbol is displayed in the FeatureManager.

In the BOX assembly, develop InPlace Mates for the MOTHERBOARD, POWERSUPPLY and CABINET components.

Create External references with the Convert Entities Sketch tool and extracting geometry from the sketch of the BOX assembly.

Working with In-Context features requires practice and time. Planning and selecting the correct reference and understanding how to incorporate changes are important. Explore various techniques using InPlace mates and External references developed In-Context of another component.

Assembly Modeling Techniques with InPlace Mates:
Plan the Top-down assembly method. Start from a sketch or with a component in the assembly.
Prepare the references. Utilize descriptive feature names for referenced features and sketches.
Utilize InPlace Mates sparingly. Load all related components into memory to propagate changes. Do not use InPlace Mates for purchased parts or hardware.
Group references. Select references from one component at a time.
Ask questions! Will the part be used again in a different assembly? If the answer is yes, do not use InPlace Mates. If the answer is no, use InPlace Mates.
Will the part be used in physical dynamics or multiple configurations? If the answer is yes, do not use InPlace mates.
Work in the Edit Component tool to obtain the required External references in the assembly. Create all non-referenced features in the part, not in the assembly.
Obtain knowledge of your company's policy on InPlace mates or develop one as part of an engineering standard.

Part Template and Assembly Template

The parts and assemblies in this project require Templates created in Project 1 and Project 2. The MY-TEMPLATES tab is located in the New dialog box. If your MY-TEMPLATES folder contains the ASM-MM-ANSI and PART-MM-ANSI Templates, skip the next few steps.

Activity: Create an Assembly and Part Template

Create the Assembly Template for the BOX.

1) Click **New** ▢ from the Menu bar.

2) Double-click the **Assembly** icon from the defaults Templates tab.

3) Click **Cancel** ✖ from the Begin Assembly PropertyManager. The Assembly FeatureManager is displayed.

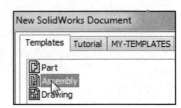

Set the Assembly Document Template options.

4) Click **Options** 🗐, **Document Properties** tab from the Menu bar.

5) Select **ANSI [ISO]** for the Overall drafting standard.

6) Click **Units**. Click **MMGS** for Units system.

7) Select **.12** (two decimal places) for Length Basic Units.

8) Select **None** for Angular units Decimal places.

9) Click **OK** from the Document Properties - Units dialog box.

Save the Assembly template.

10) Click **Save As** from the Menu bar.

11) Click the **Assembly Templates (*.asmdot)** from the Save As type box.

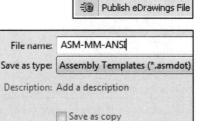

12) Select **ENGDESIGN-W-SOLIDWORKS\MY-TEMPLATES** for Save in folder.

13) Enter **ASM-MM-ANSI, [ASM-MM-ISO]** for File name. Click **Save**. The ASM-MM-ANSI FeatureManager is displayed.

Create the Part Template for the BOX.

14) Click **New** ▢ from the Menu bar.

15) Double-click the **Part** icon from the default Templates tab. The Part FeatureManager is displayed.

16) Click **Options** 📧, **Document Properties** tab from the Menu bar.

17) Select **ANSI, [ISO]** for the Overall drafting standard.

Set the Units.
18) Click **Units**. Click **MMGS** for Unit system

19) Select **.12** (two decimal places) for Length Basic Units.

20) Select **None** for Angular units Decimal places.

21) Click **OK** from the Document Properties - Units dialog box.

Save the Part Template.
22) Click **Save As** from the Menu bar.

23) Click **Part Templates (*.prtdot)** from the Save As type box.

24) Select **ENGDESIGN-W-SOLIDWORKS\MY-TEMPLATES** for Save in folder.

25) Enter **PART-MM-ANSI, [PART-MM-ISO]** in the File name text box.

26) Click **Save**.

Close all active documents.
27) Click **Windows**, **Close All** from the Menu bar.

BOX Assembly and Sketch

The BOX assembly utilizes both the Top-down assembly design approach and the Bottom-up assembly design approach. Begin the BOX assembly with a sketch.

Create a New assembly named BOX. Insert a sketch to develop component space allocations and relations in the BOX assembly.

Add dimensions and relations to the sketch. Components and assemblies reference the sketch.

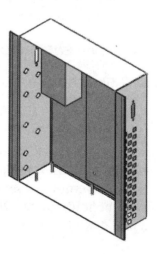

☼ In Layout-based assembly design, you can switch back and forth between Top-down and Bottom-up design methods.

The BOX assembly contains the following key components:

- POWERSUPPLY

- MOTHERBOARD

The minimum physical spatial gap between the MOTHERBOARD and the POWERSUPPLY is 25mm. The minimum physical spatial gap between the MOTHERBOARD, POWERSUPPLY and the internal sheet metal BOX wall is 20mm.

After numerous discussions with the electrical engineer, you standardize on a POWERSUPPLY size: 150mm x 75mm x 50mm. You know the overall dimensions for the BOX, POWERSUPPLY and MOTHERBOARD. You also know the dimensional relationship between these components.

Now you must build these parameters into the design intent of the sketch. There is no symmetry between the major components. Locate the sketch with respect to the BOX assembly Origin.

Activity: Box Assembly and Sketch Layout

Create the BOX assembly.

28) Click **New** ⬜ from the Menu bar.

29) Click the **MY-TEMPLATES** tab.

30) Double-click **ASM-MM-ANSI**.

31) Click **Cancel** ✖ from the Begin Assembly PropertyManager. The Assembly FeatureManager is displayed.

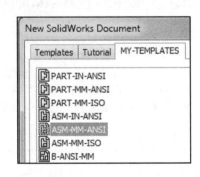

Save the BOX assembly.

32) Click **Save** 💾.

33) Select **ENGDESIGN-W-SOLIDWORKS\PROJECTS** for folder.

34) Enter **BOX** for File name. Click **Save**. The BOX FeatureManager is displayed.

Each part in the BOX assembly requires a template. Utilize the PART-MM-ANSI [PART-MM-ISO] and ASM-MM-ANSI [ASM-MM-ISO] templates. Under the Option, System Options, Default Templates section prompts you to select a different document template. Otherwise, SolidWorks utilizes the default templates.

Set the System Options.

35) Click **Options** 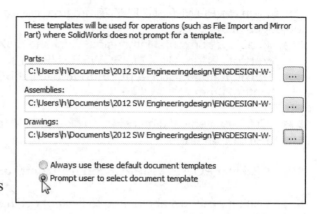, **Default Templates** from the Menu bar.

36) Click the **Prompt user to select document template** button.

37) Click **OK**.

The 2D sketch contains the 2D relationships between all major components in the BOX assembly.

Select the Sketch plane for the sketch.
38) Right-click **Front Plane** from the FeatureManager.

Sketch the profile of the BOX.
39) Click **Sketch** ✏ from the Context toolbar. The Sketch toolbar is displayed. Click **Corner Rectangle** ▢ from the Sketch toolbar.

40) Click the **Origin**. Sketch a **rectangle** as illustrated. The first point is coincident with the Origin.

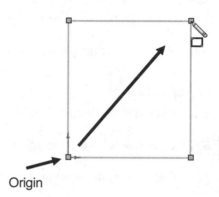

Origin

Add dimensions.
41) Click **Smart Dimension** ✏ from the Sketch toolbar.

42) Click the **left vertical line**. Click a **position** to the left of the profile. Enter **400**mm.

43) Click the **bottom horizontal line**. Click a **position** below the profile. Enter **300**mm.

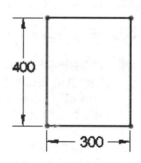

Fit the model to the Graphics window.
44) Press the **f** key.

Sketch the 2D profile for the POWER SUPPLY.
45) Click **Corner Rectangle** ▢ from the Sketch toolbar.

46) Sketch a small **rectangle** inside the BOX as illustrated.

Add dimensions.
47) Click **Smart Dimension** ✏ from the Sketch toolbar.

48) Create a vertical dimension. Enter **150**mm.

49) Create a horizontal dimension. Enter **75**mm.

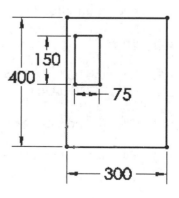

Dimension the POWERSUPPLY 20mm from the left and top edges of the BOX.

50) Create a horizontal dimension. Click the BOX left **vertical edge**. Click the POWERSUPPLY left **vertical edge**. Click a **position** above the BOX. Enter **20**mm.

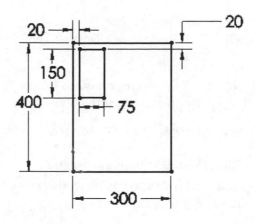

51) Create a vertical dimension. Click the BOX **top horizontal edge**.

52) Click the POWERSUPPLY **top horizontal edge**.

53) Click a **position** above and to the right of the BOX. Enter **20**mm.

Sketch the profile for the MOTHERBOARD.

54) Click **Corner Rectangle** ☐ from the Sketch toolbar.

55) Sketch a **rectangle** to the right of the POWERSUPPLY as illustrated.

Add dimensions.

56) Click **Smart Dimension** ◇ from the Sketch toolbar.

57) Click the **left vertical edge** of the MOTHERBOARD.

58) Click the **right vertical edge** of the POWERSUPPLY.

59) Click a **position** above the BOX.

60) Enter **25**mm.

61) Click the horizontal **top edge** of the MOTHERBOARD. Click the horizontal **top edge** of the BOX. Click a position off the profile. Enter **20**mm.

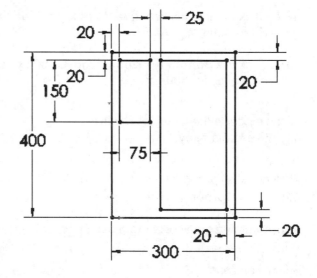

62) Click the vertical **right edge** of the MOTHERBOARD.

63) Click the vertical **right edge** of the BOX. Click a **position below** the Box.

64) Enter **20**mm.

65) Click the horizontal **bottom edge** of the MOTHERBOARD.

66) Click the horizontal **bottom edge** of the BOX.

67) Click a **position** to the right of the BOX. Enter **20**mm.

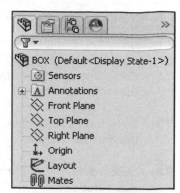

Sketch1 is fully defined and is displayed in black. Insert a Collinear relation between the top lines of the two inside rectangles if required.

Close and rename the sketch.

68) Right-click **Exit Sketch** .

69) Rename **Sketch1** to **Layout**.

Save the BOX assembly.

70) Click **Save** .

What happens when the size of the MOTHERBOARD changes? How do you ensure that the BOX maintains the required 20mm spatial gap between the internal components and the BOX boundary? How do you design for future revisions? Answer: Through Link Values.

Note: You can no longer create linked values (also referred to as shared values or linked dimensions) in SolidWorks 2012 and later. Instead, you use global variables for the same purpose.

Global Values and Equations

In previous versions of SolidWorks, linked or shared values were used to link two or more dimensions without using equations or relations. Changing any one of the linked values would change the other to which it was linked. Existing linked values are still supported, but you cannot create new linked values in SolidWorks 2012 and later. Instead, you use global variables for the same purpose as linked values. Global variables are much easier to find, change, and manage than linked values.

To use a global variable to link dimensions: 1.) Create a global variable in the Equations dialog box or the Modify dialog box for dimensions, and 2.) Set two or more dimensions equal to the global variable.

The project goal is to create three boxes of different sizes. Ensure that the models remain valid when dimensions change for various internal components. This is key! Apply Global values and equations.

Activity: Create Global variables and Equations

Display the Equations folder in the FeatureManager and display the dimensions in the Graphics window.

71) Click **Options**, **System Options** tab from the Main menu.

72) Click the **FeatureManager** folder from the System Options column.

73) Select **Show** from the Equations drop-down menu as illustrated.

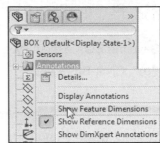

74) Click **OK** from the Systems Option - FeatureManager. View the Equations folder in the FeatureManager.

75) Right-click the **Annotations** folder in the FeatureManager.

76) Click **Show Feature Dimensions**.

77) Double-click **Layout** from the FeatureManager.

Insert a Global variable.

78) Right-click the **Equations** folder from the FeatureManager.

79) Click **Manager Equations**. The Equations, Global Variables, and Dimensions dialog box is displayed.

80) Click a **position** in the first cell under Global Variables.

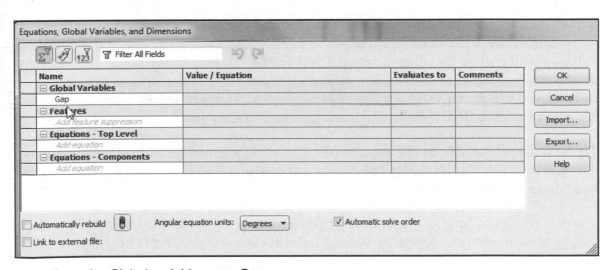

81) Enter the Global variable name **Gap**.

82) Press the **Tab** key.

83) Click the **lower right horizontal dimension 20** in the Graphics window. The dimension name is displayed in the Value / Equation column. Note: Your dimension name may be different. A green check mark indicates the value can be calculated.

84) Press the **Tab** key. The Evaluates to column displays 20.

85) Enter **Gap from the Cabinet wall** for comment. The first Global variable Gap is defined. Next define equations to equate the other variables to Gap.

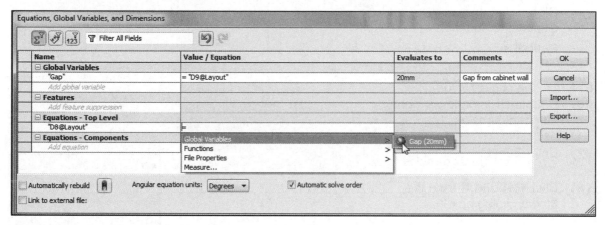

86) Click **inside** the first cell below Equations-Top Level.

87) Click the **lower right vertical dimension 20** in the Graphics window.

88) Slide the **mouse pointer** to the right of Global Variables to display Gap(20mm).

89) Click **Gap(20mm)**.

90) Press the **Enter** key to create a new row under Equations-Top Level.

91) Repeat the **above procedure** to enter the rest of the 20mm dimensions on the Layout sketch.

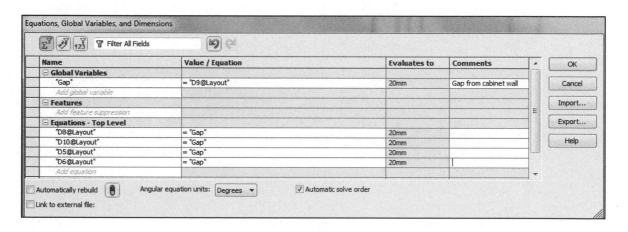

92) Click **OK** from the Equations, Global Variables, and Dimensions dialog box.

Test the Global Variables.
93) Double-click the lower right **20**mm dimension as illustrated.

94) Enter **10**mm.

95) Click **Rebuild** 🔴 from the Modify dialog box. The four Global Variables change and display the equation symbol Σ.

Return to the original value.
96) Double-click the lower right **10**mm dimension.

97) Enter **20**mm.

98) Click **Rebuild** 🔴 from the Modify dialog box.

99) Click the **Green Check mark** ✅ from the Modify dialog box. Note: All Global Variables are equal to 20mm.

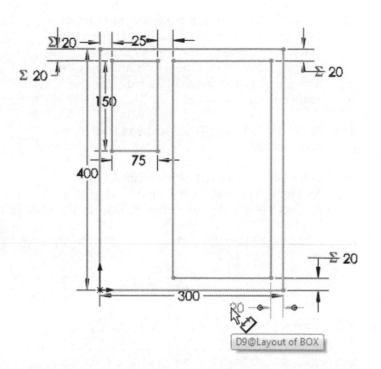

Additional dimensions are required for Equations. Each dimension has a unique variable name. The names are used as Equation variables. The default names are based on the Sketch, Feature or Part. Feature names do not have to be changed. Rename variables for clarity when creating numerous Equations.

Rename the BOX assembly width and height. The full variable name is: "box-width@ Layout". The system automatically appends Layout. If features are created or deleted in a different order, your variable names will be different.

Rename for overall box width.

100) Click the horizontal dimension **300** in the Graphics window. The Modify dialog box is displayed.

101) Delete **D2@Layout**.

102) Enter **box_width**.

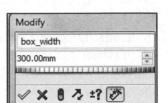

103) Click **OK** ✅ from the Modify dialog box.

Rename the overall box height.

104) Click the vertical dimension **400** in the Graphics window. The Modify dialog box is displayed.

105) Delete **D1@Layout**.

106) Enter **box height**.

107) Click **OK** ✅ from the Modify dialog box.

Display the Isometric view.

108) Click **Isometric view** 🧊 from the Heads-up View toolbar.

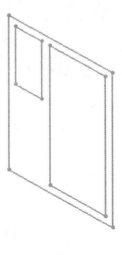

Hide all dimensions.

109) Right-click the **Annotations** folder.

110) Uncheck **Show Feature Dimensions**.

Save the BOX assembly.

111) Click **Save** 💾.

The sketch is complete. Develop the MOTHERBOARD part and the POWERSUPPLY part In-context of the BOX assembly Layout sketch.

Select the BOX assembly Front Plane to create InPlace Mates. An InPlace Mate is a Coincident Mate developed between the Front Plane of the component and the selected plane in the assembly. Utilize Convert Entities and extract geometry from the Layout Sketch to create external references for both the MOTHERBOARD part and POWERSUPPLY part. Utilize Ctrl-Tab to switch between part and assembly windows.

MOTHERBOARD-Insert Component

The MOTHERBOARD requires the greatest amount of lead-time to design and manufacture. The outline of the MOTHERBOARD is created. Create the MOTHERBOARD component from the Layout Sketch. An electrical engineer develops the Logic Diagram and Schematic required for the Printed Circuit Board (PCB). The MOTHERBOARD rectangular sketch represents the special constraints of a blank Printed Circuit Board (PCB).

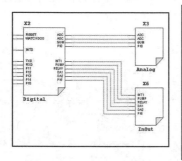
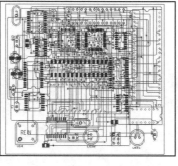
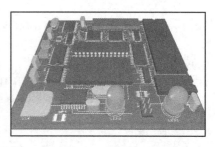

| Logic Diagram (partial) | Schematic | PCB |

Courtesy of Electronic Workbench
(www.electronicworkbench.com)

A rough design of a critical connector is located on the MOTHERBOARD in the upper right corner.

CircuitWorks Add-in is a fully integrated data interface between SolidWorks and PCB Design systems.

As the project manager, your job is to create the board outline with the corresponding dimensions from the Layout Sketch. Export the MOTHERBOARD data in an industry-standard Intermediate Data Format (IDF) from SolidWorks.

The IDF file is sent to the PCB designer to populate the board with the correct 2D electronic components.

Board Outline

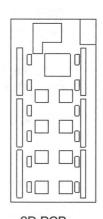

2D PCB

CircuitWorks utilizes industry-standard IDF files or PADS files, and produces the 3D SolidWorks assembly of the MOTHERBOARD. IDF and PADS are common file formats utilized in the PCB industry.

The MOTHERBOARD is fully populated with the components at the correct height. Your colleagues use the MOTHERBOARD assembly to develop other areas of the BOX assembly. An engineer develops the wire harness from the MOTHERBOARD to electrical components in the BOX.

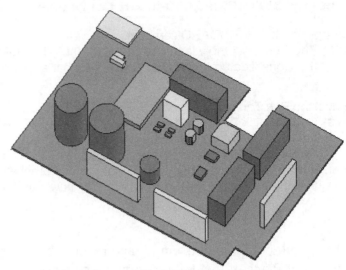

SolidWorks assembly developed with CircuitWorks
Courtesy of Computer Aided Products, Inc.

A second engineer uses the 3D geometry from each electrical component on the MOTHERBOARD to create a heat sink. As the project manager, you move components that interfere with other mechanical parts, cables and or wire harness. You distribute the updated information to your colleagues and manufacturing partners.

To save a virtual component to its own external file, right-click the component and select **Save Part (in External File)** or **Save Assembly (in External file)**.

| Hidden Tree Items |
| Rename Part |
| Save Part (in External File) |
| Add to Library |

To rename the virtual component in the assembly, right-click the **virtual component** and click **Rename Part**.

Remember, the **Prompt user to select document template** option is selected in System Options, Default Templates.

If you Do NOT want to create any External references in your part, click **Options**, **External References** and check the **Do not create references external to the model** box.

Drawings
C:\Documents and Settings\All Users\Application Data\SolidWorks   [...]

○ Always use these default document templates
◉ Prompt user to select document template

Update out-of-date linked design tables to: Prompt ▾
Assemblies
☐ Automatically generate names for referenced geometry
☑ Update component names when documents are replaced
☐ Do not create references external to the model

Activity: MOTHERBOARD-Insert Component

Create the MOTHERBOARD component.

112) The BOX assembly is the open document. Click **New Part** from the Consolidated Insert Components toolbar as illustrated.

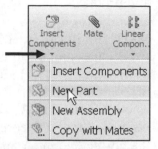

Select the Part Template.

113) Double-click **PART-MM-ANSI** from the MY-TEMPLATES tab in the New SolidWorks Document dialog box. The new part is displayed in the FeatureManager design tree with a name in the form **[Partn^assembly_name]**. The square brackets indicate that the part is a virtual component. The new Component pointer 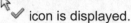 icon is displayed.

The default component is empty and requires a Sketch plane. The Front Plane of the default component is mated with the Front Plane of the BOX.

Select the Sketch plane for the default component.

114) Click **Front Plane** in the BOX assembly FeatureManager.

115) **Expand** the new inserted component, [Part1^BOX] in the FeatureManager. The FeatureManager is displayed in light blue.

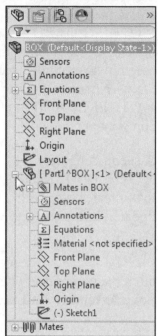

Components added In-Context of the assembly automatically received an InPlace Mate within the Mates entry in the FeatureManager and under the component.

The InPlace mate is a Coincident mate between the Front Plane of the [Part1^BOX] component and the Front Plane of the BOX.

The [Part1^BOX] entry is added to the FeatureManager. The system automatically selects the Edit Component mode.

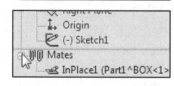

The [Part1^BOX] text is displayed in blue to indicate that the component is actively being edited. The current Sketch plane is the Front Plane. The system automatically selects the Sketch icon.

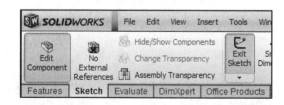

The current sketch name is 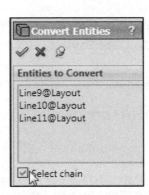 (-) Sketch1. The document name
`Sketch1 of Part1^BOX -in- BOX *` displays:

Create the sketch.
116) Click the **right vertical edge** of the right inside rectangle as Illustrated.

117) Click **Convert Entities**  from the Sketch toolbar.

118) Click the **three other sides** of the inside large rectangle. Check the **Select chain** box.

119) Click **OK** ✔ from the Convert Entities PropertyManager.

120) Click **OK** ✔ . The outside perimeter of the Layout Sketch is the current sketch.

Insert an Extruded Boss/Base feature for the [Part1^BOX] component.

121) Click **Extruded Boss/Base** from the Features toolbar. The Boss-Extrude PropertyManager is displayed.

122) Enter **10**mm for Depth in Direction 1. Accept the default settings.

123) Click **OK** ✔ from the Boss-Extrude PropertyManager. Boss-Extrude1 is displayed in the FeatureManager.

124) Rename **Boss-Extrude1** to **Base Extrude**.

The default component, "[Part1^BOX]<#> ->", Base Extrude ->, and Sketch1 -> all contain the "->" symbol indicating External references to the BOX assembly.

The Edit Component feature acts as a switch between the assembly and the component edited In-Context.

If you Do NOT want to create any External references in your model, click **Options, External References** and check the **Do not create references external to the model** box.

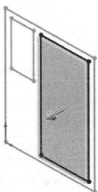

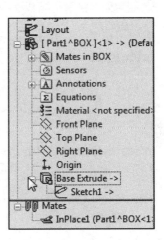

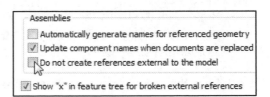

Return to the BOX assembly.

125) Click **Edit Component** 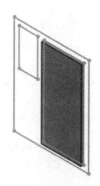 from the Assembly toolbar. The FeatureManager is displayed in black. **Rebuild** the model.

Fit and Save the BOX to the Graphics window.

126) Press the **f** key. Click **Save** 💾. Click **OK** to save internally.

Save and rename the default virtual component [Part1^BOX]<#> -> to MOTHERBOARD.

127) Right-click **[Part1^BOX]<#> ->**. Click **Open Part**.

128) Click **Save As** from the Menu bar.

129) Enter **MOTHERBOARD** in the File name box. Note: Save in the **VENDOR-COMPONENTS** folder. Click **Save**. **Close** the part.

130) **Return** to the BOX assembly. View the updated FeatureManager.

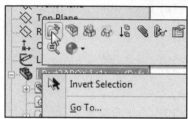

The Reference models are the MOTHERBOARD and the BOX assembly. Additional features are required that do not reference the Layout Sketch or other components in the BOX assembly. An Extruded Boss feature indicates the approximate position of an electrical connector. The actual measurement of the connector or type of connector has not been determined.

Perform the following steps to avoid unwanted assembly references:

1. Open the part from the assembly. 2. Add features. 3. Save the part. 4. Return to the assembly. 5. Save the assembly.

Insert an Extruded Boss/Base feature for the MOTHERBOARD.

131) Right-click **MOTHERBOARD** in the FeatureManager.

132) Click **Open Part**. The MOTHERBOARD FeatureManager is displayed. Press the **f** key.

Insert the Sketch.

133) Right-click the **front face** of Base Extrude. This is your Sketch plane. Click **Sketch** ✎ from the Context toolbar. The Sketch toolbar is displayed.

134) Click the outside **right vertical edge**. Hold the **Ctrl key** down.

135) Click the **top horizontal edge**. Release the **Ctrl key**.

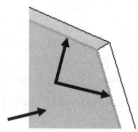

Front face

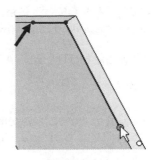

136) Click **Convert Entities** ⬚ from the Sketch toolbar. Click **OK** ✓ from the Convert Entities PropertyManager.

137) Drag the **bottom end point** of the converted line three quarters upward.

138) Drag the **left end point** of the converted line three quarters of the way to the right. Click **Front view** 📄 from the Heads-up View toolbar.

139) Click **Line** ＼ from the Sketch toolbar.

140) Sketch a **vertical line** as illustrated.

141) Sketch a **horizontal line** to complete the rectangle.

Add dimensions.
142) Click **Smart Dimension** ✐ from the Sketch toolbar.

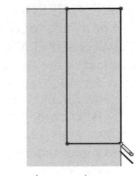

143) Enter **30**mm for the horizontal dimension. Enter **50**mm for the vertical dimension.

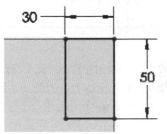

Extrude the sketch.
144) Click **Extruded Boss/Base** 🗐 from the Features toolbar. The Boss-Extrude PropertyManager is displayed.

145) Enter **10**mm for Depth in Direction 1.

146) Click **OK** ✓ from the Boss-Extrude PropertyManager.

147) Rename **Boss-Extrude2** to **Connector1**.

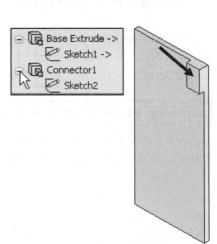

Display an Isometric view and save the model.
148) Click **Isometric view** 🔲. Click **Save** 💾.

Use color to indicate electrical connectors. Color the face of the Extruded Boss feature.

Color the Front face of Connector1.
149) Right-click the **Front face** of Connector1.

150) Click the **Appearances** drop-down arrow.

151) Click **Connector1** as illustrated.

152) Select **yellow** as the color.

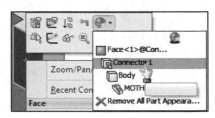

153) Click **OK** ✔ from the Color PropertyManager. The selected face is displayed in yellow.

Save the MOTHERBOARD.
154) Click **Save** 🖫.

Return to the BOX assembly.
155) Press **Ctrl Tab**. Click **Yes** to rebuild the assembly.

The MOTHERBOARD contains the new Extruded Boss feature. For now - the MOTHERBOARD is complete. The POWERSUPPLY is the second key component to be defined In-Context of the BOX assembly.

💡 A learning goal of this book is to expose the new user to various tools, techniques and procedures. It may not always use the most direct tool or process.

💡 Locate External references. Open the BOX assembly before opening individual components referenced by the assembly when you start a new session of SolidWorks. The components load and locate their external references to the original assembly.

POWERSUPPLY-Insert Component

Create a simplified POWERSUPPLY part In-Context of the BOX assembly. Utilize the Layout Sketch in the BOX assembly. Open the POWERSUPPLY part and insert an Extruded Boss to represent the connection to a cable.

Activity: POWERSUPPLY-Insert Component

Insert the POWERSUPPLY component.
156) Click **New Part** from the Consolidated Insert Components toolbar as illustrated.

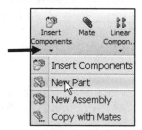

Select the Part Template.
157) Double-click the **PART-MM-ANSI** template from the MY-TEMPLATES tab. A new default virtual component is added to the FeatureManager. Note: Save the new component and name it POWERSUPPLY later in the procedure.

The Component Pointer ✔ is displayed. The new default component is empty and requires a Sketch plane.

Select the Sketch plane.

158) Click the **Front Plane** of the BOX in the FeatureManager. An InPlace mate is added to the BOX FeatureManager.

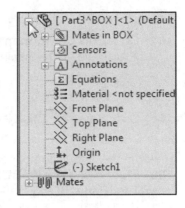

The Front Plane of the default component [Part#^BOX]<#> is mated to the Front Plane of the BOX. Sketch1 of the [Part#^BOX]<1> is the active sketch.

The system automatically selects the Edit Component tool. The default component [Part#^BOX]<#> entry in the FeatureManager is displayed in blue. The current Sketch plane is the Front Plane. The current sketch name is Sketch1. The name is indicated on the current document window title:

Sketch1 of Part3^BOX -in- BOX *

[Part#^BOX]<#> is the default name of the component created In-Context of the BOX assembly. Create the first Extruded Base feature In-Context of the Layout Sketch.

Create the sketch.

159) Click the **right vertical edge** of [Part#^BOX]<1> as illustrated.

160) Click **Convert Entities** from the Sketch toolbar.

161) Click the **three other sides** of the inside small rectangle. Check the **Select chain** box

162) Click **OK** from the Convert Entities PropertyManager.

163) Click **OK** .

Extrude the sketch.

164) Click **Extruded Boss/Base** from the Features toolbar. Blind is the default End Condition in Direction 1.

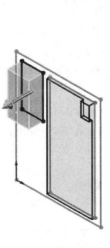

165) Enter **50**mm for Depth in Direction 1. Accept the default settings.

166) Click **OK** from the Boss-Extrude PropertyManager. Boss-Extrude1 is displayed in the FeatureManager under [Part#^BOX]<#>.

The [Part#^BOX]<#> default name is displayed in blue. The component is being edited In-Context of the BOX assembly.

Return to the BOX assembly.

167) Click **Edit Component** 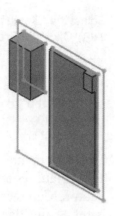 from the Assembly toolbar. The [Part#^BOX]<#> displayed in black.

Save the BOX assembly.

168) Click **Save** 🖫 . Click **Save All**. Click **OK** to save internally.

Rename the default component [Part2^BOX]<1> to POWERSUPPLY.
169) Right-click **[Part#^BOX]<1> ->**. Click **Open Part**. The [Part#^BOX]<1> -> FeatureManager is displayed.

170) Click **Save As** from the Menu bar. Enter **POWERSUPPLY** in the File name box. Note: Save in the **VENDOR-COMPONENTS** folder.

171) Click **Save**. **Close** the part. **Return** to the BOX assembly. View the updated FeatureManager.

The Extruded Boss/Base feature represents the location of the cable that connects to the POWERSUPPLY. Think about where the cables and wire harness connects to key components. You do not have all of the required details for the cables.

In a concurrent engineering environment, create a simplified version early in the design process. No other information is required from the BOX assembly to create additional features for the POWERSUPPLY.

The design intent for the POWERSUPPLY is for the cable connection to be centered on the top face of the POWERSUPPLY. Utilize a centerline and Midpoint relation to construct the Extruded Boss/Base feature centered on the top face. Open the POWERSUPPLY from the BOX assembly.

Open the POWERSUPPLY.
172) Right-click **POWERSUPPLY** in the Graphics window.

173) Click **Open Part**. The POWERSUPPLY FeatureManager is displayed. Press the **f** key.

174) Click **Top view** ⬚ from the Heads-up View toolbar.

Create the sketch.
175) Right-click the **top face** of Boss-Extrude. Click **Sketch** ✎ from the Context toolbar.

176) Click **Centerline** ┊ from the Sketch toolbar. The Insert Line PropertyManager is displayed.

177) Sketch a diagonal **centerline** from the top left to the bottom right as illustrated. Align the endpoints with the corners of the POWERSUPPLY.

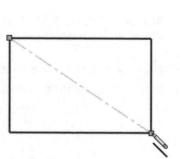

Sketch a circle.

178) Click **Circle** 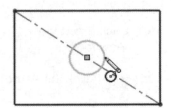 from the Sketch toolbar.

179) Click the **Midpoint** of the centerline. Sketch a **circle** as illustrated. The centerline midpoint is the center of the circle.

Add a dimension.

180) Click **Smart Dimension** from the Sketch toolbar.

181) Click the **circumference** of the circle. Click a **position** off the profile.

182) Enter **15**mm for diameter. The circle is displayed in black.

Extrude the sketch.

183) Click **Extruded Boss/Base** from the Features toolbar. The Boss-Extrude PropertyManager is displayed. Blind is the default End Condition in Direction 1.

184) Enter **10**mm for End Condition in Direction 1.

185) Click **OK** from the Boss-Extrude PropertyManager. The feature is displayed in the FeatureManager.

186) Rename the **Boss-Extrude2** to **Cable1**.

187) Click **Isometric view** from the Heads-up View toolbar.

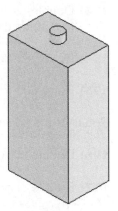

Save the POWERSUPPLY.

188) Click **Save** .

Return to the BOX assembly.

189) Press **Ctrl Tab**. Click **Yes** to update the assembly.

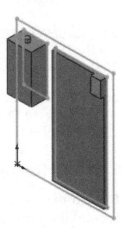

The POWERSUPPLY contains the new Extruded Boss feature, Cable1. Recall the initial design parameters. The requirement calls for three different size boxes. Test the Layout Sketch dimensions for the three configurations: *Small, Medium, and Large*.

Do the Layout Sketch, MOTHERBOARD and POWERSUPPLY reflect the design intent of the BOX assembly? Review the next steps to confirm the original design intent.

Display all dimensions.

190) Click **Front view** 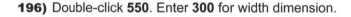 from the Heads-up View toolbar.

191) Right-click the **Annotations** folder.

192) Click **Show Feature Dimensions**.

193) Double-click **300**. Enter **550** for the horizontal dimension.

194) Double-click **400**.

195) Enter **600** for the vertical dimension.

Return to the original dimensions.

196) Double-click **550**. Enter **300** for width dimension.

197) Double-click **600**. Enter **400** for height dimension.

198) Click **Isometric view** from the Heads-up View toolbar.

199) Click **Save**.

Hide all dimensions.
200) Right-click the **Annotations folder**.

201) Un-check **Show Feature Dimensions**.

Suppress the MOTHERBOARD and POWERSUPPLY.
202) Click **MOTHERBOARD** from the FeatureManager.

203) Hold the **Ctrl** key down.

204) Click **POWERSUPPLY** from the FeatureManager.

205) Release the **Ctrl** key.

206) Right-click **Suppress**. Both components and their Mates are Suppressed. They are displayed in light gray in the FeatureManager.

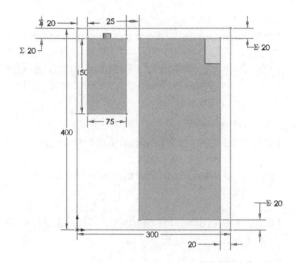

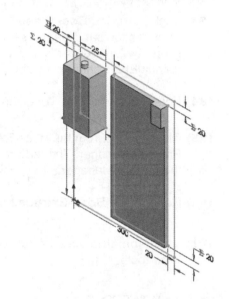

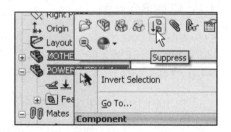

Electro-mechanical assemblies contain hundreds of parts. Suppress components saves rebuild time and simplifies the model. Only display the components and geometry required to create a new part or to mate a component.

Improve rebuild time and display performance for large assemblies. Review the System Options listed in Large Assembly Mode. Adjust the large assembly threshold to match your computer performance.

Check Remove detail during zoom/pan/rotate. Check Hide all planes, axes and sketches.

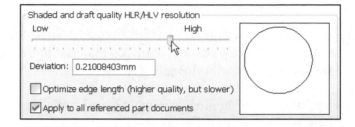

Review the System Option, Image Quality. Drag the resolution slider towards Low to improve computer performance.

Review the BOX Assembly

The BOX assembly utilized the Top-down assembly approach by incorporating a Layout Sketch. The Layout Sketch consisted of 2D profiles and dimensions of the BOX, MOTHERBOARD and POWERSUPPLY.

The MOTHERBOARD was developed as a new part In-Context of the BOX assembly. An InPlace mate was created between the MOTHERBOARD Front Plane and the BOX assembly Front Plane. In the Edit Component mode, you extracted the Layout Sketch geometry to create the first feature of the MOTHERBOARD part. The first feature, sketch and Sketch plane contained External references to the BOX assembly.

You opened the MOTHERBOARD from the BOX assembly. Additional features were added to the MOTHERBOARD part. The POWERSUPPLY was developed as a new part In-Context of the BOX assembly. The BOX assembly design intent and requirements were verified by modifying the dimensions to the different sizes of the BOX configurations.

The MOTHERBOARD and POWERSUPPLY are only two of numerous components in the BOX assembly. They are partially complete but represent the overall dimensions of the final component. Electro-Mechanical assemblies contain parts fabricated from sheet metal. Fabricate the CABINET and BRACKET for the BOX assembly from Sheet metal.

Sheet Metal Overview

Sheet metal manufactures create sheet metal parts from a flat piece of raw material. To produce the final part, the material is - cut, formed and folded. In SolidWorks, the material thickness does not change.

Talk to colleagues. Talk to sheet metal manufacturers. Review other sheet metal parts previously designed by your company. You need to understand a few basic sheet metal definitions before starting the next parts, CABINET and BRACKET.

The CABINET begins as a solid part. The Shell feature determines the material thickness. The Rip feature removes material at the edges to prepare for sheet metal bends. The Insert Sheet Metal Bends feature transforms the solid part into a sheet metal part.

The BRACKET begins as a sheet metal part. The Base Flange feature is the first feature in a sheet metal part. Create the Base Flange feature from a sketch. The material thickness and bend radius of the Base Flange are applied to all other sheet metal features in the part.

There are two design states for sheet metal parts:

- Formed

- Flat

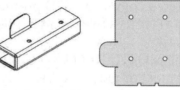

Design the CABINET and BRACKET in the formed state. The Flatten feature creates the Flat Pattern for the manufactured flat state.

Formed State   Flat State
BRACKET

Sheet metal parts can be created from the flat state. Sketch a line on a face to indicate bend location. Insert the Sketched Bend feature to create the formed state.

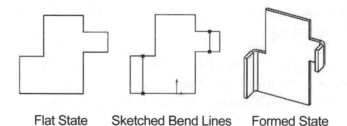

Flat State   Sketched Bend Lines   Formed State

Bends

Example: Use a flexible eraser, 50mm or longer. Bend the eraser in a U shape. The eraser displays tension and compression forces.

The area where there is no compression or tension is called the neutral axis or neutral bend line.

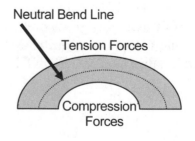

Neutral Bend Line

Tension Forces

Compression Forces

Eraser Example

Assume the material has no thickness. The length of the material formed into a 360° circle is the same as its circumference.

The length of a 90° bend would be ¼ of the circumference of a circle.

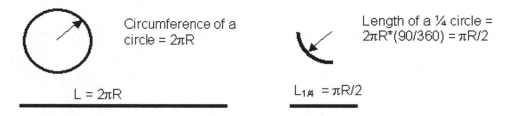

Circumference of a circle = $2\pi R$

$L = 2\pi R$

Length of a ¼ circle = $2\pi R*(90/360) = \pi R/2$

$L_{1/4} = \pi R/2$

Length of 90° bend

In the real world, materials have thickness. Materials develop different lengths when formed in a bend depending on their thickness.

There are three major properties which determine the length of a bend:

- Bend radius

- Material thickness

- Bend angle

The distance from the inside radius of the bend to the neutral bend line is labeled δ.

The symbol 'δ' is the Greek letter delta. The amount of flat material required to create a bend is greater than the inside radius and depends on the neutral bend line.

The true developed flat length, L, is measured from the endpoints of the neutral bend line.

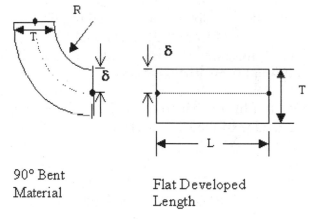

90° Bent Material

Flat Developed Length

True developed flat length

Example:

T is the Material thickness.

Create a 90° bend with an inside radius of R:

$L = \frac{1}{2}\pi R + \delta T$

The ratio between δ and T is called the K-factor. Let K = 0.41 for Aluminum:

$K = \delta/T$

$\delta = KT = 0.41T$

$L = \frac{1}{2}\pi R + 0.41T$

U.S. Sheet metal shops use their own numbers.

Example:

One shop may use K = 0.41 for Aluminum, versus another shop uses K = 0.45.

Use tables, manufactures specifications or experience.

Save design time and manufacturing time. Work with your Sheet metal shop to know their K-factor. Material and their equipment produce different values. Build these values into the initial design.

Where do you obtain the material required for the Bend? The answer comes from the Sheet metal Flange position option. The four options are: Material Inside, Material Outside, Bend Outside and Bend from Virtual Sharp.

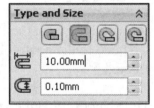

Material Inside: The outside edge of the Flange coincides with the fixed edge of the sheet metal feature.

Material Outside: The inside edge of the Flange coincides with the fixed edge of the sheet metal feature.

Bend Outside: The Flange is offset by the bend radius.

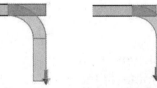

Material Inside   Material Outside   Bend Outside

Bend from Virtual Sharp: The Flange maintains the dimension to the original edge and varies the bend material condition to automatically match with the flange's end condition.

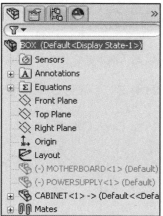

Relief

Sheet metal corners are subject to stress. Excess stress will tear material. Remove material to relieve stress with Auto Relief. The three options are: Rectangular, Tear and Obround.

- Rectangular: Removes material at the bend with a rectangular shaped cut.

- Tear: Creates a rip at the bend, a cut with no thickness.

- Obround: Removes material at the bend with a rounded shaped cut.

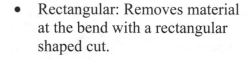

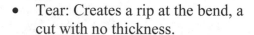

Rectangular   Tear   Obround

CABINET-Insert Component

Create the CABINET component inside the assembly and attach it to the Front Plane. The CABINET component references the Layout sketch.

Create the CABINET component as a solid box. Shell the box to create the constant sheet metal thickness. Utilize the Rip feature to cut the solid edges. Utilize the Insert Bends feature to create a sheet metal part from a solid part.

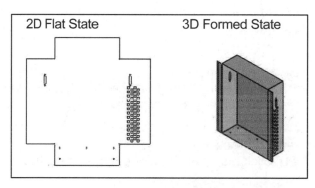

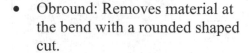

Add additional sheet metal Edge Flange and Hem features. Add dies, louvers, and cuts to complete the CABINET.

Utilize configurations to create the 3D formed state for the BOX assembly and the 2D flat state for the CABINET drawing.

Activity: CABINET-Insert Component

Insert the CABINET Component.
207) Open the BOX assembly.

208) Click **New Part** from the Consolidated Insert Components toolbar as illustrated.

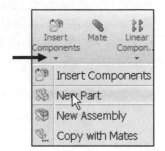

Select the Part Template.
209) Double-click the **PART-MM-ANSI** template from the MY-TEMPLATES tab. A new virtual default component is added to the FeatureManager. Note: Save the new component and name it CABINET later in the procedure.

The Component Pointer  is displayed. The default component [Part3^BOX]<1> is empty and requires a Sketch plane. The Front Plane of the [Part3^BOX]<1>, component is mated with the Front Plane of the BOX. [Part3^BOX]<1> is the name of the component created In-Context of the BOX assembly.

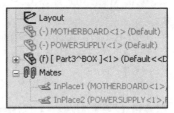

Select the Sketch plane.
210) Click **Front Plane** from the BOX FeatureManager. The system automatically selects the Sketch toolbar.

Create the sketch.
211) Click the **right vertical edge** of the outside BOX as illustrated.

212) Click the **Convert Entities** Sketch tool. The Convert Entities PropertyManager is displayed.

213) Click the **three other sides** of the outside BOX.

214) Check the **Select chain** box.

215) Click **OK** from the Convert Entities PropertyManager.

216) Click **OK**. The outside perimeter is the current sketch.

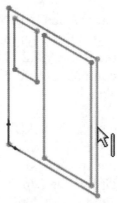

Extrude the sketch.

217) Click **Extruded Boss/Base** 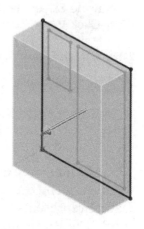 from the Features toolbar. The Boss-Extrude PropertyManager is displayed.

218) Enter **100**mm for Depth in Direction 1. Accept the default settings.

219) Click **OK** ✓ from the Boss-Extrude PropertyManager.

Return to the BOX assembly.

220) Click **Edit Component** 🗊 from the Assembly toolbar.

Rename the default component [Part3^BOX]<1> to CABINET.

221) Right-click **[Part3^BOX]<1> ->**. Click **Open Part**. The [Part3^BOX]<1> -> FeatureManager is displayed.

222) Click **Save As** from the Menu bar. Click **OK**. Enter **CABINET** in the File name box. Note: Save in the VENDOR-COMPONENTS folder.

223) Click **Save. Close** the part.

224) Return to the BOX assembly. View the updated FeatureManager.

Caution: Do not create unwanted geometry references. Open the part when creating features that require no references from the assembly. The features of the CABINET require no additional references from the BOX assembly, MOTHERBOARD or POWERSUPPLY.

Open the CABINET part.

225) Right-click **CABINET** from the FeatureManager.

226) Click **Open Part**. The CABINET FeatureManager is displayed.

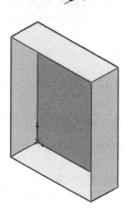

Fit the model to the Graphics window.

227) Press the **f** key. Click the **Isometric view** 🔲 from the Heads-up View toolbar.

Create the Shell.

228) Click the **front face** of the Boss-Extrude1 feature.

229) Click **Shell** 🔲 from the Features toolbar. The Shell1 PropertyManager is displayed.

230) Check the **Shell outward** box. Enter **1**mm for Thickness.

231) Click **OK** ✓ from the Shell1 PropertyManager. Shell1 is displayed in the FeatureManager.

Display the Sheet Metal toolbar.
232) Click **View**, **Toolbars** from the Main menu.

233) Check **Sheet Metal**. The Sheet Metal toolbar is displayed.

234) Click and drag a **location** inside the Graphics window for the toolbar.

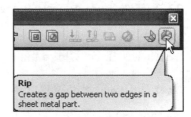

Maintain constant thickness. Sheet metal features only work with constant wall thickness. The Shell feature maintains constant wall thickness. The solid Extruded Base feature represents the overall size of the sheet metal CABINET. The CABINET part is solid. The Rip feature and Insert Sheet Metal Bends feature convert a shelled solid part into a sheet metal part.

CABINET-Rip Feature and Sheet Metal Bends

The Rip feature creates a cut of no thickness along the edges of the Extruded Base feature. Rip the Extruded Base feature along the four edges. The Bend feature creates sheet metal bends.

Specify bend parameters: bend radius, bend allowance and relief. Select the bottom face to remain fixed during bending and unbending. In the next example, you will create the Rip and the Insert Bend features in two steps. The Bends PropertyManager contains the Rip parameter. Utilize the Rip feature to create a simple Rip and Bend in a single step.

Activity: CABINET-Rip Feature and Sheet Metal Bends

Fit the model to the Graphics window.
235) Press the **f** key.

Insert a Rip feature. A Rip feature creates a gap between two edges in a sheet metal part.
236) Click **Rip** from the Sheet Metal toolbar. The Rip PropertyManager is displayed.

237) **Rotate** the part to view the inside edges.

238) Click the **inside lower left edge** as illustrated.

239) Click the **inside upper left edge**.

240) Click the **inside upper right edge**.

241) Click the **inside lower right edge**. The selected entities are displayed in the Edges to Rip box. Accept the default Rip Gap.

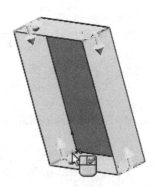

242) Click **OK** ✔ from the Rip PropertyManager. Rip1 is displayed in the FeatureManager.

Insert the Sheet Metal Bends.
243) Click the inside **bottom face** to remain fixed.

244) Click **Insert Bends** ⌦ from the Sheet Metal toolbar. The Bends PropertyManager is displayed. Face<1> is displayed.

245) Enter **2.00**mm for Bend radius.

246) Enter **.45** for K-Factor.

247) Enter **.5** for Rectangular Relief.

248) Click **OK** ✔ from the Bends PropertyManager.

249) Click **OK** to the message, "Auto relief cuts were made for one or more bends".

250) Zoom in 🔍 on the Rectangular relief in the upper back corner.

Display the full view.

251) Click **Isometric view** 🧊 from the Heads-up View toolbar.

252) Click **Save** 💾. View the created features in the FeatureManager.

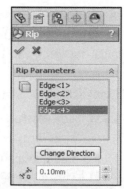

You just created your first sheet metal part. The .45 K-Factor is based on your machine shops parameters for Aluminum.

💡 Save manufacturing cost and reduce setup time. A sheet metal manufacturer maintains a turret of standard relief tools for Rectangular and Obround relief. Obtain the dimensions of these tools to utilize in your design.

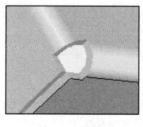

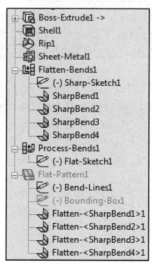

The CABINET part is in its 3D formed state. Display its 2D flat manufactured state. Test every feature to determine if the part can be manufactured as a sheet metal part.

Alternate between 3D formed and 2D flat for every additional sheet metal feature you create. The Flatten feature alternates a sheet metal part between the flat state and formed state.

Display the Flat State.

253) Click **Flatten** from the Sheet Metal toolbar.

Fit the model to the Graphics window.
254) Press the **f** key.

Display the part in its fully formed state.

255) Click **Flatten** from the Sheet Metal toolbar.

Save the CABINET.
256) Click **Save**.

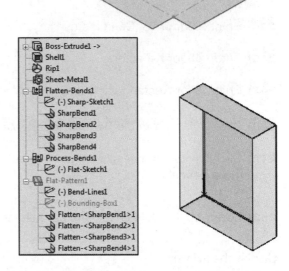

Note: To return to the solid part, utilize the No Bends feature to roll back the model before the first sheet metal bend.

CABINET-Edge Flange

Create the right flange wall and left flange wall with the Edge Flange feature. The Edge Flange feature adds a wall to a selected edge of a sheet metal part.

Select the inside edges when creating bends. Create the right hem and left hem with the Hem feature. The Hem feature folds back at an edge of a sheet metal part. A Hem can be open, closed, double or tear-drop.

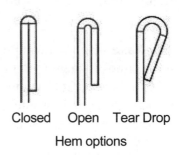

Closed   Open   Tear Drop

Hem options

In the preliminary design stage, review Hem options with your sheet metal manufacturer.

Activity: CABINET-Edge Flange

Insert the front right Flange.

257) Click the **front vertical right edge** of the CABINET as illustrated.

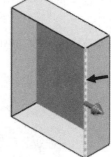

258) Click **Edge Flange** from the Sheet Metal toolbar. The Edge-Flange PropertyManager is displayed.

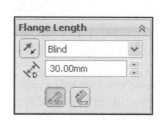

259) Select **Blind** for Length Edge Condition.

260) Enter **30**mm for Length. Click the **Reverse Direction** button. The Direction arrow points towards the right. Accept all other defaults.

261) Click **OK** from the Edge-Flange PropertyManager. Edge-Flange1 is displayed in the FeatureManager.

Insert the right Hem.

262) Click **Front view** from the Heads-up View toolbar.

263) Click the **right edge** of the right flange.

264) Click **Hem** from the Sheet Metal toolbar. The Hem PropertyManager is displayed.

265) Click **Isometric view** from the Heads-up View toolbar.

266) Click the **Reverse Direction** button.

267) Enter **10**mm for Length.

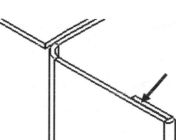

268) Enter **.10**mm for Gap Distance.

269) Click **OK** from the Hem PropertyManager. Hem1 is displayed in the FeatureManager.

Insert the left Edge Flange wall.

270) Click **Front view** from the Heads-up View toolbar.

271) Click the **front vertical left edge** as illustrated.

272) Click **Edge Flange** 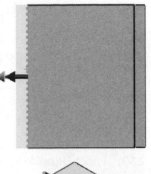 from the Sheet Metal toolbar. The Edge-Flange PropertyManager is displayed.

273) Select **Blind** for Length End Condition.

274) Enter **30**mm for Length. The feature direction arrow points towards the left. Reverse the arrow direction if required. Accept the default settings.

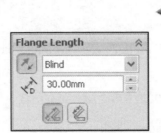

275) Click **OK** ✓ from the Edge-Flange PropertyManager. Edge-Flange2 is displayed in the FeatureManager.

Insert the left edge Hem.

276) Select the **left edge** of the left flange as illustrated.

277) Click **Isometric view** 🔲 from the Heads-up View toolbar.

278) Click **Hem** 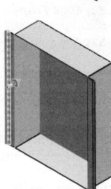 from the Sheet Metal toolbar. The Hem PropertyManager is displayed.

279) Enter **.10**mm for Gap Distance.

280) Click the **Reverse Direction** button.

281) Click **OK** ✓ from the Hem PropertyManager. Hem2 is displayed in the FeatureManager.

Display the part in its flat manufactured state and formed state.

282) Click **Flatten** 🔳 from the Sheet Metal toolbar.

283) Click **Top view** 🔲 from the Heads-up View toolbar.

284) Click **Flatten** 🔳 to display the part in its fully formed state.

285) Click **Isometric view** 🔲 from the Heads-up View toolbar.

286) Rename **Edge-Flange1** to **Edge-Flange1-Right**.

287) Rename **Edge-Flange2** to **Edge-Flange2-Left**.

288) Rename **Hem1** to **Hem1-Right**.

289) Rename **Hem2** to **Hem2-Left**.

Save the CABINET.
290) Click **Save** 💾.

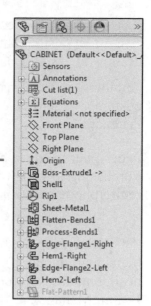

☀ Simplify the FeatureManager. Sheet metal parts contain numerous flanges and hems. Rename features with descriptive names.

☀ Save design time. Utilize Mirror Feature to mirror sheet metal features about a plane. Example: In a second design iteration, the Edge Flange1-Right and Hem1-Right are mirrored about Plane1.

The Mirror feature was not utilized in the CABINET. A Mirror Plane and Equations are required to determine the flange and hem locations.

☀ Maintain the design intent. In the Top-down design process return to the Layout Sketch after inserting a new component to verify the assembly design intent.

Increase and decrease dimension values in the Layout Sketch. To avoid problems in a top level assembly, select the same face to remain fixed for the Flat Pattern and Bend/Unbend features. Save sheet metal parts in their 3D formed state.

The CABINET requires additional solid and sheet metal features. These features require no references in the BOX assembly. Work in the CABINET part.

CABINET - Hole Wizard and Linear Pattern features

Sheet metal holes are created through a punch or drill process. Each process has advantages and disadvantages: cost, time and accuracy.

Investigate a Linear Pattern feature of holes. Select a self-clinching threaded fastener that is inserted into the sheet metal during the manufacturing process. The fastener requires a thru hole in the sheet metal.

Holes should be of equal size and utilize common fasteners. Why? You need to ensure a cost effect design that is price competitive. Your company must be profitable with their designs to ensure financial stability and future growth.

Another important reason for fastener commonality and simplicity is the customer. The customer or service engineer does not want to supply a variety of tools for different fasteners.

A designer needs to be prepared for changes. You proposed two fasteners. Ask Purchasing to verify availability of each fastener. Select a Ø5.0mm hole and wait for a return phone call from the Purchasing department to confirm. Design flexibility is key!

Dimension the hole position. Do not dimension to sheet metal bends. If the sheet metal manufacturer modifies the bend radius, the hole position does not maintain the design intent. Reference the hole dimension by selecting the Origin in the FeatureManager.

Utilize the Hole Wizard feature to create a Simple Through All hole. The hole is positioned on the inside bottom face based on the selection point.

Activity: CABINET - Hole Wizard and Linear Pattern feature

Insert the first hole. Open the Hole Wizard.

291) Click **Hole Wizard** from the Features toolbar. The Hole Specification PropertyManager is displayed.

292) Click the **Hole** tab.

293) Select **Ansi Metric** for Standard.

294) Select **Drill sizes** for Type.

295) Select **Ø5.0** for Size.

296) Select **Through All** for End Condition.

297) Click the **Positions** tab.

298) Click the **inside bottom face** as illustrated. The Point Sketch tool is active.

299) Click **again** to place the center point of the hole.

300) Right-click **Select** to deselect the Point Sketch tool.

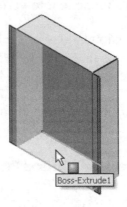

Dimension the hole. Create a horizontal and vertical dimension from the Origin.

301) Click **Smart Dimension** 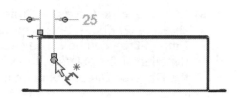.

302) Click **Top view** ⊞ from the Heads-up View toolbar.

303) Expand CABINET from the fly-out FeatureManager.

304) Click the **Origin** from the FeatureManager.

305) Click the **center point** of the hole in the Graphics window.

306) Enter **25**mm for the horizontal dimension.

307) Expand CABINET from the fly-out FeatureManager.

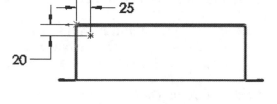

308) Click the **Origin** from the FeatureManager.

309) Click the **center point** of the hole in the Graphics window.

310) Enter **20**mm for the vertical dimension.

311) Click **OK** ✔ from the Dimension PropertyManager.

312) Click **OK** ✔ from the Hole Position PropertyManager. The Hole is displayed in the FeatureManager.

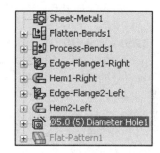

313) Click **Isometric view** ▢ from the Heads-up View toolbar.

314) Click **Save** 💾.

Insert a Linear Pattern of holes.

315) Click **Linear Pattern** ⣿ from the Features toolbar. The Linear Pattern PropertyManager is displayed.

316) Click the **back inside edge** for Direction 1. Edge<1> is displayed in the Pattern Direction box. The direction arrow points to the right. Click the Reverse Direction button if required.

317) Enter **125**mm for Spacing.

318) Enter **3** for Number of Instances.

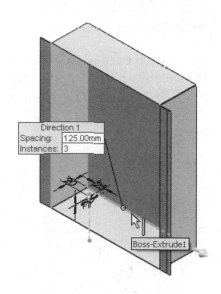

319) Click the **left edge** as illustrated for Direction 2. Edge<2> is displayed in the Pattern Direction box. The second direction arrow points to the front. Click the Reverse Direction button if required.

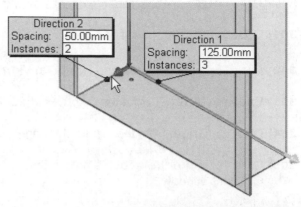

320) Enter **50**mm for Spacing.

321) Enter **2** for Number of Instances.

Remove an Instance.

322) Click inside the **Instances to Skip** box.

323) Click the **front middle hole**, (2,2) in the Graphics window as illustrated. (2,2) is displayed in the Instances to Skip box.

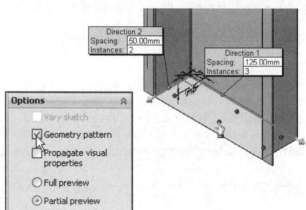

324) Check the **Geometry pattern** box. Accept the default settings.

325) Click **OK** ✅ from the Linear Pattern PropertyManager. LPattern1 is displayed in the FeatureManager.

326) Click **Isometric view** 🧊 from the Heads-up View toolbar.

Use the Geometry pattern option to improve system performance for sheet metal parts. The Geometry pattern option copies faces and edges of the seed feature. Type options such as Up to Surface are not copied.

The Instances to Skip option removes a selected instance from the pattern. The value (2, 2) represents the instance position in the first direction - second instance, second direction - second instance.

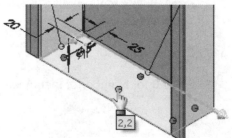

CABINET-Sheetmetal Design Library Feature

Sheet metal manufacturers utilize dies and forms to create specialty cuts and shapes. The Design Library contains information on dies and forms. The Design Library, features folder contains examples of predefined sheet metal shapes. Insert a die cut on the right wall of the CABINET. The die cut is for a data cable. The team will discuss sealing issues at a later date.

Activity: CABINET-Sheetmetal Design Library Feature

Insert a Sheet metal Library Feature.

327) Click **Design Library** from the Task Pane.

328) **Expand** the Design Library folder.

329) **Expand** the features folder.

330) Click the **Sheetmetal** folder.

331) Click and drag the **d-cutout** feature into the Graphics window.

332) Release the mouse button on the **right flange outside face** of the CABINET as illustrated. The d-cutout PropertyManager is displayed.

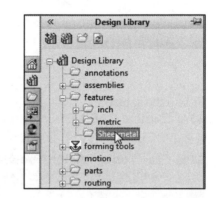

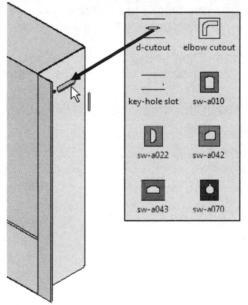

Edit the sketch.

333) Click the **Edit Sketch** button from the d-cutout PropertyManager. The Library Feature Profile dialog box is displayed. DO NOT SELECT THE FINISH BUTTON.

Rotate the d-cutout.

334) Expand CABINET in the fly-out FeatureManager.

335) Click **Tools**, **Sketch Tools**, **Modify** Modify... from the Menu bar. The mouse pointer displays the modify move/rotate icon . The Modify Sketch dialog box is displayed.

336) Enter **90deg** in the Rotate box. Note: Press the **Enter** key. View the d-cutout feature.

337) Click **Close** from the Modify Sketch dialog box.

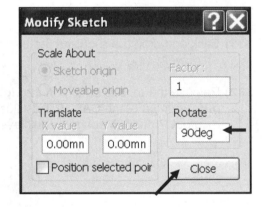

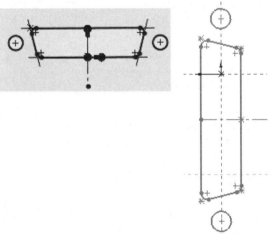

Display the Right view.

338) Click **Right view** 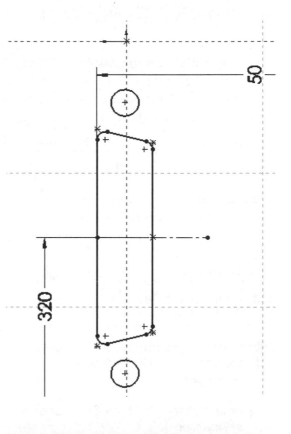 from the Heads-up View toolbar.

339) Zoom to Area on the d-cutout.

Dimension the d-cutout.

340) Click **Smart Dimension** from the Sketch toolbar.

Create a vertical dimension.

341) Click the **midpoint** of the d-cutout. The Point PropertyManager is displayed.

342) Expand CABINET from the fly-out FeatureManager.

343) Click the **Origin** of the CABINET.

344) Enter **320**mm.

Create a horizontal dimension.

345) Click the **left vertical line** of the d-cutout.

346) Expand the CABINET from the fly-out FeatureManager.

347) Click the **Origin** of the CABINET.

348) Enter **50**mm.

349) Click **Finish** from the Library Feature Profile dialog box. The d-cutout entry is displayed in the FeatureManager.

350) Click **Isometric view** from the Heads-up View toolbar.

351) Click **Save**.

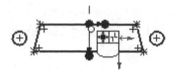

💡 With thin sheet metal parts, select the dimension references with care. Use the Zoom and Rotate commands to view the correct edge. Use reference planes or the Origin to create dimensions. The Origin and planes do not change during the flat and formed states.

💡 Save time with the Modify Sketch tool. Utilize the left and right mouse point positioned on the large black dots. The left button Pans and the right button Rotates. The center dot reorients the Origin of the sketch and Mirrors the

sketch about y, x or both.

The d-cutout goes through the right and left side. Through All is the current End Condition Type. This is not the design intent. Redefine the end condition through the right flange.

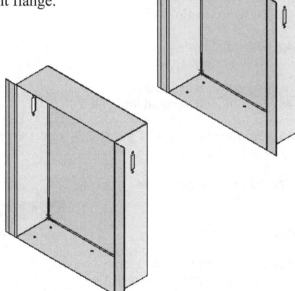

If needed, modify the D Cut End Condition.

352) Right-click **D Cut** ⊞ 🔟 D Cut in the FeatureManager.

353) Click **Edit Feature** from the Context toolbar. The D Cut PropertyManager is displayed.

354) Select **Through All** for End Condition in Direction 1.

355) Click **Isometric view** 🔲 from the Heads-up View toolbar.

356) Click **OK** ✅ from the D Cut PropertyManager. View the results in the Graphics window.

The D Cut feature is positioned before the Flat Pattern1 icon in the FeatureManager. The D Cut is incorporated into the Flat Pattern. The sheet metal manufacturing process is cost effective to perform cuts and holes in the flat state.

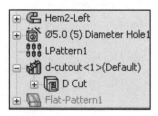

CABINET-Louver Forming Tool

Form features, such as louvers are added in the formed state. Formed features are normally more expensive than cut features. Louvers are utilized to direct air flow. The louvers are used to dissipate the heat created by the internal electronic components.

The forming tools folder contains numerous sheet metal forming shapes. In SolidWorks, the forming tools are inserted after the Bends are processed. Suppress forming tools in the Flat Pattern.

Activity: CABINET-Louver Forming Tool

Insert a Louver Forming tool.

357) Rotate the model to display the inside right flange.

358) Expand the **forming tools** folder in the Design Library.

359) Click the **louvers** folder. A louver is displayed.

360) Click and drag the **louver** to the inside right flange of the CABINET as illustrated. The Form Tool Feature PropertyManager is displayed.

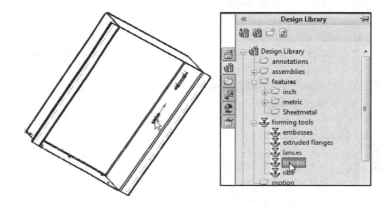

Rotate the Louver.

361) Rotate the Louver as illustrated.

362) Click **OK** ✔ from the Form Tool Feature PropertyManager.

363) Click **Right view** ⬚ from the Heads-up View toolbar.

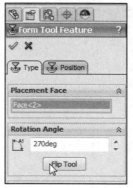
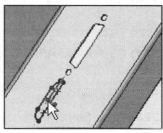

Dimension the Louver.

364) Click **Smart Dimension** ✐ from the Sketch toolbar.

365) Click the **center point** of the Louver as illustrated.

366) Click the **Origin** of the CABINET from the FeatureManager.

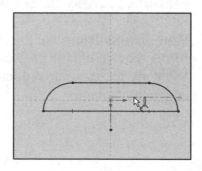

367) Enter **100**mm for the vertical dimension.

368) Click the **Origin** of the CABINET from the fly-out FeatureManager. Click the **vertical centerline** of the Louver.

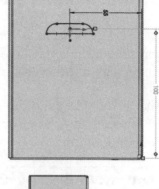

369) Enter **55**mm for the horizontal dimension.

Display the Louver.
370) Click **Finish** from the Position form feature dialog box.

371) Press the **f** key.

Add a Linear Pattern of louvers.
372) Click **Linear Pattern** from the Features toolbar.

373) Click the CABINET **back vertical edge** for Direction 1 as illustrated. Edge<1> is displayed in the Pattern Direction box.

374) Enter **25**mm for Spacing.

375) Enter **6** for Number of Instances. The direction arrow points upward. Click the Reverse Direction button if required.

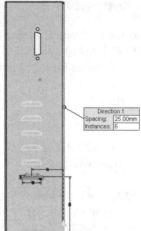

376) Click **OK** from the Linear Pattern PropertyManager. LPattern2 is displayed in the FeatureManager.

377) Click **Isometric view** from the Heads-up View toolbar.

Save the CABINET.
378) Click **Save** .

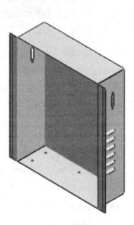

Manufacturing Considerations

How do you determine the size and shape of the louver form? Are additional die cuts or forms required in the project? Work with a sheet metal manufacturer. Ask questions. What are the standards? Identify the type of tooling in stock? Inquire on form advantages and disadvantages.

One company that has taken design and form information to the Internet is Lehi Sheetmetal, Westboro, MA. (www.lehisheetmetal.com).

Standard dies, punches and manufacturing equipment such as brakes and turrets are listed in the Engineering helpers section.

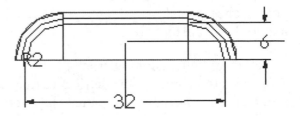

Dimensions are provided for their standard forms. The form tool you used for this project creates a 32mm x 6mm louver. The tool is commercially available.

Your manufacturer only stocks 3in., (75mm) and 4in., (100mm) louvers. Work with the sheet metal manufacturer to obtain a cost effective alternative. Create a pattern of standard square cuts to dissipate the heat.

If a custom form is required, most custom sheet metal manufacturers can accommodate your requirement. Example: Wilson Tool, Great Bear Lake, MN (www.wilsontool.com).

However, you will be charged for the tool. How do you select material? Consider the following:

- Strength

- Fit

- Bend Characteristics

- Weight

- Cost

All of these factors influence material selection. Raw aluminum is a commodity. Large manufacturers such as Alcoa and Reynolds sell to a material supplier, such as *Pierce Aluminum.

**ALUMINUM SHEET
NON HEAT TREATABLE, 1100-0
QQ-A-250/1 ASTM B 209**

All thickness and widths available from coil for custom blanks

SIZE IN INCHES	WGT/SHEET
.032 x 36 x 96	11.05
.040 x 36 x 96	13.82
.040 x 48 x 144	27.65
.050 x 36 x 96	17.28
.050 x 48 x 144	34.56
.063 x 36 x 96	21.77
.063 x 48 x 144	43.55
.080 x 36 x 96	27.65
.080 x 48 x 144	55.30
.090 x 36 x 144	46.66
.090 x 48 x 144	62.26
.100 x 36 x 96	34.56
.125 x 48 x 144	86.40
.125 x 60 x 144	108.00
.190 x 36 x 144	131.33

*Courtesy of Piece Aluminum Co, Inc.
Canton, MA

Pierce Aluminum in turn, sells material of different sizes and shapes to distributors and other manufacturers. Material is sold in sheets or cut to size in rolls. U.S. Sheet metal manufacturers work with standard 8ft - 12ft stock sheets. For larger quantities, the material is usually supplied in rolls. For a few cents more per pound, sheet metal manufacturers request the supplier to shear the material to a custom size.

Check Dual Dimensioning from Tools, Options Document Properties to display both Metric and English units. Do not waste raw material. Optimize flat pattern layout by knowing your sheet metal manufacturers equipment.

Discuss options for large cabinets that require multiple panels and welds. In Project 1, you were required to be cognizant of the manufacturing process for machined parts.

In Project 4 and 5 you created and purchased parts and recognized the Draft feature required to manufacture plastic parts.

Whether a sheet metal part is produced in or out of house, knowledge of the materials, forms and layout provides the best cost effective design to the customer.

Save manufacturing cost and time. Obtain a list of standard dies and forms from your sheet metal manufacturer. Utilize their standard dies and forms at the start of your design. Prepare for lead-time if custom dies and forms are required.

You require both a protective and cosmetic finish for the Aluminum BOX. Review options with the sheet metal vendor. Parts are anodized. Black and clear anodized finishes are the most common. In harsh environments, parts are covered with a protective coating such as Teflon® or Halon®.

The finish adds thickness to the material. A few thousandths could cause problems in the assembly. Think about the finish before the design begins.

Your sheet metal manufacturer suggests a pattern of square cuts replace the Louvers for substantial cost savings. Suppress the Louver. The Linear Pattern is a child of the Louver. The Linear Pattern is suppressed. Utilize the Sketch Linear Step and Repeat option.

Activity: Manufacturing Considerations-Sketch Linear Pattern

Suppress the louver.
379) Right-click **louver1** from the FeatureManager. Click **Suppress**.

Sketch the profile of squares.
380) Click **Right view**  from the Heads-up View toolbar.

381) Right-click the right face of the **CABINET** in the Graphics window. Shell1 is highlighted in the FeatureManager.

382) Click **Sketch** ✏ from the Context toolbar.

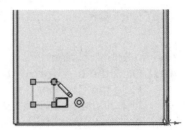

Create the first square.
383) Click **Corner Rectangle** ☐ from the Sketch toolbar. Sketch a **square** in the lower left corner as illustrated.

Add dimensions.
384) Click **Smart Dimension** ✧ from the Sketch toolbar.

385) Create a vertical and horizontal **10**mm dimension.

386) Click the **Origin**.

387) Click the **bottom horizontal line**, of the square.

388) Click a **position** to the left of the profile.

389) Enter **10**mm.

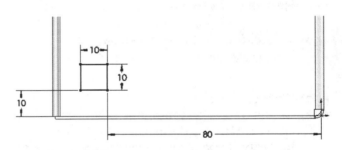

Create a horizontal dimension.
390) Click the **Origin**.

391) Click the **right vertical line** of the square.

392) Click a position **below** the profile.

393) Enter **80**mm.

Create a second square.
394) Click **Corner Rectangle** ☐ from the Sketch toolbar.

395) Sketch a **square diagonal** to the right of the first square as illustrated.

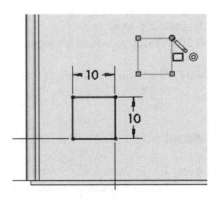

Create a horizontal dimension.

396) Click **Smart Dimension** ✎ from the Sketch toolbar.

397) Click the **right vertical** line of the first box.

398) Click the **left vertical** line of the second box. Enter **5**mm.

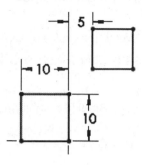

Add an Equal relation.

399) Right-click **Select**. Click the **top line** of the second square. Hold the **Ctrl key** down. Click the **left line** of the second square. The selected entities are displayed in the Selected Entities box. Release the **Ctrl** key. Click **Equal** from the Add Relations box.

400) Click **OK** ✔ from the Properties PropertyManager.

Add an Equal relation.

401) Click the **top horizontal line** of the first square. Hold the **Ctrl key** down. Click the **top horizontal line** of the second square. The selected entities are displayed in the Selected Entities box. Release the **Ctrl** key. Click **Equal** from the Add Relations box.

402) Click **OK** ✔ from the Properties PropertyManager.

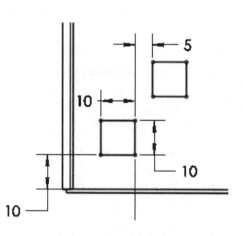

Add a Collinear relation.

403) Click the **top horizontal line** of the first square. Hold the **Ctrl** key down. Click the **bottom horizontal line** of the second square. Release the **Ctrl** key.

404) Click **Collinear** from the Add Relations box.

405) Click **OK** ✔ from the Properties PropertyManager.

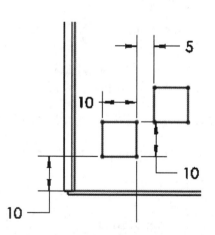

Repeat the sketch.

406) Window-Select all **eight lines** of the two squares.

407) Click **Linear Sketch Pattern** from the Sketch toolbar. The Linear Pattern PropertyManager is displayed.

408) Enter **2** for Number in Direction 1.

409) Enter **30**mm for Spacing in Direction 1. The pattern direction arrow points to the right.

410) Check the **Add angle dimension between axes** box.

411) Enter **13** for Number in Direction 2. The pattern direction arrow points upward.

412) Enter **20**mm for Spacing in Direction 2.

413) Click **OK** ✅ from the Linear Pattern PropertyManager.

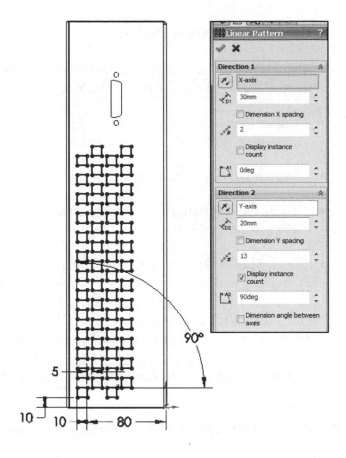

Extrude the sketch.

414) Click **Extruded Cut** 🔲 from the Features toolbar. The Cut-Extrude PropertyManager is displayed.

415) Check the **Link to thickness** box. Accept the default settings.

416) Click **OK** ✅ from the Cut-Extrude PropertyManager.

417) Click **Isometric view** 🔲 from the Heads-up View toolbar.

418) Rename the **Cut-Extrude2** to **Vent**.

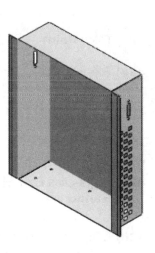

Save the CABINET in its formed state.

419) Click **Save** 💾.

💡 Before you utilize the Sketch Linear Pattern feature, determine the design intent of the pattern in the part and in a future assembly. If there are mating components, use Pattern features. Pattern features provide additional options compared to a simple sketch tool.

Additional Pattern Options

Three additional pattern options to discuss in this book are: Curve Driven Pattern, Sketch Driven Pattern and Table Driven Pattern.

▦	Linear Pattern
✤	Circular Pattern
⧈	Mirror
⊞	Curve Driven Pattern
⋮⋮⋮	Sketch Driven Pattern
▦	Table Driven Pattern
⊛	Fill Pattern

- Curve Driven Pattern creates instances of the seed feature through a sketched curve or edge of a sketch.

- Sketch Driven Pattern creates instances of the seed feature though sketched points.

- Table Driven Pattern creates instances of the seed feature through X-Y coordinates in an existing table file or text file.

Utilize a Sketch Driven Pattern for a random pattern. A Sketch Driven Pattern requires a sketch and a feature. The Library Feature Sheetmetal folder contained the d-cutout and other common profiles utilized in sheet metal parts. Create a sketch with random Points on the CABINET left face. Insert sw-b212 and create a Sketch Pattern. View SolidWorks help for additional information on Pattern features.

Activity: Additional Pattern Options-Sketch Driven Pattern

Create the sketch.

420) Click **Left view** ⬚ from the Heads-up View toolbar.

421) Right-click the left face of the **CABINET** for the Sketch plane.

422) Click **Sketch** ✏ from the Context toolbar.

423) Click **Point** ＊Point from the Sketch toolbar.

424) Sketch a random series of **points** on the left face as illustrated. The Point PropertyManager is displayed. Remember the first point you select. This point is required to constrain the sw-b212 sheet metal feature.

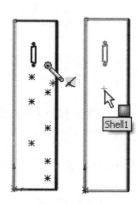

Close the sketch.

425) Right-click **Exit Sketch** ✏. Rename **Sketch** to **Sketch-Random**. **Rotate** the CABINET to view the inside left face.

426) Click **Save** 💾.

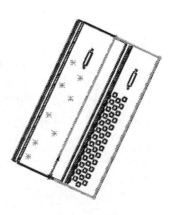

Insert the sw-b212 sheet metal tool.

427) **Expand** the Design Library 📚 folder in the Task Pane.

428) **Expand** the features folder. Click the **Sheetmetal** folder.

429) Drag the **sw-b212** feature to the inside left face of the CABINET. The sw-b212 PropertyManager is displayed.

430) Click the **Edit Sketch** button. DO NOT SELECT THE FINISH BUTTON AT THIS TIME.

Add a Midpoint relation between the center point of the sw-b212 feature and the first random sketch point.
431) Click the **vertical centerline** of sw-b212.

432) Hold the **Ctrl** key down.

433) Click the **Point** of the first random sketch point. The selected entities are displayed in the Selected Entities box. Release the **Ctrl** key.

434) Click **Midpoint** / from the Add Relations box. Click the **Finish** button.

Edit the End Condition.
435) Right-click **Cut-Extrude1** in the FeatureManager.

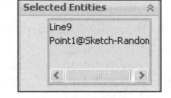

436) Click **Edit Feature**.

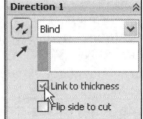

437) Select **Blind** for End Condition in Direction 1.

438) Check the **Link to thickness** box.

439) Click **OK** ✓ from the Cut-Extrude1 PropertyManager.

Insert the Sketch Driven Pattern feature.
440) Click **Insert**, **Pattern/Mirror**, **Sketch Driven Pattern** from the Menu bar. The Sketch Driven Pattern PropertyManager is displayed.

441) Click inside the **Reference Sketch** box.

442) **Expand** the CABINET fly-out FeatureManager.

443) Click **Sketch-Random**. If required, click inside the **Features to Pattern** box.

444) Click **Cut-Extrude1** from the fly-out FeatureManager.

445) Click **OK** ✓ from the Sketch Driven Pattern PropertyManager. Sketch-Pattern1 is displayed in the FeatureManager.

446) Click **Isometric view** ▢. Click **Save** ▣.

The Sketch Driven Pattern is complete. Suppress the patterns when not required for future model development.

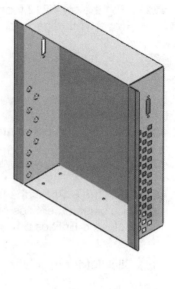

Utilize the Suppress option with configurations to save the formed and flat state of the CABINET.

CABINET-Formed and Flat States

How do you control the 3D formed state in the assembly and the 2D flat state in the drawing? Answer: Utilize configurations. A configuration is a variation of a part or assembly within a single document. Variations include different dimensions, features, and properties.

The BOX assembly requires the CABINET in the 3D formed state. Utilize the Default Configuration in the assembly.

The CABINET drawing requires the 2D flat state. The drawing requires a view with the Linear Pattern of square cuts and the random sketch pattern suppressed. Create a new configuration named NO-VENT. Suppress the patterns. Create a Derived configuration named NO-VENT-FLAT. Un-suppress the Flat Pattern feature to maintain the CABINET in its flat state. Create two configurations in the 3D formed and 2D flat state.

Activity: CABINET-Formed and Flat States

Display the ConfigurationManager and add a Configuration.

447) Drag the **Split bar** ⇌ downward to divide the FeatureManager in half.

448) Click the **ConfigurationManager** tab.

449) Right-click **CABINET**.

450) Click **Add Configuration**. The Add Configuration PropertyManager is displayed.

451) Enter **NO-VENT** for Configuration name.

452) Enter **SUPPRESS VENT CUTS** for Description.

453) Click **OK** ✓ from the Add Configuration PropertyManager.

Suppress the Vent Extruded Cut feature.
454) Right-click **Vent** from the FeatureManager.

455) Click **Suppress** ↓🗄 from the Context toolbar.

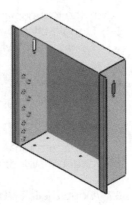

The NO-VENT Configuration contains the suppressed Vent Extruded Cut feature. A Flat Pattern exists for the Default Configuration. The Flat Pattern Configuration is called a Derived Configuration.

Create a Flat Pattern Derived Configuration for the NO-VENT configuration. Un-suppress the Flat Pattern feature.

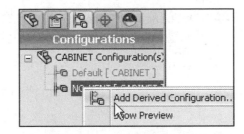

Create a Derived Configuration.
456) Right-click **NO-VENT [CABINET]** from the ConfigurationManager.

457) Click **Add Derived Configuration**. The Add Configuration PropertyManager is displayed.

458) Enter **NO-VENT-FLAT-PATTERN** for Configuration name.

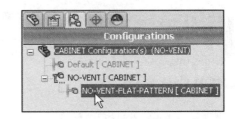

459) Click **OK** ✅ from the Add Configuration PropertyManager.

Un-suppress the Flat-Pattern1 feature.
460) Right-click **Flat-Pattern1** from the CABINET FeatureManager.

461) Click **Unsuppress** ↑🗄 from the Context toolbar to display the Flat State.

Suppress the Random Sketch points.
462) Right-click **Sketch Random** from the CABINET FeatureManager.

463) Click **Suppress** ↓🗄 from the Context toolbar.

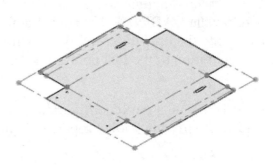

Display the four configurations.

464) Double-click **Default [CABINET]** from the ConfigurationManager.

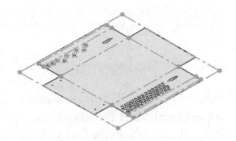

465) UnSuppress Flat-Pattern1 from the FeatureManager.

466) Double-click **NO-VENT [CABINET]** from the ConfigurationManager. View the results in the Graphics window.

467) Double-click **NO-VENT-FLAT-PATTERN [CABINET]** from the ConfigurationManager.

468) Double-click **Default [CABINET]** from the ConfigurationManager to return to the Default Configuration.

469) Suppress Flat-Pattern1 from the CABINET FeatureManager.

470) Return to a single CABINET FeatureManager.

471) Click the **FeatureManager** tab.

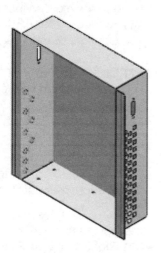

Save the CABINET.

472) Click **Save** .

CABINET-Sheet Metal Drawing with Configurations

Sheet metal drawings are produced in the flat state. Create a new C-size [A2] drawing size for the CABINET. Utilize the default SolidWorks Drawing Template. Insert the Top view into the drawing. Modify the Drawing View Properties from the Default configuration to the Flat configuration.

To create a multi configuration drawing, insert the Default configuration for Model View. Modify the View Properties in the drawing to change the configuration.

Activity: CABINET-Sheet Metal Drawing with Configurations

Create a new drawing.

473) Click **Make Drawing from Part/Assembly** ⊞ from the Menu bar.

474) Double-click **Drawing** from the Templates tab.

475) Select **C (ANSI) Landscape [A2-Landscape]** for Sheet Format/Size.

476) Click **OK**. The View Palette is displayed on the right side of the Graphics window.

Insert an Isometric view using the View Palette.

477) Click and drag the **Isometric view** from the View Palette to the right side of the drawing as illustrated.

478) Click **OK** ✔ from the Drawing View1 PropertyManager.

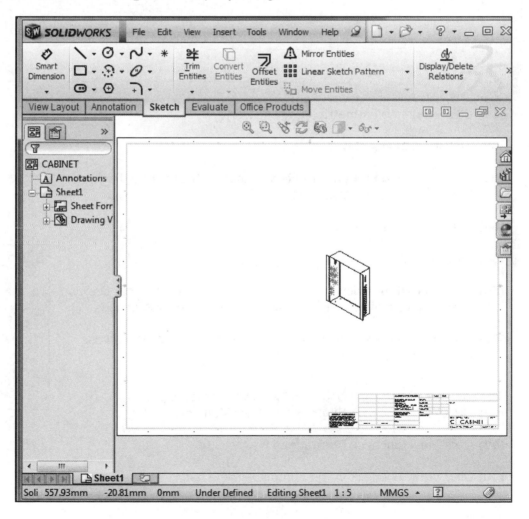

Insert the Default Flat pattern view.

479) Click and drag the **Flat pattern view** on the left side of the drawing.

480) Click **OK** ✅ from the Projected View PropertyManager.

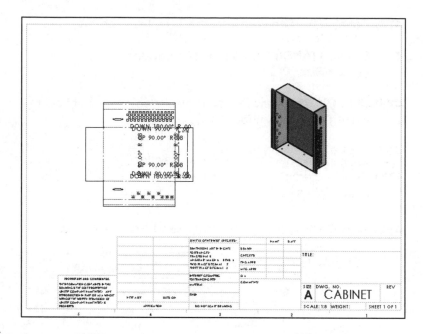

Copy and Paste the two views.
481) Click inside the **Isometric view** boundary.

482) Hold the **Ctrl** key down.

483) Click inside the **Flat pattern view** boundary. The Multiple View PropertyManager is displayed.

484) Release the **Ctrl** key.

485) Press **Ctrl C** to copy.

486) Click a **position** below the views. Click **Ctrl V** to paste. Four Drawing Views are displayed in the CABINET Drawing FeatureManager.

487) Click and drag the **views** under each other as illustrated. Address the drawing view scale if needed.

Hide Sketches, Points, and Origins in the drawing if required.
488) Click **View**; uncheck **Sketches** from the Menu bar.

489) Click **View**; uncheck **Points** from the Menu bar.

490) Click **View**; uncheck **Origins** from the Menu bar.

Modify the configurations.

491) Click inside the **lower left Flat pattern** view in Sheet1.

492) Right-click **Properties**.

493) Select **NO-VENT-FLAT-PATTERN** from the Configuration information drop-down menu.

494) Click **OK** from the Drawing View Properties dialog box.

495) Click inside the **lower right Isometric** view.

496) Right-click **Properties**.

497) Select **NO-VENT "SUPPRESS VENT CUTS"** from the Configuration information drop down menu.

498) Click **OK** from the Drawing View Properties dialog box.

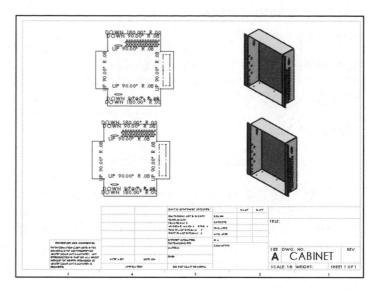

Hide Origins, Planes, and Sketches in drawing views. Insert centerlines to represent sheet metal bend lines.

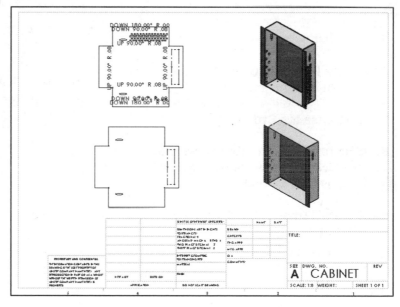

Save and close the CABINET drawing.

499) Click **Save** 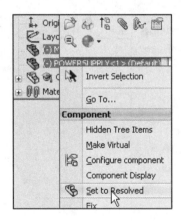.

500) Enter **CABINET** For File name.

501) Click **Save**.

502) Click **Yes**.

503) Click **File**, **Close** from the Menu bar.

The CABINET is in the Default configuration.

Return to the BOX assembly. Review the BOX FeatureManager. The MOTHERBOARD and POWERSUPPLY are suppressed. Restore the MOTHERBOARD and POWERSUPPLY. Collapse entries in the FeatureManager.

Return to the BOX assembly.
504) **Open** the BOX assembly.

Display the components.
505) Click **MOTHERBOARD** in the BOX FeatureManager.

506) Hold the **Ctrl** key down.

507) Click **POWERSUPPLY** in the FeatureManager.

508) Release the **Ctrl** key.

509) Right-click **Set to Resolved**. The selected entities are displayed in the Graphics window.

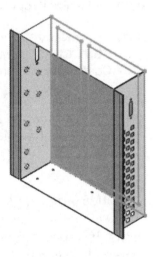

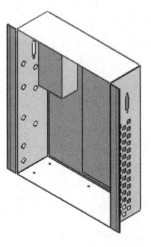

Save the BOX assembly.

510) Click **Save** 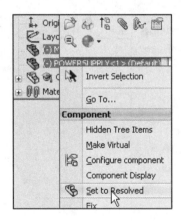.

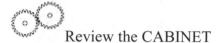 Work in SolidWorks from the design through the manufacturing process to avoid data conversion issues and decreased shop time. The SolidWorks Manufacturing Partner Network (www.solidworks.com) lists Sheet Metal manufacturers that utilize SolidWorks.

Additional details on Sheet Metal, Rip, Insert Bends, Flat Pattern, Forming Tools, Edge Flange, Hem, Library Feature, Linear Step and Repeat and Configuration Properties, are available in SolidWorks Help.

Review the CABINET

You created the CABINET component inside the BOX assembly. The InPlace Mate inserted a Coincident Mate between the Front Plane of the CABINET and the Front Plane of the BOX. The CABINET component referenced the outside profile of the Layout Sketch.

The first feature of the CABINET was the Extruded Boss/Base (Boss-Extrude1) feature that created a solid box. You utilized the Shell feature to create the constant wall thickness.

The Rip feature cut removed the solid edges. The Insert Bends feature created a sheet metal part from a solid part. The Flat Pattern feature displayed the sheet metal part in the manufactured 2D flat state.

The Edge Flange and Hem features were added to the right and left side of the CABINET. Sheet metal dies and form louvers were inserted from the Library Features.

The Sketch Linear Pattern was utilized to replace the form louvers with more cost effective cuts. You created a configuration to represent the 2D flat state for the CABINET drawing. The 3D formed state was utilized in the BOX assembly.

The Vent feature located in the Sheet Metal toolbar allows you to create a vent by selecting a closed profile as the outer vent boundary.

The Fill Pattern feature allows you to fill a closed boundary to create a grid for a sheet-metal perforation - style pattern.

PEM® Fasteners and IGES Components

The BOX assembly contains an additional support BRACKET that is fastened to the CABINET. To accommodate the new BRACKET, the Linear Pattern of five holes is added to the CABINET.

How do you fasten the BRACKET to the CABINET?

Answer: Utilize efficient and cost effective self-clinching fasteners. Example: PEM® Fasteners.

PEM® Fastening Systems, Danboro, PA USA manufactures self-clinching fasteners, called PEM® Fasteners.

PEM® Fasteners, Examples

PEM® Fasteners are used in numerous applications and are provided in a variety of shapes and sizes. IGES standard PEM® models are available at (www.pennfast.com) and are used in the following example.

Answer the following questions in order to select the correct PEM® Fastener for the assembly:

QUESTION:	ANSWER:
What is the material type?	Aluminum
What is the material thickness of the CABINET?	1mm
What is the material thickness of the BRACKET?	19mm
What standard size PEM® Fastener is available?	FHA-M5-30ZI
What is the hole diameter required for the CABINET?	5mm ± 0.08
What is the hole diameter required for the BRACKET?	5.6mm maximum
Total Thickness = CABINET Thickness + BRACKET Thickness + NUT Thickness Total Thickness = 1mm + 19mm + 5mm = 25mm (minimum)	25mm (minimum)

The selected PEM® Fastener for the project is the FHA-M5-30ZI.

- **FHA** designates the stud type and material. Flush-head Aluminum.

- **M5** designates the thread size.

- **30** designates length code: 30mm.

- **ZI** designates finish.

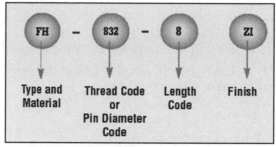

Part Number General Designation for PEM®

The overall dimensions of the FH PEM® Fasteners series are the same for various materials. Do not utilize PEM® Fasteners as a feature in the Flat State of a component. In the sheet metal manufacturing process, PEM® Fasteners are added after the material pattern has been fabricated.

Proper PEM® Fasteners material selection is critical. There are numerous grades of aluminum and stainless steel. Test the installation of the material fastener before final specification.

Note: When the fastener is pressed into ductile host metal, it displaces the host material around the mounting hole, causing metal to cold flow into a designed annular recess in the shank.

Ribs prevent the fastener from rotating in the host material once it has been properly inserted. The fasteners become a permanent part of the panel in which they are installed.

During the manufacturing process, a stud is installed by placing the fastener in a punched or drilled hole and squeezing the stud into place with a press. The squeezing action embeds the head of the stud flush into the sheet metal.

Obtain additional manufacturing information that directly affects the design.

Manufacturing Information:	**FH-M4ZI:**	**FH-M5ZI:**
Hole Size in sheet metal (CABINET).	4.0mm ± 0.8	5.0mm ± 0.8
Maximum hole in attached parts (BRACKET).	4.6mm	5.6mm
Centerline to edge minimum distance. (from center point of hole to edge of CABINET)	7.2mm	7.2mm

---

Activity: PEM® Fasteners and IGES Components

Obtain the PEM Fasteners for this project either from the Internet at; www.pennfast.com, (FH-M5-30ZI, FH-M4-25ZI) or from the enclosed DVD in the book in the COMPONENTS folder.

🔆 Part geometry for Aluminum and Stainless Steel fasteners is the same.

Open the files from the COMPONENTS folder in the book DVD.
511) Copy the files, **FH-M5-30ZI** and **FH-M4-25ZI** to the ENGDESIGN-W-SOLIDWORKS\VENDOR COMPONENTS folder.

512) Open the **FH-M5-30ZI** part from the ENGDESIGN-W-SOLIDWORKS\VENDOR COMPONENTS folder.

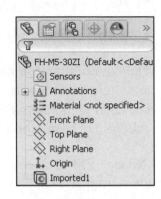

Insert the FH-M5-30ZI fastener into the BOX assembly.
513) Open the **BOX** assembly.

514) Click **Insert Components** from the Menu bar. The Insert Component PropertyManager is displayed.

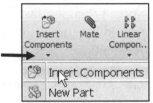

515) Double-click **FH-M5-30ZI** in the Open documents box.

516) Click inside the **Graphics window** in the inside bottom left corner near the seed feature of the Linear Pattern of holes.

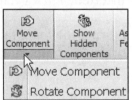

517) Zoom to Area 🔍 on the hole and fastener.

Rotate the FH-M5-30ZI component.
518) Click **Rotate Component** 🔄 from the Assembly toolbar. The Rotate Component PropertyManager is displayed.

519) Rotate the **FH-M5-30ZI** component until the head faces downward as illustrated.

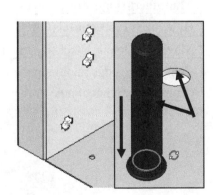

520) Click **OK** ✔ from the Rotate Component PropertyManager.

Insert a Concentric mate.

521) Click **Mate** 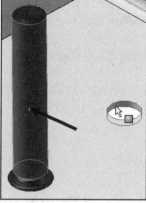 Mate from the Assembly toolbar. The Mate PropertyManager is displayed.

522) Click the cylindrical **shaft** face of the FH-M5-30ZI fastener.

523) Click the cylindrical **face** of the back left hole. The selected faces are displayed in the Mate Selections box. Concentric is selected by default.

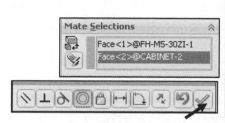

524) Click **OK** ✅ from the Pop-up dialog box.

Insert a Coincident mate.

525) Click the **head face** of the FH-M5-30ZI fastener as illustrated.

526) Click the **outside bottom face** of the CABINET. The selected faces are displayed in the Mate Selections box. Coincident is selected by default.

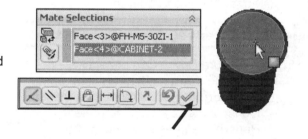

527) Click **OK** ✅ from the Pop-up dialog box.

Insert a Parallel mate.

528) **Expand** BOX from the fly-out FeatureManager.

529) Click the **Front Plane** of the FH-M5-30ZI fastener.

530) Click the **Right Plane** of the CABINET. The selected planes are displayed in the Mate Selections box.

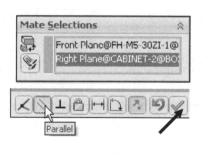

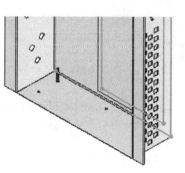

531) Click **Parallel** from the Pop-up menu.

532) Click **OK** ✅ from the Pop-up dialog box.

533) Click **OK** ✅ from the Mate PropertyManager.

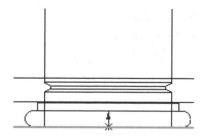

Fit to the Graphics window.

534) Press the **f** key. Click **Isometric view** .

535) Click **Save** .

Three Mates fully define the FH-M5-30ZI fastener. The PEM® fastener creates an interference fit with hole1.

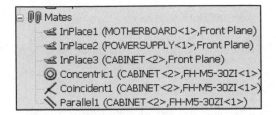

Feature Driven Component Pattern

A Feature Driven Component Pattern is a pattern that utilizes an existing component in an assembly and a driving feature from another component in the same assembly. Utilize the Feature Driven Component Pattern tool to create multiple copies of a component in an assembly. The FH-M5-30ZI references the hole seed feature. The Feature Driven Component Pattern displays the five FH-M5-30ZI fasteners based on the CABINET LPattern1 feature.

Activity: Derived Component Pattern

Insert a Feature Driven Component Pattern.

536) Click the **Feature Driven Component Pattern** tool from the Consolidated Linear Component Pattern menu as illustrated. The Feature Driven PropertyManager is displayed.

537) Click **FH-M5-30ZI** from the fly-out FeatureManager.

538) Click **inside** the Driving Feature box.

539) Click **CABINET\LPattern1** from the BOX fly-out FeatureManager.

540) Click **OK** from the Feature Driven PropertyManager. DerivedLPattern1 is displayed in the FeatureManager.

Expand DerivedLPattern1.

541) **Expand** DerivedLPattern1 in the FeatureManager. View the results.

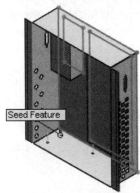

Seed Feature

Save the BOX.

542) Click **Isometric view** 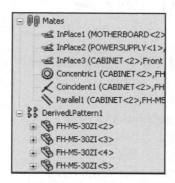 from the Heads-up View toolbar.

543) Click **Save** 💾.

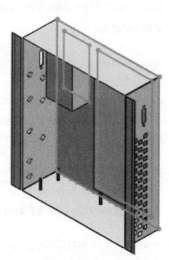

The additional FH-M5-30ZI instances are located under DerivedLPattern1 in the FeatureManager. The purchasing manager for your company determines that the FH-M5-30ZI fastener has a longer lead-time than the FH-M4-25ZI fastener.

The FH-M4-25 fastener is in stock. Time is critical. You ask the engineer creating the corresponding BRACKET if the FH-M4-25ZI fastener is a reliable substitute.

You place a phone call. You get voice mail. You send email. You wait for the engineer's response and create the next assembly feature.

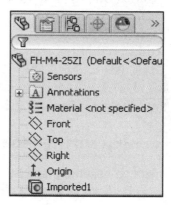

🔅 The common PEM® fastener library is available in SolidWorks Toolbox.

How do you locate a hole for a fastener that has to go through multiple components in an assembly? Answer: Utilize an Assembly Hole feature.

MOTHERBOARD-Assembly Hole Feature

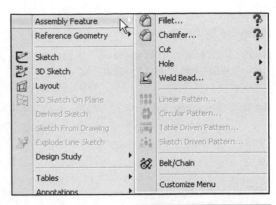

Assembly features are holes and cuts created in the assembly. Utilize the Assembly Feature option from the Menu bar or from the Assembly tab after components have been assembled. The Assembly Feature option includes Cut, Hole, Linear Pattern, Circular Pattern, Table Driven Pattern, Sketch Driven Pattern, Belt/Chain and Weld Bead. Create a Cut or Hole Assembly feature to activate the Assembly Feature Pattern tool.

Insert an Assembly feature into the BOX assembly. When creating an Assembly feature, determine the components to be affected by the feature. Assembly features are listed at the bottom of the FeatureManager. They are not displayed in the components affected by the feature.

Additional holes are created through multiple components in the BOX assembly. A through hole is required from the front of the MOTHERBOARD to the back of the CABINET. Since the hole is an Assembly feature, it is not displayed in the CABINET part.

Activity: MOTHERBOARD-Assembly Hole Feature

Insert an Assembly feature with the Hole Wizard.

544) Click **Hole Wizard** from the Assembly toolbar. The Hole Specification PropertyManager is displayed.

545) Click the **Hole** icon type.

546) Select **Ansi Metric** for Standard.

547) Select **Drill sizes** for Type.

548) Select **Ø5.0** for Size.

549) Click **Through All** for End Condition.

550) Click the **Positions** tab.

551) Click the **front face** of the MOTHERBOARD as illustrated. The Point Sketch tool is displayed.

552) Click **again** to position the center point of the hole.

553) Right-click **Select** to deselect the Sketch Point tool.

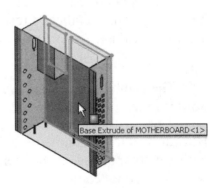

554) Click **Front view** from the Heads-up View toolbar.

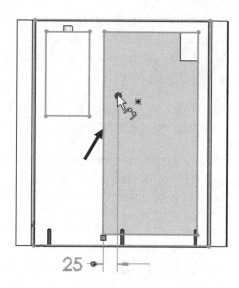

Dimension the hole on the MOTHERBOARD.
555) Click **Smart Dimension** from the Sketch toolbar.

Add a horizontal dimension.
556) Click the **left edge** of the MOTHERBOARD. Click the **centerpoint** of the hole. Click a **position** below the profile. Enter **25**mm.

Add a vertical dimension.
557) Click the **Origin** of the CABINET from the fly-out FeatureManager.

558) Click the **centerpoint** of the hole.

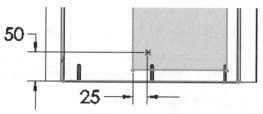

559) Click a **position** to the left of the profile. Enter **50**mm.

560) Click **OK** from the Dimension PropertyManager.

561) Click **OK** from the Hole Position PropertyManager. The hole is added to the FeatureManager.

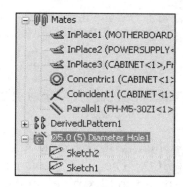

Save the BOX.
562) Click **Isometric view** . Click **Save** .

Close all parts and assemblies.
563) Click **Window**, **Close All** from the Menu bar.

Assembly FeatureManager and External References

The FeatureManager contains numerous entries in an assembly. Understanding the organization of the components and their Mates is critical in creating an assembly without errors. Errors occur within features such as a radius that is too large to create a Fillet feature. When an error occurs in the assembly, the feature or component is labeled in red.

Mate errors occur when you select conflicting geometry such as a Coincident Mate and a Distance Mate with the same faces.

In the initial mating, undo a mate that causes an error. Mate errors occur later on in the design process when you suppress required components and features. Plan ahead to avoid a problem.

Use reference planes for Mates that will not be suppressed.

When a component displays a '->' symbol, an External reference exists between geometry from the assembly or from another component.

When the system does not locate referenced geometry, the part name displays a '->?' symbol to the right of the component name in the FeatureManager.

Explore External references in the next activity. Open the CABINET without opening the BOX assembly. The '->?' symbol is displayed after the part name.

Activity: Assembly FeatureManager and External References

Open the CABINET part.

564) Click **Open** from the Menu bar.

565) Browse to the **ENGDESIGN-W-SOLIDWORKS\PROJECTS** folder.

566) Click **Part** for Files of type.

567) Double-click **CABINET**. The CABINET FeatureManager is displayed.

The CABINET cannot locate the referenced geometry BOX. The "->?" is displayed to the right of the CABINET name and Extrude1 in the FeatureManager.

Open the BOX assembly.

568) Open the BOX assembly. The BOX FeatureManager is displayed.

Reload the CABINET

569) Right-click **CABINET** in the FeatureManager.

570) Click the **More arrow** at the bottom of the pop-up menu.

571) Click **Reload**. Click **OK** from the Reload dialog box.

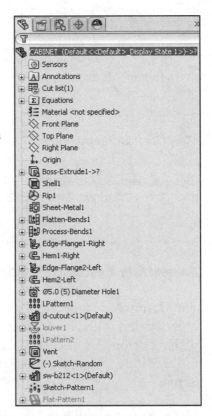

572) Return to the CABINET assembly.

Save the CABINET.
573) Click **Save** 🔳.

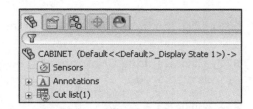

The CABINET references the BOX assembly. The External references are resolved and the question marks are removed.

Replace Components

The Replace Components option replaces a component in the current assembly with another component. Replace a part with a new part or assembly or an assembly with a part. If multiple instances of a component exist, select the instances to replace.

The BRACKET engineer contacts you. You proceed to replace the FH-M5-30ZI fastener with the FH-M4-25ZI fastener. Use the Replace Components option to exchange the FH-M5-30ZI fastener with FH-M4-25ZI fastener in the BOX assembly.

Edit the Mates and select new Mate references. The Derived Component Pattern updates with the new fastener. Edit the CABINET and change the hole size of the seed feature from Ø5.0mm to Ø4.0mm. The CABINET Linear Pattern updates with the new hole.

Before you replace Component-A in and assembly with Component-B, close Component-B. The Replace Component option produces an error message with open components. Example; Close the FH-M4-25ZI part before replacing the FH-M5-30ZI part in the current assembly.

Activity: Replace Components

Replace the FH-M5-30ZI component with the FH-M4-25ZI component.
574) Open the **BOX** assembly.

575) Right-click **FH-M5-30ZI**
⊟ 🐾 FH-M5-30ZI<1> from the
FeatureManager.

Replace five FH-M5-30ZI with FH-M4-25ZI by replacing the seed feature.

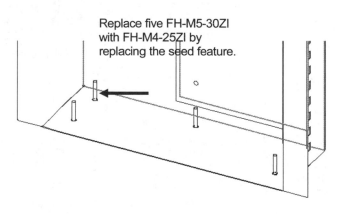

576) Click the **More arrow** ❯ at the bottom of the Pop-up menu.

577) Click **Replace Components** . The Replace PropertyManager is displayed.

578) Click the **Browse** button.

579) Double-click **FH-M4-25ZI** from the ENGDESIGN-W-SOLIDWORKS\VENDOR COMPONENTS folder.

580) Click **OK** ✔ from the Replace PropertyManager. The What's Wrong box is displayed.

581) Click **Close** from the What's Wrong dialog box.

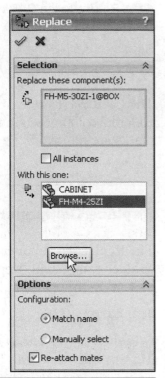

The five FH-M5-30ZI fasteners are replaced with the five FH-M4-25ZI fasteners. Mate Entity errors are displayed.

A red X is displayed on the Mated Entities PropertyManager. Redefine the Concentric Mate and Coincident Mate. Select faces from the FH-M4-25ZI component.

Edit the Concentric Mate.

582) **Expand** the first Mate entity. ❓ Face of FH-M4-25ZI-1

583) Double-click **Concentric1**.

584) Click the **FH-M4-25ZI cylindrical face** as illustrated in the Graphics window. View the results in the Mate Entities box.

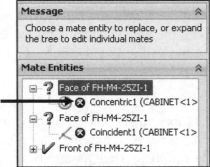

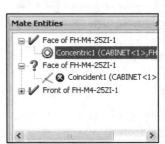

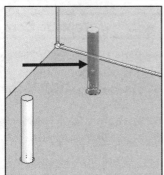

Edit the Coincident mate.

585) **Expand** the second Mate entity.

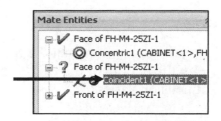

586) Click **Bottom view** ⌗.

587) Double-click **Coincident1**.

588) Click the **FH-M4-25 bottom face**. View the results in the Mate Entities box.

589) Click **OK** ✔ from the Mate Entities PropertyManager.

Edit the Ø5.0mm hole to a Ø4.0mm hole

590) Click **CABINET** from the FeatureManager.

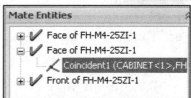

591) Click **Edit Component** from the Assembly toolbar. The CABINET entries are displayed in blue.

592) **Expand** CABINET in the FeatureManager.

593) Right-click **(5) Diameter Hole1** from the FeatureManager.

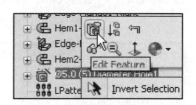

594) Click **Edit Feature**. The Hole Specification PropertyManager is displayed.

595) Select **Ø4.0** for Size.

596) Click **OK** ✔ from the Hole Specification PropertyManager. The (4) Diameter Hole1 is displayed in the FeatureManager. Click **OK**.

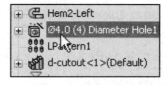

597) Click **Edit Component** from the Assembly toolbar to return to the BOX assembly. The CABINET entries are displayed in black.

Save the BOX.

598) Click **Save** 💾. Click **Save All**.

🔆 The hole of the CABINET requires an Ø4mm ± 0.8. Add a limit dimension in the drawing and a revision note to document the change from an Ø5.0mm to an Ø4.0mm.

🔆 IGES imported geometry requires you to select new mate references during Replace Components. Utilize reference planes for Mate entities to save time. The Replace Components option attempts to replace all Mate entities. Reference planes provide the best security in replacing the appropriate Mate entity during the Replace Components option.

Equations

How do you ensure that the fasteners remain symmetrical with the bottom of the BOX when the box-width dimension varies?

Answer: With an Equation. Equations use shared names to control dimensions. Use Equations to connect values from sketches, features, patterns and various parts in an assembly.

Create an Equation. Each dimension has a unique variable name. The names are used as Equation variables. The default names are based on the Sketch, Feature or Part. Rename default names for clarity.

Resolve all Mate Errors before creating Equations. Mate Errors produce unpredictable results with parameters driven by equations.

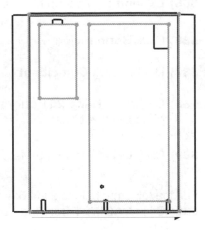

Increase BOX width.

Fasteners are not symmetrical with the bottom of the BOX.

Activity: Create an Equation

Display Feature dimensions.
599) Right-click the CABINET **Annotations** folder.

600) Check the **Show Feature Dimensions** box.

601) Click **Front view** from the Heads-up View toolbar.

Edit the LPattern1 dimension name.
602) Click the **125**mm dimension in the Graphics window. The Dimension PropertyManager is displayed.

Rename the D3@LPattern1 dimension.
603) Click inside the **D3@LPattern1** box.

604) Enter **lpattern-bottom** for Name.

605) Click **OK** from the Dimension PropertyManager.

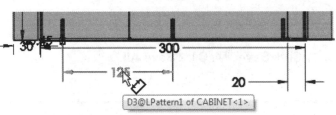

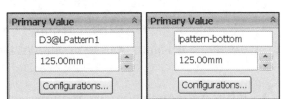

Edit the 4.0mm hole dimension name.
606) **Rotate** to display the back hole.

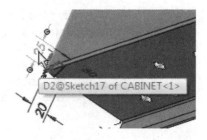

607) Click the **25**mm dimension. 25 is the Ø4.0mm hole horizontal dimension in the lower left corner. The Dimension PropertyManager is displayed.

Rename the Sketch dimension name.
608) Enter **hole-bottom** for Name.

609) Click **OK** ✅ from the Dimension PropertyManager.

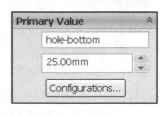

610) Click **Isometric view** 🔲 from the Heads-up View toolbar.

Display the BOX feature dimensions.
611) Right-click the BOX **Annotations** folder.

612) Check the **Show Feature Dimensions** box.

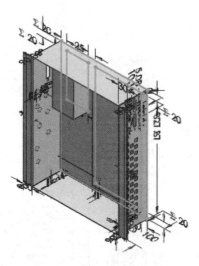

Create an Equation.

613) Click **Front view** ⬛ from the Heads-up View toolbar.

614) Click **Tools**, **Equations** Σ Equations... from the Menu bar. The Equations, Global Variables, and Dimensions dialog box is displayed.

Create the first half of equation1.
615) Click a **position** in the fifth cell under Equations - Top Level.

616) Click the lpattern-bottom @Pattern1 horizontal dimension, **125**mm in the Graphics window. The variable "lpattern-bottom @LPattern1@ CABINET" is added.

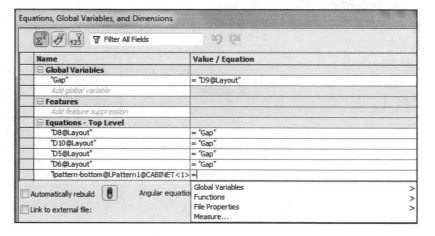

Create the second half of the Equation.

617) Enter **0.5*(** from the keypad.

Name	Value / Equation
⊟ **Global Variables**	
"Gap"	= "D9@Layout"
Add global variable	
⊟ **Features**	
Add feature suppression	
⊟ **Equations - Top Level**	
"D8@Layout"	= "Gap"
"D10@Layout"	= "Gap"
"D5@Layout"	= "Gap"
"D6@Layout"	= "Gap"
"lpattern-bottom@LPattern1@CABINET<1>	=0.5*

618) Click the box_width horizontal dimension, **300** from the Graphics window. The variable "box_width@layout" is added to the equation text box.

619) Enter **−2*** from the keypad.

620) Click **Bottom**

	"D6@Layout"	= "Gap"
	"lpattern-bottom@LPattern1@CABINET<1>	=0.5*("box_width@Layout"

view ⬚ from the Heads-up View toolbar.

621) Click the M4 hole-bottom dimension, **25**. The variable "hole-bottom@Sketch12@CABINET.Part" is added to the equation text box.

622) Enter **)** from the keypad.

623) Press the **Tab** key. View the results.

	"D6@Layout"	= "Gap"
	"lpattern-bottom@LPattern1@CABINET<1>	_width@Layout"-2*"hole-bottom @Sketch17@CABINET<1>.Part")

624) Click **OK** from the dialog box.

Name	Value / Equation	Evaluates to
Add global variable		
⊟ **Features**		
Add feature suppression		
⊟ **Equations - Top Level**		
"D8@Layout"	= "Gap"	20mm
"D10@Layout"	= "Gap"	20mm
"D5@Layout"	= "Gap"	20mm
"D6@Layout"	= "Gap"	20mm
Add equation		
⊟ **Equations - Components**		
-bottom@LPattern1@CABINET<1>.Part"	= 0.5 * ("box_width@Layout" - 2 * "hole-bottom @Sketch17@CAB	125mm

The Equations dialog box contains the complete equation. A green check mark indicates that the Equation is solved. The Equation evaluates to 125mm. The

Equation $^{\Sigma}$ icon, placed in front of the 125 dimension, indicates that an equation drives the value.

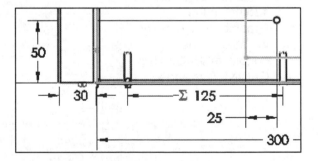

Remember the initial design parameters. There are three different size boxes. Remember the design intent. You build a Layout Sketch to control the different sizes.

Test the equation by modify the Layout Sketch dimensions.

Verify the Equation. Modify the dimensions.

625) Click **Front view** 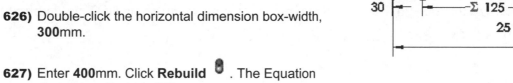 from the Heads-up View toolbar.

626) Double-click the horizontal dimension box-width, **300**mm.

627) Enter **400**mm. Click **Rebuild** . The Equation dimension displays 175.

628) Double-click the dimension box-height, **400**mm.

629) Enter **500**mm. Click **Rebuild** .

Return to the original dimensions.
630) Double-click on the horizontal dimension, **400**mm.

631) Enter **300**mm.

632) Double-click the vertical dimension, **500**mm.

633) Enter **400**mm. Click **Rebuild** .

Hide the dimensions.
634) Right-click the CABINET **Annotations** folder; uncheck the **Show Feature Dimension** box.

635) Right-click the BOX **Annotations** folder; uncheck the **Show Feature Dimension** box. View the Equations folder.

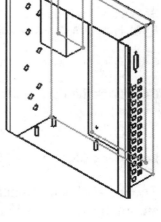

Save the BOX.
636) Click **Isometric view** from the Heads-up View toolbar.

637) Click **Save** .

The BOX assembly requires three configurations. Recall the CABINET drawing. You had two configurations that controlled the Flat Pattern feature state.

Now control dimensions, features and parts in an assembly. A Design Table is an efficient tool to control variations in assemblies and parts.

Design Tables

A Design Table is an Excel spreadsheet used to create multiple configurations in a part or assembly. Utilize the Design Table to create three configurations of the BOX assembly:

- Small

- Medium

- Large

Activity: Design Tables

Create a Design Table.
638) Click **Insert**, **Tables**, **Design Table** from the Main menu. The Design Table PropertyManager is displayed. Accept the Auto-create default setting.

639) Click **OK** ✅ from the Design Table PropertyManager.

Select the input dimension.
640) Hold the **Ctrl** key down.

641) Click the **box_width@Layout** dimension.

642) Click the **box_height@Layout** dimension.

643) Release the **Ctrl** key.

644) Click **OK** from the Dimension dialog box.

The input dimension names and default values are automatically entered into the Design Table.

The value Default is entered in Cell A3. The value 400 is entered in Cell B3 for box_height and the value 300 is entered in Cell C3 for box_width.

Enter the three configuration names.
645) Click Cell **A4**. Enter **Small**.

646) Click Cell **A5**. Enter **Medium**.

647) Click Cell **A6**. Enter **Large**.

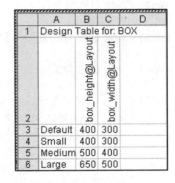

Enter the dimension values for the Small configuration.
648) Click Cell **B4**. Enter **400** for box_height.

649) Click Cell **C4**. Enter **300** for box_width.

Enter the dimension values for the Medium configuration.
650) Click Cell **B5**. Enter **600**. Click Cell **C5**. Enter **400**.

Enter the dimension values for the Large configuration.
651) Click Cell **B6**. Enter **650**.

652) Click Cell **C6**. Enter **500**.

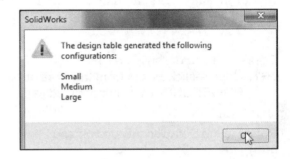

Build the three configurations.
653) Click a **position** outside the EXCEL Design Table in the Graphics window.

654) Click **OK** to generate the three configurations.

Display the configurations.
655) Click the **Configuration Manager** tab. The Design Table icon is displayed.

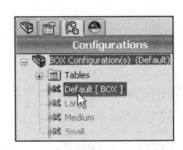

656) Double-click **Small**. Double-click **Medium**.

657) Double-click **Large**.

Return to the Default configuration.
658) Double-click **Default**.

Fit the model to the Graphics window.
659) Press the **f** key.

Edit to the Design Table.
660) Right-click **Design Table** from the BOX FeatureManager.

661) Click **Edit Table**.

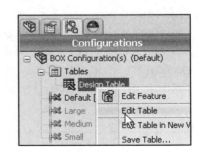

662) Click **OK** from the Add Rows and Columns dialog box.

663) Click Cell **D2**. Enter **$State@POWERSUPPLY<1>**.

664) Click Cell **D3**. Enter **R** for Resolved.

665) Copy Cell **D3**. Select Cell **D4** and Cell **D5**. Click **Paste**. Click Cell **D6**. Enter **S** for Suppressed. Note: If needed remove any hyperlinks.

666) Update the configurations. Click a **position** outside the EXCEL Design Table.

	A	B	C	D	E
1	Design Table for: BOX				
2		box_height@Layout	box_width@Layout	$State@POWERSUPPLY<1>	
3	Default	400	300	R	
4	Small	400	300	R	
5	Medium	500	400	R	
6	Large	650	500	S	

The Design Table component variables must match the FeatureManager names in order to create configurations. The Design Table variable $State@POWERSUPPLY<1> is exactly as displayed in the BOX FeatureManager. If the name in the BOX FeatureManager was P-SUPPLY<2> then the Design Table variable is $State@ P-SUPPLY<2>.

Display the Large Configuration.
667) Double-click **Large** from the ConfigurationManager. The POWERSUPPLY is suppressed in the Large configuration.

Return to the Default configuration and FeatureManager.
668) Double-click **Default**.

669) Click the **FeatureManager** tab.

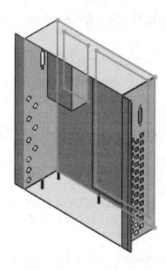

Fit the model to the Graphics window.
670) Press the **f** key.

Save the BOX assembly.
671) Click **Save** .

672) Click **Save All**.

Dimensions, Color, Configurations, Custom Properties, and other variables are controlled through Design Tables. The engineer working on the sheet metal bracket had to visit another customer. As the project manager, it is up to you to get the job done. Create the sheet metal BRACKET in the next activity. The sheet metal BRACKET is created In-Context to illustrate the Top-down Assembly Modeling process. Sheet metal parts are created in a Bottom-up approach and then inserted and mated in the assembly.

BRACKET Part-Sheet Metal Features

The CABINET was converted from a solid part to a sheet metal part by using Rip and Insert Bends feature. The sheet metal BRACKET part starts with the Base Flange feature. The Base Flange feature utilizes a U-shaped sketch profile. The material thickness and bend radius are set in the Base Flange.

Create the BRACKET In-Context of the BOX assembly. The BRACKET references the inside bottom edge and left corner of the CABINET.

 Orient the Flat Pattern. The first sketched edge remains fixed with the Flatten tool.

Sketch the first line from left to right Coincident with the inside bottom edge. The BRACKET is not symmetrical and requires a right hand and left hand version. Utilize Mirror Component feature to create a right-hand copy of the BRACKET.

Activity: BRACKET Part-Sheet Metal Features

Insert the BRACKET component.
673) If required, open the **BOX** assembly.

674) Click **New Part** for the Consolidated Insert Components drop-down menu. Click **OK** from the New SolidWorks Document dialog box. A new virtual default component is added to the FeatureManager. Note: Save the new component and name it BRACKET later in the procedure

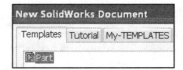

The Component Pointer 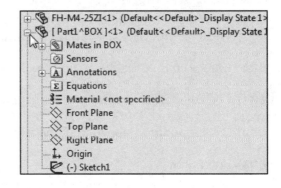 is displayed on the mouse pointer.

675) Click **Front Plane** from the BOX FeatureManager.

676) **Expand** the new inserted component, [Part1^BOX]<1> in the FeatureManager. The FeatureManager is displayed in light blue.

Sketch the profile for the Base Flange.

677) Click **Front view** from the Heads-up View toolbar.

678) Check **View**, **Temporary Axis** from the Menu bar. Click **Hidden Lines Visible** from the Heads-up View toolbar.

679) **Zoom to Area** on the front lower inside edge of the BOX as illustrated.

680) Click **Line** from the Sketch toolbar.

681) Sketch a **horizontal line** Collinear with the inside edge of the CABINET. Note: Sketch the horizontal line from left to right.

682) Sketch a **vertical line** and a **horizontal line** to complete the U-shaped profile.

Add an Equal relation.
683) Right-click **Select**. Click the **top horizontal** line. Hold the **Ctrl** key down.

684) Click the **Bottom horizontal** line.

685) Release the **Ctrl** key. Click **Equal** = from the Add Relations box.

686) Click **OK** ✔ from the PropertyManager.

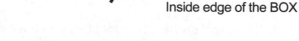

Inside edge of the BOX

The bottom horizontal line is displayed in black. The bottom line is Collinear with the inside bottom edge. Add a Collinear relation between the bottom line and the inside bottom edge if required.

Add dimensions.
687) Click **Smart Dimension** ✎ from the Sketch toolbar.

688) Click the **top horizontal line**. Click the **Top** of the fastener. Click a **position** to the right of the fastener. Enter **2mm**.

689) Click the **centerline** of the fastener. Click the **vertical line**. Click a **position** below the fastener.

690) Enter **10**mm. Click the **left edge** of the CABINET.

691) Click the **left end point** of the top horizontal line.

692) Click a **position** above the profile.

693) Enter **5**mm.

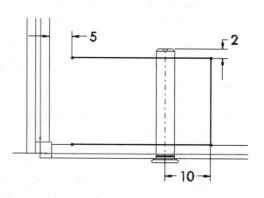

Insert the Base Flange feature.
694) Click **Isometric view** 🔲 from the Heads-up View toolbar.

695) Click **Base Flange/Tab** 🔖 from the Sheet Metal toolbar. The Base Flange PropertyManager is displayed.

696) Select **Up to Vertex** for End Condition in Direction 1.

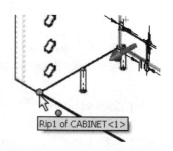

697) Click the CABINET **front left vertex** as illustrated.

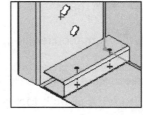

698) Enter **1**mm for Thickness. If required, click the Reverse direction box. Enter **2**mm for Bend Radius. Enter **.45** for K factor. If needed, check the Reverse direction box. Accept the default settings.

699) Click **OK** ✓ from the Base Flange PropertyManager.

700) Click **Front view** ⬛. Note the location of the BRACKET.

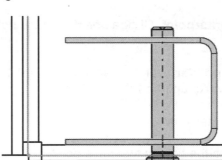

The PEM® fasteners protrude through the BRACKET in the BOX assembly.

Save the BOX assembly.

701) Click **Edit Component** ⬛ from the Assembly toolbar.

702) Click **Save** 💾. Click **Save All**.

Rename the default component [Part1^BOX]<1> to BRACKET.

703) Right-click **[Part1^BOX]<1>** from the FeatureManager. Click **Open Part**. Click **Save As** from the Menu bar. Enter **BRACKET**. Click **Save**. The BRACKET FeatureManager is displayed. **Close** the BRACKET part.

704) Return to **BOX** assembly.

The BRACKET part contains no holes. The BRACKET requires two holes that reference the location of the fasteners. Utilize In-Context Extruded Cut feature to locate the holes. Display Wireframe mode to view the fastener position. Allow for a 0.8mm clearance.

Activity: BRACKET Part-In-context Features

Insert the Extruded Cut feature In-Context of the BOX assembly.

705) Click **BRACKET** from the FeatureManager.

706) Right-click **Edit Part**. The Edit Component icon is activated and the BRACKET entry in the FeatureManager is displayed in blue.

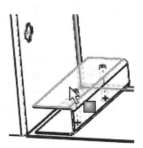

707) Press the **f** key. **Rotate** to view the top face of the BRACKET.

708) Right-click the **top face** of the BRACKET.

709) Click **Sketch** ⌐ from the Context toolbar.

710) Click **Top view** ⬜ from the Heads-up View toolbar.

711) Click **Wireframe** ⬜ from the Heads-up View toolbar.

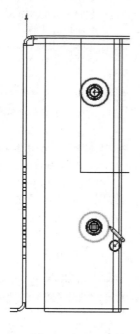

Sketch two circular profiles referencing the PEM center points.
712) Click **Circle** ⊘ from the Sketch toolbar.

713) Drag the **mouse pointer** over the centerpoint of the top PEM circular edge to "wake up" the PEM center point.

714) Click the **PEM centerpoint**. Click a **position** to the right of the center point.

715) Drag the **mouse pointer** over the centerpoint of the bottom PEM circular edge to "wake up" the PEM centerpoint.

716) Click the **PEM center point**.

717) Click a **position** to the right of the centerpoint.

Deselect the Circle Sketch tool and add an Equal relation.
718) Right-click **Select**. Click the **circumference** of the top circle.

719) Hold the **Ctrl** key down. Click the **circumference** of the bottom circle.

720) Release the **Ctrl** key. Click **Equal** from the Add Relations box. Click **OK** ✔ from the Properties PropertyManager.

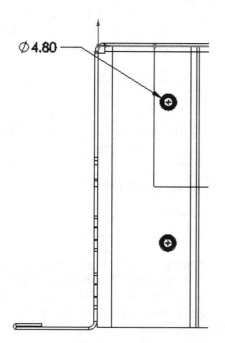

Add dimensions.
721) Click **Smart Dimension** ✎ from the Sketch toolbar.

722) Click the **circumference** of the top circle.

723) Click a **position** off the profile.

724) Enter **4.80**mm.

725) Click **Extruded Cut** ▤ from the Features toolbar. The Cut-Extrude PropertyManager is displayed.

726) Select **Through All** for End Condition in Direction1. Accept the default settings,

727) Click **OK** ✔ from the Cut-Extrude PropertyManager.

Return to the BOX assembly.

728) Click **Edit Component** 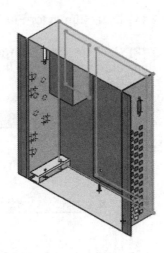 from the Assembly toolbar.

729) Click **Isometric view** from the Heads-up View toolbar.

730) Click **Shaded With Edges** from the Heads-up View toolbar.

Save the Box assembly and its referenced components.

731) Click **Save** . Click **Save All**. Click **Yes**.

No additional references from the BOX components are required. The Edge Flange/Tab feature, Break Corner feature, and Miter Flange feature are developed in the BRACKET part.

The Edge Flange feature adds a flange on the selected edge. Edit the Flange Profile to create a 30mm tab, 20mm from the front face. The tab requires fillet corners. The Break Corner feature adds fillets or chamfers to sheet metal edges.

> **Activity: BRACKET Part-Edge Flange Tab, Break Corner, and Miter Flange Features**

Open the BRACKET.
732) Right-click **BRACKET** in the Graphics window.

733) Click **Open Part** from the Context toolbar.

Display the 2D flat state.
734) Right-click **Flat-Pattern1** from the FeatureManager.

735) Click **Unsuppress** from the Context toolbar.

Display the 3D formed state.
736) Right-click **Flat-Pattern1** from the FeatureManager.

737) Click **Suppress** from the Context toolbar.

Insert a tab using the Edge Flange feature.
738) Click the **top left edge** of the BRACKET as illustrated.

739) Click **Edge Flange** from the Sheet Metal toolbar. The Edge Flange PropertyManager is displayed.

740) Click a **position** above the BRACKET for direction.

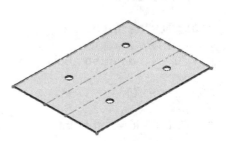

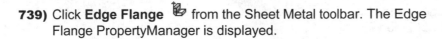

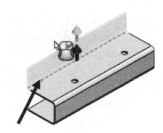

741) Enter **20**mm for Flange Length.

742) Click the **Edit Flange Profile** button. Do not click the Finish button at this time. Move the Profile Sketch dialog box off to the side of the Graphics window.

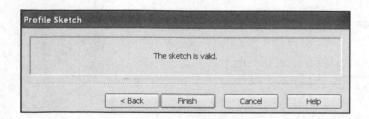

743) Drag the **front edge** of the tab towards the front hole as illustrated.

744) Drag the **back edge** of the tab towards the front hole as illustrated.

Add dimensions.

745) Click **Smart Dimension** from the Sketch toolbar.

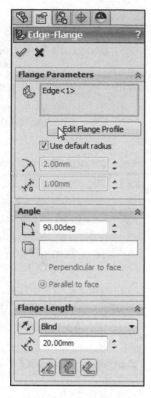

746) Click the **front left vertex** of Base-Flange1.

747) Click the **front vertical** line as illustrated.

748) Click a **position** above the profile.

749) Enter **20**mm.

750) Click the **top horizontal** line.

751) Click a **position** above the profile.

752) Enter **30**mm.

753) Click **Finish** from the Profile Sketch box.

754) Click **OK** from the PropertyManager. Edge-Flange1 is displayed in the FeatureManager.

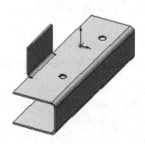

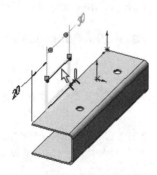

Insert the Break Corner/Corner Trim Feature.

755) Zoom in 🔍 on the tab.

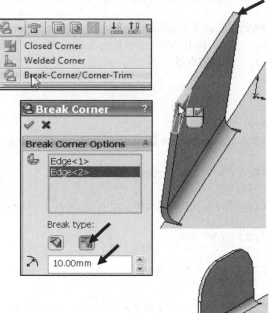

756) Click **Break Corner/Corner Trim** 🖃 from the Sheet Metal toolbar. The Break Corner PropertyManager is displayed.

757) Click the small **front edge** of the flange as illustrated. Edge<1> is displayed in the Corner Edges box.

758) Click the small **back edge** of the flange. Edge<2> is displayed in the Corner Edges box.

759) Select **Fillet** for Break type.

760) Enter **10**mm for Radius.

761) Click **OK** ✔ from the Break Corner PropertyManager. Break Corner1 is displayed in the FeatureManager.

762) Click **Save** 💾.

The Miter Flange feature inserts a series of flanges to one or more edges of a sheet metal part. A Miter Flange requires a sketch. The sketch can contain lines or arcs and can contain more than one continuous line segment.

The Sketch plane is normal (perpendicular) to the first edge, closest to the selection point. The geometry you create must be physically possible. New flanges cannot cross or deform. The current 2mm bend radius of the BRACKET is too small to insert a Miter Flange feature. Leave the area open for stress relief and utilize a Gap distance of 0.5mm.

Gap for stress relief

Insert the Miter Flange Feature.

763) Click **Miter Flange** ▧ from the Sheet Metal toolbar.

764) Click the BRACKET **front inside edge** as illustrated. The new sketch Origin is displayed in the left corner.

765) Zoom to Area 🔍 on the left corner.

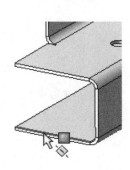

766) Click **Line** ＼ from the Sketch toolbar.

767) Sketch a **small vertical line** at the front inside left corner. The first point is coincident with the new sketch Origin as illustrated.

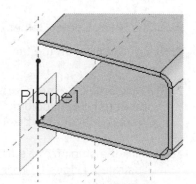

Add dimensions.
768) Click **Smart Dimension** from the Sketch toolbar.

769) Click the **vertical line**.

770) Click a **position** to the front of the BRACKET.

771) Enter **2**mm.

772) Click **Exit Sketch**.

773) Click the **Propagate** ⌐ icon to select inside tangent edges.

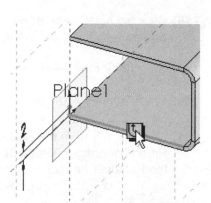

774) Uncheck the **Use default radius** box.

775) Enter **.5**mm for Bend Radius.

776) Check the **Trim side bends** box.

777) Enter **.5**mm for Rip Gap distance.

778) Click **OK** ✔ from the Miter Flange PropertyManager. Miter Flange1 is displayed in the FeatureManager.

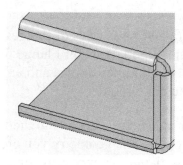

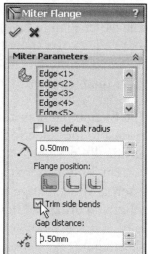

Display the flat and formed states.

779) Click **Isometric view** 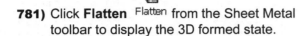 from the Heads-up View toolbar.

780) Click **Flatten** Flatten from the Sheet Metal toolbar to display the 2D flat state.

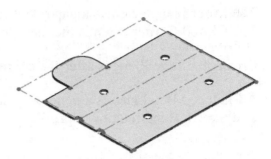

781) Click **Flatten** Flatten from the Sheet Metal toolbar to display the 3D formed state.

Save the BRACKET.
782) If required, click **View**; uncheck **Planes** from the Menu bar.

783) Click **Save** .

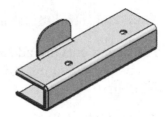

The Trim side bends option creates a Chamfer at the inside corners of the Miter Flange feature. Miter profiles produce various results. Combine lines and arcs for more complex miters.

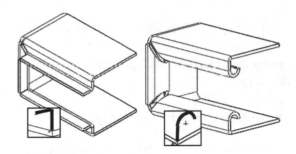

Covers are very common sheet metal parts. Covers protect equipment and operators from harm.

Utilize the Base Flange, Miter Flange, Close Corner, and Edge Flange to create simple covers. Add holes for fasteners. The Base Flange is a single sketched line that represents the area to be covered.

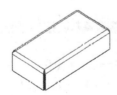

Base Flange
Sketch Line on Right
Plane

Miter Flange
Select 4 edges around
the top face.

Insert the Miter Flange feature and select the 4 edges on the top face. Utilize the Material Outside option for Flange position to add the walls to the outside of the Base Flange.

The Miter Flange creates a gap between the perpendicular walls. The Closed Corner extends the faces to eliminate the gap.

Create the Edge Flanges after the corners are closed.

A Sketched Bend inserts a bend on a flat face or plane. The Sketch Bend requires a profile.

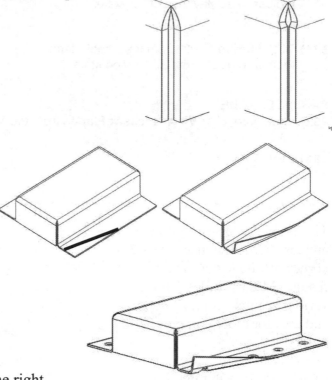

Holes and cuts are added to the cover. A hole or cut can cross a bend line.

The Base Flange utilized the MidPlane option. Utilize the same plane to mirror the sets of holes.

There is plenty of time to develop additional sheet metal parts in the provided exercises at the end of the chapter.

Return to the BOX assembly.

A second BRACKET is required for the right side of the CABINET. Design for the three configurations and reuse geometry.

BRACKET-Mirror Component

Insert a reference plane through the center PEM® fastener. The right side BRACKET requires a right hand version of the left side BRACKET. Utilize the Mirror Component option to create the mirrored right side BRACKET.

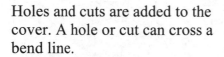
Activity: BRACKET-Mirror Component

784) Open the **BOX** assembly.

Display Sketches and Temporary Axes.
785) Click **View**; check **Sketches** from the Menu bar.

786) Click **View**; check **Temporary Axes** from the Menu bar.

787) **Zoom to Area** on the top circular face of the middle PEM® fastener.

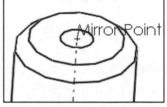

Insert a Reference Point.
788) Click **Insert**, **Reference Geometry**, **Point** from the Menu bar. The Point PropertyManager is displayed.

789) Click the **top circular edge** of the back middle fastener. The Reference Point is located at the center point of the top circle.

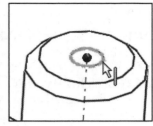

790) Click **OK** ✔ from the Point PropertyManager. Point1 is displayed.

791) Press the **f** key.

792) Rename **Point1** to **Mirror Point**.

Insert a Reference Plane.
793) Click **Insert**, **Reference Geometry, Plane** from the Menu bar. The Plane PropertyManager is displayed.

794) Click **Mirror Point** from the fly-out FeatureManager. Mirror Point is displayed in the First Reference box.

795) Click **Right Plane** in the BOX assembly from the fly-out FeatureManager. Right Plane is displayed in the Second Reference box.

796) Click **OK** ✔ from the Plane PropertyManager. PLANE1 is displayed in the FeatureManager.

Fit the model to the Graphics window.
797) Press the **f** key.

798) Rename **PLANE1** to **Mirror PLANE**.

799) Click **Save** 🖫.

800) Click **Save All**.

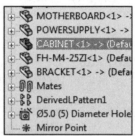

Mirror the Component.

801) Click **Mirror PLANE** from the FeatureManager.

802) Click **Insert**, **Mirror Components** from the Menu bar. The Mirror PLANE is selected.

803) Click **BRACKET** from the BOX fly-out FeatureManager.

804) Click **Next** .

805) Click the **Create opposite hand version** button. Accept the default conditions.

806) Click **OK** from the Mirror Components PropertyManager.

807) Click **OK**. MirrorComponent1 is displayed in the FeatureManager.

808) **Expand** MirrorComponent1 in the FeatureManager. View the new name of the mirror component.

The right hand version of the BRACKET is named MirrorBRACKET<#>. An interference exists between the MOTHERBOARD and MirrorBRACKET<#>. Address the interference in the next section.

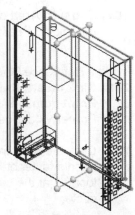

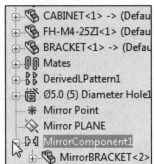

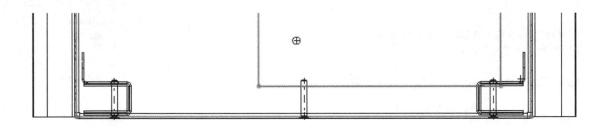

MirrorBRACKET Part-Bends, Unfold and Fold and Jog Features

Open the MirrorBRACKET part. The MirrorBRACKET is not a sheet metal part. The MirrorBRACKET references only the BRACKET's geometry. The MirrorBRACKET must be manufactured in a 2D Flat State. Insert Sheet Metal Bends to create the Flat State.

Control the unfold/fold process. Insert an Extruded Cut across a bend radius. Utilize the Unfold and Fold features to flatten and form two bends. Perform cuts across bends after the Unfold. Utilize unfold/fold steps for additional complex sheet metal components.

Sketch the cut profile across the Bend Radius. The Extruded Cut is linked to the Thickness. A Jog feature creates a small, short bend. Insert a Jog feature to offset the right tab and to complete the MirrorBRACKET.

Activity: MirrorBRACKET Part-Bends, Unfold and Fold and Jog Features

Open MirrorBRACKET and Insert Bends.
809) Right-click **MirrorBRACKET** in the FeatureManager.

810) Click **Open Part** from the Context toolbar. The MirrorBRACKET FeatureManager is displayed.

Fit the Model to the Graphics window.
811) Press the **f** key.

812) Click the **inside bottom face** of MirrorBRACKET.

813) Click **Insert Bends**  from the Sheet Metal toolbar. The Bends PropertyManager is displayed. Face<1> is displayed in the Fixed Face box. Accept the default conditions.

814) Click **OK** ✅ from the Bends PropertyManager. Process-Bends1 is displayed in the FeatureManager.

Unfold two Bends.

815) Click **Unfold**  from the Sheet Metal toolbar. The Unfold PropertyManager is displayed.

816) **Zoom to Area** on the right inside bottom face.

817) Click the right **inside bottom face** as illustrated.

818) Rotate the **model** to view the two long bends. Click the two long **bends** to unfold as illustrated. RoundBend5 and RoundBend6 is displayed in the Bends to unfold box.

819) Click **OK** ✓ from the Unfold PropertyManager. Unfold1 is displayed in the PropertyManager.

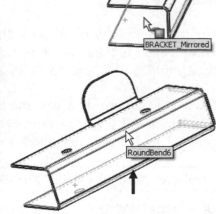

820) Click **Top view** ⬚ from the Heads-up View toolbar.

Note: Only the selected two bends unfold. The tab remains folded.

Create the Sketch for an Extruded Cut.

821) Right-click the **right bottom face** as illustrated. This is your Sketch plane.

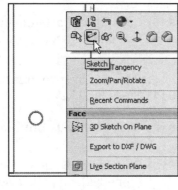

822) Click **Sketch** ✏ from the Context toolbar. The Sketch toolbar is displayed.

823) Click **Corner Rectangle** ▢ from the Sketch toolbar.

824) Click the **lower right corner** of the Mirror BRACKET.

825) Click a **position** coincident with the left bend line as illustrated.

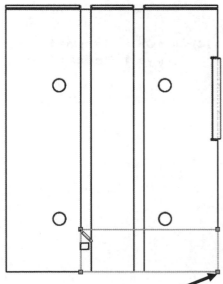

Add a dimension.

826) Click **Smart Dimension** 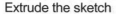 from the Sketch toolbar.

827) Click the **top horizontal edge** of the MirrorBRACKET.

828) Click the **bottom horizontal line** of the rectangle.

829) Enter **15**mm.

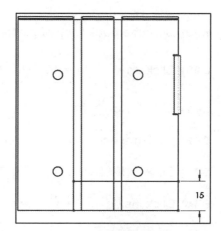

Extrude the sketch

830) Click **Extruded Cut** from the Sheet Metal toolbar. The Cut-Extrude PropertyManager is displayed.

831) Check the **Link to Thickness** box.

832) Click **OK** from the Cut-Extrude PropertyManager. The Cut-Extrude feature is displayed in the FeatureManager.

833) Click **Wireframe** from the Heads-up View toolbar.

Fold the bends.

834) Click **Fold** from the Sheet Metal toolbar. The Fold PropertyManager is displayed.

835) Click the **Collect All Bends** button. All Unfolded bends are displayed in the Bends to fold box.

836) Click **OK** from the Fold PropertyManager. Fold1 is displayed in the FeatureManager. View the Cut-Extrude1 feature in the Graphics window.

837) Click **Shaded With Edges** .

838) Click **Isometric view** . from the Heads-up View toolbar.

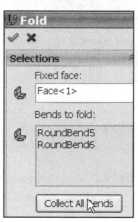

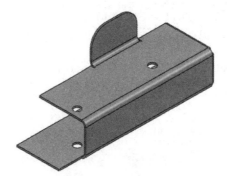

Insert a Jog feature.
839) Click the **back face** of the tab as illustrated.

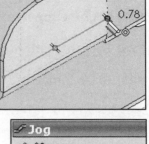

840) Click **Jog** ✐ from the Sheet Metal toolbar.

841) Click **Line** ╲ from the Sketch toolbar.

842) Sketch a **horizontal line** across the midpoints of the tab.

843) Click **Exit Sketch**. The Jog PropertyManager is displayed.

844) Click inside the **Fixed Face** box.

845) Click the **back face** of the tab below the horizontal line. The direction arrow points to the right.

846) Enter **5**mm for Offset Distance.

847) Click the **Reverse Direction** button. The direction arrow points to the left.

848) Click **OK** ✔ from the Jog PropertyManager. Jog1 is displayed in the FeatureManager.

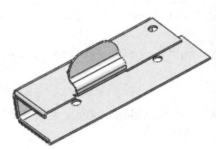

Display the Jog.
849) Use the **arrow keys** to rotate and view the Jog feature.

Display the flat/formed state.
850) Click **Flatten** ▱ from the Sheet Metal toolbar to display the flat state.

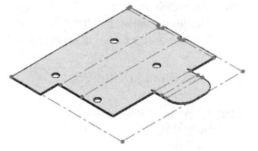

851) Click **Flatten** Flatten from the Sheet Metal toolbar to display the formed state.

852) Click **Isometric view** ⬛ from the Heads-up View toolbar.

853) Click **Save** 💾.

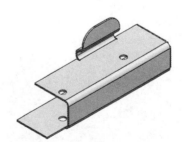

Return to the BOX assembly.
854) Open the **BOX** assembly.

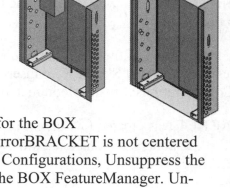
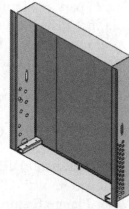

Save the BOX assembly.
855) Click **Save** 💾.

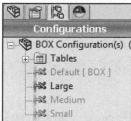

Note: The BRACKET and
MirrorBRACKET remain
centered around the Mirror Plane for the BOX
assembly configurations. If the MirrorBRACKET is not centered
for the Small, Medium, and Large Configurations, Unsuppress the
Mirror Point and Mirror Plane in the BOX FeatureManager. Un-
suppress the Mates created for the MirrorBRACKET part.

The ConfigurationManager, Add Configuration, Advanced,
Suppress new features and mates option controls the state of new
features and mates in multiple configuration assemblies.

The BOX configurations are complete.

Fully define all sketches. Rebuild errors occur in a Top Down design approach if
the sketches are not fully defined.

Additional details on IGES, Replace Component, Equations, Design Table and
Configurations, Mirror Component, and Sheet Metal features (Base Flange, Edge Flange,
Miter Flange, Extruded Cut, Sketch Bend, Break Corner, Closed Corner, Hem, Fold,
Unfold, and Jog) are available in SolidWorks Help. Index all Sheet metal features under
the keyword, sheet metal.

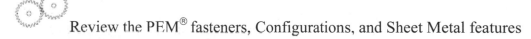

Review the PEM® fasteners, Configurations, and Sheet Metal features

You obtained IGES format PEM® self-clinching fasteners either from the Internet or from
the enclosed DVD in this book. IGES files are opened in SolidWorks and saved as part
files. The PEM® fasteners were inserted into the BOX assembly seed hole. You
developed a Component Pattern for the fasteners that referenced the CABINET Linear
Pattern of Holes.

You added an Assembly Cut feature to the BOX assembly. The Cut feature created a hole
through the MOTHERBOARD and CABINET components.

Due to part availability, you utilized the Replace Components feature to exchange the Ø5.0mm fastener with the Ø4.0mm fastener. The new mates were redefined to update the Ø 4.0mm fastener. The Ø 4.0mm fastener required the CABINET holes to be modify from the Ø5.0mm to the Ø4.0mm.

Equations controlled the location of the CABINET holes based on the Layout Sketch dimensions. Variables in the Layout Sketch were inserted into a Design Table to create the BOX assembly Configurations: Small, Medium and Large.

The sheet metal BRACKET was created In-Context of the BOX assembly and referenced the location of the PEM® fasteners. You inserted the Base Flange, Edge Flange, and Miter Flange features.

The Mirror Components option was utilized to create a right hand version of the BRACKET. The MirrorBRACKET required Insert Bends, Fold/Unfold, and an Extruded Cut across the Bends for manufacturing. The Jog feature provided an offset for the tab by inserting two additional bends from and single sketched line.

Project Summary

You created three different BOX sizes utilizing the Top down assembly modeling approach. The Top down design approach is a conceptual approach used to develop products from within the assembly.

The BOX assembly began with a Layout Sketch. The Layout Sketch built in relationships between 2D geometry that represented the MOTHERBOARD, POWERSUPPLY, and CABINET components. These three components were developed In-Context of the Layout Sketch. The components contain InPlace Mates and External references to the BOX assembly.

Additional Extruded features were inserted in the MOTHERBOARD and POWERSUPPLY parts.

The CABINET part developed an Extruded Boss/Base feature In-Context of the BOX assembly. The Shell feature created a constant wall thickness for the solid part.

The Rip feature and Insert Bends feature were added to convert the solid part into a sheet metal part.

Sheet metal components utilized the Edge, Hem, Extruded Cut, Die Cuts, Forms, and Flat Pattern features. Model sheet metal components in their 3D formed state. Manufacture sheet metal components in their 2D flatten state.

The Component Pattern maintained the relationship between the Cabinet holes and the PEM® self-clinching fasteners. You utilized Replace Components to modify the Ø5.0mm fastener to the Ø 4.0mm fastener.

The new mates were redefined to update the Ø4.0mm fastener.

Equations controlled the location of the Cabinet holes. Variables in the Layout Sketch were inserted into a Design Table to create the BOX assembly Configurations: Small, Medium and Large.

The BRACKET was created utilizing sheet metal features: Base Flange, Edge Flange, and Miter Flange. You utilized Mirror Components to create a right hand version, MirrorBRACKET. The MirrorBRACKET required Insert Bends and an Extruded Cut feature through the sheet metal bends.

The Mirror Plane built in the design intent for the BRACKET and MirrorBRACKET to remain centered for each BOX assembly configuration.

Sheet metal parts can be created individually or In-Context of an assembly.

Project Terminology

Assembly Feature: Features created in the assembly such as a hole or extrude. The Hole assembly feature was utilized to create a hole through 2 different components in the BOX assembly.

Bend Allowance (Radius): The arc length of the bend as measured along the neutral axis of the material.

Derived Component Pattern: A pattern created inside the assembly. A Derived Pattern utilizes a component from the assembly. There are two methods to reference a Derived Pattern. Method 1 - Derived Pattern references a Linear or Circular feature pattern. Method 2 - Derived Pattern references a local pattern created in the assembly.

Design Table: A Design Table is an Excel spreadsheet used to create multiple configurations in a part or assembly. Utilize a Design Table to create the small, medium and large BOX configurations.

Equations: Mathematical expressions that define relationships between parameters and or dimensions are called Equations. Equations use shared names to control dimensions. Use Equations to connect values from sketches, features, patterns and various parts in an assembly. Use Equations in different parts and assemblies.

Extruded Boss/Base: Add material to the part. Utilize to create the CABINET as a solid part.

Flange: Adds a flange (wall) to a selected edge of a sheet metal part.

Flat Pattern: The flat manufactured state of a sheet metal part. It contains no bends.

Flattened: Creates the flat pattern for a sheet metal part. The Flattened feature toggles between the flat and formed state. A Flat configuration is created. Utilize the Flat configuration in the drawing to dimension the flat pattern.

Fold/Unfold: Flatten one or more bends to create an extruded cut across a bend.

Hem: Adds a hem to a selected edge of a sheet metal part.

Insert Sheet Metal Bends: Converts a solid part into a sheet metal part. Requires a fixed face or edge and a bend radius.

Jog: Adds an offset to a flange by creating 2 bends from one sketch line.

Layout Sketch: Specifies the location of the key components. Components and assemblies reference the Layout Sketch.

Link Values: Are used to define equal relations. Create an equal relation between two or more sketched dimensions and or features with a Link Value. Link Values require a shared name. Use Link Values within the same part.

Mirror Components: Creates a right/left hand version of component about a plane in an assembly.

Reload: Refreshes shared documents.

Replace: Substitutes one or more instances of a component with a closed different component in an assembly.

Rip: A Rip feature is a cut of 0 thickness. Utilize a Rip feature on the edges of a shelled box before inserting sheet metal bends

Shell: Creates a constant wall thickness for the CABINET Extruded Base. The Shell feature represents the thickness of the sheet metal utilized for the cabinet.

Suppressed: A feature that is not displayed. Hide features to improve clarity. Suppressed features to improve model Rebuild time.

Top Down design approach: A conceptual approach used to develop products from within the assembly. Major design requirements are translated into sub-assemblies or individual components and key relationships

Questions

1. Explain the Top-down assembly approach? When do you create a Layout Sketch?

2. How do you create a new component In-Context of the assembly?

3. What is the difference between a Link Value and an Equation?

4. Name three characteristics unique to Sheet metal parts.

5. Identify the indicator for a part in an Edit Component state.

6. Where should you position the Layout Sketch?

7. For a solid extruded block, what features do you add to create a Sheet metal part?

8. Name the two primary states of a Sheet metal part.

9. Explain the procedure on how to insert a formed Sheet metal feature such as a dimple or louver?

10. Identify the type of information that a Sheet metal manufacturer provides.

11. Explain what is an Assembly Feature?

12. What features are required before you create the Component Pattern in the assembly?

13. True or False. If you utilize Replace Component, you have to modify all the mate references.

14. Define a Design Table. Provide an example.

Exercises

Exercise 6.1: Insert a Extruded Cut feature In-Context of an assembly.

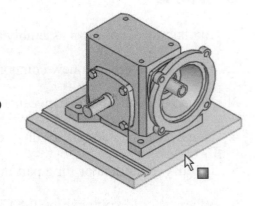

- Open the MOTOR assembly from the Chapter6 - Homework folder located in the DVD of the book. Rotate the assembly and view the bottom plate. Insert mounting holes through the plate In-Context of the assembly.

- Edit the b5g-plate In-Context of the assembly. Right-click the right face of b5g-plate from the Graphics window.

- Click Edit Part.

- Right-click the bottom face of b5g-plate for the sketch plane. Create a sketch.

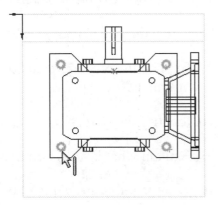

- Display a Bottom, Wireframe view.

- Ctrl-select the 4 circular edges of the f718b-30-b5-g mounting holes.

- Apply the Convert Entities sketch tool.

- Apply an Extruded Cut feature for the 4 selected circular holes thru the plate.

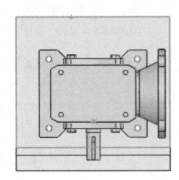

- Return to the assembly.

- Rebuild the model and view the results.

Exercise 6.2: Create a New Part In-Context of an assembly.

- Open the PLATE1 assembly from the Chapter6 - Homework folder located in the DVD of the book.

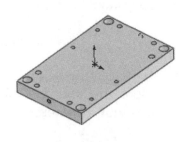

- Insert a new part into the assembly. Use the default part template. The mouse pointer displays the 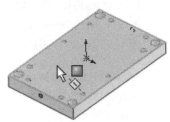 icon.

- The new virtual component name is displayed in the FeatureManager. Click the top face.

- Click the Convert Entities Sketch tool.

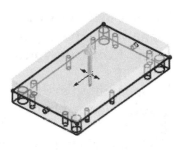

- Create a top plate from the bottom plate. Extrude upwards 50mm.

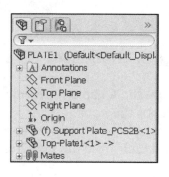

- Return to the assembly.

- Name the virtual component Top-Plate1.

Exercise 6.3: Layout Sketch.

Create components from a Layout sketch utilizing blocks.

- Open Block-Layout from the Chapter6 - Homework folder located in the DVD of the book. Three blocks are displayed in the Graphics window as illustrated.

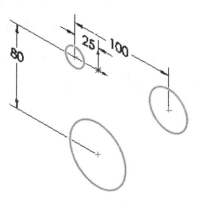

Blocks can include the following items: Text (Notes), Dimensions, Sketch entities, Balloons, Imported entities and text, and Area hatch.

- Activate the Make Part from Block tool.

- Click the three blocks in the Graphics window.

- Click the On Block option. Click OK. View the results.

- Edit the 20MM-6 block In-Context of the assembly.

- Extruded the 20MM-6 block 10mms.

- Return to the assembly.

- Extruded the 40MM-4 block In-Context of the assembly 10mms.

- Extruded the 60MM-4 block In-Context of the assembly 10mms.

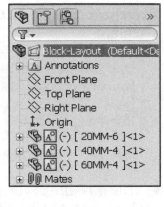

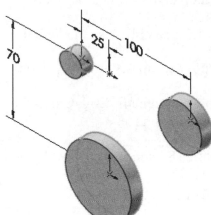

- Modify the 80mm dimension to 70mms. View the results.

Exercise 6.4: Create an assembly from a Layout sketch. Utilize the Entire assembly method.

- Open Entire assembly from the Chapter6 - Homework folder located in the DVD of the book.

- Double-click the Layout sketch from the FeatureManager to display the dimensions.

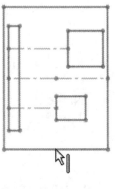

- Insert a New Part In-Context to the assembly. Select the PART-IN-ANSI template.

- Select the Top Plane and the bottom horizontal line as illustrated.

- Right-click Select chain.

- Select the Convert Entities Sketch tool.

- Extruded the Layout sketch in a downward direction 20mms.

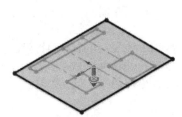

- Return to the assembly.

- Rename the new virtual part, Bottom. View the results.

Exercise 6.5: L-BRACKET Part.

Create the L-BRACKET Sheet metal parts.

- Create a family of sheet metal L-BRACKETS. L-BRACKETS are used in the construction industry.

- Create an eight hole L-BRACKET.

- Use Equations to control the hole spacing.

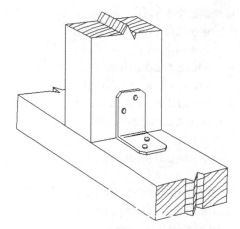

Stong-Tie Reinforcing Brackets
Courtesy of Simpson Strong Tie
Corporation of California

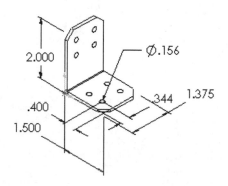

Review Design Tables in On-line Help and the SolidWorks Tutorial.

	A	B	C	D	E
1	Design Table for: lbracket				
2		height@Sketch1	width@Sketch1	depth@Base-Extrude-Thin	hole1dia@Sketch2
3	First Instance	2	1.5	1.375	0.156
4	Small2x1.5x1	2	1.5	1	0.156
5	Medium3x2x2	3	2	2	0.156
6	Large4x3x4	4	3	4	0.25

L-BRACKET DESIGN TABLE

- Create a Design Table for the L-Bracket.

- Rename the dimensions of the L-Bracket.

- Create a drawing that contains the flat state and formed state of the L-BRACKET.

Exercise 6.6: SHEET METAL SUPPORTS.

Create the SHEET METAL SUPPORT parts. The following examples are courtesy of Strong Arm Corporation.

- Obtain additional engineering information regarding dimensions and bearing loads. Visit www.strongtie.com.

Similar parts can be located in local hardware and lumber stores. An actual physical model provides a great advantage in learning SolidWorks Sheet metal functionality.

- Estimate the length, width and depth dimensions. Beams are 2" x 4" (50mmx100mm), posts are 4" x 4" (100mmx100mm).

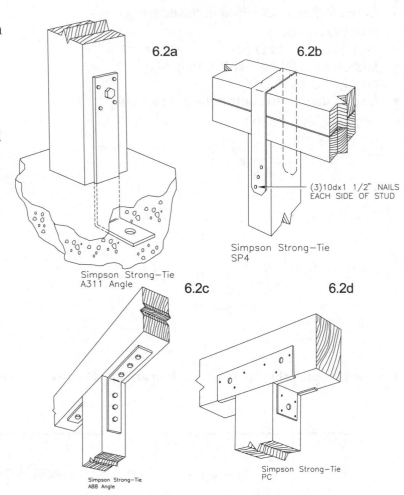

6.2a

Simpson Strong-Tie
A311 Angle

6.2b

(3)10dx1 1/2" NAILS
EACH SIDE OF STUD

Simpson Strong-Tie
SP4

6.2c

Simpson Strong-Tie
A88 Angle

6.2d

Simpson Strong-Tie
PC

Exercise 6.7: Create a lofted bend feature in a Sheet Metal part.

- Open the Lofted part from the Chapter6 - Homework folder located in the DVD of the book. View the FeatureManager.

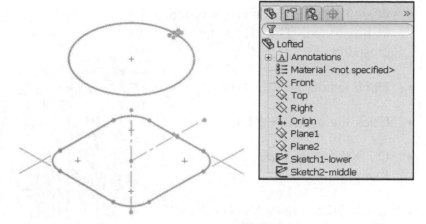

- Apply a Lofted Bend feature on both Sketches as illustrated.

- Thickness = .01in. The direction arrow is inward,

- View the results.

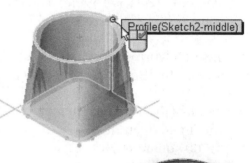

Exercise 6.8: Create a Vent in a Sheet Metal box.

- Open Vent from the Chapter6 - Homework folder located in the DVD of the book. View the FeatureManager.

- Create the illustrated sketch for the Vent on the Right top side of the box.

- Click the Vent 🔲 tool.

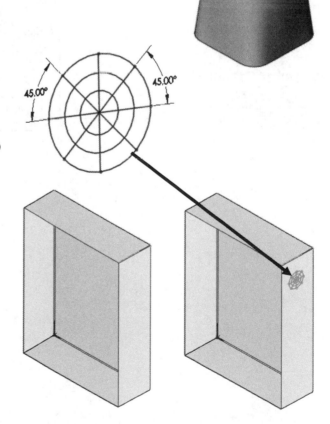

- Click the large circle circumference as illustrated. The 2D Sketch is a closed profile.

- Click inside the Ribs box.

- Click the horizontal sketch line.

- Click the three remaining sketch lines.

- Click OK to create Vent1. The Area of Vent1 is 1590.67 square mms. The Open area is 51/13%.

- Insert a Mirror Vent on the left side of the Box. Create Mirror1.

- Modify the vent design and obtain an open flow area of 1,700 square mms.

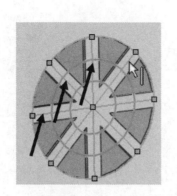

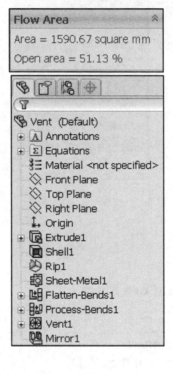

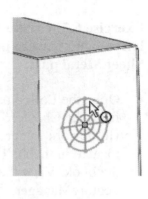

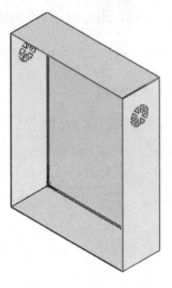

Project 7

SolidWorks SimulationXpress, Sustainability and DFMXpress

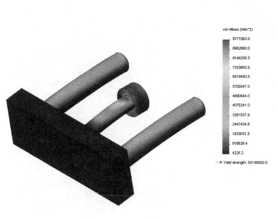

Below are the desired outcomes and usage competencies based on the completion of Project 7.

Project Desired Outcomes:	Usage Competencies:
• Understand and navigate SolidWorks SimulationXpress analysis.	• Operate SolidWorks SimulationXpress Wizard. • Apply SimulationXpress to a simple part.
• Awareness of SolidWorks Sustainability and SustainabilityXpress.	• Display and understand the four environmental impact factors. • Generate a customer report and find suitable alternative materials.
• Knowledge of SolidWorks DFMXpress.	• Apply SolidWorks DFMXpress Wizard.

Notes:

Project 7 - SolidWorks SimulationXpress, Sustainability and DFMXpress

Project Objective

Execute a SolidWorks SimulationXpress analysis on the MGPM12-1010 assembly. Determine if the MGPM12-1010 assembly can support an applied load under a static load condition.

Perform a SolidWorks SustainabilityXpress analysis on a part. View the environmental impact calculated in four key areas: *Carbon Footprint*, *Energy Consumption*, *Air Acidification* and *Water Eutrophication*. Material and Manufacturing process region and Usage region are used as input variables. Compare similar materials and environmental impacts on the base line design.

Implement DFMXpress on a part. DFMXpress is an analysis tool that validates the manufacturability of SolidWorks parts. Use DFMXpress to identify design areas that *may cause problems* in fabrication or increase the costs of production.

On the completion of this project, you will be able to:

- Implement a SolidWorks SimulationXpress analysis on a component in the MGPM12-1010 assembly.
- Apply SolidWorks SustainabilityXpress to a part.
- View the four key environmental impact areas:
  - o Carbon Footprint
  - o Energy Consumption
  - o Air Acidification
  - o Water Eutrophication
- Generate a customer sustainability report and locate suitable alternative materials
- Perform a DFMXpress analysis on a simple part

This project uses short step-by-step tutorials to practice and reinforce the subject matter and objectives. All needed models for the short tutorials are located on the DVD in the book.

SolidWorks SimulationXpress

SimulationXpress is a Finite Element Analysis (FEA) tool incorporated into SolidWorks. SimulationXpress calculates the displacement and stress in a part based on material, restraints and static loads.

When loads are applied to a part, the part tries to absorb its effects by developing internal forces. Stress is the intensity of these internal forces. Stress is defined in terms of Force per unit Area: $Stress = \dfrac{f}{A}$.

Different materials have different stress property levels. Mathematical equations derived from Elasticity theory and Strength of Materials are utilized to solve for displacement and stress. These analytical equations solve for displacement and stress for simple cross sections. Example: Bar or Beam. In complicated parts, a computer based numerical method such as Finite Element Analysis is used.

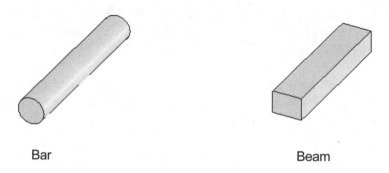

Bar   Beam

SolidWorks SimulationXpress utilizes linear static analysis based on the Finite Element Method. The Finite Element Method is a numerical technique used to analyze engineering designs. FEM divides a large complex model into numerous smaller models. A model is divided into numerous smaller segments called elements.

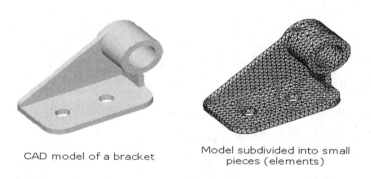

CAD model of a bracket   Model subdivided into small pieces (elements)

SolidWorks SimulationXpress utilizes a tetrahedral element containing 10 nodes. Each node contains a series of equations. SimulationXpress develops the equations governing the behavior of each element. The equations relate displacement to material properties, restraints, "boundary conditions" and applied loads.

SimulationXpress organizes a large set of simultaneous algebraic equations.

The Finite Element Analysis (FEA) equation is:

$[K]\{U\} = \{F\}$ where:

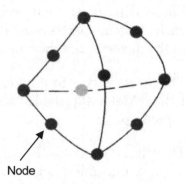

1. [K] is the structural stiffness matrix.

2. {U} is the vector of unknown nodal displacements.

3. {F} is the vector of nodal loads.

Node

SimulationXpress determines the X, Y and Z displacement at each node. This displacement is utilized to calculate strain.

Tetrahedral Element

Strain is defined as the ratio of the change in length, δL to the original length, L.

Stress is proportional to strain in a Linear Elastic Material.

The Elastic Modulus (Young's Modulus) is defined as stress divided by strain.

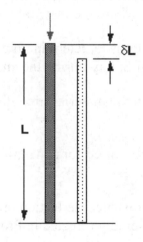

Strain = $\delta L / L$

Compression Force Applied
Original Length L
Change in Length δL

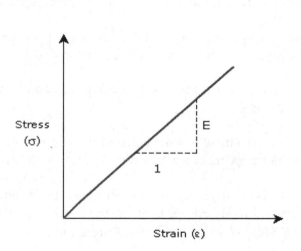

Elastic Modulus is the ratio of Stress to Strain for Linear Elastic Materials.

SimulationXpress determines the stress for each element based on the Elastic Modulus of the material and the calculated strain.

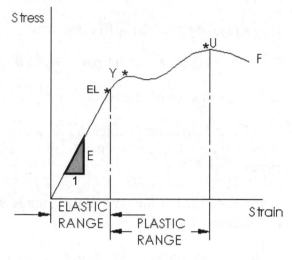

Stresss versus Strain Plot
Linearly Elastic Material

The Stress versus Strain Plot for a Linearly Elastic Material provides information about a material.

The Elastic Modulus, E is the stress required to cause one unit of strain. The material behaves linearly in the Elastic Range.

The material remains in the Elastic Range until it reaches the elastic limit.

The point EL is the elastic limit. The material begins Plastic deformation.

The point Y is called the Yield Point. The material begins to deform at a faster rate. The material behaves non-linearly in the Plastic Range. The point U is called the ultimate tensile strength. Point U is the maximum value of the non-linear curve. Point U represents the maximum tensile stress a material can handle before a facture or failure. Point F represents where the material will fracture.

Designers utilize maximum and minimum stress calculations to determine if a part is safe. SimulationXpress reports a recommended Factor of Safety during the analysis.

The SimulationXpress Factor of Safety is a ratio between the material strength and the calculated stress.

The von Mises stress is a measure of the stress intensity required for a material to yield. The SimulationXpress Results plot displays von Mises stress.

Will the ROD that you created in Project 1 experience unwanted deflection or stress? Insure a valid ROD component for the GUIDE-ROD assembly (created in Project 2) by using the SolidWorks SimulationXpress tool.

In Project 1, you assigned AISI 304 material to the GUIDE. You also calculated the mass in grams of the part with the Mass Properties tool.

A Newton is defined as the force acting on a mass of one kilogram at a location where the acceleration due to gravity is $1 m/s^2$. Weight equals mass * gravity. Weight is a Force.

$$1\ newton = \frac{1\ kg - m}{s^2}$$

How do you determine if the MGPM12-1010 assembly supports the weight under static load conditions? Answer: Utilize SimulationXpress. Determine the engineering data required for the simplified static analysis.

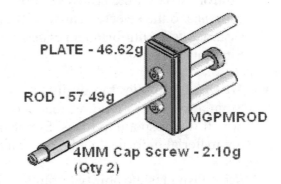

PLATE - 46.62g

ROD - 57.49g

MGPMROD

4MM Cap Screw - 2.10g
(Qty 2)

Analyze the 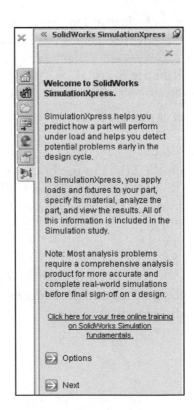 MGPM12-1010_MGPRod.SLDPRT part.

Component:	MATERIAL:	Mass:
PLATE	AISI 304	46.62g
ROD	AISI 304	56.65g
2 - 4MMCAPSCREWS	AISI 304	4.20g
Customer Component Fasten to Rod	Unknown	1500.00g
Total	**Total**	1607.47g ~1.6kg

Assume $g_c = 9.81 \text{ m/s}^2$.

$$Weight = mg = 1.6 \ kg \ x \ 9.81 \frac{m}{s^2}$$

$$Weight = 16.1 \frac{kg - m}{s^2}$$

$$Weight = 16N$$

SimulationXpress guides you through various default steps to define fixtures, loads, material properties, analyze the model, view the results and the optional Optimization process.

The SimulationXpress interface consists of the following Menus:

- *Welcome*: Informs the user about SolidWorks SimulationXpress. Provides the ability to set units and a save in folder location for the results.

- *Fixtures*: Provides the ability to apply restraints or fixtures to selected entities of the part. This keeps the part from moving when loads are applied. Faces with fixtures are treated as perfectly rigid. This may cause unrealistic results when the fixture is in the vicinity of: *Fixed Holes*, *Fixed vs. Supported* and *Fixed vs. Attached Parts*.

« SolidWorks SimulationXpress

Welcome to SolidWorks SimulationXpress.

SimulationXpress helps you predict how a part will perform under load and helps you detect potential problems early in the design cycle.

In SimulationXpress, you apply loads and fixtures to your part, specify its material, analyze the part, and view the results. All of this information is included in the Simulation study.

Note: Most analysis problems require a comprehensive analysis product for more accurate and complete real-world simulations before final sign-off on a design.

Click here for your free online training on SolidWorks Simulation fundamentals.

→ Options

→ Next

- *Loads*: Provides the ability to apply either a *force* or *pressure* to the selected entity of the part. Loads are assumed to be uniform and constant when using SimulationXpress.

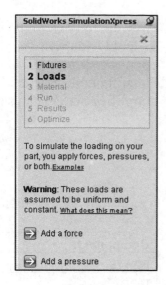

- *Material*: Provides the ability to assign material to the part. SimulationXpress requires that part's material to predict how it will respond to loads. SimulationXpress assumes that the material deforms in a linear fashion with increasing load.

- *Run*: Provides the ability to run the simulation and to mesh the study. The results are based on the specified study criterion.

- *Result*: Examine the animation of the part's response to verify that the correct loads and fixtures were applied. The Results folder in the SimulationXpress Study displays the following:

  o *Stress (-vonMises-)*

  o *Displacment (-Res disp-)*

  o *Deformation (-Displacement-)*

  o *Factor of Safety (-Max von Mises Stress-)*

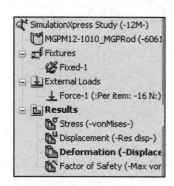

- *Optimize*: Optimizes a model dimension based on a specified criterion. This is an optional step.

Run the SimulationXpress wizard: Analyzes the MGPM12-1010_MGPRod.SLDPRT part. The part is a reference component in the MGPM12-1010 assembly.

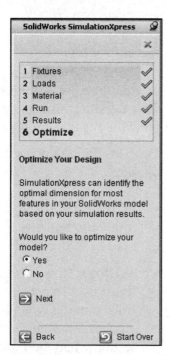

Note: *Use the models from the SimulationXpress folder* located on the DVD in the book. Add three Restraints to the back circular faces of the part. The three faces are fixed. Apply a force on the front face of the part. The force is perpendicular to Plane 2.

The book is design to expose the new user to many tools, techniques and procedures. It may not always use the most direct tool or process.

Activity: SolidWorks SimulationXpress - Analyze the MGPMRod Part

Close all documents.
1) Click **Window**, **Close All** from the Menu bar.

Open the MGPM12-1010 assembly from the SimulationXpress folder.
2) Click **Open** 📂 from the Menu bar.

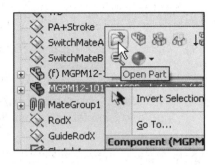

3) Double-click the **MGPM12-1010** assembly from the ENGDESIGN-W-SOLIDWORKS\SimulationXpress folder. The MGPM12-1010 Assembly FeatureManager is displayed. Note: Copy the folder from the DVD in the book. Copy all models to your local hard drive.

Open the Rod component from the assembly.
4) Right-click 🔩 MGPM12-1010_MGPRod.SLDPRT from the FeatureManager.

5) Click **Open Part**. The Rod is displayed in the Graphics window. This is an older part without imported geometry.

6) Click **Isometric view** 🔲 from the Heads-up View toolbar.

7) Click **SimulationXpress Analysis Wizard** 📐 from the Evaluate tab in the CommandManager or click **Tools**, **SimulationXpress** from the Menu bar. The SimulationXpress Welcome screen is displayed.

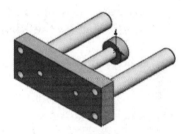

Define the Options.
8) Click the **Options** button from the Welcome screen.

Set units and Study location.
9) Select **SI** for Units.

10) Click the **Browse** button.

11) Select your desired **file folder** location.

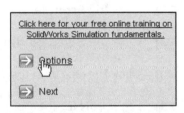

12) Click **OK** from the dialog box.

13) Click **Next**. The Fixtures screen is displayed. View your options.

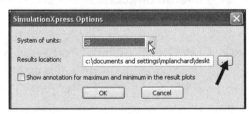

Apply Fixtures to the model.
14) Click **Add a fixture**. The Fixture PropertyManager is displayed.

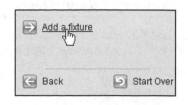

15) Click the **back middle circular face**. Small hatch symbols are displayed on the circular face.

16) Click the **back left circular face**.

17) Click the **back right circular face**. The three selected faces are displayed.

18) Click **OK** ✔ from the Fixture PropertyManager. View the updated SimulationXpress Study tree.

Apply Loads to the model.
19) Click **Next**.

Apply a distributed load set to the part front face. The applied Force is 16N. The downward Force is Normal (Perpendicular) to the Plane2 (Top) Reference Plane.

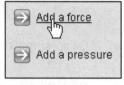

20) Click **Add a force**. The Force PropertyManager is displayed.

21) Click **Isometric view** 🔲 from the Heads-up View toolbar.

22) Click the **Front face** as illustrated.

23) Click the **Selected direction** box.

24) Click **Plane2** from the Fly-out FeatureManager.

25) Enter **16**N for the Force value. If required, click the **Reverse direction** check box. The force symbols point downward.

26) Click **OK** ✔ from the Force PropertyManager. View the updated SimulationXpress Study tree.

Add Material to the part.
27) Click **Next**.

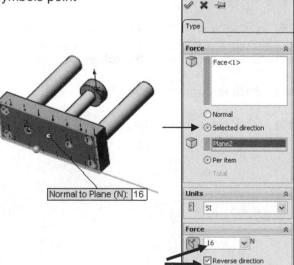

28) Click **Choose Material**. The Material dialog box is displayed.

29) Select **6061 Alloy**.

30) Click **Apply**.

31) Click **Close** from the Material dialog box. View the updated Study tree. A green check mark is displayed on the part folder. This indicates that material is applied.

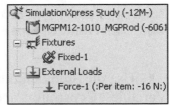

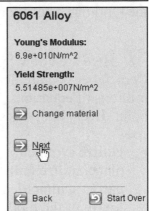

Run the analysis.
32) Click **Next**.

33) Click **Run Simulation**. View the results. View the animation of the model in the Graphics window and the updated Study tree.

34) Click **Stop animation**.

View the results.
35) Double-click on the **Stress** folder. View the results.

36) Double-click on the **Displacement** folder. View the results.

37) Double-click on the **Deformation** folder. View the results.

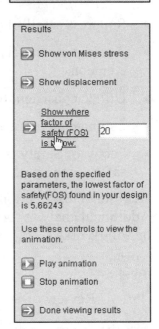

38) Double-click on the **Factor of Safety** folder. View the results. Red indicates areas (FOS) less than 1. Blue indicates areas (FOS) greater than 1.

39) Click **Yes, continue**. The present FOS is approximately 5.66.

40) Enter **20**.

41) Click **Show where factor of safety (FOS) is below**: View the results. View the red areas on the piston.

42) Click **Done viewing results**. View the new Results screen. You can choose between: *Generate report* or *Generate eDrawing* file. Skip this section.

43) Click **Next**.

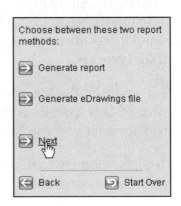

44) Click **No**. As an exercise, click yes and follow the wizard to optimize the part. Note: You can not optimize a part with imported geometry. Use FeatureWorks to create the needed features and geometry.

45) Click **Next**. You have completed the SimulationXpress Wizard. That was easy! Think about the steps and input values.

Close all parts and assemblies.
46) Click **Window**, **Close All** from the Menu bar.

 Use the Study Optimization table to select Variables, Constraints and Goals when performing the Optimize procedure in SimuationXpress.

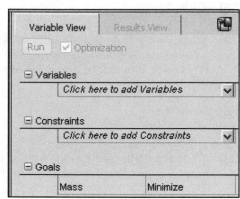

 Here are a few tips in performing the analysis. Remember you are dealing with thousands or millions of equations. These tips are a starting point. Every analysis situation is unique.

- Utilize symmetry. If a part is symmetric about a plane, utilize half of the model for analysis. If a part is symmetric about two planes, utilize a fourth of the model for analysis.

- Suppress small fillets and detailed features in the part.

- Avoid parts that have aspect ratios over 100, use SolidWorks Simulation with Beam elements.

- Utilize consistent units.

- Estimate an intuitive solution based on the fundamentals of stress analysis techniques.

- Factor of Safety is a guideline for the designer. The designer is responsible for the safety of the part.

Additional information on SolidWorks SimulationXpress is located in SolidWorks Help. Additional analysis tools for static, dynamic, thermal and fluid analysis are also available in SolidWorks.

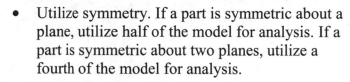 Review of SolidWorks SimulationXpress

SolidWorks SimulationXpress is a Finite Element tool that calculates stress, displacement and FOS based on Fixtures, Loads, and material on a part. You utilized SolidWorks SimulationXpress to analysis the stress, displacement and FOS on a part early in the design process.

You assigned the Material to be Aluminum, added fixtures to three faces and applied a force of 16N. Based on the inputs, the results provided visual representation of displacement and von Mises stress.

Use the Optimize tool to redesign the part as an exercise. See SolidWorks Help for additional information.

SolidWorks Sustainability

Sustainable engineering is the integration of social, environmental, and economic conditions into a product or process. Soon all design will be Sustainable Design.

SolidWorks Sustainability allows students and designers to be environmentally conscious about their designs.

Every license of SolidWorks 2012 provides access to SolidWorks SustainabilityXpress. SustainabilityXpress calculates environmental impact on a part in four key areas: *Carbon Footprint, Energy Consumption, Air Acidification and Water Eutrophication.*

Material and Manufacturing process region and Transportation Usage region are used as input variables. Two SolidWorks Sustainability products are available: *SolidWorks SustainabilityXpress* and *SolidWorks Sustainability.*

SolidWorks SustainabilityXpress: Handles part documents and is included in the core software.

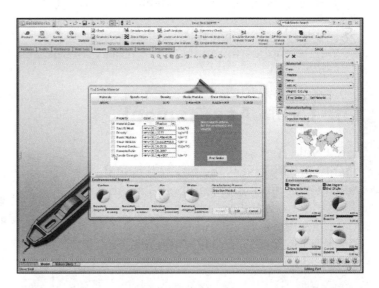

SolidWorks Sustainability: Provides the following functions:

- Same functions as SustainabilityXpress

- Life Cycle Assessment (LCA) of assemblies

- Configuration support:

    o Save inputs and results per configuration

- Expanded reporting capabilities for assemblies

- Specify amount and type of energy consumed during use

- Specify method of transportation

- Support for Assembly Visualization

SolidWorks Sustainability provides real-time feedback on key impact factors in the Environmental Impact Dashboard, which updates dynamically with any changes by the user. You can generate customize reports to share the results.

Run SustainabilityXpress and analyze a part. Display the four environmental impact factors, view suitable alternative materials, compare their environmental impact factors and generate a customer report.

☀ SolidWorks Sustainability is included in the SolidWorks Educational edition and is a separate application in the SolidWorks commercial edition.

Life Cycle Assessment

Life Cycle Assessment is a method to quantitatively assess the environmental impact of a product throughout its entire lifecycle, from the procurement of the raw materials, through the production, distribution, use, disposal and recycling of that product.

- **Raw Material Extraction:**
  - o Planting, growing, and harvesting of trees
  - o Mining of raw ore (example: bauxite)
  - o Drilling and pumping of oil
- **Material Processing** - The processing of raw materials into engineered materials:
  - o Oil into Plastic
  - o Iron into Steel
  - o Bauxite into Aluminum

- **Part Manufacturing** - Processing of material into finished parts:
  o Injection molding
  o Milling and Turning
  o Casting
  o Stamping
- **Assembly** - Assemble all of the finished parts to create the final product.
- **Product Use** - End consumer uses product for intended lifespan of product.
- **End of Life** - Once the product reaches the end of its useful life, how is it disposed of:
  o Landfill
  o Recycled
  o Incinerated

Life Cycle Assessment Key Elements

SolidWorks Sustainability provides the ability to assess the following key elements on the life cycle:

- **Identify and quantify the environmental loads involved:**
  o Energy and raw materials consumed
  o Emissions and wastes generated
- **Evaluate the potential environmental impacts of these loads.**
- **Assess the options available for reducing these environmental impacts.**

SustainabilityXpress Wizard

Materials: Provides the ability to select various material classes and Names for the part from the drop-down menu.

Manufacturing: Provides the ability to select both the process and usage region for the part. At this time, there are four different areas to choose from: North America, Europe, Asia, and Japan.

Process: Provides the ability to select between multiple different production techniques to manufacture the part.

Environmental Impact: This area included four quantities: *Carbon Footprint*, *Total Energy*, *Air Acidification*, and *Water Eutrophication*. Each graph presents a graphical breakdown in: Material Impact, Transportation and Use, Manufacturing, and End of Life.

- **Carbon Footprint:** A measure of carbon-dioxide and other greenhouse gas emissions such as methane (in CO_2 equivalents units, CO_{2e}) which contribute to emissions, predominantly caused by burning fossil fuels. Global warming Potential (GWP) is also commonly referred to as a carbon footprint.

- **Energy Consumption:** A measure of the non-renewable energy sources associated with the part's lifecycle in nits of Mega Joules (MJ). This impact includes not only the electricity of fuels used during the product's lifecycle, but also the upstream energy required to obtain and process these fuels, and the embodied energy of materials which would be released if burned. Energy Consumed is expressed as the net calorific value of energy demand from non-renewable resources (petroleum, natural gas, etc.).

Efficiencies in energy conversion (power, heat, steam, etc.) are taken into account.

- **Air Acidification:** Sulfur dioxide, nitrous oxides and other acidic emissions to air cause an increase in the acidity of rain water, which in turn acidifies lakes and soil. These acids can make the land and water toxic for plants and aquatic life. Acid rain can also slowly dissolve man-made building materials such as concrete. This impact is typically measured in nits of either kg sulfur dioxide equivalent SO_{2e} or moles H+ equivalent.

- **Water Eutrophication:** When an over abundance of nutrients are added to a water ecosystem, Eutrophication occurs. Nitrogen and phosphorous from waste water and agricultural fertilizers causes an overabundance of algae to bloom, which then depletes the water of oxygen and results in the death of both plant and animal life. This impact is typically measured in either kg phosphate equivalent (PO_{4e}) or kg nitrogen (N) equivalent.

Generate a Report: Provides the ability to generate a customer report that captures your baseline design and comparison between the various input parameters.

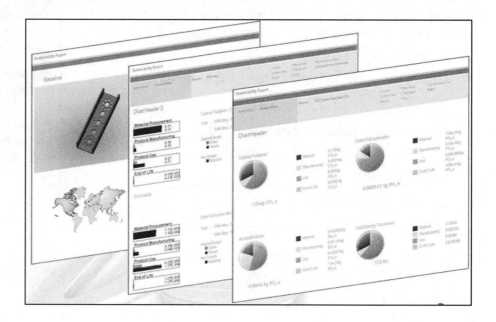

References

It is very important to understand definitions and how SolidWorks information is obtained on the above areas for baseline calculations and comparative calculations. The following standards and agencies were used in the development cycles of this tool:

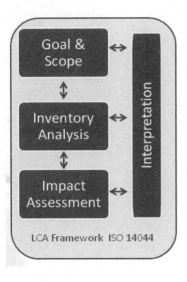

- **Underlying LCA Technology: PE International:**

  - 20 years of LCA experience

  - LCA international database

  - GaBi4 - leading software application for product sustainability

  - www.pe-international.com

- **International LCA Standards:**

  - Environmental Management Life Cycle Assessment Principles and Framework ISO 14040/44 www.iso.org

- **US EPA LCA Resources:**

  - http://www.epa.gov/nrmrl/lcaccess/

SolidWorks Sustainability Methodology

The following chart was created to provide a visualization of the Methodology used in SolidWorks Sustainability. Note the Input variables and the Output areas along with the ability to create and send a customer report are based on your selected decisions.

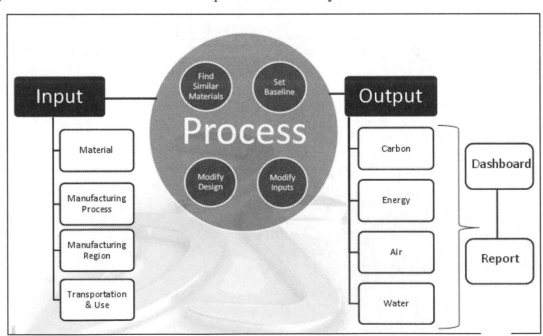

Tutorial: SustainabilityXpress 7.1

Close all parts, assemblies and drawings. Run SustainabilityXpress. Perform a simple analysis on a part.

1)   Open the **CLAMP** part from the ENGDESIGN-W-SOLIDWORKS\SustainabilityXpress folder.

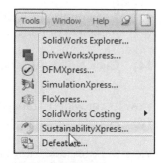

Activate SustainabilityXpress.

2)   Click **Tools**, **SustainabilityXpress/Sustainability** from the Main menu. SustainabilityXpress is displayed in the Task Pane area.

View the SolidWorks Sustainability dialog box.

3)   Click **Continue**. View the Sustainability Wizard.

Select the Material Class.

4)   Click the **Drop-down arrow** under Class. View your options.

5)   Select **Steel** from the drop-down menu as illustrated.

Select the Material Name. Material Name is Class dependent.

6)   Click the **Drop-down arrow** under Name. View your options.

7)   Select **Stainless Steel (ferritic)** from the drop-down menu as illustrated.

Select the Manufacturing Region. Each region produces energy by different method combinations. Impact of a kWh is different for each region. Example methods include: Fossil Fuels, Nuclear and Hydro-electric.

8)   Position the **mouse pointer** over the regions. View your available options.

9)   Click **Asia**.

Select Built to last period.

10)   Enter **10** years.

Select the Manufacturing Process. Manufacturing Process is Material Name dependent.

11)   Click the **Drop-down arrow** under Process. View your options.

12)   Select **Milled** from the drop-down menu as illustrated.

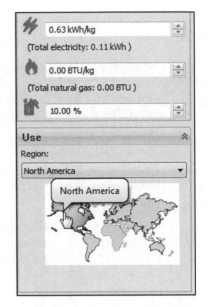

View the provided information on energy and scrap rate. Select Use Region.

13) Position the **mouse pointer** over the regions. View your available options.

14) Click **North America**.

View the provided Transportation modes.
15) Accept the default: **Boat**.

View the provided information on End of Life.

16) Accept the defaults.

Enter Duration of Use.
17) Enter **10** for years.

Set your design Baseline. A Baseline is required to comprehend the environmental impact of the original design.

18) Click the **Set Baseline**  tool from the bottom of the Environmental Impact screen. The Environmental Impact of this part is displayed. The Environmental Impact is calculated in four key areas: *Carbon Footprint, Energy Consumption, Air Acidification* and *Water Eutrophication*.

19) Position the **mouse pointer** over the Carbon box. View Factor percentage. 74.31% represents the Material percentage of the total Carbon footprint of the part (0.87kg). Material has the largest impact in the example.

 74.31% represents the Material percentage of the Energy footprint of the part.

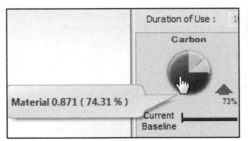

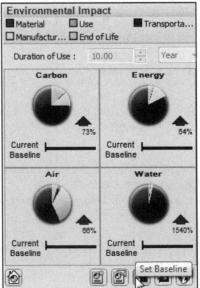

20) Click inside the **Carbon** Consumption impact screen to display a Baseline bar chart of the Carbon Footprint. View the results.

21) Click the **Home** icon to return to the Environmental Impact display.

22) Click inside the **Energy** Consumption impact screen to display a Baseline bar chart of Energy Consumption. View the results.

23) Click the **Home** icon to return to the Environmental Impact display.

24) Click inside the **Air** Consumption impact screen to display a Baseline bar chart of Air Acidification. View the results.

25) Click the **Home** icon to return to the Environmental Impact display.

26) Click inside the **Water** Consumption impact screen to display a Baseline bar chart of Water Eutrophication. View the results.

27) Click the **Home** icon to return to the Environmental Impact display.

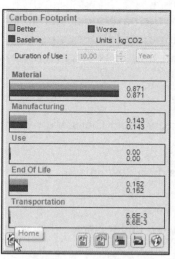

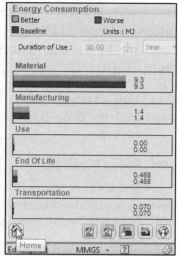

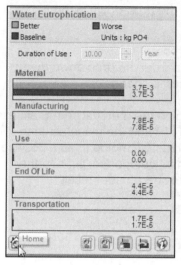

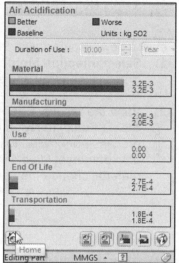

In the next section, compare the baseline design to a different Material Class, Name, and Process. Let's compare the present material Stainless Steel (ferritic) to Nylon 6/10.

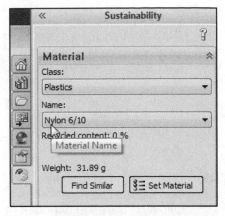

Select a new Material Class.
28) Click the **Drop-down arrow** under Class. View your options.

29) Select **Plastics** from the drop-down menu.

Select a new Material Name.
30) Click the **Drop-down arrow** under Name. View your options.

31) Select **Nylon 6/10** from the drop-down menu.

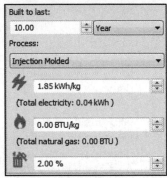

Select a new Manufacturing Process.
32) Select **Injection Molded** from the drop-down menu. Note: The energy and scrap change.

33) **View** the results. Changing the material from Stainless Steel (ferritic) to Nylon 6/10 and the manufacturing process from milled to injection molded had a positive environmental impact in all categories, but a further material change may provide a better result.

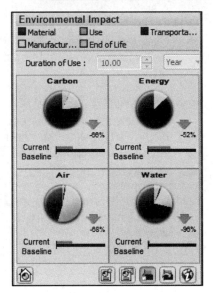

Find a similar material and compare the Environmental Impact to Nylon 6/10. This function is a real time saver.

34) Click the **Find Similar** button as illustrated. The Find Similar Material dialog box is displayed.

35) Click the **Value (-any-)** drop-down arrow. View your options.

36) Select **Plastics**. You can perform a general search or customize your search on physical properties of the material.

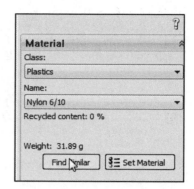

Select a Similar Material from the provided list. You can find similar materials based on the following Properties: *Material Class, Thermal Expansion Coefficient, Specific Heat, Density, Elastic Modulus, Shear Modulus, Thermal Conductivity, Poissons Ratio, Tensile Strength and Yield Strength.*

SolidWorks Simulation provides the ability to search on various materials properties. Select either Greater than, Less than or approximately from the drop-down menu.

37) Click the **Find Similar** button as illustrated. SolidWorks provides a full list of comparable materials that you can further refine.

38) In this exercise, select a Material Class. Check the **ABS PC** box.

39) Check the **Nylon 101** box.

40) Click **inside the top left box (Show Selected Only)** as illustrated. The selected materials are displayed with their properties.

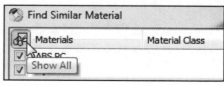

View the Environment Impact for the alternative materials.

41) Click the **ABS PC** material row as illustrated. View the results. The material is lower in Carbon Footprint, Energy Consumption, Air Acidification and Water Eutrophication than Nylon 6/10.

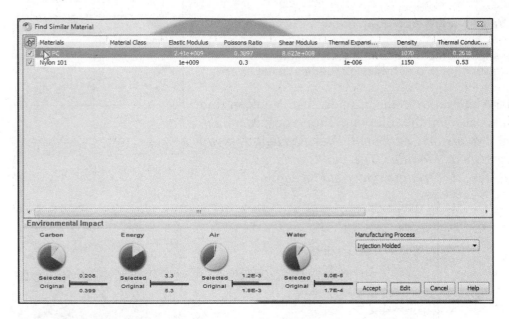

42) Click the **Nylon 101** material row. The material is lower in Carbon Footprint, Air Acidification and Water Eutrophication but higher in Energy Consumption than Nylon 6/10. You decide to stay with Nylon 6/10 due to cost and other design issues.

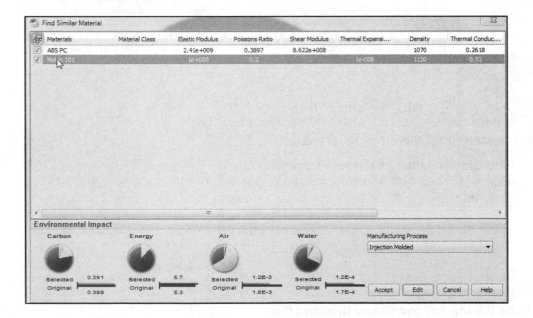

43) Click the **Cancel** button from the Find Similar Material dialog box.

Run a Report.

44) Click the **Generate Report** button as illustrated. SolidWorks provides the ability to communicate this report information throughout your organization. Sustainability generates a report that will compare designs (material, regions) and explain each category of Environmental Impact and show how each design compares to the Base line.

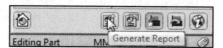

☀ You can't generate a report if Microsoft Word is running.

45) Review the Generated report. **Ctrl - Click here for alternative units such as Miles Driven In a Car.** View the results. Note: Internet access is required

Click here for alternative units such as Miles Driven In A Car

46) **View the Glossary** in the report section and the other hyperlinks for additional information.

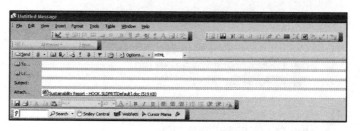

Send the Report to a colleague
47) Click the **Send Report** button as illustrated.

48) **Close** the report.

49) **Close** the part.

☀ Click the Online information icon to obtain additional information and comparisons of the model. Internet access is required.

The same process can be performed on an assembly using Sustainability as illustrated with additional inputs. Click on a component in an assembly to display the Material dialog box.

SolidWorks DFMXpress

DFMXpress is an analysis tool that validates the manufacturability of SolidWorks parts. Use DFMXpress to identify design areas that may cause problems in fabrication or increase the costs of production. The DFMXpress tool uses a Wizard. Run the DFMXpress Wizard. Analyze a simple part. View the results.

Activity: SolidWorks DFMXpress - Analyze the AXLE and ROD Part

Close all documents.
1) Click **Window**, **Close All** from the Menu bar.

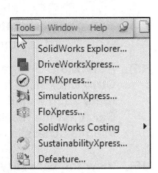

Open the AXLE part.
2) Open the AXLE part from the **ENGDESIGN-W-SOLIDWORKS\ DFMXpress** folder. The AXLE FeatureManager is displayed. Note: The folder is located on the DVD in the book. Copy all models to your local hard drive.

Activate DFMXpress.
3) Click **Tools**, **DFMXpress**. View the DFMXpress wizard in the Task Pane location.

4) Click the **Settings** button. View the optional settings in DFMXpress.

5) Click the **Back** button to return to the Main menu.

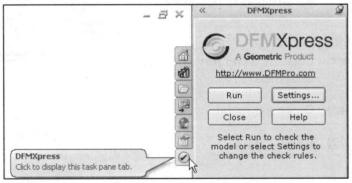

Run the DFMXpress wizard on the AXLE part.
6) Click the **Run** button.

7) **Expand** each folder. View the results.

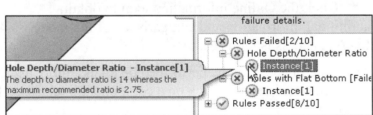

8) Click on each **red check mark**. The hole in the AXLE was created with an Extruded Cut feature (failed due to flat bottom). Use the Hole Wizard to create holes for Manufacturing. Use caution for depth to diameter ratio.

9) Click on each **green check mark**. Review the information. This is a very useful tool to use for manufacturing guidelines.

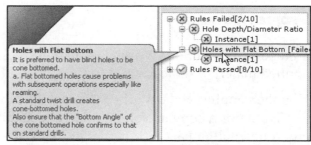

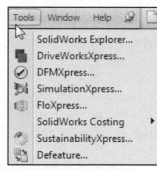

Close the DFMXpress Wizard.
10) Click the **Close** button.

Open the ROD part.
11) Open the ROD part from the **ENGDESIGN-W-SOLIDWORKS\ DFMXpress** folder. The ROD FeatureManager is displayed. Note: The folder is located on the DVD in the book. Copy all models to your local hard drive.

Activate DFMXpress.
12) Click **Tools**, **DFMXpress**. View the DFMXpress wizard in the Task Pane location.

Run the DFMXpress wizard on the AXLE part.
13) Click the **Run** button.

14) **Expand** each folder. View the results.

15) Click on each **red check mark**. The hole in the ROD was created using the Hole Wizard feature. Use caution for depth to diameter ratio.

16) Click on **each green check mark**. Review the information. This is a very useful tool to use for manufacturing guidelines.

Close the DFMXpress Wizard.
17) Click the **Close** button.

18) **Close** all models.

 Review the DFMXpress Wizard

DFMXpress is an analysis tool that validates the manufacturability of SolidWorks parts. Use DFMXpress to identify design areas that may cause problems in fabrication or increase the costs of production.

The DFMXpress tool uses a Wizard. Run the DFMXpress Wizard. Analyze a simple part. View the results.

Project Summary

You performed a SolidWorks SimulationXpress analysis on a part with a Force load. You applied material, set loads and fixtures. Run the optimization tool as an exercise.

You then applied SolidWorks SustainabilityXpress to a part. Every license of SolidWorks 2012, obtains a copy of SolidWorks SustainabilityXpress. SustainabilityXpress calculates environmental impact on a model in four key areas: *Carbon Footprint, Energy Consumption, Air Acidification and Water Eutrophication.*

Material and Manufacturing process region and Transportation Usage region are used as input variables.

You applied material, manufacturing process and selected the manufacturing process region and the part transportation and usage region. You then viewed the environmental impact of your design.

You viewed suitable alternative materials, re-ran the environmental impact study and compared their environmental impact factors in all four categories. A customer report was generated.

You used DFMXpress to identify design areas that may cause problems in fabrication or increase the costs of production. The DFMXpress tool uses a Wizard.

Exercises

Example 7.1: SimulationXpress is used to analyze?

- A = Drawings

- B = Parts and Drawings

- C = Assemblies and Parts

- D = Parts

- E = All of the above

The correct answer is _____.

Example 7.2: How many degrees of freedom does a physical structure have?

- A = Zero.

- B = Three - Rotations only

- C = Three - Translations only

- D = Six - Three translations and three rotations

The correct answer is _____.

Example 7.3: Brittle materials has little tendency to deform (or strain) before fracture and does not have a specific yield point. It is not recommended to apply the yield strength analysis as a failure criterion on brittle material. Which of the following failure theories is appropriate for brittle materials?

- A = Mohr-Columb stress criterion

- B = Maximum shear stress criterion

- C = Maximum von Mises stress criterion

- D = Minimum shear stress criterion

The correct answer is _____.

Example 7.4: You are performing an analysis on your model. You select three faces and apply a 40 lb load. What is the total force applied to the model?

- A = 40 lbs

- B = 20 lbs

- C = 120 lbs

- D: Additional information is required.

The correct answer is _____.

Example 7.5: In an engineering analysis, you select a face to restrain. What is the affect?

- A = The face will not translate but can rotate

- B = The face will rotate but can not translate

- C = You can not apply a restraint to a face

- D = The face will not rotate and will not translate

The correct answer is _____.

Example 7.6: A material is orthotropic if its mechanical or thermal properties are not unique and independent in three mutually perpendicular directions.

- A = True

- B = False

The correct answer is _____.

Exercise 7.7: SimulationXpress Engineering Analysis

Determine the Factor of Safety, the von Mises Stress Plot and Deflection Plot utilizing SimulationXpress.

- Open the AXLE-100 part. The AXLE-100 part is available on the Multi-media DVD in the Chapter 7 - Homework folder.

- Active SimulationXpress.

- Select SI for system of units.

- Enter 6061 Alloy for material.

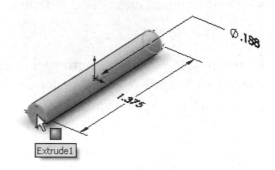

- Fix the AXLE-100 part at both ends.

- Apply a 100N force load along the entire cylindrical face normal to the Top plane.

- Calculate and explain the FOS.

- View the stress distribution in the model. Note the units.

- View the displacement distribution in the model. Note the uints.

- Determine the modifications to the material, geometry, restraints and loading conditions that will increase the Factor of Safety in your design.

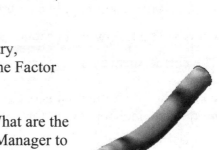

- Optimize the FOS design with the Axle diameter. What are the effects? Double-click Boss-Extrude1 in the FeatureManager to view the new diameter.

- Experiment the different options and features.

Project 8

Intelligent Modeling Techniques

 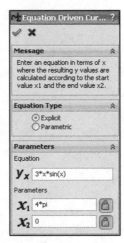 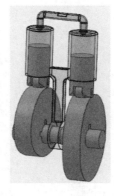 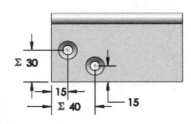

Below are the desired outcomes and usage competencies based on the completion of Project 8.

Project Desired Outcomes:	Usage Competencies:
• Utilize and understand various Design Intent tools and Intelligent Modeling Techniques.	• Apply Design Intent and Intelligent Modeling Techniques in a Sketch, Feature, Part, Plane, Assembly and Drawing.
• Awareness of the Equation Driven Sketch Curve tool.	• Create Explicit Dimension Driven and Parametric Equation Driven Curves. • Generate curves using the Curve Through XYZ Points tool for a NACA aerofoil and CAM data text file.
• Knowledge of the SolidWorks Xpert tools.	• Apply SketchXpert, DimXpert and FeatureXpert.

Notes:

Project 8 - Intelligent Modeling Techniques

Project Objective

Understand some of the available tools in SolidWorks to perform intelligent modeling. Intelligent modeling is incorporating design intent into the definition of the sketch, feature, part, and assembly or drawing document. Intelligent modeling is most commonly addressed through design intent using a:

- Sketch:
  - Geometric relations
  - Fully defined Sketch tool
  - SketchXpert
  - Equations:
    - Explicit Dimension Driven
    - Parametric Driven Curve
  - Curves:
    - Curve Through XYZ Points:
    - Projected Composite
- Feature:
  - End Conditions:
    - Blind, Through All, Up to Next, Up to Vertex, Up to Surface, Offset from Surface, Up to Body and Mid Plane
  - Along a Vector
  - FeatureXpert (Constant Radius)
  - Symmetry:
    - Mirror
- Plane
- Assembly:
  - Symmetry (Mirror / Pattern)
  - Assembly Visualization
  - SolidWorks Sustainability
  - MateXpert
- Drawing:
  - DimXpert (Slots, Pockets, Machined features, etc.)

This project uses short step-by-step tutorials to practice and reinforce the subject matter and objectives. Learn by doing, not just by reading! All needed models for this chapter are located on the DVD in the book.

Design Intent

What is design intent? All designs are created for a purpose. Design intent is the intellectual arrangement of features and dimensions of a design. Design intent governs the relationship between sketches in a feature, features in a part and parts in an assembly or drawing document.

The SolidWorks definition of design intent is the process in which the model is developed to accept future modifications. Models behave differently when design changes occur.

Design for change! Utilize geometry for symmetry, reuse common features, and reuse common parts. Build change into the following areas that you create: sketch, feature, part, assembly and drawing.

Design Intent is how your part reacts as parameters are modified. Example: If you have a hole in a part that must always be .125≤ from an edge, you would dimension to the edge rather than to another point on the sketch. As the part size is modified, the hole location remains .125≤ from the edge.

Sketch

In SolidWorks, relations between sketch entities and model geometry, in either 2D or 3D sketches, are an important means of building in design intent. In this chapter - we will only address 2D sketches.

Apply design intent in a sketch as the profile is created. A profile is determined from the Sketch Entities. Example: Rectangle, Circle, Arc, Point, Slot etc.

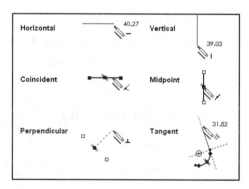

Develop design intent as you sketch with Geometric relations. Sketch relations are geometric constraints between sketch entities or between a sketch entity and a plane, axis, edge, or vertex. Relations can be added automatically or manually.

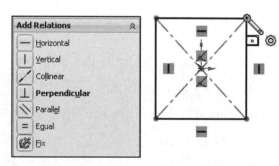

As you sketch, allow SolidWorks to automatically add relations. Automatic relations rely on:

- Inferencing

- Pointer display

- Sketch snaps and Quick Snaps

After you sketch, manually add relations using the Add Relations tool, or edit existing relations using the Display/Delete Relations tool.

Fully Defined Sketch

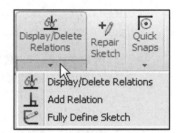

Sketches are generally in one of the following states:

- Under defined

- Fully defined

- Over defined

Although you can create features using sketches that are not fully defined, it is a good idea to always fully define sketches for production models. Sketches are parametric, and if they are fully defined, changes are predictable. However, sketches in drawings, although they follow the same conventions as sketches in parts, do not need to be fully defined since they are not the basis of features.

SolidWorks provides a tool to help the user fully define a sketch. The Fully Defined Sketch tool provides the ability to calculate which dimensions and relations are required to fully define under defined sketches or selected sketch entities. You can access the Fully Define Sketch tool at any point and with any combination of dimensions and relations already added. See Chapter 5 for additional information.

Your sketch should include some dimensions and relations before you use the Fully Define Sketch tool.

The Fully Defined Sketch tool uses the Fully Define Sketch PropertyManager. The Fully Define Sketch PropertyManager provides the following selections:

- ***Entities to Fully Define***. The Entities to Filly Define box provides the following options:

  - **All entities in sketch**. Fully defines the sketch by applying combinations of relations and dimensions.

  - **Selected entities**. Provides the ability to select sketch entities.

    - **Entities to Fully Define**. Only available when the Selected entities box is checked. Applies relations and dimensions to the specified sketch entities.

  - **Calculate**. Analyzes the sketch and generates the appropriate relations and dimensions.

- *Relations*. The Relations box provide the following selections:

  - **Select All**. Includes all relations in the results.

  - **Deselect All**. Omits all relations in the results.

  - **Individual relations**. Include or exclude needed relations. The available relations are: **Horizontal, Vertical, Collinear, Perpendicular, Parallel, Midpoint, Coincident, Tangent, Concentric** and **Equal radius/Length**.

- *Dimensions*. The Dimensions box provides the following selections:

  - **Horizontal Dimensions**. Displays the selected Horizontal Dimensions Scheme and the entity used as the Datum - Vertical Model Edge, Model Vertex, Vertical Line or Point for the dimensions. The available options are: **Baseline, Chain** and **Ordinate**.

  - **Vertical Dimensions.** Displays the selected Vertical Dimensions Scheme and the entity used as the Datum - Horizontal Model Edge, Model Vertex, Horizontal Line or Point for the dimensions. The available options are: **Baseline, Chain** and **Ordinate**.

  - **Dimension**. Below sketch and Left of sketch is selected by default. Locates the dimension. There are four selections: **Above sketch, Below the sketch, Right of sketch** and **Left of sketch**.

Activity: Fully Defined Sketch

Close all parts, assemblies and drawings. Apply the Fully Defined Sketch tool. Modify the dimension reference location in the sketch profile with control points.

1) Open **Fully Defined Sketch 8-1** from the ENGDESIGN-W-SOLIDWORKS/Chapter 8 - Intelligent Modeling folder. View the FeatureManager. Sketch1 is under defined.

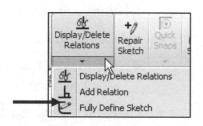

2) Edit 🖉 Sketch1. The two circles are equal and symmetrical about the y axis. The rectangle is centered at the origin.

3) Click the **Fully Define Sketch** ✎ tool from the Consolidated Display/Delete Relations drop-down menu. The Fully Defined Sketch PropertyManager is displayed.

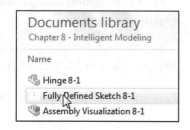

4) The All entities in the sketch are selected by default. Click **Calculate**. View the results. Sketch1 is fully defined to the origin.

5) Click **OK** ✓ from the PropertyManager. Drag all dimensions off the profile.

6) Modify the **vertical dimension to 50mms** and the **diameter dimension to 25mms** as illustrated. SolidWorks suggests a dimension scheme to create a fully defined sketch.

7) Click **View**; uncheck **Sketch Relations** from the Main menu.

Modify the dimension reference location in the sketch profile with control points.

8) Click the **90**mm dimension in the Graphic window.

9) Click and drag the **left control point** as illustrated.

10) Release the mouse pointer on the **left vertical line** of the profile. View the new dimension reference location of the profile.

11) Repeat the same procedure for the horizontal **50mm** dimension. Select the new reference location to the left hole diameter as illustrated.

12) **Close** the model. View the results.

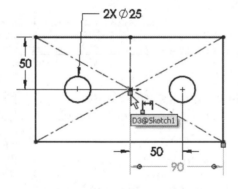

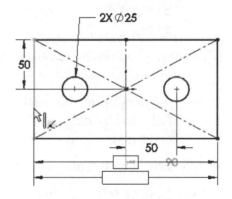

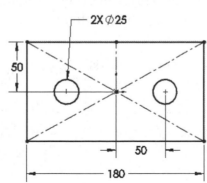

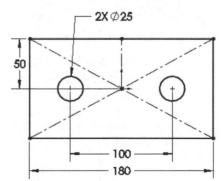

SketchXpert

SketchXpert resolves conflicts in over defined sketches and proposes possible solution sets. Color codes are displayed in the SolidWorks Graphics window to represent the sketch states. The SketchXpert tool uses the SketchXpert PropertyManager. The SketchXpert PropertyManager provides the following selections:

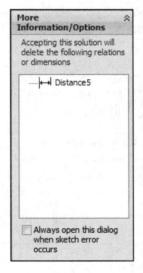

- *Message*. The Message box provides access to the following selections:

  - **Diagnose**. The Diagnose button generates a list of solutions for the sketch. The generated solutions are displayed in the Results section of the SketchXpert PropertyManager.

  - **Manual Repair**. The Manual Repair button generates a list of all relations and dimensions in the sketch. The Manual Repair information is displayed in the Conflicting Relations/ Dimensions section of the SketchXpert PropertyManager.

- *More Information/Options*. Provides information on the relations or dimensions that would be deleted to solve the sketch.

  - **Always open this dialog when sketch error occurs**. Selected by default. Opens the dialog box when a sketch error is detected.

- *Results*. The Results box provides the following selections:

  - **Left or Right arrows**. Provides the ability to cycle through the solutions. As you select a solution, the solution is highlighted in the Graphics window.

  - **Accept**. Applies the selected solution. Your sketch is no longer over-defined.

- *More Information/Options*. The More Information/Options box provides the following selections:

  - **Diagnose**. The Diagnose box displays a list of the valid generated solutions.

  - **Always open this dialog when sketch error occurs**. Selected by default. Opens the dialog box when a sketch error is detected.

- *Conflicting Relations/Dimensions*. The Conflicting Relations/Dimensions box provides the ability to select a displayed conflicting relation or dimension. The select item is highlight in the Graphics window. The options include:

  - **Suppressed**. Suppresses the relation or dimension.

  - **Delete**. Removes the selected relation or dimension.

  - **Delete All**. Removes all relations and dimensions.

  - **Always open this dialog when sketch error occurs**. Selected by default. Opens the dialog box when a sketch error is detected.

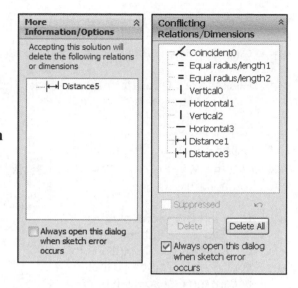

Activity: SketchXpert

Close all parts, assemblies and drawings. Create an over defined sketch. Apply SketchXpert to select a solution.

1) Open **SketchXpert 8-1** from the ENGDESIGN-W-SOLIDWORKS/Chapter 8 - Intelligent Modeling folder. View the FeatureManager. Sketch1 is under defined.

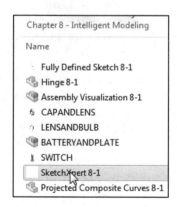

2) **Edit** Sketch1. The two circles are equal and symmetrical about the y axis. The rectangle is centered at the origin. Insert a dimension to create an over defined sketch.

3) Click **Smart Dimension**.

4) Add a dimension to the **left vertical line**. This makes the sketch over-defined. The Make Dimension Driven dialog box is displayed.

5) Check the **Leave this dimension driving** box option.

6) Click **OK**. The Over Defined warning is displayed.

7) Click the **red Over Defined** message. The SketchXpert PropertyManager is displayed.

Color codes are displayed in the Graphics window to represent the sketch states.

8) Click the **Diagnose** button. The Diagnose button generates a list of solutions for your sketch. You can either accept the first solution or click the Right arrow key in the Results box to view the section solution. The first solution is to delete the vertical dimension of 105mm.

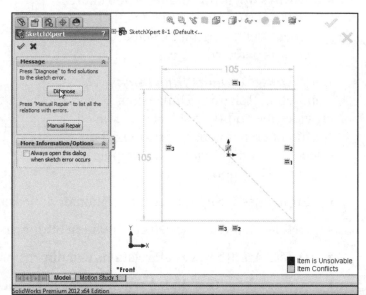

View the Second solution.

9) Click the **Right arrow** key in the Results box. The second solution is displayed. The second solution is to delete the horizontal dimension of 105mm.

View the Third solution.

10) Click the **Right arrow** key in the Results box. The third solution is displayed. The third solution is to delete the Equal relation between the vertical and horizontal lines.

Accept the Second solution.

11) Click the **Left arrow** key to obtain the second solution.

12) Click the **Accept** button. The SketchXpert tool resolves the over-defined issue. A message is displayed.

13) Click **OK** ✔ from the SketchXpert PropertyManager.

14) **Rebuild** 🔘 the model. View the results.

15) **Close** the model.

Equations

Dimension driven by equations

You want to design a hinge assembly that you can modify easily to make similar assemblies. You need an efficient way to create two matching hinge pieces and a pin for a variety of hinge assembly sizes.

Some analysis and planning can help you develop a design that is flexible, efficient, and well defined. You can then adjust the size as needed, and the hinge assembly still satisfies the design intent.

In the following example, one hole is fixed, one is driven by an equation (Equations create a mathematical relations between dimensions), and the other two are mirrored. As the size of the hinge changes, the holes remain properly spaced along the length and width of the model.

If you modify the height or length hinge dimension, the screw holes maintain the mathematical relation design intent.

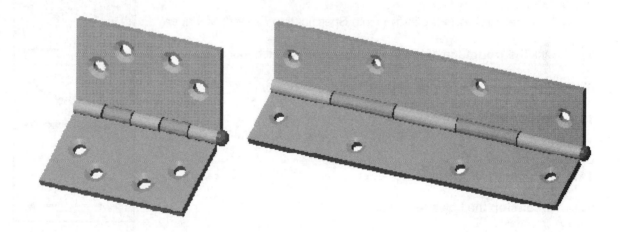

You can add comments to equations to document your design intent. Place a single quote (') at the end of the equation, then enter the comment. Anything after the single quote is ignored when the equation is evaluated. Example: "D2@Sketch1" = "D1@Sketch1" / 2 'height is 1/2 width. You can also use comment syntax to prevent an equation from being evaluated. Place a single quote (') at the beginning of the equation. The equation is then interpreted as a comment, and it is ignored. See SolidWorks Help for additional information.

Activity: Equation

Close all parts, assemblies and drawings. Create two driven equations. Insert two Countersink holes. To position each hole on the hinge, one dimension is fixed, and the other is driven by an equation.

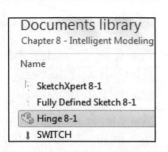

1) Open the **Hinge 8-1** part from the ENGDESIGN-W-SOLIDWORKS/Chapter 8 - Intelligent Modeling folder.

Add Screw holes using the Hole Wizard.

2) Click **Hole Wizard** from the Features toolbar. The Hole Specification PropertyManager is displayed.

3) Click **Countersink** for Hole Type as illustrated.

4) Select **Ansi Metric** for Standard. Accept the default Type.

5) Select **M8** for Size from the drop-down menu.

6) Select **Through All** for End Condition.

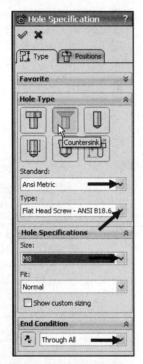

Place the location of the holes.

7) Click the **Positions** tab in the Hole Specification PropertyManager.

8) Click the **front face** of the model. The Point Sketch tool is displayed.

9) Click to **place the holes** approximately as shown.

10) **De-select** the Point Sketch tool.

Dimension the holes.

11) Click **Smart Dimension**.

12) **Dimension** the holes as illustrated.

13) Click **OK** from the Hole Position PropertyManager.

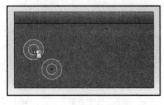

14) Click **OK** from the Hole Specification PropertyManager. View the new feature in the FeatureManager.

Add equations to control the locations of the screw Countersink holes. First - add an equation to control the location of one of the points. Create an equation that sets the distance between the point and the bottom edge to one-half the height of the hinge.

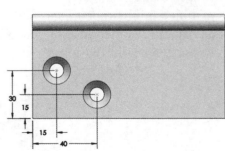

15) Right-click **Sketch5**. Click **Edit Sketch**.

16) Double-click **Extruded-Thin1** from the FeatureManager. Dimensions are displayed in the Graphics window.

17) Click **Tools**, **Equations** from the Menu bar. The Equations, Global Variables, and Dimensions dialog box is displayed.

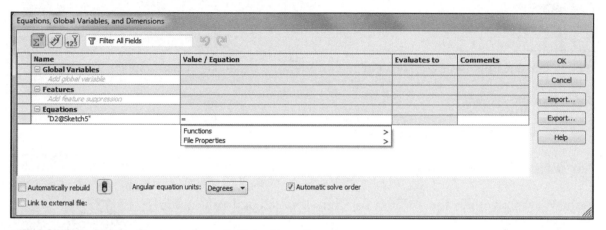

18) Click inside the **first empty** cell under Equations.

19) Click the **30mm** dimension in the Graphics window. An = sign is displayed in the Value / Equation cell.

20) Click the **60mm** dimension in the Graphics window.

21) Enter **/2** in the dialog box to complete the dimension.

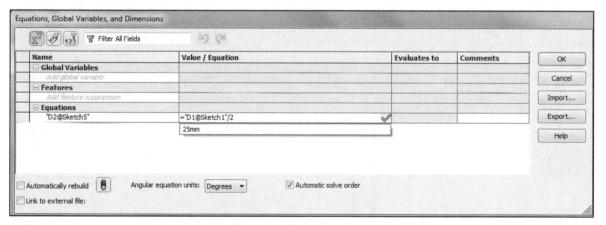

22) Press the **Tab** key. This equation sets the distance between the point and the bottom edge to one-half the height of the hinge.

Create the second equation. Add an equation to control the location of the other point.

23) Click inside the **first empty** cell under Equations.

24) Click the **40mm** dimension from the Graphics window. An = sign is displayed in the Value / Equation cell.

25) Click the **120mm** dimension for the base.

26) Enter **/3** in the dialog box to complete the dimension.

27) Press the **Tab** key. View the active equations. This sets the distance between the point and the side edge to one-third the length of the hinge.

⊟ Equations		
"D1@Sketch5"	= "D1@Sketch1" / 2	30mm
"D3@Sketch5"	= "D1@Extrude-Thin1" / 3	40mm
Add equation		

28) Click **OK** from the Equations dialog box. View the equations in the Graphics window.

29) **Exit** the Sketch.

30) **Close** the model.

Equation Driven Curve

SolidWorks provides the ability to address Explicit and Parametric equation types.

Explicit Equation Driven Curve

Explicit provides the ability to define x values for the start and endpoints of the range. Y values are calculated along the range of x values.

Mathematical equation can be inserted into a sketch. Utilize parenthesis to manage the order of operations. For example, calculate the volume of a solid bounded by a curve as illustrated. The region bounded by the equation $y = 2+x\cos x$, and the x-axis, over the interval $x = -2$ to $x = 2$, is revolved about the x-axis to generate a solid.

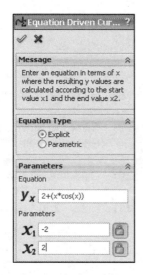

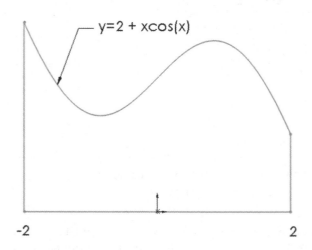

$y = 2 + x\cos(x)$

-2   2

Activity: Explicit Equation Driven Curve

Create an Explicit Equation Driven Curve on the Front plane. Revolve the curve. Add material. Calculate the volume of the solid.

1)   Create a **New** ▯ part. Use the default ANSI, IPS Part template.

Create a 2D Sketch on the Front Plane.

2)   Right-click **Sketch** from the Front Plane in the FeatureManager. The Sketch toolbar is displayed. Front Plane is your Sketch plane.

3)   Click the **Equation Driven Curve Sketch** ⌁ tool from the Consolidated drop-down menu. The Equation Driven Curve PropertyManager is displayed.

4)   Enter the **Equation y$_x$** as illustrated.

5)   Enter the **parameters x$_1$, x$_2$** that defines the lower and upper bound of the equation as illustrated.

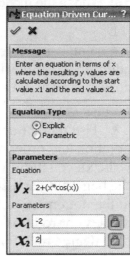

6)   Click **OK** ✔ from the Equation Driven Curve Sketch PropertyManager. View the curve in the Graphics window. Size the curve in the Graphics window. The Sketch is under defined.

7)   Insert **three lines to close** the profile as illustrated.

8)   Insert the **required dimensions** and any needed geometric relations to fully define the sketch.

Create a Revolved feature.

9)   Click the **horizontal line**.

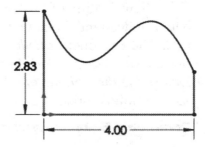

10)   Click the **Revolved Boss/Base** ⚙ tool from the Feature toolbar. 360 degrees is the default.

11)   Click **OK** ✔ from the PropertyManager. View the results in the Graphics window. Revolve1 is displayed. Utilize the Section tool, parallel with the Right plane to view how each cross section is a circle.

12)   Apply **Brass** for material.

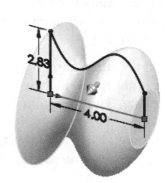

13) Calculate the **volume** of the part using the Mass Properties tool. View the results. Also note the surface area and the Center of mass.

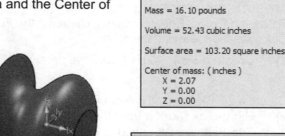

Density = 0.31 pounds per cubic inch

Mass = 16.10 pounds

Volume = 52.43 cubic inches

Surface area = 103.20 square inches

Center of mass: (inches)
X = 2.07
Y = 0.00
Z = 0.00

14) **Save** the model.

15) Name the model **Explicit Equation Driven Curve 8-1**.

16) **Close** the model.

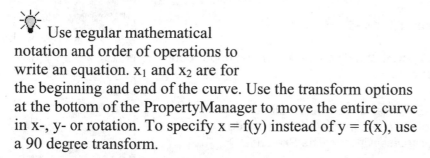

You can create parametric (in addition to explicit) equation-driven curves in both 2D and 3D sketches.

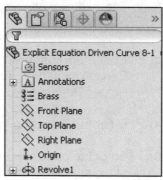

Use regular mathematical notation and order of operations to write an equation. x_1 and x_2 are for the beginning and end of the curve. Use the transform options at the bottom of the PropertyManager to move the entire curve in x-, y- or rotation. To specify $x = f(y)$ instead of $y = f(x)$, use a 90 degree transform.

View the .mp4 files located on the DVD in the book to better understand the potential of the Equation Driven Curve tool. The first one: *Calculating Area of a region bounded by two curves (secx)^2 and sin x* in SolidWorks; the second one: *Determine the Volume of a Function Revolved Around the x Axis* in SolidWorks.

Parametric Equation Driven Curve

The Parametric option of Equation Driven Curve can be utilize to represent two parameters, x- and y-, in terms of a third variable, t.

In the illustration below, a string is wound about a fixed circle of radius 1, and is then unwound while being held taught in the x-y-plane. The end point of the string, P traces an involute of the circle. The initial point (1,0) is on the x- axis and the string is tangent to the circle at Q. The angle, t is measured in radians from the positive x- axis.

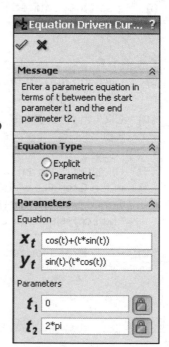

Equation Driven Cur... ?

Message

Enter a parametric equation in terms of t between the start parameter t1 and the end parameter t2.

Equation Type

○ Explicit
⦿ Parametric

Parameters

Equation

x_t cos(t)+(t*sin(t))

y_t sin(t)-(t*cos(t))

Parameters

t_1 0

t_2 2*pi

Use the Equation Driven Curve, Parametric option to illustrate how the parametric equations x = cos t + t sin t and y = sin t - t cos t represents the involute of the circle from t = 0 to t = 2π.

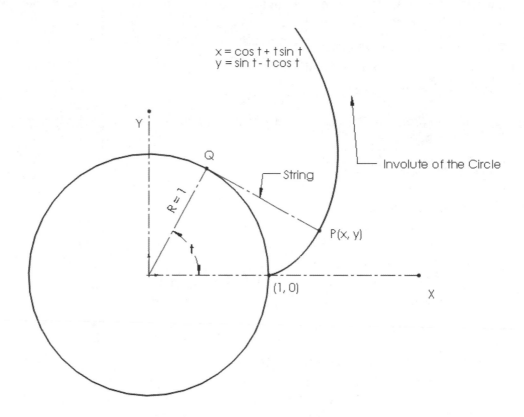

☀️ Rename a feature or sketch for clarity. Slowly click the feature or sketch name twice and enter the new name when the old one is highlighted.

Activity: Parametric Equation Driven Curve

Close all parts, assemblies and drawings. Create a Parametric Equation Driven Curve.

1) Open the **Parametric Equation 8.1** part from the ENGDESIGN-W-SOLIDWORKS/ Chapter 8 - Intelligent Modeling folder. The circle of radius 1 is displayed.

2) **Edit** Sketch1 from the FeatureManager.

Activate the Equation Driven Curve tool.

3) Click **Tools**, **Sketch Entities**, **Equation Driven Curve** from the Menu bar. The Equation Driven Curve PropertyManager is displayed.

4) Click **Parametric** in the Equation Type dialog box.

5) Enter the parametric equations for x_t and y_t as illustrated.

6) Enter t_1 and t_2 as illustrated.

You can also utilize sketch parameters in the equation. For example, the diameter dimension DIA2 is represented as "D2@Sketch2". The equation can be multiplied by the diameter.

X = "D2@Sketch2"*(cos(t)+(t*sin(t)))

Y = "D2@Sketch2"*(sin(t)-(t*cos(t)))

7) Click **OK** ✓ from the PropertyManager. View the results in the Graphics window.

8) **Close** the model.

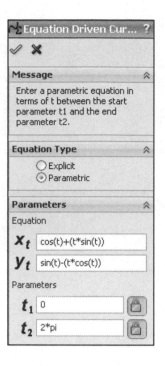

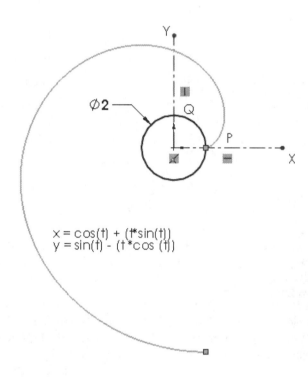

$$x = \cos(t) + (t^*\sin(t))$$
$$y = \sin(t) - (t^*\cos(t))$$

Curves

Another intelligent modeling technique is to apply the Curve Through XYZ Points feature. This feature provides the ability to either type in (using the Curve File dialog box) or click Browse and import a text file with x-, y-, z-, coordinates for points on a curve.

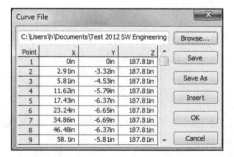

Generate the text file by any program which creates columns of numbers. The Curve 𝒰 feature reacts like a default spline that is fully defined.

🔆 It is highly recommend that you insert a few extra points on either end of the curve to set end conditions or tangency in the Curve File dialog box.

Imported files can have an extension of either *.sldcrv or *.text. The imported data: x-, y-, z-, must be separated by a space, a comma, or a tab.

The National Advisory Committee for Aeronautics (NACA) developed airfoil shapes for aircraft wings. The shape of the aerofoil is defined by parameters in the numerical code that can be entered into equations to generate accurate cross-sections.

Activity: Curve Through XYZ Points

Create a curve using the Curve Through XYZ Points tool and the Composite curve tool.
Import the x-, y-, z- data for an NACA aerofoil file for various cross sections.

1) Create a **New** 🗋 part. Use the default ANSI, IPS Part template.

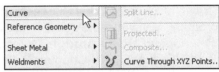

Browse to select the imported x-, y-, z- data.

2) Click **Insert**, **Curve**, **Curve Through XYZ Points** from the Menu bar. The Curve File dialog box is displayed.

3) **Browse** to the ENGDESIGN-W-SOLIDWORKS/Chapter 8 - Intelligent Modeling folder.

4) Select **Text Files (*.txt)** for file type.

5) Double-click **LowerSurfSR5.txt**. View the results in the Curve File dialog box.

6) Click **OK** from the Curve File dialog box. Curve1 is displayed in the FeatureManager.

7) **Repeat** the above procedure to create Curve2 from the UpperSurfSR5.txt file. Both the Lower and Upper curves are on the same Z plane. They are separate entities.

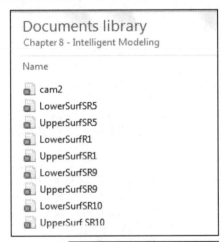

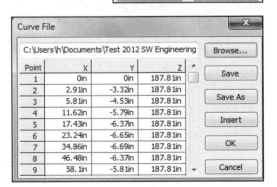

Use the Composite curve tool to join the Lower and Upper curves.

8) Click **Insert**, **Curve**, **Composite** from the Menu bar.

9) Select **Curve1** and **Curve2** from the FeatureManager as illustrated. Both curves are displayed in the Entities to Join box.

10) Click **OK** ✓ from the PropertyManager. View the results in the FeatureManager.

11) **Close** the model.

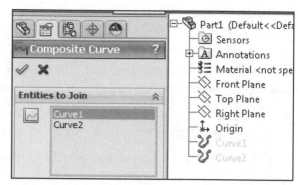

| **Activity: Curve Through XYZ Points - Second Tutorial** |

Create a curve using the Curve Through XYZ Points tool. Import the x-, y-, z- data from a CAM program. Verify that the first and last points in the curve file are the same for a closed profile.

1) Open **Curve Through XYZ points 8-2** from the ENGDESIGN-W-SOLIDWORKS/Chapter 8 - Intelligent Modeling folder.

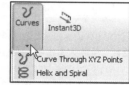

2) Click the **Curve Through XYZ Points** 𝒰 tool from the Features CommandManager. The Curve File dialog box is displayed.

Import curve data. Additional zero points were added on either end of the curve to set end conditions.

3) Click **Browse** from the Curve File dialog box. Browse to the ENGDESIGN-W-SOLIDWORKS/Chapter 8 - Intelligent Modeling folder.

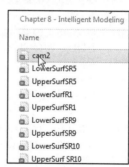

4) Set file type to **Text Files**.

5) Double-click **cam2.text**. View the data in the Curve File dialog box. View the sketch in the Graphics window. Review the data points in the dialog box.

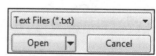

6) Click **OK** from the Curve File dialog box. Curve1 is displayed in the FeatureManager. Use the curve to create a fully defined base sketch of a CAM. Let's view the final results.

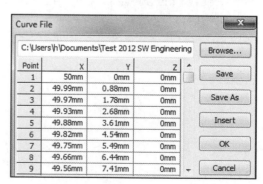

7)   **Close** the model.

8)   Open **Curve Through XYZ points 8-3**
from the ENGDESIGN-W-SOLIDWORKS/
Chapter 8 - Intelligent Modeling folder to
view the final Cam model. Curve1 is used
to fully defined Sketch1. **Close** the model.

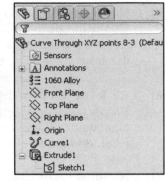

Activity: Projected Composite Curves

Create a plane and sketch for two composite curves. Note: Before you create a Loft feature, you must
insert a plane for each curve and add additional construction geometry to the sketch.

1)   Open the **Projected Composite Curves 8-1** part from the ENGDESIGN-W-SOLIDWORKS/
Chapter 8 - Intelligent Modeling folder. Two NACA aerofoil profiles are displayed created
from x-, y-, z- data.

Create a reference plane through each point of the composite curve.

2)   Click **Insert**, **Reference**
Geometry, **Plane** from the
Menu bar. The Plane
PropertyManager is displayed.

3)   Click **Front Plane** for First
Reference from the fly-out
FeatureManager.

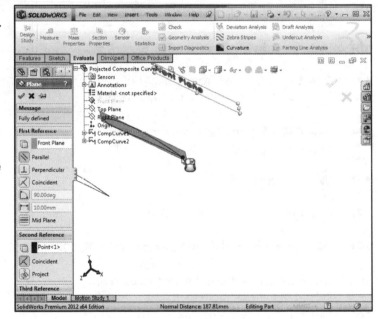

4)   Click the **end point** of
CompCurve1 in the Graphics
window as illustrated for
Second Reference. The Plane
is fully defined.

5)   Click **OK** ✅ from the Plane
PropertyManager. Plane1 is
displayed in the
FeatureManager.

Create a Reference Plane through
CompCurve 2, parallel to Plane1.

6)   Click **Insert**, **Reference**
Geometry, **Plane** from the
Menu bar. The Plane
PropertyManager is displayed.
Plane1 is the First Reference.

7) Click the **end point** of CompCurve2 in the Graphics window as illustrated,

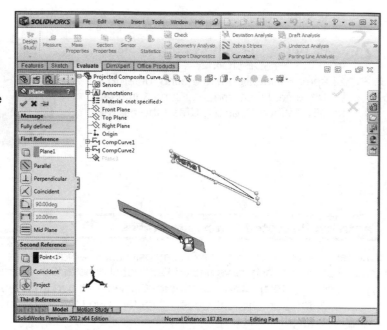

8) Click **OK** ✔ from the Plane PropertyManager. Plane2 is displayed in the FeatureManager.

9) **Rename** Plane1 to PlaneSR5.

10) **Rename** Plane2 to PlaneSR9. When preparing multiple cross sections it is important to name the planes for clarity and future modification in the design.

Convert CompCurve1 to Plane SR5 and CompCurve2 to Plane SR9.

11) Right-click **PlaneSR5** from the FeatureManager.

12) Click **Sketch** from the Context toolbar.

13) Click the **Convert Entities** Sketch tool.

14) Click **CompCurve1** from the fly-out FeatureManager in the Graphics window.

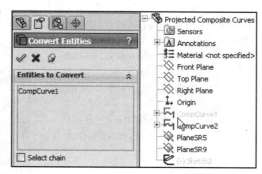

15) Click **OK** ✔ from the Convert Entities PropertyManager.

16) Click **Exit Sketch**.

17) **Rename** Sketch1 to SketchSR5.

18) Right-click **PlaneSR9** from the FeatureManager.

19) Click **Sketch** from the Context toolbar.

20) Click the **Convert Entities** Sketch tool.

21) Click **CompCurve2** from the fly-out FeatureManager in the Graphics window.

22) Click **OK** ✔ from the Convert Entities PropertyManager.

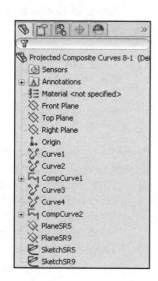

23) Click **Exit Sketch**.

24) **Rename** Sketch2 to SketchSR9.

25) Display an **Isometric** view.

26) **Show** Front Plane. View the results in the Graphics window.

27) **Close** the model.

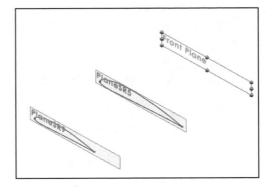

To save time in developing a series of curves, sketches can be copied and rotated.

Feature - End Conditions

Build design intent into a feature by addressing End Conditions (Blind, Through All, Up to Next, Up to Vertex, Up to Surface, Offset from Surface, Up to Body and Mid Plane) symmetry, feature selection, and the order of feature creation.

Example A: The Extruded Base feature remains symmetric about the Front Plane. Utilize the Mid Plane End Condition option in Direction 1. Modify the depth, and the feature remains symmetric about the Front Plane.

Example B: Create 34 teeth in the model. Do you create each tooth separately using the Extruded Cut feature? No.

Create a single tooth and then apply the Circular Pattern feature. Modify the Circular Pattern from 32 to 24 teeth. Think about Design Intent when you apply an End Condition in a feature during modeling. The basic End Conditions are:

- *Blind* - Extends the feature from the selected sketch plane for a specified distance; default End Condition.

- *Through All* - Extends the feature from the selected sketch plane through all existing geometry.

- *Up to Next* - Extends the feature from the selected sketch plane to the next surface that intercepts the entire profile. The intercepting surface must be on the same part.

- *Up To Vertex* - Extends the feature from the selected sketch plane to a plane that is parallel to the sketch plane and passing through the specified vertex.

- *Up To Surface* - Extends the feature from the selected sketch plane to the selected surface.

- *Offset from Surface* - Extends the feature from the selected sketch plane to a specified distance from the selected surface.

- *Up To Body* - Extends the feature up to the selected body. Use this option with assemblies, mold parts, or multi-body parts.

- *Mid Plane* - Extends the feature from the selected sketch plane equally in both directions.

Activity: End Condition options

Create Extruded Cut features using various End Condition options. Think about Design Intent of the model.

1) Open **End Condition 8-1** from the ENGDESIGN-W-SOLIDWORKS/ Chapter 8 - Intelligent Modeling folder. The FeatureManager displays two Extrude features, a Shell feature, and a Linear Pattern feature.

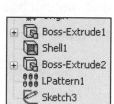

2) Click the **circumference of the front most circle**. Sketch3 is highlighted in the FeatureManager.

Create an Extruded Cut feature using the Selected Contours and Through All option.

3) Click **Extruded Cut** 🔲 from the Features toolbar. The Cut-Extrude PropertyManager is displayed.

4) **Expand** the Selected Contours box.

5) Click the **circumference of the front most circle**. Sketch3-Contour<1> is displayed in the Selected Contours box. The direction arrow points downward, if not click the Reverse Direction button.

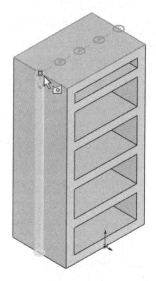

6) Select **Through All** for End Condition in Direction 1. Only the first circle of your sketch is extruded.

7) Click **OK** ✓ from the Cut-Extruded PropertyManager. Cut-Extrude1 is displayed in the FeatureManager.

Create an Extruded Cut feature using the Selected Contours and the Up To Next option. Think about Design Intent!

8) Click **Sketch3** from the FeatureManager.

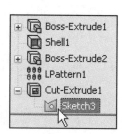

9) Click **Extruded Cut** 🔲 from the Features toolbar. The Cut-Extrude PropertyManager is displayed.

10) **Expand** the Selected Contours box.

11) Click the **circumference of the second circle** from the left. Sketch3-Contour<1> is displayed in the Selected Contours box.

12) Select **Up To Next** for End Condition in Direction 1. Only the second circle of your sketch is extruded.

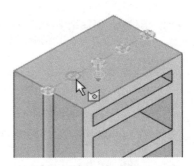

13) Click **OK** ✔ from the Cut-Extrude PropertyManager. Cut-Extrude2 is displayed in the FeatureManager.

14) **Rotate** your model and view the created cut through the first plate.

Create an Extruded Cut feature using the Selected Contours and the Up To Vertex option.
15) Click **Sketch3** from the FeatureManager.

16) Click the **Extruded Cut** 🔲 Features tool. The Cut-Extrude PropertyManager is displayed.

17) **Expand** the Selected Contours box.

18) Click the **circumference of the third circle** from the left. Sketch3-Contour<1> is displayed in the Selected Contours box.

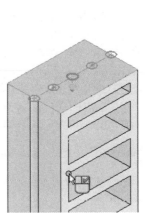

19) Select **Up To Vertex** for End Condition in Direction 1. Only the third circle of your sketch is extruded.

20) Select a **vertex point below** the second shelf as illustrated. Vertex<1> is displayed in the Vertex box in Direction 1.

21) Click **OK** ✔ from the PropertyManager. Cut-Extrude3 is displayed in the FeatureManager. The third circle has an Extruded Cut feature through the top two shelves.

Create an Extruded Cut feature using the Selected Contours and the Offset From Surface option.
22) Click **Sketch3** from the FeatureManager.

23) Click the **Extruded Cut** 🔲 Features tool.

24) **Expand** the Selected Contours box.

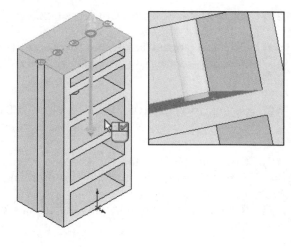

25) Click the circumference of the **fourth circle** from the left. Sketch3-Contour<1> is displayed in the Selected Contours box.

26) Select **Offset From Surface** for End Condition in Direction 1.

27) Click the **face** of the third shelf. Face<1> is displayed in the Face/Plane box in Direction1.

28) Enter **60**mm for Offset Distance.

29) Click the **Reverse offset** box if required.

30) Click **OK** ✔ from the PropertyManager. Cut-Extrude4 is displayed in the FeatureManager.

31) Click **Isometric** view. View the created features.

32) **Close** the model.

Along a Vector

In engineering design, vectors are utilized in modeling techniques. In an extrusion, the sketch profile is extruded perpendicular to the sketch plane. The Direction of Extrusion Option allows the profile to be extruded normal to the vector.

Activity: Along a Vector

Utilize a vector, created in a separate sketch, and the Direction of Extrusion to modify the extruded feature.

1) Open **Along Vector 8-1** from the ENGDESIGN-W-SOLIDWORKS/Chapter 8 - Intelligent Modeling folder. View the model in the Graphics window. The current sketch profile is extruded normal to the Top Sketch Plane.

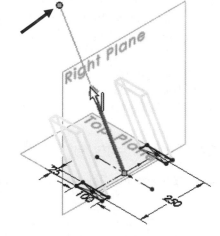

2) Edit **Boss-Extrude1** from the FeatureManager. The Boss-Extrude1 PropertyManager is displayed.

3) Select **Up to Vertex** for End Condition in Direction 1.

4) Click the **top endpoint** of Sketch Vector.

5) Click inside the **Direction of Extrusion** box.

6) Click **Sketch Vector** from the Graphics window as illustrated. The feature is extruded along the Sketch Vector normal to the Sketch Profile.

7) Click **OK** ✔ from the Boss-Extrude1 PropertyManager. View the results in the Graphics window.

8) **Close** the part.

FeatureXpert (Constant Radius)

FeatureXpert manages the interaction between fillet and draft features when features fail. The FeatureXpert, manages fillet and draft features for you so you can concentrate on your design. See Xperts Overview in SolidWorks Help for additional information.

When you add or make changes to constant radius fillets and neutral plane drafts that cause rebuild errors, the **What's Wrong** dialog appears with a description of the error. Click **FeatureXpert** in the dialog to run the FeatureXpert to attempt to fix the error.

The FeatureXpert can change the feature order in the FeatureManager design tree or adjust the tangent properties so a part successfully rebuilds. The FeatureXpert can also, to a lesser extent, repair reference planes that have lost references.

Supported features:

- *Constant radius fillets.*
- *Neutral plane drafts.*
- *Reference planes.*

Unsupported items:

- *Other types of fillets or draft features.*

- *Mirror or pattern features. When mirror or pattern features contain a fillet or draft feature, the FeatureXpert cannot manipulate those features in the mirrored or patterned copies.*

- *Library features. Fillet or draft features in a library feature are ignored by the FeatureXpert and the entire Library feature is treated as one rigid feature.*

- *Configurations and Design Tables. The FeatureXpert is not available for parts that contain these items.*

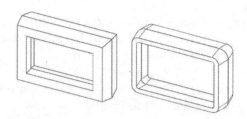

Utilize symmetry, feature order and reusing common features to build design intent into a part. Example A: Feature order. Is the entire part symmetric? Feature order affects the part.

Apply the Shell feature before the Fillet feature and the inside corners remain perpendicular.

Symmetry

An object is symmetrical when it has the same exact shape on opposite sides of a dividing line (or plane) or about a center or axis. The simplest type of Symmetry is a "Mirror" as we discussed above in this chapter.

Symmetry can be important when creating a 2D sketch, a 3D feature or an assembly. Symmetry is important because:

- Mirrored shapes have symmetry where points on opposite sides of the dividing line (or mirror line) are the same distance away from the mirror line.

- For a 2D mirrored shape, the axis of symmetry is the mirror line.

- For a 3D mirrored shape, the symmetry is about a plane.

- Molded symmetrical parts are often made using a mold with two halves, one on each side of the axis of symmetry.

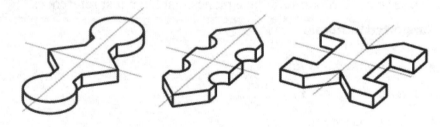

- The axis or line where two mold parts join is called a parting line.

- When items are removed from a mold, sometimes a small ridge of material is left on the object. Have you ever notice a parting line on a molded object such as your toothbrush or a screwdriver handle?

Parting line

Bodies to Mirror

When a model contains single point entities, the pattern and mirror features may create disjoint geometry and the feature will fail. For example, a cone contains a single point at its origin. To mirror the cone about a plane would create disjoint geometry. To resolve this issue, utilize Bodies to Pattern option.

Activity: Bodies to Mirror

Create a Mirror feature with the Top plane and utilize the Body to Mirror option.

1) Open **Bodies to Mirror 8-1** from the ENGDESIGN-W-SOLIDWORKS/Chapter 8 - Intelligent Modeling folder. View the model in the Graphics window.

2) Select **Mirror** from the Features toolbar. The Mirror PropertyManager is displayed.

3) **Expand** the Bodies to Mirror dialog box.

4) Click the **face of the Cone** in the Graphics window. Note the icon feedback symbol.

5) Click inside the **Mirror Face/Plane** dialog box.

6) Click **Top Plane** from the fly-out FeatureManager.

7) Uncheck the **Merge solids** box. (You cannot merge these two cones at a single point).

8) Click **OK** from the Mirror PropertyManager. View the results in the Graphics window. Mirror1 is displayed in the FeatureManager.

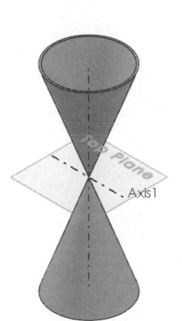

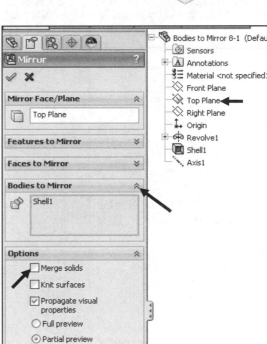

Planes

Certain types of models are better suited to design automation than others. When setting up models for automation, consider how they fit into an assembly and how the parts might change when automated.

Create planes so that sketches and features can be referenced to them (coincident, up to surface, etc.) This provides the ability to dimension for the plane to be changed and the extrusion extended with it. When placed in an assembly, other components can be mated to a plane so that they move with consideration to the parts altered.

Incorporating planes into the design process prepares the model for future changes. As geometry becomes more complex, additional sketch geometry(construction lines, circles) or reference geometry (axis, planes) may be required to construct planes.

For example, a sketched line coincident with the silhouette edge of the cone provides a reference to create a plane through to the outer cone's face. An axis, a plane and an angle creates a new plane through the axis at a specified angle.

Activity: Angle Plane

Create a plane at a specified angle through an axis.

1)   Open **Angle Planes 8-1** from the ENGDESIGN-W-SOLIDWORKS/Chapter 8 - Intelligent Modeling folder. View the model in the Graphics window.

Create an Angle Plane.

2)   Click **Insert, Reference Geometry, Plane** from the Menu bar. The Plane PropertyManager is displayed.

3)   Click **Axis1** in the Graphics window as illustrated. Axis1 is displayed in the First Reference box.

4)   Click **Right Plane** from the fly-out FeatureManager. Right Plane is displayed in the Second Reference box.

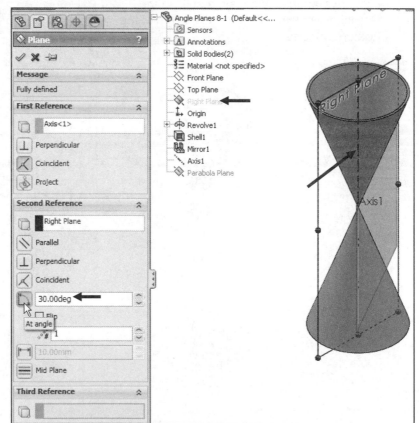

Set the angle of the Plane.
5) Click the **At angle** box.

6) Enter **30** for Angle. The Plane is fully defined.

7) Click **OK** ✓ from the Plane PropertyManager. Plane1 is displayed in the FeatureManager.

8) **Close** the Model.

As geometry becomes more complex in mechanical design, planes and conical geometry can be combined to create circular, elliptical, parabolic and hyperbolic sketches with the Intersection Curve Sketch Curve.

Conic Sections and Planes

Conic sections are paths traveled by planets, satellites, electrons and other bodies whose motions are driven by inverse-square forces. Once the path of a moving body is known, information about its velocity and forces can be derived. Using planes to section cones creates circular, elliptical, parabolic, hyperbolic and other cross sections are utilized in engineering design.

The Intersection Curve tool provides the ability to open a sketch and create a sketched curve at the following kinds of intersections: a plane and a surface or a model face, two surfaces, a surface and a model face, a plane and the entire part, and a surface and the entire part.

Activity: Conic Section

Apply the Intersection Curve tool. Create a sketched curve
1) Open **Conic Section 8-1** from the ENGDESIGN-W-SOLIDWORKS/Chapter 8 - Intelligent Modeling folder. View the model in the Graphics window. Create a Sketch.

2) Right-click **Plane1 Circle** from the FeatureManager.

3) Click **Sketch** ⌐ from the Context toolbar. The Sketch toolbar is displayed in the CommandManager.

4) Click **Tools, Sketch Tools, Intersection Curve** from the Menu bar. The Intersection Curve PropertyManager is displayed.

5) Click the **top face** of the cone as illustrated. Face<1> is displayed in the Selected Entities dialog box.

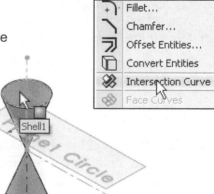

6) Click **OK** ✓ from the Intersection Curve PropertyManager. Click **OK** ✓ .

7) Click **Exit Sketch**.

8) **Hide** Plane1 Circle. The intersection of the cone with the plane creates a circle. **Close** the model. You can use the resulting sketched intersection curve in the same way that you use any sketched curve.

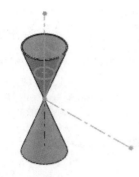

Assembly

Utilizing symmetry, reusing common parts and using the Mate relation between parts builds the design intent into an assembly. For example: Reuse geometry in an assembly. The assembly contains a linear pattern of holes. Insert one screw into the first hole. Utilize the Component Pattern feature to copy the machine screw to the other holes.

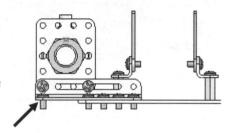

Assembly Visualization

In an assembly, the designer selects material based on cost, performance, physical properties, manufacturing processes, and sustainability. The Assembly Visualization tool provides the ability to rank components based on the values of their custom properties, and activate a spectrum of colors that reflects the relative values of the properties for each component. The default custom properties are:

- *SW-Density*
- *SW-Mass*
- *SW-Material*
- *SW-Rebuild Time*
- *SW-Surface Area*
- *SW-Volume*
- *Description*
- *PartNo*
- *Quantity*

To reduce the weight in the overall assembly, first review the component with the greatest mass. Alternative materials or changes to the geometry that still meet safely requirements may save in production and shipping costs. When comparing mass, volume and other properties with assembly visualization, utilize similar units.

Activity: Assembly Visualization

Apply the Assembly Visualization tool to an assembly. View your options

1) Open **Assembly Visualization 8-1** from the ENGDESIGN-W-SOLIDWORKS/Chapter 8 - Intelligent Modeling folder. View the assembly in the Graphics window. Material was assigned to each component in the assembly.

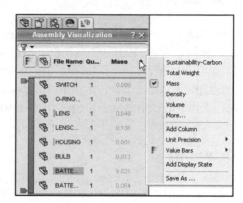

2) Click the **Assembly Visualization** tool from the Evaluate Tab in the CommandManager. The Assembly Visualization PropertyManager is displayed.

3) Click the **expand arrow** to the right of Mass as illustrated to display the default visualization properties. Mass is selected by default.

4) **Explore** the available tabs and SolidWorks Help for additional information. **Close** the model.

SolidWorks Sustainability - Assembly

With the SolidWorks Sustainability you can determine Life Cycle Assessment (LCA) properties for the components in an assembly. The LCA Sustainability properties are:

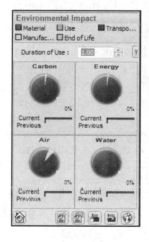

- *Air*

- *Carbon*

- *Energy*

- *Manufacturing Location*

- *Material Class*

- *Material Specific*

- *Total Carbon*

- *Total Energy*

- *Total Water*

- *Use Location*

By combining SolidWorks Sustainability and Assembly Visualization, you can determine the components in the assembly that contain the greatest carbon footprint and review materials with similar properties to reduce CO_2 emissions. See chapter on SolidWorks Sustainability and SolidWorks Help for additional information.

You need SolidWorks Professional or SolidWorks Premium to access SolidWorks Sustainability for an assembly. Every license of SolidWorks 2011 provides access to SolidWorks SustainabilityXpress for a part.

MateXpert

The MateXpert is a tool that provides the ability to identify mate problems in an assembly. You can examine the details of mates that are not satisfied, and identify groups of mates which over define the assembly. If the introduction of a component leads to multiple mate errors, it may be easier to delete the component, review the design intent, reinsert the component and then apply new mates. See SolidWorks Help for additional information.

Drawing

Utilize dimensions, tolerance and notes in parts and assemblies to build the design intent into a drawing.

Example A: Tolerance and material in the drawing. Insert an outside diameter tolerance +.000/-.002 into the Pipe part. The tolerance propagates to the drawing.

Define the Custom Property Material in the Part. The Material Custom Property propagates to your drawing.

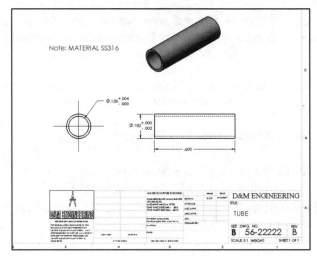

DimXpert

DimXpert for parts is a set of tools you use to apply dimensions and tolerances to parts according to the requirements of the ASME Y14.41-2003 standard.

DimXpert dimensions show up in a different color to help identify them from model dims and reference dims. DimXpert dims are the dimensions that are used when calculating tolerance stack-up using TolAnalyst.

DimXpert applies dimensions in drawings so that manufacturing features (patterns, slots, pockets, etc.) are fully-defined.

DimXpert for parts and drawings automatically recognize manufacturing features. What are manufacturing features? Manufacturing features are *not SolidWorks features*. Manufacturing features are defined in 1.1.12 of the ASME Y14.5M-1994 Dimensioning and Tolerancing standard as: "The general term applied to a physical portion of a part, such as a surface, hole or slot.

The DimXpertManager provides the following selections:
Auto Dimension Scheme ✤, **Show Tolerance Status** ⚭,
Copy Scheme ✤, **TolAnalyst Study** ⊡.

💡 Care is required to apply DimXpert correctly on complex surfaces or with some existing models. See SolidWorks help for detail information on DimXpert with complex surfaces.

Activity: DimXpert 8-1

Apply the DimXpert tool. Apply Prismatic and Geometric options

1) Open **DimXpert 8-1** from the ENGDESIGN-W-SOLIDWORKS/ Chapter 8 - Intelligent Modeling folder.

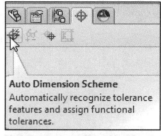

2) Click the **DimXpertManager** tab as illustrated.

3) Click the **Auto Dimension Scheme** tab from the DimXpertManager. The Auto Dimension Scheme PropertyManager is displayed.

4) Click the **Prismatic** box.

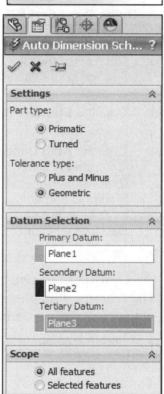

5) Click the **Geometric** box.

Select the Primary Datum.
6) Click the **front face** of the model. Plane1 is displayed in the Primary Datum box.

Select the Secondary Datum.
7) Click **inside** the Secondary Datum box.

8) Click the **left face** of the model. Plane2 is displayed in the Secondary Datum box.

Select the Tertiary Datum.
9) Click **inside** the Tertiary Datum box.

10) Click the **bottom face** of the model. Plane3 is displayed in the Tertiary Datum box. Accept the default options.

11) Click **OK** ✓ from the Auto Dimension Scheme PropertyManager. View the results in the Graphics window and in the DimXpertManager.

12) **Close** all models. Do not save the updates.

🔆 Right-click Delete to delete the Auto Dimension Scheme in the DimXpert PropertyManager.

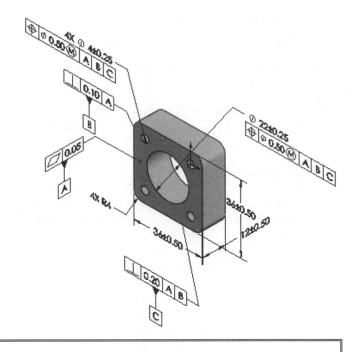

Activity: DimXpert 8-2

Apply DimXpert: Geometric option. Edit a Feature Control Frame.

1) Open **DimXpert 8-2** from the ENGDESIGN-W-SOLIDWORKS/Chapter 8 - Intelligent Modeling folder.

2) Click the **DimXpertManager** ✛ tab.

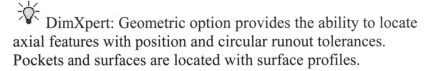

3) Click the **Auto Dimension Scheme** ✛ tool from the DimXpertManager. The Auto Dimension PropertyManager is displayed. Prismatic and Plus and Minus is selected by default. In this section, you will select the Geometric option.

🔆 DimXpert: Geometric option provides the ability to locate axial features with position and circular runout tolerances. Pockets and surfaces are located with surface profiles.

4) Click the **Geometric** box as illustrated. Note: the Prismatic box is checked.

Create the Primary Datum.
5) Click the **back face** of the model. Plane1 is displayed in the Primary Datum box.

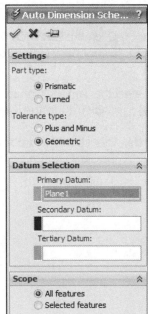

Select the Secondary Datum.

6) Click **inside** the Secondary Datum box.

7) Click the **left face** of the model. Plane2 is displayed in the Secondary Datum box.

Select the Tertiary Datum.

8) Click **inside** the Tertiary Datum box.

9) Click the **top face** of the model. Plane3 is displayed in the Tertiary Datum box.

10) Click **OK** ✅ from the Auto Dimension PropertyManager.

11) Click **Isometric view** from the Heads-up View toolbar. View the Datum's, Feature Control Frames, and Geometric tolerances. All features are displayed in green.

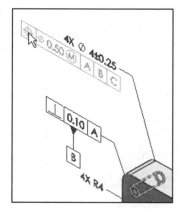

Edit a Feature Control Frame.

12) **Double-click** the illustrated Position Feature Control Frame. The Properties dialog box is displayed.

Modify the 0.50 tolerance.

13) Click **inside** the Tolerance 1 box.

14) **Delete** the existing text.

15) Enter **0.25**.

16) Click **OK** from the Properties dialog box.

17) **Repeat** the above procedure for the second Position Feature Control Frame. View the results.

18) **Close** the model. Do not save the model.

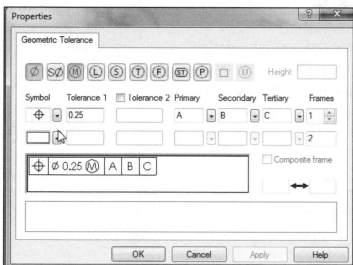

Project Summary

In this project, you performed short step by step tutorials to understanding some of the available tools in SolidWorks to perform intelligent modeling. Intelligent modeling is incorporating design intent into the definition of the Sketch, Feature, Part and Assembly or Drawing document.

All designs are created for a purpose. Design intent is the intellectual arrangement of features and dimensions of a design. Design intent governs the relationship between sketches in a feature, features in a part and parts in an assembly or drawing document.

The SolidWorks definition of design intent is the process in which the model is developed to accept future modifications. Models behave differently when design changes occur.

You create Explicit Dimension Driven and Parametric Equation Driven Curves and generated curves using the Curve Through XYZ Points tool for a NACA aerofoil and CAM data text file.

You also applied SketchXpert, DimXpert and FeatureXpert.

View the .mp4 files located on the DVD in the book to better understand the potential of the Equation Driven Curve tool.

Definitions

DimXpert: DimXpert speeds the process of adding reference dimensions by applying dimensions in drawings so that manufacturing features, such as patterns, slots, and pockets, are fully-defined. You select a feature's edge to dimension, then DimXpert applies all associated dimensions in that drawing view for the feature.

Feature: The first feature you create in a part is the base. This feature is the basis on which you create the other feature. The base feature can be an extrusion, a revolve, a sweep, a loft, thickening of a surface, or a sheet metal flange. However, most base features are extrusions.

FeatureXpert: The FeatureXpert, manages fillet and draft features. The FeatureXpert can change the feature order in the FeatureManager design tree or adjust the tangent properties so a part successfully rebuilds.

Involute: The path of the endpoint of a cord as it is pulled straight (held taut) and unwrapped from a circular disk.

Sketch: The sketch is the basis for a 3D model. Create a sketch on any of the default planes (**Front Plane**, **Top Plane**, and **Right Plane**), or a created plane.

SketchXpert: SketchXpert resolves over-defined sketches and proposes possible solution sets.

Xperts: The SolidWorks Xperts are tools that help novices use SolidWorks like experts, without having to understand the details of the software. The Xperts take care of the details of software functionality. Using Xperts, you can focus on your design intent while the Xperts focus on what the software should do to achieve it.

Notes:

Appendix

Engineering Changer Order (ECO)

D&M	Engineering Change Order			ECO # _____ Page 1 of __
	☐ Hardware			Author
	☐ Software			Date
Product Line	☐ Quality			Authorized Mgr.
	☐ Tech Pubs			Date
Change Tested By				
Reason for ECO(Describe the existing problem, symptom and impact on field)				

D&M Part No.	Rev From/To	Part Description	Description	Owner

ECO Implementation/Class		Departments	Approvals	Date
All in Field	☐	Engineering		
All in Test	☐	Manufacturing		
All in Assembly	☐	Technical Support		
All in Stock	☐	Marketing		
All on Order	☐	DOC Control		
All Future	☐			
Material Disposition		ECO Cost		
Rework	☐	DO NOT WRITE BELOW THIS LINE(ECO BOARD ONLY)		
Scrap	☐	Effective Date		
Use as is	☐	Incorporated Date		
None	☐	Board Approval		
See Attached	☐	Board Date		

This text follows the ASME Y14 Engineering Drawing and Related Documentation Practices for drawings. Display of dimensions and tolerances are as follows:

TYPES of DECIMAL DIMENSIONS (ASME Y14.5M)			
Description:	**UNITS: MM**	**Description:**	**UNITS: INCH**
Dimension is less than 1mm. Zero precedes the decimal point.	0.9 0.95	Dimension is less than 1 inch. Zero is not used before the decimal point.	.5 .56
Dimension is a whole number. Display no decimal point. Display no zero after decimal point.	19	Express dimension to the same number of decimal places as its tolerance. Add zeros to the right of the decimal point.	1.750
Dimension exceeds a whole number by a decimal fraction of a millimeter. Display no zero to the right of the decimal.	11.5 11.51	If the tolerance is expressed to 3 places, then the dimension contains 3 places to the right of the decimal point.	

TABLE 1
TOLERANCE DISPLAY FOR INCH AND METRIC DIMENSIONS (ASME Y14.5M)

DISPLAY:	UNITS: INCH:	UNITS: METRIC:
Dimensions less than 1	.5	0.5
Unilateral Tolerance	$1.417^{+.005}_{-.000}$	$36^{\;0}_{-0.5}$
Bilateral Tolerance	$1.417^{+.010}_{-.020}$	$36^{+0.25}_{-0.50}$
Limit Tolerance	.571 .463	14.50 11.50

SolidWorks Keyboard Shortcuts

Listed below are the pre-defined keyboard shortcuts in SolidWorks:

Action:	Key Combination:
Model Views	
Rotate the model horizontally or vertically:	**Arrow** keys
Rotate the model horizontally or vertically 90 degrees.	**Shift** + **Arrow** keys
Rotate the model clockwise or counterclockwise	**Alt** + left of right **Arrow** keys
Pan the model	**Ctrl** + **Arrow** keys
Magnifying Glass	**g**
Zoom in	**Shift + z**
Zoom out	**z**
Zoom to fit	**f**
Previous view	**Ctrl + Shift + z**
View Orientation	
View Orientation menu	**Spacebar**
Front view	**Ctrl + 1**
Back view	**Ctrl + 2**
Left view	**Ctrl + 3**
Right view	**Ctrl + 4**
Top view	**Ctrl + 5**
Bottom view	**Ctrl + 6**
Isometric view	**Ctrl + 7**
NormalTo view	**Ctrl + 8**
Selection Filters	
Filter edges	**e**
Filter vertices	**v**
Filter faces	**x**
Toggle Selection Filter toolbar	**F5**
Toggle selection filters on/off	**F6**
File menu items	
New SolidWorks document	**Ctrl + n**
Open document	**Ctrl + o**
Open From Web Folder	**Ctrl + w**
Make Drawing from Part	**Ctrl + d**
Make Assembly from Part	**Ctrl + a**
Save	**Ctrl +s**
Print	**Ctrl + p**
Additional shortcuts	
Access online help inside of PropertyManager or dialog box	**F1**
Rename an item in the FeatureManager design tree	**F2**
Rebuild the model	**Ctrl + b**
Force rebuild – Rebuild the model and all its features	**Ctrl + q**
Redraw the screen	**Ctrl + r**

Cycle between open SolidWorks document	**Ctrl + Tab**
Line to arc/arc to line in the Sketch	**a**
Undo	**Ctrl + z**
Redo	**Ctrl + y**
Cut	**Ctrl + x**
Copy	**Ctrl + c**
Additional shortcuts	
Paste	**Ctrl + v**
Delete	**Delete**
Next window	**Ctrl + F6**
Close window	**Ctrl + F4**
Selects all text inside an Annotations text box	**Ctrl + a**

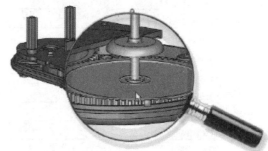

 In a sketch, the Esc key un-selects geometry items currently selected in the Properties box and Add Relations box. In the model, the Esc key closes the PropertyManager and cancels the selection.

 Press the **g** key to activate the Magnifying glass tool. Use the Magnifying glass tool to inspect a model and make selections without changing the overall view.

 Press the **s** key to view/access previous command tools in the Graphics window.

Windows Shortcuts

Listed below are the pre-defined keyboard shortcuts in Microsoft Windows:

Action:	Keyboard Combination:
Open the Start menu	Windows Logo key
Open Windows Explorer	Windows Logo key + E
Minimize all open windows	Windows Logo key + M
Open a Search window	Windows Logo key + F
Open Windows Help	Windows Logo key + F1
Select multiple geometry items in a SolidWorks document	Ctrl key (Hold the Ctrl key down. Select items.) Release the Ctrl key.

CSWA Certification - Introduction

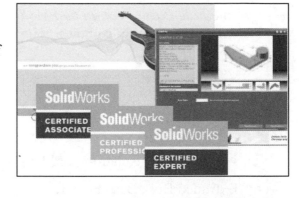

DS SolidWorks Corp. offers various stages of certification representing increasing levels of expertise in 3D CAD design as it applies to engineering: *Certified SolidWorks Associate CSWA, Certified SolidWorks Professional CSWP and Certified SolidWorks Expert CSWE* along with specialty fields in Simulation, Sheet Metal, and Surfacing.

The CSWA Certification indicates a foundation in and apprentice knowledge of 3D CAD design and engineering practices and principles. The main requirement for obtaining the CSWA certification is to take and pass the on-line proctored 180 minute exam (minimum of 165 out of 240 points). The new CSWA exam consists of fourteen questions in the following categories and subject areas:

- *Drafting Competencies*: (Three questions - multiple choice - 5 points each).

  A00006: Drafting Competencies - To create drawing view 'B' it is necessary to select drawing view 'A' and insert which SolidWorks view type?

  - Questions on general drawing views: Projected, Section, Break, Crop, Detail, Alternate Position, etc.

- *Basic Part Creation and Modification*: (Two questions - one multiple choice / one single answer - 15 points each).

  B22001: Basic Part (Hydraulic Cylinder Half) - Step 1
  Build this part in SolidWorks.
  (Save part after each question in a different file in case it must be reviewed)

  Unit system: MMGS (millimeter, gram, second)
  Decimal places: 2
  Part origin: Arbitrary
  All holes through all unless shown otherwise.
  Material: Aluminium 1060 Alloy

  - Sketch Planes:
    - Front, Top, Right
  - 2D Sketching:
    - Geometric Relations and Dimensioning
  - Extruded Boss/Base Feature
  - Extruded Cut Feature
  - Modification of the basic part (sketch dimension and or feature)

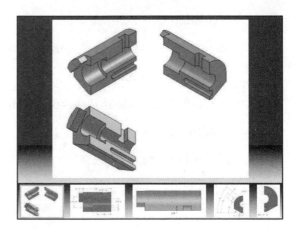

- *Intermediate Part Creation and Modification*: (Two questions - one multiple choice / one single answer - 15 points each).

  - Sketch Planes:

    - Front, Top, Right

  - 2D Sketching:

    - Geometric Relations and Dimensioning

  - Extruded Boss/Base Feature

  - Extruded Cut Feature

  - Revolved Boss/Base Feature

  - Mirror and Fillet Feature

  - Circular and Linear Pattern Feature

  - Plane Feature

  - Modification of the intermediate part:

    - Sketch, Feature, Pattern, etc.

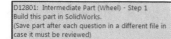

D12801: Intermediate Part (Wheel) - Step 1
Build this part in SolidWorks.
(Save part after each question in a different file in case it must be reviewed)

Unit system: MMGS (millimeter, gram, second)
Decimal places: 2
Part origin: Arbitrary
All holes through all unless shown otherwise.
Material: Aluminium 1060 Alloy

A = 134.00
B = 890.00

Note: All geometry is symmetrical about the plane represented by the line labeled F'' in the M-M Section View.

What is the overall mass of the part (grams)?

- *Advanced Part Creation and Modification:* (Three questions - one multiple choice / two single answers - 15 points each).

  - Sketch Planes:

    - Front, Top, Right, Face, Created Plane, etc.

    - 2D Sketching or 3D Sketching

  - Sketch Tools:

    - Offset Entities, Convert Entitles, etc.

  - Extruded Boss/Base Feature

  - Extruded Cut Feature

  - Revolved Boss/Base Feature

  - Mirror and Fillet Feature

  - Circular and Linear Pattern Feature

  - Shell Feature

  - Plane Feature

C12801: Advanced Part (Bracket) - Step 1
Build this part in SolidWorks.
(Save part after each question in a different file in case it must be reviewed)

Unit system: MMGS (millimeter, gram, second)
Decimal places: 2
Part origin: Arbitrary
All holes through all unless shown otherwise.
Material: AISI 1020 Steel
Density = 0.0079 g/mm^3

A = 64.00
B = 20.00
C = 26.50

What is the overall mass of the part (grams)?

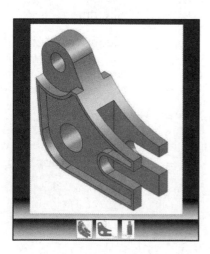

- More Difficult Geometry Modifications

- *Assembly Creation and Modification*: (Two different assemblies - four questions - two multiple choice / two single answers - 30 points each).

  - Insert the first (fixed) component

  - Insert all needed components

  - Insert Standard Mates

  - Modification of key parameters in the assembly

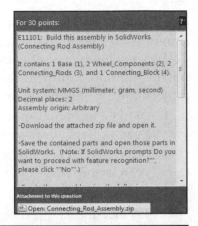

Note: Download the needed components during the exam for create the assembly.

A total score of 165 out of 240 or better is required to obtain your CSWA Certification.

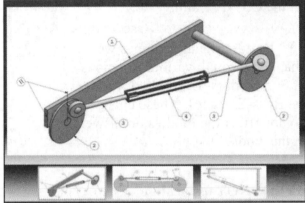

☀ You are allowed to answer the questions in any order you prefer. Use the Summary Screen during the CSWA exam to view the list of all questions you have or have not answered.

During the exam, use the control keys at the bottom of the screen to:

- *Show the Previous the Question.*

- *Reset the Question.*

- *Show the Summary Screen.*

- *Move to the Next Question.*

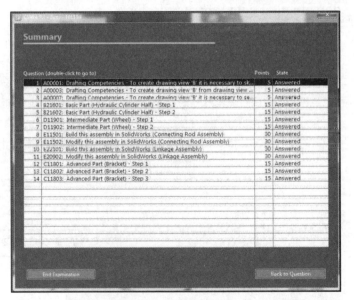

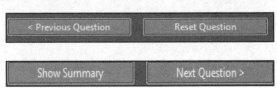

During the exam, SolidWorks provides the ability to click on a detail view below (as illustrated) to obtain additional information and dimensions during the exam.

💡 No SimulationXpress (CosmosXpress) questions are on the CSWA exam.

💡 No Sheetmetal questions are on the CSWA exam.

💡 FeatureManager names were changed through various revisions of SolidWorks. Example: Extrude1 vs. Boss-Extrude1. These changes do not affect the models or answers in this book.

💡 Do NOT use feature recognition when you open the downloaded components for the assembly in the CSWA exam.

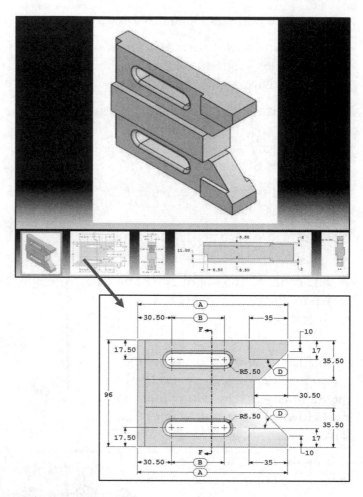

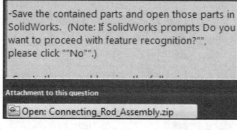

-Save the contained parts and open those parts in SolidWorks. (Note: If SolidWorks prompts Do you want to proceed with feature recognition?"", please click ""No"".)

Attachment to this question

Open: Connecting_Rod_Assembly.zip

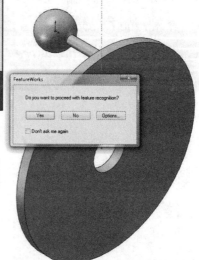

💡 View the .pdf on the DVD for a sample CSWA exam.

CSWASampleExam2010.pdf

Intended Audience

The intended audience for the CSWA exam is anyone with a minimum of 6 - 9 months of SolidWorks experience and basic knowledge of engineering fundamentals and practices. SolidWorks recommends that you review their SolidWorks Tutorials on Parts, Assemblies, and Drawings as a prerequisite and have at least 45 hours of classroom time learning SolidWorks or using SolidWorks with basic engineering design principles and practices.

To prepare for the CSWA exam, it is recommended that you first perform the following:

- Take a CSWA exam preparation class or review a text book written for the CSWA exam.

- Complete the SolidWorks Tutorials

- Practice creating models from the isometric working drawings sections of any Technical Drawing or Engineering Drawing Documentation text books.

- Complete the sample CSWA exam in a timed environment, available at www.solidworks.com or visit www.dmeducation.net for exam details and models.

Additional reference to help you prepare:

- Planchard & Planchard, Official Certified SolidWorks® Professional (CSWP) Certification Guide with Multimedia DVD, 2011, 2010, SDC Pub., Mission, KS.

- SolidWorks website.

About the CSWA exam

Most CAD professionals today recognize the need to become certified to prove their skills, prepare for new job searches, and to learn new skill, while at their existing jobs.

Specifying a CSWA or CSWP certification on your resume is a great way to increase your chances of landing a new job, getting a promotion, or looking more qualified when representing your company on a consulting job.

Helpful On-Line Information

The SolidWorks URL: http://www.solidworks.com contains information on Local Resellers, Solution Partners, Certifications, SolidWorks users groups, and more.

Access 3D ContentCentral using the Task Pane to obtain engineering electronic catalog model and part information.

Use the SolidWorks Resources tab in the Task Pane to obtain access to Customer Portals, Discussion Forums, User Groups, Manufacturers, Solution Partners, Labs, and more.

Helpful on-line SolidWorks information is available from the following URLs:

- http://www.dmeducation.net

   Helpful tips, tricks and what's new in SolidWorks.

- http://www.mechengineer.com/snug/

   News group access and local user group information.

- http://www.nhcad.com

   Configuration information and other tips and tricks.

- http://www.dmeducation.net

   Helpful tips, tricks and what's new in SolidWorks.

- http://www.topica.com/lists/SW

   Independent News Group for SolidWorks discussions, questions and answers.

*On-line tutorials are for educational purposes only. Tutorials are copyrighted by their respective owners.

Index

*asmdot, 5-66
*sldlfp, 2-34
*sldprt, 2-34
*stl, 5-83
.drwdot, 3-20
.prtdot, 1-32
.slddrw, 3-25
.sldprt, 1-37
3 Point Arc Sketch tool, 4-44
3D ContentCentral, 1-22, 2-69
3MMCAPSCREW, 2-60
4MMCAPSCREW Part, 2-39
99-FBM8-1-25, 3-72
Accelerator key, 1-21

A

Add a dimension, 1-71
Add Derived Configuration, 6-63
Add Dimension, 1-48
Add File Location - Design Library, 2-33
Add Ins, 1-22
Add Revision - Drawing, 3-60
Advanced Mate, 2-17
  Angle mate, 2-18, 2-95
  Distance mate, 2-17, 2-93
  Mate Alignment option, 2-18
  Linear/Linear Coupler mate, 2-17
  Path mate, 2-17
  Symmetric mate, 2-17
  Width mate, 2-17
Advanced Mode - New Document, 1-27
Air Acidification, 7-13, 7-16
Along A Vector, 8-26
Angle mate, 2-17, 2-95
Angular Dimension, 1-92, 4-29
ANGULAR Tolerance, 3-15
Animate Exploded view, 2-54
Animation - eDrawing, 5-84
Animation Controller, 2-54
Annotation tools, 3-48
  Balloons, 3-69
  Bill of Materials, 3-71
  Center Mark, 3-48

  Centerline, 3-48
  Insert Magnetic line, 3-70
  Note, 3-57
Annotations - Drawing, 3-10
Annotations folder, 6-32
ANSI - Drafting Standard, 1-6, 1-7, 1-32, 1-33
ANSI - IPS Assembly Template, 5-66
ANSI - MMGS Assembly Template, 5-66
Appearance Callout tool, 6-27
Appearance, 1-87
Appearances command, 4-56
Apply Fixtures, 7-10
Apply Loads, 7-10
Arc Condition - Drawing, 3-52
ASME Y14.3 Standard, 1-40
ASM-IN-ANSI, 5-72
ASM-MM-ANSI, 6-12
ASM-MM-ISO Assembly Template, 5-67
Assembly - Fixed Component, 2-9
Assembly - Float Component, 2-9
Assembly FeatureManager Syntax, 2-13
Assembly Modeling Approach, 2-5
Assembly Template, 5-65
Assembly tool, 2-9
  Collision Detection, 2-30
  Exploded View, 2-51
  Feature Driven Component Pattern, 2-62
  Insert Component, 2-9, 2-11, 5-67
  Insert Components, 5-71
  Mate, 5-73, 5-75, 5-78
  Rotate Component, 2-24, 2-33, 5-79
  View Mates, 2-27
  Assembly Visualization, 8-32
Auto Balloon Annotation tool, 3-69
Auxiliary View - Drawing, 3-29, 3-30

B

B Landscape size Drawing Template, 3-5
Back - View Orientation, 1-13
Balloon - Drawing, 3-69
B-ANSI-MM, 3-20, 3-23
B-ANSI-MM, 3-68
Base Feature, 1-26, 1-38

Base Sketch, 1-26
BATTERANDPLATE Sub-assembly, 5-70
  Insert BATTERY Part, 5-70
  Insert BATTERYPLATE Part, 5-71
BATTERY Part, 4-15
  Extruded Base feature, 4-19
  Extruded Boss feature, 4-26
  Extruded Cut feature, 4-23, 4-24
  Fillet feature, 4-21
  Measure tool, 4-30
  Second Fillet feature, 4-25
BATTERYPLATE Part, 4-33
  Extruded Boss feature, 4-36
  Fillet feature, 4-37
  Offset Entities, 4-36
  Save As command, 4-34
Bends PropertyManager, 6-102
Bill of Materials - Drawing, 3-71
Binary .stl file, 5-84
Binder Clip Assembly, 2-91
Blind - End Condition, 1-51, 1-71, 5-23
Boss-Extrude PropertyManager, 1-51, 4-19
Bottom - View Orientation, 1-13
Bottom-up Assembly Modeling, 2-5, 2-6
Boundary Condition, 7-4
BOX Assembly overview, 6-8, 6-13
Box-Select, 1-76
BRACKET Part, 6-89
  Base Flange/Tab feature, 6-90
  Edge Flange feature, 6-93
  Extruded Cut feature, 6-92
  In-Context feature, 6-91
  Mirror Component feature, 6-98, 6-100
  Miter Flange feature, 6-95
BULB Part, 4-57
  Circular Pattern feature, 4-65, 4-68
  Dome feature, 4-64
  Extruded Cut feature, 4-67
  Revolved Base feature, 4-58, 4-60, 4-61
  Revolved Cut Thin feature, 4-62
  Spline feature, 4-61

C

CABINET Part, 6-37
  Additional Pattern options, 6-60
  Edge Flange feature, 6-42, 6-44
  Extruded Base feature, 6-39
  Extruded Cut feature, 6-59
  Flat State feature, 6-42, 6-44, 6-62
  Formed state feature, 6-62

Fully formed state feature, 6-42, 6-44
Hem feature, 6-43
Hole Wizard feature, 6-45, 6-46
Insert Bends feature, 6-41
Linear Pattern feature, 6-45, 6-47
Linear Pattern sketch, 6-57
Louver Forming feature, 6-53
Manufacturing Considerations, 6-54
RIP feature, 6-40
Sheet metal drawing, 6-64
Sheet metal Library feature, 6-49
Shell feature, 6-39
Sketch Driven Pattern feature, 6-61
Cam mate, 2-18
CAPANDLENS Sub-assembly, 5-72
  Insert LENSANDBULB Assembly, 5-73
  Insert LENSCAP Sub-assembly, 5-72
  Insert ORING Part, 5-72
Carbon Footprint, 7-13, 7-16
Center Mark Drawing tool, 3-48
Center Mark PropertyManager, 3-48, 3-49
Center Rectangle Sketch tool, 4-18, 5-15
Centerline Drawing tool, 3-48
Centerline Sketch tool, 1-92, 2-40, 3-34
Centerpoint Arc Sketch tool, 4-51
Certification - CSWA, 1-30
Certified SolidWorks Associate CSWA, 1-30
Chamfer feature, 1-73, 1-75, 1-84, 2-42
  Angle distance, 1-73
  Distance distance, 1-73
  Vertex, 1-73
Chamfer PropertyManager, 1-73, 2-42
Circle Sketch tool, 1-58, 1-71
Circular Pattern feature, 4-65, 4-68
C-Landscape, 6-65
Clear All Filters Command, 2-24, 2-26, 5-69
Clearance fit, 2-48
Close All Command, 1-34, 2-8
Close Profile, 3-33
Closed Contour, 6-25, 6-29
Coincident mate, 2-16, 2-25, 2-37, 5-68
Coincident Relation, 1-72, 1-81, 4-50, 5-48
Coincident Sketch Relation, 1-46
Coincident SmartMate, 2-44
Coincident/Concentric SmartMate, 2-44, 2-62
Collapse Display Pane, 1-88
Collinear Relation, 3-34
Collision Detection tool, 2-30
Color tool, 6-27
Color, 1-87

CommandManager, 1-15
  Assembly Document, 1-17
  Custom tab, 1-16
  Drawing Document, 1-16
  Part Document, 1-15
Component Properties, 2-15
  Fixed sign, 2-15
  Minus sign, 2-15
  None, 2-15
  Plus sign, 2-15
  Question mark, 2-15
Component State, 2-14
  Hidden, 2-14
  Lightweight, 2-14
  Out-of-Date Lightweight, 2-14
  Rebuild, 2-14
  Resolved Assembly, 2-14
  Resolved part, 2-14
  Smart Component, 2-14
  Suppressed, 2-14
Concentric mate, 2-16, 2-19, 2-25, 5-71
ConfigurationManager tab, 2-54, 5-11
Confirmation Corner, 1-12
  Active Sketch, 1-12
  Exit Sketch, 1-12
Conic Sections and Planes, 8-31
Consolidated Slot toolbar, 1-95
Consolidated toolbar, 1-11, 1-12
Context-Sensitive toolbar, 1-15, 1-16, 1-17
Context toolbar, 1-70
Convert Entities Sketch tool, 1-75, 4-48
Copy Command, 1-57, 1-59
Copy/Paste Command, 1-80
Coradial Relation, 1-86
Corner Rectangle feedback icon, 1-45
Corner Rectangle Sketch tool, 1-45
Cosmetic Thread feature, 1-222
Counberbore Hole, 4-48
Counter Weight Assembly, 2-90
Create a Sketch, 1-75
Crop View Drawing tool, 3-33
CSWA Certification, 1-30
Ctrl-Tab, 2-10
Curve Through XYZ Points feature, 8-18
Curves, 8-16, 8-18
  Explicit Equation Driven, 8-14
  Parametric Equation Driven, 8-16
  Projected Composite, 8-21
  Through XYZ Points, 8-18

Custom Document Templates, 1-31
Custom Properties - Drawing Template, 3-6
Custom Properties - Drawing, 3-64, 3-65
Custom Property - Drawing, 3-6, 3-64, 3-14
CUSTOM-B, 3-19
CUSTOMER Assembly, 2-5, 2-71
Customize CommandManager, 1-15, 1-16
Customize Dialog box, 1-16
Customize FeatureManager, 1-20
Customizing Toolbar, 4-69
Cut-Extrude PropertyManager, 1-61
Cut-Revolve PropertyManager, 4-64, 5-25

D
Datum Planes, 1-40
d-cutout feature, 6-49
Deactivate a tool, 1-46
Defeature tool, 5-4
Deformation (-Displacement-), 7-8, 7-11
Degrees of Freedom, 2-6, 2-12
Delete a mate, 2-65
Delete feature, 4-34
Derived Configuration, 6-63
Design Checklist, 4-71
Design Intent, 1-47, 4-6, 8-4
  Assembly, 4-8
  Drawing, 4-8
  Feature, 4-7
  Part, 4-8
  Sketch, 4-6, 8-4
Design Library, 1-22, 2-32
  3D ContentCentral, 1-22
  Add File Location, 2-33
  SolidWorks Content, 1-22
  Toolbox, 1-22
Design Table - Configurations, 5-10, 5-11
Design Table, 5-9, 5-10, 5-11
Detail Drawing - 3-37
Detail View - Drawing, 3-29, 3-32, 3-37
DFMXpress Wizard, 7-26
DFMXpress, 1-127, 7-26
Dimension - Angular, 1-92, 4-29
Dimension - Dual, 4-14
Dimension - Fillet, 3-55, 3-56
Dimension - Modify, 1-49
Dimension - Radial, 1-96
Dimension - Vertical, 1-101
Dimension driven by equations, 8-11
Dimension Feedback icon, 1-12

Dimension Holes - Drawing, 3-45
Dimension PropertyManager, 1-48
   Leaders tab, 1-48
   Other tab, 1-48
   Value tab, 1-48
Dimension Text Dialog box, 3-47
Dimension the Sketch, 4-19
Dimetric - View Orientation, 1-13
DimXpert, 8-34
Displacement (-Res disp-), 7-8, 7-11
Display Modes - Drawing, 3-35
Display Pane, 1-88
Display Style, 1-13
   Hidden Lines Removed, 1-13
   Hidden Lines Visible, 1-13
   Shaded With Edges, 1-13
   Shaded, 1-13
   Wireframe, 1-13
Display Temporary Axes, 1-100
Display/Delete Relations PropertyManager,
   1-85
Display/Delete Relations Sketch tool, 1-85
Distance mate, 2-17, 2-26, 5-74, 5-77
Dock CommandManager, 1-16, 1-18
Document Properties, 1-33, 2-10, 3-9, 4-11
Dome feature, 4-64, 5-18
Dome PropertyManager, 4-64, 5-18
Double - View Orientation, 1-13
Draft Analysis - Mold Design, 4-75
Draft Angle Command, 4-29
Draft feature, 5-41
Draft option, 5-22
Draft PropertyManager, 5-42, 5-43
Drafting Standard, 1-6, 1-32
DraftXpert PropertyManager, 5-42
DraftXpert, 4-75, 5-42
Drawings, 3-1
   Add Revision, 3-60
   ANGULAR Tolerance, 3-15
   Annotations, 3-10
   Arc Condition, 3-52
   As Phantom lines, 3-35, 3-36
   Auto Balloon, 3-69
   Auxiliary View, 3-29, 3-30
   Center Marks, 3-48
   Centerline, 3-48
   Custom Properties, 3-14, 3-64
   Detail View, 3-29, 3-32
   Detail, 3-37
   Dimension Holes, 3-45
   Display Modes, 3-35

Edge Properties, 3-27
Exploded View, 3-67
Font, 3-14
Formatting toolbar, 3-15
Gap, 3-40, 3-41
General Note, 3-56
Hide View dimension, 3-43
Hole Callouts, 3-45
Insert Dimension into a view, 3-42
Insert Magnetic lines, 3-70
Leader line, 3-40, 3-41
Logo, 3-17
Move Dimension to a Different View, 3-44
Move Dimensions, 3-40, 3-41
Move views, 3-26, 3-28
Parametric Note, 3-56
Reference Dimension, 3-53
Revision Table, 3-59
Section View, 3-29, 3-32
Sheet Format, 3-19
Sheet Scale, 3-68
Show in exploded state command, 3-69
System Properties, 3-12, 3-26
Tangent Edges, 3-36
Title Block, 3-14
Tolerance Note, 3-15
View Properties, 3-27
Drawing Layers, 3-11
Drawing mode - Edit Sheet Format, 3-5
Drawing mode - Edit Sheet, 3-5
Drawing Sheet Format, 3-5
Drawing Template, 3-5, 3-21
Drawing Template - B (ANSI) Landscape, 3-5
Drawing Template - Custom Properties, 3-6
Drawing tool - Auxiliary View, 3-30
Drawing tool - Detail View, 3-32
Drawing tool - Model Items, 3-39
Drawing tool - Model View, 3-24
Drawing tool - Section View, 3-31
DrawnBy Custom Property, 3-63
DrawnDate Custom Property, 3-63
Drop-down menu, 1-11
Dual Dimension, 4-14
Dynamic Mirror Sketch tool, 1-92, 1-94, 4-66

E

Edge Feedback icon, 1-12
Edit - Menu bar menu, 1-11
Edit Component tool, 6-24, 6-91
Edit feature tool, 1-82, 1-87, 4-19

Edit Material Command, 1-104
Edit Part Color, 1-87
Edit Sheet Format mode, 3-5
Edit Sheet mode, 3-5
Edit Sketch Command, 1-80
Edit Sketch Plane Sketch tool, 1-84
Edit Sketch tool, 1-85
eDrawing, 5-83
  Animation, 5-84
  Save, 5-84, 5-85
Education Edition, 1-14
Elastic Modulus, 7-5, 7-6
Elasticity Theory, 7-4
Elements, 7-4
End Condition, 8-23
  Blind, 1-51, 1-71, 5-23, 8-23
  Mid Plane, 1-94, 8-24
  Offset from Surface, 8-24
  Through All, 1-61, 4-67, 8-23
  Up To Body, 8-24
  Up To Surface, 5-54, 8-24
  Up To Vertex, 8-23, 8-26
Energy Consumption, 7-13, 7-16
ENGDESIGN-W-SOLIDWORKS folder, 1-8
Equal Relation, 1-59, 4-28
Equation Driven Curve Sketch tool, 8-14
Equations - Dimension by equations, 8-11
Equations, 6-19, 6-82, 8-11
Exit Sketch tool, 1-81
Explicit Equation Driven Curve, 8-14
Explode PropertyManager, 2-51
Exploded View - Drawing, 3-67
Exploded View Assembly tool, 2-51
Export File, 5-83
External References, 6-25, 6-28, 6-78
Extruded Boss/Base feature, 1-51, 1-71, 4-19
Extruded Cut feature, 1-57, 1-61, 1-74, 1-77

F

Face Feedback icon, 1-12
Face Filter, 2-24
Factor of Safety, 7-6, 7-11
Fasteners, 1-56
FEA tool, 7-3
FEA, 7-5
Feature - Cosmetic Thread, 1-122
Feature - End Condition, 8-23
Feature - Fillet, 3-55
Feature - Rename, 1-30, 1-47, 1-53
Feature Driven Component Pattern tool, 2-62

Feature Driven Component Pattern, 6-74
Feature Statistics, 5-19
Feature tools, 1-26
  Chamfer, 1-75, 1-84, 2-42
  Circular Pattern, 4-65, 4-68
  Curve Through XYZ Points, 8-18
  Dome, 4-64, 5-18, 5-41
  Edit, 1-82, 4-19
  Extruded Boss/Base, 1-51, 1-71, 4-19
  Extruded Cut, 1-57, 1-74, 1-77
  Fillet Multiple radius, 4-37, 4-39
  Fillet, 1-63, 1-64, 4-22
  Freeform, 5-18
  Helix/Spiral, 5-27, 5-45
  Hole Wizard, 1-65, 1-72, 4-48
  Instant3D, 1-53
  Linear Component Pattern, 2-64
  Linear Pattern 1-102, 5-56
  Lofted Boss/Base, 5-13, 5-17, 5-34, 5-37
  Mirror, 1-98, 5-61, 8-29
  Plane, 5-14, 5-25, 8-21
  Revolved Boss/Base, 2-42, 4-46
  Revolved Cut, 4-63
  Revolved Thin Cut feature, 5-24, 5-25
  Rib, 5-56
  Shell, 4-46, 5-23, 5-39
  Swept Boss/Base, 5-7, 5-30
FeatureManager Design Tree, 1-19
  Customize, 1-20
  Fly-out, 1-21
  New Folder, 2-68
  Split, 1-20
FeatureManager tabs, 1-19
  ConfigurationManager, 1-19
  DimXpertManager, 1-19
  PropertyManager, 1-19
FeatureXpert, 4-26, 8-27
FEM, 7-4
FH-M4-25ZI, 6-70
FH-M5-30ZI, 6-70
File - Menu bar menu, 1-11
File Locations - Add Folder, 3-20
File Locations - SolidWorks, 1-31
File Management, 1-6, 1-7
Fillet Feature - Multiple radius, 4-37, 4-39
Fillet feature, 1-63, 1-64, 3-55, 4-21, 4-22
Fillet PropertyManager, 1-64, 4-22
FilletXpert PropertyManager, 1-64
FilletXpert, 1-64
Filter icon, 2-27

Finish Custom Property, 3-65, 3-66
Finite Element Analysis tool, 7-3
First Angle Projection, 1-42, 1-43
Fit the Model, 1-52
Fixed Component, 2-9
Fixtures, 7-10
FLASHLIGHT Assembly, 4-4, 5-64, 5-76
   Insert BATTERYANDPLATE, 5-79
   Insert CAPANDLENS, 5-78
   Insert HOUSING, 5-76
   Insert SWITCH, 5-76
   Interference, 5-82
Float CommandManager, 1-16, 1-18
Float Component, 2-9
Fly-out FeatureManager, 1-21
Force Normal, 7-10
Four View - View Orientation, 1-13
Freeform feature, 5-18
Front - View Orientation, 1-13
Fully Defined - Sketch state, 1-50, 1-113
Fully Defined Sketch tool, 8-5
Fully Defined Sketch, 1-50, 1-113

G
Gap - Drawing, 3-40, 3-41
Gear mate, 2-18
General Note - Drawing, 3-56
Geometric Pattern box, 1-98, 1-103
Geometric Relation - Coincident, 1-72, 1-81,
   4-50, 5-48
Geometric Relation - Collinear, 3-34
Geometric Relation - Coradial, 1-85
Geometric Relation - Equal, 1-59, 4-28
Geometric Relation - Horizontal, 1-60, 2-43
Geometric Relation - Intersection, 4-52, 5-60
Geometric Relation - Perpendicular, 4-67
Geometric Relation - Pierce, 5-8, 5-29
Geometric Relation - Tangent, 5-16
Geometric Relation - Vertical, 1-99, 4-52
Geometry Relation - Pierce, 5-46, 5-52
Getting Started, 1-30
Global Values, 6-17
Graphics Window, 1-28
Grid/Snaps, 1-14, 1-28
GUIDE Part, 1-89
   First Extruded Cut feature, 1-94
   Hole Wizard feature, 1-101
   Interference Problem, 2-58
   Linear Pattern feature, 1-102
   Mass Properties, 1-104, 1-105

Materials Dialog box, 1-104
Mirror feature, 1-98
Modify, 3-22
Second Extruded Cut feature, 1-99
Section View, 2-56
Vertical Relation, 1-99
GUIDE Drawing, 3-23
   Additional Feature, 3-54
   Auxiliary View, 3-30.3-43
   Balloons, 3-69
   Bill of Materials, 3-71
   Center Marks, 3-48
   Centerlines, 3-48,   3-50
   Crop View, 3-33
   Custom Properties, 3-14
   Detail View, 3-32, 3-37
   Display Modes, 3-35
   Document Properties, 3-61
   Drawing Template, 3-7
   Exploded View, 3-67
   General Note, 3-56
   Insert Dimensions, 3-39
   Insert Drawing Views, 3-24
   Layers, 3-11
   Logo, 3-17
   Modify Dimensions, 3-50
   Move Dimension, 3-44
   Move Views, 3-26
   Parametric Note, 3-56
   Part Number, 3-61
   Partial Auxiliary View, 3-33
   Revision Table, 3-59
   Section View, 3-31
   Sheet Format, 3-12
   Sheet Properties, 3-9
Guided Actuators,
GUIDE-ROD Assembly, 2-5, 2-5, 2-11
   Coincident mate, 2-25
   Collision Detection, 2-30
   Design Library, 2-32
   Exploded View, 2-51
   Feature Driven Component Pattern, 2-62
   First Concentric mate, 2-19
   Insert Component, 2-11, 2-23
   Insert Flange Bolt mates, 2-35
   Linear Component Pattern, 2-11, 2-23, 2-64
   Mate Errors, 2-27
   Mate the PLATE Component, 2-23
   Mate the ROD Component, 2-18
   Modify Component Dimensions, 2-31

PLATE - Coincident mate, 2-25
PLATE - Parallel mate, 2-26
Redefining mates, 2-64
ROD - Parallel mate, 2-22
Second Concentric mate, 2-25
Tolerance and Fit, 2-49

H
Heads-up View toolbar, 1-13
  3D Drawing View, 1-14
  Apply Scene, 1-14
  Display Style, 1-13
  Edit Appearance, 1-13
  Hide/Show Items, 1-13
  Previous View, 1-13
  Rotate View, 1-14
  Section View, 1-13
  View Orientation, -13
  View Setting, 1-14
  Zoom to Area, 1-13
  Zoom to Fit, 1-13
Helix/Spiral feature, 5-27, 5-45
Helix/Spiral PropertyManager, 5-27, 5-45
Help - Menu bar menu, 1-11
Help - SolidWorks, 1-29
Hidden Lines Removed - Display Style, 1-13
Hidden Lines Visible - Display Style, 1-13
Hide Annotations Command, 3-56
Hide Command, 1-70
Hide Component, 2-22
Hide Feature Dimensions Command, 5-19
Hide FeatureManager, 1-19
Hide Plane, 4-29
Hide Reference Geometry, 2-66
Hide View Dimension, 3-43
Hinge Mate, 2-18
Hole Callout - Drawing, 3-45
Hole Specification PropertyManager, 4-49
Hole Wizard - 2D Sketch, 1-125
Hole Wizard - 3D Sketch, 1-125
Hole Wizard feature, 1-65, 1-72, 4-48
HOOK Part, 5-90
Horizontal Dimension, 5-36
Horizontal Relation, 1-60, 2-43
Horizontal Sketch Relation, 1-46
HOUSING Part, 5-31
  Draft feature, 5-41
  Extruded Base feature, 5-33
  Extruded Boss feature, 5-38

Extruded Cut feature, 5-53
First Rib and Linear Pattern feature, 5-55
Handle with Swept feature, 5-48, 5-52
Lofted Boss feature, 5-34, 5-37
Mirror feature, 5-61
Second Extruded Boss feature, 5-40
Second Rib feature, 5-58
Shell feature, 5-39
Thread with Swept feature, 5-43, 5-47

I
IGES Components, 6-70
Image Quality, 6-33
In-Context features, 6-10, 6-24, 6-28, 6-29
Injection Molded Process, 4-32
InPlace Mates, 6-10, 6-24
Insert a mate, 2-19
Insert Component Assembly tool, 2-9, 2-11, 2-23, 5-67, 6-89
Insert Design Table, 5-10
Insert Dimension into a Drawing View, 3-42
Insert Dimension, 1-48
Insert Drawing Views, 3-24, 3-26
Insert magnet lines, 3-70
Instant3D feature, 1-53
Intelligent Modeling Techniques, 8-1
  Along A Vector, 8-26
  Assembly Visualization, 8-32
  Curve Through XYZ Points feature, 8-18
  DimXpert, 8-34
  End Conditions, 8-23
  Equations, 8-11
  Explicit Equation Driven Curve, 8-14
  FeatureXpert, 8-27
  Fully Defined Sketch tool, 8-5
  MateXpert, 8-34
  Parametric Equation Driven Curve, 8-16
  Planes, 8-30
  SketchXpert, 8-18
  Sustainability, 8-33
  Symmetry, 8-28
Interference Detection tool, 5-82
Interference fit, 2-48
Interference Issues, 5-82
Interference Problem, 2-58
Internal Force, 7-4
Intersection Relation, 4-52, 5-49, 5-60
IPS, 1-6, 1-32, 4-12
ISO - Drafting Standard, 1-6, 1-7, 1-32
Isometric - View Orientation, 1-13

K - L

K factor, 6-91
Layer - Modify, 3-50
Layer Color, 3-11
Layers toolbar, 3-11
Layout sketch, 6-13, 6-29
LCA, 7-13, 7-14
Leader line - Drawing, 3-40, 3-41
Left - View Orientation, 1-13
LENS Part, 4-42
    Extruded Boss/Base feature, 4-47
    Hole Wizard feature, 4-48
    Revolved Base feature, 4-43
    Revolved Boss Thin feature, 4-51
    Second Extruded Boss/Base feature, 4-53
    Shell feature, 4-46
    Third Extruded Boss/Base feature, 4-56
    Transparent Optical Property, 4-55
LENSANDBULB Sub-assembly, 5-65
    Insert LENS Part, 5-67
LENSCAP Part, 5-20
    Extruded Base feature, 5-22
    Extruded Cut feature, 5-23
    Helix/Spiral feature, 5-27
    Plane feature, 5-26
    Revolved feature
    Revolved Thin Cut feature, 5-24, 5-25
    Shell feature, 5-23
    Swept Boss/Base feature, 5-30
Life Cycle Assembly, 7-13, 7-14
Limit mate Assembly, 2-93
Line Properties PropertyManager, 1-47
Line Segments, 1-92
Line Sketch tool, 1-75, 1-92, 4-66
Linear Component Pattern Feature, 2-64
Linear Motion, 2-6
Linear Pattern - Geometry pattern, 6-48
Linear Pattern - Instance, 6-48
Linear Pattern feature, 1-102, 5-56
Linear Pattern sketch tool, 6-58
Linear/Linear Coupler mate, 2-17
Linearly Elastic Material, 7-6
Link to Property Command, 3-64
Loads, 7-4, 7-10
Locked Reference, 2-15
Lofted Boss/Base feature, 5-13, 5-17, 5-34
Logo - Drawing, 3-17

M

Magnet lines tool, 3-70
Magnifying glass, 1-14
Manufacturing Considerations, 1-106
Manufacturing Design Issues, 4-83
Mass Properties tool, 1-104, 1-105
Mate Alignment option, 2-17
Mate Assembly tool, 2-19, 2-25, 5-73, 5-79
Mate Error message, 2-27
Mate Pop-up toolbar, 2-20
Mate PropertyManager, 2-16
Mate Types, 2-16
Material Custom Property, 3-65
Material Dialog box, 1-105
Material Editor, 1-104
MateXpert, 8-34
Measure tool, 4-30
Mechanical Mates, 2-18
    Screw mate assembly, 2-94
    Cam mate, 2-18
    Gear mate, 2-18
    Hinge mate, 2-18
    Rack Pinion mate, 2-18
    Screw mate, 2-18
    Universal Joint mate, 2-18
Menu bar menu, 1-10
    Edit, 1-11
    File, 1-11
    Help, 1-11
    Insert, 1-11
    Pin, 1-11
    Tools, 1-11
    View, 1-11
    Window, 1-11
Menu bar toolbar, 1-10
    New, 1-10
    Open, 1-10
    Options, 1-10
    Print, 1-10
    Rebuild, 1-10
    Save, 1-10
    Select, 1-10
    Undo, 1-10
MGPM12-1010 Assembly, 2-70, 2-71, 7-7
MGPM12-1010_ROD Component, 7-9
Mid Plane End Condition, 1-94
Mirror Components feature, 6-100

Mirror Components PropertyManager, 6-100
Mirror Entities sketch tool, 5-36, 5-51
Mirror feature, 1-98, 5-61, 8-29
Mirror PropertyManager, 1-98
Mirror sketch tool, 1-92
MirrorBRACKET Part, 6-101
　Fold feature, 6-104
　Insert Bends feature, 6-102
　Jog feature, 6-105
　Unfold feature, 6-103
Miter Flange PropertyManager, 6-96
MMGS, 1-6, 1-32, 1-33, 4-12
Model Items Drawing tool, 3-39, 3-43
Model Items PropertyManager, 3-39, 3-43
Model View Drawing tool, 3-24
Model View PropertyManager, 3-7, 3-24
Modify Dimension, 1-49
Modify Sketch Dialog box, 6-50
Modify Sketch tool, 6-50, 6-52, 6-53
Mold - Parting Line feature, 4-77
Mold - Parting Surface feature, 4-78
Mold - Scale feature, 4-77
Mold - Shut-off Surfaces, 4-78
Mold - Tooling Split feature, 4-79
Mold Base, 4-73
Mold Design - Draft analysis, 4-74
Mold Design - Parting Lines, 4-74
Mold Design - Parting Surfaces, 4-74
Mold Design - Ruled Surface, 4-74
Mold Design - Tooling Split, 4-74
Mold toolbar, 4-75
Mold Tooling Design, 4-73
MOLD-TOOLING Part, 4-76
　Parting Lines feature, 4-77
　Parting Surface feature, 4-78
　Scale tool feature, 4-77
　Shut-off Surface feature, 4-78
　Tooling Split feature, 4-79
MOTHERBOARD - Hole Wizard feature, 6-76
MOTHERBOARD - Insert Component, 6-22
MOTION LINKAGE Assembly, 2-101
Motion Study, 1-25
　Animation, 1-25
　Basic Motion, 1-25
　MotionManager, 1-25
Mouse Gesture Wheel, 1-20
Mouse Pointer, 1-12, 1-45
Move a component, 2-21
Move Component PropertyManager, 2-30

Move Dimension in View, 3-40, 3-41
Move Dimension to Different View, 3-44
Move views - Drawing, 3-26, 3-28
Multi-body Parts, 4-40
MY-TEMPLATES folder, 1-8, 4-12
MY-TEMPLATES tab, 1-37, 3-19, 3-23

N
Named View, 1-79
New Assembly Document, 2-8
New Document Command, 1-10
New Drawing, 3-7
New Folder - FeatureManager, 2-68
New Folder, 1-7
New Motion Study, 1-26
New Part Document, 1-26, 1-27, 1-37, 1-70
New View tool, 1-14, 1-79
Newton, 7-6
Node, 7-4, 7-5
Non-Uniform Rational B-Spline, 4-57, 4-60
Normal To View tool, 1-99
Note Annotation tool, 3-57
Novice Mode - New Document, 1-27
NURB, 4-57, 4-60

O
Offset Entities Sketch tool, 4-23, 4-36
Offset from Surface - End Condition, 8-24
Open Document Command, 1-10
Options Command, 1-10
Orientation Dialog box, 1-79
Origin, 1-28, 1-45, 1-46, 1-71, 5-33, 5-36
O-RING Part, 5-7
　Design Table, 5-9, 5-10, 5-11
　Large, 5-10, 5-11
　Medium, 5-10, 5-11
　Small, 5-10, 5-11
　Swept Base feature, 5-7
Orthographic Projection, 1-40
Out of Context, 2-15
Over defined - Sketch state, 1-50, 1-113

P - Q
Pack and Go, 2-77
Parallel mate, 2-16, 2-22, 2-26, 2-36
Parametric Driven Curve Sketch tool, 8-16
Parametric Note - Drawing, 3-56, 3-57
Part Template, 1-32
Part Template - PART-IN-ANSI, 4-11

Part Template - PART-MM-ISO, 4-11
Partial Auxiliary View, 3-33
PART-IN-ANSI - Part Template, 4-11
PART-MM-ANSI, 1-32, 1-33
PART-MM-ISO - Part Template, 4-11
Path mate, 2-17
PEM Fasteners, 6-70, 6-71
Perpendicular Relation, 4-67
Phantom Lines - Drawing, 3-35, 3-36
PhotoView 360, 1-127
Pierce Relation, 5-8, 5-29, 5-46, 5-52
Pin - Menu bar menu, 1-11
Plane feature, 5-14, 5-25, 8-21
Plane feature tool, 5-14, 5-25, 8-21
Plane PropertyManager, 5-14, 5-25
PLATE Part Overview, 1-35
  Add Dimension, 1-49
  Equal Relation, 1-59
  Extruded Base feature, 1-38, 1-44, 1-51
  Extruded Cut feature, 1-57, 1-61
  Fillet feature, 1-63
  Hole Wizard feature, 1-65
  Horizontal Relation, 1-60
  Modify Dimension, 1-49, 1-52
  Rename, 1-53
Point Sketch tool, 1-65, 5-17, 6-60
Polygon Sketch tool, 2-43
Power Trim Option, 1-76
Power Trim tool, 4-66
POWERSUPPLY Part, 6-28
Primary Datum Plane, 1-40
Print Document command, 1-10
Profile Lines, 4-59
Projected Composite Curve, 8-21
PROJECTS folder, 1-8, 2-10
Propagate icon, 6-96
Publish eDrawing, 5-84

R
Rack Pinion mate, 2-18
Radial dimension, 1-96
Rapid Prototype, 5-83
Rebuild Document Command, 1-10
Rebuild Errors, 1-83
Redefining Mates, 2-64
Reference Dimension - Drawing, 3-53
Reference is Broken, 2-15
Reference Plane, 5-14

Reference Planes, 1-40
Reference Planes - Default, 1-28
Reference Planes - Front, 1-28
Reference Planes - Right, 1-28
Reference Planes - Top, 1-28
Reference Point, 6-99
Registry, 1-31
Remove an Instance, 6-42
Remove Trailing zeros, 3-46
Rename a Drawing View, 3-30
Rename Feature, 1-30, 1-47, 1-53, 4-22
Rename Sketch, 1-30, 1-47, 1-53
Replace Component, 6-79
Resolved Component, 2-13
Revision Table - Drawing, 3-59
Revision Table PropertyManager, 3-59
Revolve PropertyManager, 2-42
Revolved Base feature, 2-42
Revolved Boss Thin Feature, 4-51
Revolved Boss/Base feature, 4-43, 4-46
Revolved Cut Feature, 4-63, 5-25
Revolved Cut Thin Feature, 4-62
Revolved Thin Cut feature, 5-24, 5-25
Rib feature, 5-56
Right - View Orientation, 1-13
Right-Click Select Command, 1-46, 1-49
Right-Click, 1-11
ROD Part Overview, 1-68, 1-70
  Chamfer feature, 1-73, 1-74, 1-84
  Coincident Relation, 1-73
  Convert Entities, 1-74, 1-75
  Copy/Paste Command, 1-80
  Coradial Relation, 1-85
  Design Changes / Rollback bar, 1-81, 1-84
  Edit Color, 1-87
  Extruded Base feature, 1-71
  Extruded Cut feature, 1-74, 1-77, 1-83
  Hole Wizard feature, 1-72
  Recover for Rebuild errors, 1-83
  Trim Entities, 1-76
  View Orientation, 1-79
Rollback Bar, 1-81, 1-84
Rotate a component, 2-21, 2-24
Rotate Component Assembly tool, 2-24, 5-79
Rotate Component tool, 2-33, 6-72
Rotate Part, 4-47
Rotate View, 1-65
Rotational Motion, 2-6

S

Save - eDrawing, 5-84, 5-85
Save a Drawing Document, 3-16
Save As Command, 1-33, 2-10, 4-34
Save As Copy option, 2-59
Save Command, 1-37
Save Document Command, 1-10
Save Drawing Document, 3-25
Save Part Template, 1-33
Save Sheet Format, 3-19
Screw mate Assembly, 2-94
Screw mate, 2-18
Secondary Datum Plane, 1-40
Section View, 5-80
  Assembly, 2-56
  Drawing, 3-29, 3-31
  Part, 4-50
  Drawing tool, 3-31
Seed feature, 4-65
Select Command, 1-46, 1-49
Select Other tool, 2-75, 2-94
Select Other, 2-75, 2-94
Select Sketch Entities Command, 1-10
Selection Filter, 2-24, 4-31
Selection Filter toolbar, 2-24, 5-69
Sensors tool, 1-19
Set Document Properties, 4-11, 4-12
Set the Layer Color, 3-11
Set to Resolved, 6-68
Shaded - Display Style, 1-13
Shaded With Edges - Display Style, 1-13
Sheet Format, 3-12, 3-19
Sheet Format/Size Dialog box, 3-7
Sheet Metal, 6-34
  Base Flange/Tab feature, 6-94
  Bends feature, 6-34
  Break Corner/Corner Trim feature, 6-95
  CABINET Part, 6-37
  Design Library, 6-49
  Design Table, 6-86
  Edge Flange feature, 6-42, 6-44, 6-93
  Flat State feature, 6-42
  Fold feature, 6-104
  Fully formed state feature, 6-42
  Hem feature, 6-43
  Insert Bends feature, 6-41, 6-102
  Jog feature, 6-105
  K factor, 6-36, 6-41

  Louver Forming tool feature, 6-53
  Manufacturing Consideration, 6-55
  Miter Flange feature, 6-95
  Relief feature, 6-37
  RIP feature, 6-40
  Unfold feature, 6-102
Sheet Properties Dialog box, 3-9
Sheet Properties, 3-9, 3-26
Sheet Scale - Drawing, 3-68
Sheet Scale, 3-9
Shell feature, 4-46, 5-23, 5-39
Short Cut Key, 4-69
Shortcut toolbar, 1-21
Show Annotations Command, 3-56
Show Feature Dimension, 6-17
Show Feature Dimensions Command, 5-19
Show in exploded state - Drawing, 3-69
Silhouette Edge, 5-24, 5-60
SimulationXpress, 7-3
Single - View Orientation, 1-13
Sketch - Design Intent, 8-4
Sketch - Edit Sketch, 1-80
Sketch - Fully Defined, 1-50
Sketch - Rename, 1-30, 1-47, 1-53
Sketch Driven Pattern feature, 6-61
Sketch Fillet Sketch tool, 5-35
Sketch Plane, 1-70
Sketch Relations - View, 1-46
Sketch state - Fully defined, 1-50, 1-113
Sketch state - Over defined, 1-50, 1-113
Sketch state - Under defined, 1-50, 1-113
Sketch states, 8-5
Sketch status - Black, 4-20
Sketch status - Blue, 4-20
Sketch status - Light Blue, 4-20
Sketch status - Red, 4-20
Sketch tool - 1-58
  3 point Arc, 4-44
  Center Rectangle, 4-18, 5-15
  Centerline, 1-82, 2-40, 3-34
  Centerpoint Arc, 4-51
  Circle, 1-71
  Convert Entities, 1-75, 4-48
  Corner Rectangle, 1-45
  Dimension driven by equation, 8-11
  Display/Delete Relations, 1-85
  Dynamic Mirror, 1-92, 1-94, 4-66
  Edit Sketch Plane, 1-84

Edit Sketch, 1-85
Equation Driven Curve, 8-14
Exit Sketch, 1-81
Fully Defined, 8-5
Line, 1-75, 1-92, 4-66
Linear Pattern, 6-58
Mirror Entities, 5-36, 5-51
Mirror, 1-92
Modify, 6-50, 6-52
Offset Entities, 4-23, 4-26, 4-36, 4-39
Parametric Equation Sketch tool, 8-16
Point, 1-65, 5-17, 6-60
Polygon, 2-43
Sketch Fillet, 5-35
SketchXpert, 8-8
Smart Dimension, 1-48, 1-71
Spline, 3-33, 4-61
Straight Slot, 1-95
Tangent Arc, 5-50, 5-60
Trim Entities, 1-76, 4-66
SketchXpert, 1-50, 8-8
Smart Dimension Sketch tool, 1-48, 1-71
SmartMates, 2-44
Socket Head Cap Screw Part, 2-39
3MMCAPSCREW, 2-60
4MMCAPSCREW - Chamfer, 2-42
4MMCAPSCREW - Extruded Cut 2-44
4MMCAPSCREW - Revolved Boss, 2-42
4MMCAPSCREW - Save As Copy, 2-60
Solid Features, 5-20
SolidWorks - File Locations, 1-31
SolidWorks CommandManager, 1-15
SolidWorks DFMXpress, 7-26
SolidWorks Document - Advanced Mode, 1-17
SolidWorks Document - Novice Mode, 1-27
SolidWorks eDrawing, 5-83
SolidWorks Help, 1-29
SolidWorks Resources tab, 1-9
SolidWorks SimulationXpress, 7-3
Deformation (-Displacement-), 7-8
Displacement (-Res disp-), 7-8
Factor of Safety, 7-8
Fixtures tab, 7-7
Loads tab, 7-8
Material tab, 7-8
Results tab, 7-8
Run tab, 7-8
Stress (-vonMises-), 7-8

Welcome tab, 7-7
SolidWorks Sustainability, 7-13
SolidWorks SustainabilityXpress, 7-13
SolidWorks Tutorials, 1-30
SolidWorks Web Help, 1-29
Spline Sketch tool, 3-33, 4-61
Split FeatureManager, 1-20, 2-54
Standard Mates, 2-16
Angle mate, 2-17
Coincident mate, 2-16, 2-25, 2-37, 5-68
Concentric mate, 2-16, 2-19, 2-25, 5-71
Distance mate, 2-17, 2-26, 5-74, 5-77
Lock mate, 2-17
Mate Alignment option, 2-17
Parallel mate, 2-16, 2-22, 2-26, 2-36
Tangent mate, 2-16
Start a SolidWorks Session, 1-9
Stereo Lithography File Format, 5-83
STL File Format, 5-83
Straight Slot Sketch tool, 1-95
Strain, 7-5
Strength of Materials, 7-4
Stress (-vonMises-), 7-11, 7-8
Stress vs. Strain Plot, 7-6
Stress, 7-4, 7-5
Sub-folders, 1-7
Supply Content, 2-70
Suppress Components, 6-32
Suppress Folder, 2-68
Suppressed Components, 2-68
Sustainability, 7-13
Air Acidification, 7-13
Carbon Footprint, 7-13
Energy Consumption, 7-13
Water Eutrophication, 7-13
SustainabilityXpress, 7-13
Wizard - Air Acidification, 7-20
Carbon Footprint, 7-20
Energy Consumption, 7-20
Environmental Impact, 7-19
Wizard - Manufacturing, 7-19
Wizard - Materials, 7-19
Wizard - Process, 7-19
Wizard - Report, 7-25
Wizard - Water Eutrophication, 7-21
Swept Boss/Base feature, 5-7, 5-30
Swept Path, 5-8
Swept Profile, 5-8, 5-8

SWEPT-CUT CASE Part, 5-92
SWITCH Part, 5-13
   Dome feature, 5-18
   Lofted Base feature, 5-13, 5-17
Symmetric mate, 2-17
Symmetry - Bodies to mirror, 8-28
System Feedback icons, 1-12
   Dimension, 1-12
   Edge, 1-12
   Face, 1-12
   Vertex, 1-12
System Options, 1-31
System Options - File Location, 4-13

T
Tags, 1-28
Tangent Arc Sketch tool, 5-50, 5-60
Tangent Edges, 1-78, 4-13
Tangent Edges - Drawing, 3-36
Tangent mate, 2-16
Tangent Relation, 5-16
Task Pane, 1-22
   Appearances, Scenes, Decals 1-22, 1-24
   Custom Properties, 1-22, 1-24
   Design Library, 1-22
   Document Recovery, 1-22, 1-24
   File Explorer, 1-22, 1-23
   Search, 1-22, 1-23
   SolidWorks Resources, 1-22
   View Palette, 1-22, 1-23
Templates - Custom, 1-31, 1-32
Temporary Axes, 1-100, 3-34
Tertiary Datum Plane, 1-40
Tetrahedral Element, 7-4
Third Angle Projection, 1-41, 1-42
Three Rotational, 2-6
Three Translational, 2-6
Through All End Condition, 1-61, 4-67
Tip of the Day, 1-9
Title Block - Drawing, 3-14
Tolerance and Fit, 2-47
Tolerance Note, 3-15
Tolerance/Precision, 1-108
Toolbox, 1-22
Top - View Orientation, 1-13
Top-down Assembly Approach, 2-5, 2-6, 6-6
Trailing zeros, 3-46
Transition fit, 2-48

Transparency option, 4-56
Trim Entities Sketch tool, 1-76, 4-66
Trimetric - View Orientation, 1-13
Tutorials - SolidWorks, 1-30
Two View - View Orientation, 1-13

U - V
U-BLOCK Assembly, 2-104
Under defined - Sketch state, 1-50, 1-113
Undo Command, 1-10, 2-21, 4-20
Undo Document Command, 1-10, 2-21, 4-20
Unified National Coarse Thread Series, 1-122
Unit System, 1-6, 1-32, 1-33
Units - Document Properties, 2-10
Units, 4-12
Universal Joint mate, 2-18
Up To Body - End Condition, 8-24
Up To Surface - End Condition, 5-54
Up To Vertex - End Condition, 8-23, 8-26
User Library, 2-70
User Specified Name command, 3-73
VENDOR-COMPONENTS folder, 1-8
Vertex Feedback icon, 1-12
Vertical Dimension, 1-101, 5-36
Vertical Relation, 1-99, 4-52
Vertical Sketch Relation, 1-46
View - Menu bar menu, 1-11
View Mates tool, 2-27, 2-29
View Orientation, 1-13
   Back, 1-13
   Bottom, 1-13
   Dimetric, 1-13
   Double view, 1-13
   Four View, 1-13
   Front, 1-13
   Isometric, 1-13
   Left, 1-13
   Right, 1-13
   Single View, 1-13
   Top, 1-13
   Trimetric, 1-13
   Two View, 1-13
View Palette, 6-65
View Sketch Relations, 1-46, 1-85

W - X - Y - Z
Water Eutrophication, 7-13. 7-16
WEIGHT Part, 5-92

What's Wrong Dialog box, 1-84, 4-26
Width mate, 2-17
Window - Menu bar menu, 1-11
Window-Select, 5-51, 6-58
Wireframe - Display Style, 1-13

WIZARD-MOTION-LINKAGE, 2-102
Young's Modulus, 7-5
Zoom to Area tool, 1-13
Zoom to Fit tool, 1-13